Birkhäuser

Applied and Numerical Harmonic Analysis

More information about this series at http://www.springer.com/series/4968

Akram Aldroubi · Carlos Cabrelli ·
Stéphane Jaffard · Ursula Molter
Editors

New Trends in Applied Harmonic Analysis, Volume 2

Harmonic Analysis, Geometric Measure Theory, and Applications

Editors
Akram Aldroubi
Vanderbilt University
Nashville, TN, USA

Carlos Cabrelli
Department of Mathematics (FCEyN)
Universidad de Buenos Aires
Buenos Aires, Argentina

Stéphane Jaffard
Department of Mathematics (LAMA)
Université Paris-Est Créteil
Creteil, Paris, France

Ursula Molter
Department of Mathematics (FCEyN)
Universidad de Buenos Aires
Buenos Aires, Argentina

ISSN 2296-5009 ISSN 2296-5017 (electronic)
Applied and Numerical Harmonic Analysis
ISBN 978-3-030-32355-4 ISBN 978-3-030-32353-0 (eBook)
https://doi.org/10.1007/978-3-030-32353-0

Mathematics Subject Classification (2010): 42-XX, 28A80, 94A08, 94A12

This book is published under the imprint Birkhäuser, www.birkhauser-science.com by the registered company Springer Nature Switzerland AG
The registered company address is: Gewerbestrasse 11, 6330 Cham, Switzerland

ANHA Series Preface

The *Applied and Numerical Harmonic Analysis (ANHA)* book series aims to provide the engineering, mathematical, and scientific communities with significant developments in harmonic analysis, ranging from abstract harmonic analysis to basic applications. The title of the series reflects the importance of applications and numerical implementation, but richness and relevance of applications and implementation depend fundamentally on the structure and depth of theoretical underpinnings. Thus, from our point of view, the interleaving of theory and applications and their creative symbiotic evolution is axiomatic.

Harmonic analysis is a wellspring of ideas and applicability that has flourished, developed, and deepened over time within many disciplines and by means of creative cross-fertilization with diverse areas. The intricate and fundamental relationship between harmonic analysis and fields such as signal processing, partial differential equations (PDEs), and image processing is reflected in our state-of-the-art *ANHA* series.

Our vision of modern harmonic analysis includes mathematical areas such as wavelet theory, Banach algebras, classical Fourier analysis, time-frequency analysis, and fractal geometry, as well as the diverse topics that impinge on them.

For example, wavelet theory can be considered an appropriate tool to deal with some basic problems in digital signal processing, speech and image processing, geophysics, pattern recognition, biomedical engineering, and turbulence. These areas implement the latest technology from sampling methods on surfaces to fast algorithms and computer vision methods. The underlying mathematics of wavelet theory depends not only on classical Fourier analysis, but also on ideas from abstract harmonic analysis, including von Neumann algebras and the affine group. This leads to a study of the Heisenberg group and its relationship to Gabor systems, and of the metaplectic group for a meaningful interaction of signal decomposition methods. The unifying influence of wavelet theory in the aforementioned topics illustrates the justification for providing a means for centralizing and disseminating information from the broader, but still focused, area of harmonic analysis. This will be a key role of *ANHA*. We intend to publish with the scope and interaction that such a host of issues demands.

Along with our commitment to publish mathematically significant works at the frontiers of harmonic analysis, we have a comparably strong commitment to publish major advances in the following applicable topics in which harmonic analysis plays a substantial role:

Antenna theory
Biomedical signal processing
Digital signal processing
Fast algorithms
Gabor theory and applications
Image processing
Numerical partial differential equations
Prediction theory
Radar applications
Sampling theory
Spectral estimation
Speech processing
Time-frequency and time-scale analysis
Wavelet theory

The above point of view for the *ANHA* book series is inspired by the history of Fourier analysis itself, whose tentacles reach into so many fields.

In the last two centuries Fourier analysis has had a major impact on the development of mathematics, on the understanding of many engineering and scientific phenomena, and on the solution of some of the most important problems in mathematics and the sciences. Historically, Fourier series were developed in the analysis of some of the classical PDEs of mathematical physics; these series were used to solve such equations. In order to understand Fourier series and the kinds of solutions they could represent, some of the most basic notions of analysis were defined, e.g., the concept of "function". Since the coefficients of Fourier series are integrals, it is no surprise that Riemann integrals were conceived to deal with uniqueness properties of trigonometric series. Cantor's set theory was also developed because of such uniqueness questions.

A basic problem in Fourier analysis is to show how complicated phenomena, such as sound waves, can be described in terms of elementary harmonics. There are two aspects of this problem: first, to find, or even define properly, the harmonics or spectrum of a given phenomenon, e.g., the spectroscopy problem in optics; second, to determine which phenomena can be constructed from given classes of harmonics, as done, for example, by the mechanical synthesizers in tidal analysis.

Fourier analysis is also the natural setting for many other problems in engineering, mathematics, and the sciences. For example, Wiener's Tauberian theorem in Fourier analysis not only characterizes the behavior of the prime numbers, but also provides the proper notion of spectrum for phenomena such as white light; this latter process leads to the Fourier analysis associated with correlation functions in filtering and prediction problems, and these problems, in turn, deal naturally with Hardy spaces in the theory of complex variables.

Nowadays, some of the theory of PDEs has given way to the study of Fourier integral operators. Problems in antenna theory are studied in terms of unimodular trigonometric polynomials. Applications of Fourier analysis abound in signal processing, whether with the fast Fourier transform (FFT), or filter design, or the adaptive modeling inherent in time-frequency-scale methods such as wavelet theory. The coherent states of mathematical physics are translated and modulated Fourier transforms, and these are used, in conjunction with the uncertainty principle, for dealing with signal reconstruction in communications theory. We are back to the raison d'être of the *ANHA* series!

University of Maryland
College Park, MD, USA

John J. Benedetto
Series Editor

Foreword

Argentina is well known for its strong tradition of fundamental contributions to harmonic analysis, going back to the pioneering work of Alberto Calderón. Researchers from around the world gathered in Mar del Plata, Argentina in 2013 for a very successful *CIMPA Research School* devoted to new trends in applied harmonic analysis. In 2017, a second CIMPA school was held at the University of Buenos Aires, focused on aspects of mathematical analysis which have recently had a significant impact on image and signal processing. Many of these technological developments have relied on breakthroughs in both harmonic analysis and geometric measure theory that have yielded solutions to deep theoretical problems in these fields.

The 2017 CIMPA School gave researchers a chance to see new interfaces between geometric measure theory and harmonic analysis, and to apply this knowledge to real-life problems. The following eight courses were presented during the CIMPA school, especially aimed at Ph.D. students and Postdoctoral researchers in mathematics, signal processing, and image processing.

The following courses were taught at the school:

- Massimo Fornasier, *Variational quantization of measures: An harmonic analysis approach.*
- Karlheinz Gröchenig, *Gabor analysis: Applications, theory, and mysteries.*
- Stéphane Jaffard, *Multifractal analysis based on wavelet bases: Part 1. Mathematical foundations and p-leaders analysis.*
- Pertti Mattila, *The Fourier transform and Hausdorff dimension.*
- Maria Cristina Pereira, *Dyadic harmonic analysis and weighted inequalities.*
- Pablo Shmerkin, *From additive combinatorics to geometric measure theory.*
- Xavier Tolsa, *The Riesz transform, rectifiabilty, and harmonic measure.*
- Herwig Wendt, *Multifractal analysis based on wavelet bases: Part 2. Estimation, Bayesian models and multivariate data.*

The following plenary talks were also presented by leading researchers from around the globe:

- Patrice Abry, *Multivariate Self-similarity: Multiscale eigen structures for the estimation of Hurst exponents.*
- John Benedetto, *Frames and some algebraic forays.*
- Jorge Betancor, *Hardy spaces with variable exponents.*
- Emanuel Carneiro, *Regularity theory for maximal operators—An overview.*
- Maria Charina, *Regularity of refinable functions: Matrix approach.*
- Marianna Csornyei, *The Kakeya needle problem for rectifiable sets.*
- Laura De Carli, *Many questions and few answers on exponential bases.*
- Martin De Hoop, *Frame-based multi-scale Gaussian beams, wavefield approximation and boundary value problems.*
- Patrick Flandrin, *The sound of silence—Recovering signals from time-frequency zeros.*
- Alex Iosevich, *Finite point configurations: Analysis, combinatorics and number theory.*
- Tamas Keleti, *Hausdorff dimension of unions of subsets of lines or k-planes.*
- Mihalis Kolountzakis, *Measurable Steinhaus sets do not exist for finite sets or the integers in the plane.*
- Roberto Leonarduzzi, *p-leader analysis and classification of oscillating singularities.*
- Clothilde Melot, *Intertwinning wavelets or multiresolution analysis on graphs through random forests.*
- Shahaf Nitzan, *The Balian-Low theorem in the finite dimensional setting.*
- Carlos Perez, *Borderline weighted estimates for the maximal function and for rough singular integral operators.*
- José Luis Romero, *Sharp sampling density conditions for shift-invariant spaces.*
- Ville Suomala, *Patterns in random fractals.*
- Rodolfo Torres, *Smoothing properties of bilinear operators and Leibniz-type rules.*
- Yimin Xiao, *Fine Properties of Gaussian Random Fields on the Sphere.*

During the special days of birthday celebration, the following researchers delivered talks on many very interesting subjects:

- John Benedetto, *Super-resolution by means of Beurling minimal extrapolation.*
- Luis Caffarelli, *A problem of interacting obstacles.*
- Ricardo Durán, *Improved Poincaré inequalities in fractional Sobolev spaces.*
- Eleonor Harboure, *Local Calderón-Zygmund theory on proper open subsets.*
- Chris Heil, *Wavelets, Self-Similarity, and the Joint Spectral Radius: A Retrospective.*
- Eugenio Hernandez, *Greedy algorithm and embeddings.*
- Stéphane Jaffard, *Wavelets on the hunt for gravitational waves.*

- Irene MartŠnez Gamba, *Existence and uniqueness theory for binary collisional kinetic models.*
- Sheldy Ombrosi, *Weighted endpoint estimates for commutators of Calderón-Zygmund operators.*

In this volume, we present ten articles based on the courses and plenary talks of the 2017 CIMPA school.

Our Chap. 1 is by John J. Benedetto, Katherine Cordwell, and Mark Magsino, on the topic of *constant amplitude zero autocorrelation sequences*. These *CAZAC sequences* play important roles in the design of waveforms for radar and communication. The authors provide a detailed analysis of the connections between CAZAC sequences and deep mathematical results in Fourier analysis and the theory of Hadamard matrices. This gives us a unified exposition of the theory of CAZAC sequences and also introduces new techniques for constructing CAZAC sequences. Another fundamental contribution is a discussion of the unpublished results of the late Uffe Haagerup on the number of CAZAC generating cyclic N-roots, and the connection of this proof to the uncertainty principle.

Chapter 2, by Víctor Almeida, Jorge J. Betancor, Estefanía Dalmasso, and Lourdes Rodríguez-Mesa, is based on a course given by Betancor during the 2017 CIMPA school. Of course the Lebesgue spaces L^p and their analogous discrete versions ℓ^p are ubiquitous throughout harmonic analysis, and mathematical analysis in general. Versions of the Lebesgue spaces with a *variable exponent* $p(x)$ were first introduced by Orlicz in 1931. In recent years, variable exponent function spaces have attracted new attention due to applications in fluids and related areas. In this chapter, the authors study Hardy spaces with variable exponents, which were first introduced by Nakai and Sawano and, independently, by Cruz-Uribe and Wang. While the classical Hardy spaces are naturally adapted to the Laplacian operator, the authors explain how variable-exponent Hardy spaces are associated with a more general class of operators, and they also consider local versions of Hardy spaces with variable exponents.

Emanuel Carneiro reports in Chap. 3 on the recent progress on the regularity theory of maximal operators and also discusses some of the current open problems in this area. Maximal operators are a classical subject in harmonic analysis, being fundamental tools for the proofs of many types of results on pointwise convergence, including the Lebesgue differentiation theorem and Carleson's theorem on the pointwise convergence of Fourier series, and many others. Carneiro focuses on the classical Hardy–Littlewood maximal operator in this chapter, but while this operator is classical, the results he surveys are very recent, not to mention deep and mathematically elegant.

Chapter 4, by Karlheinz Gröchenig and Sarah Koppensteiner, is based on a course given by Gröchenig at the 2017 CIMPA school. The course is a broad survey of the structure and characterizations of *Gabor frames* over lattices. Such frames have been studied and applied to many scientific and engineering problems, for many decades. Yet, as the authors explain, it is still extremely difficult to determine whether a particular window function will generate a Gabor frame over a particular lattice! The various characterizations of Gabor frames over lattices all ultimately involve the issue of the invertibility of some operator, which is an intrinsically difficult problem. Gröchenig and Koppensteiner masterfully present the many distinct approaches to Gabor frames in a single streamlined exposition fundamentally based on the Poisson Summation Formula.

In Chap. 5, Alex Iosevich considers the *Approximate Unit Distance Problem*, which developed out of the Erdös unit distance conjecture, a longstanding and extremely difficult conjecture in extremal combinatorics. The best known bound for Erdös' conjecture is $Cn^{4/3}$, which was obtained by Spencer, Szemeredi, and Trotter in 1984. Iosevich considers an approximate setting for this problem and shows that in that context the bound can be significantly improved for many point sets.

Chapter 6, by Pertti Mattila, is based on the course he gave at the 2017 CIMPA school. He surveys recent results on the Hausdorff dimension of projections and intersections of general subsets of Euclidean space, focusing on integral-geometric properties of Hausdorff dimension and their relations to Kakeya-type problems. Integral-geometric properties are those related to affine subspaces of Euclidean spaces and to rigid motions, and Mattila presents estimates of the Hausdorff dimension of exceptional sets of planes and rigid motions, and projections onto restricted families of planes.

María Cristina Pereyra gives in Chap. 7 an extended survey of how sparse techniques have revolutionized the theory of dyadic harmonic analysis and weighted inequalities. In this chapter, based on the course she gave at the 2017 CIMPA school, Pereyra begins with detailed background on the Hilbert transform and the maximal function before turning to a comprehensive treatment of dyadic harmonic analysis, dyadic maximal functions, and dyadic operators. Weighted operators and sparse domination by positive dyadic operators are then explored, and the chapter concludes with a summary and a discussion of recent progress. An extensive list of references complements the exposition of the course.

Chapter 8, by Alberto Criado, Carlos Pérez, and Israel P. Rivera-Ríos, deals with quantitative weighted BMO estimates. The authors give a new, quantitative proof of an extrapolation theorem originally due to Harboure, Macías, and Segovia. They also obtain sharp weighted L^{∞}_{c}—BMO-type estimates for some Calderón—Zygmund operators.

Chapter 9, by Pablo Shmerkin, is also based on a course given at the 2017 CIMPA school. In his survey, Schmerkin presents a self-contained proof of a formula for the L^q dimensions of self-similar measures on the real line under exponential separation. This is a special case of more general results by the author but gives a simpler approach to a setting that is still very important. He also reviews some applications to the study of Bernoulli convolutions and intersections of self-similar Cantor sets.

The final chapter, Chap. 10 by Erick Herbin and Yimin Xiao, considers sample path properties of set-indexed fractional Brownian motion. The Hausdorff dimensions of inverse images and corresponding hitting probabilities are considered.

On a personal note, I had the great pleasure of attending both the 2013 and the 2017 CIMPA schools. These were extraordinary events, which allowed researchers and students from mathematics and science and engineering to learn and interact. The organizers did a terrific job of organizing two schools that will have a lasting impact for years to come.

Atlanta, Georgia
June 2019

Christopher Heil

Preface

The CIMPA school *Harmonic Analysis, Geometric Measure Theory and Applications* took place in Buenos Aires during 2 weeks, in August 2017. It was conceived as a continuation of the very successful previous school *New Trends in Applied Harmonic Analysis Sparse Representations, Compressed Sensing and Multifractal Analysis* which had also been held in Argentina, at Mar del Plata. Once again, this CIMPA school was a big success, with a large number of Ph.D. students and young PostDocs, mainly coming from all over Latin America.

Still focused on harmonic analysis taken in a broad sense, the aim of this second school was to specifically aim at the new interlaces with geometric measure theory which recently have had a huge impact, in particular, in image and signal processing, and how these new understandings can be applied to solve real-life problems. The relevance of this school was due to the fact that several technological deadlocks have been recently solved through the resolution of deep theoretical problems in these areas, and the purpose of the school was to expose their resolutions, but also their implications both for theory and in applications, and the new challenges which they raise.

In the middle of the school, a climax was reached with the special day in honor of Ursula Molter, which gave the young audience a great opportunity to revisit 30 years of pure and applied harmonic analysis and geometric measure theory through the various topics where Ursula has made seminal contributions.

The courses of this CIMPA school were taught by leaders in these areas. Though the purpose was to expose recent deep breakthroughs, these scientists managed to meet the challenge of remaining accessible to a broad audience, mainly composed of Ph.D. students and PostDocs, and of diverse scientific cultures, covering mathematics, and signal and image processing.

The school emulated new interactions between these communities, and the courses collected in this book faithfully transpose the atmosphere of feverish interdisciplinary interactions that took place during these 2 weeks.

Nashville, USA — Akram Aldroubi
CABA (Buenos Aires), Argentina — Carlos Cabrelli
Creteil, France — Stéphane Jaffard
CABA (Buenos Aires), Argentina — Ursula Molter

Acknowledgements

The editors of this volume gratefully acknowledge the local organizing committee for their extraordinary work. Without their help this school would not have been possible. They were the driving force behind the scene.

Diana Carbajal (University of Buenos Aires).
Sigrid Heineken (University of Buenos Aires).
Carolina Mosquera (University of Buenos Aires).
Andrea Olivo (University of Buenos Aires).
Victoria Paternostro (University of Buenos Aires).
Ezequiel Rela (University of Buenos Aires).
Pablo Shmerkin (Torcuato Di Tella University).
Alexia Yavícoli (University of Buenos Aires).

We also want to thank the Department of Mathematics of the FCEyN from the University of Buenos Aires and the Instituto de Matematica Luis Santaló who kindly provided the facilities and technical equipment.

Finally, we want to highlight the excellent job of our staff Monica Lucas and Liliana Grandz and the extraordinary help of the student Nahuel García.

The support of our sponsors was crucial for the realization of the meeting. We thank all of them very much for their generous financial help.

BEZOUT—Labex Bézout—Université Paris-Est.
CDC—International Mathematical Union (IMU).
CIMPA—International Center for Pure and Applied Mathematics.
CONICET—Consejo Nacional de Investigación Científica.
Departamento de Matemática—FCEyN—Universidad de Buenos Aires.
FUNDACEN—Fundación de Ciencias Exactas y Naturales.
HEXAGON Consulting.
ICTP (Trieste)—International Centre for Theoretical Physics.
IMAS—Instituto de Investigaciones Matemáticas Luis A. Santaló.
UPEC—University Paris-Est Créteil.

Contents

Chapter 1
CAZAC Sequences and Haagerup's Characterization of Cyclic N-roots

John J. Benedetto, Katherine Cordwell and Mark Magsino

Abstract Constant amplitude zero autocorrelation (CAZAC) sequences play an important role in waveform design for radar and communication theory. They also have deep and intricate connections in several topics in mathematics, including Fourier analysis, Hadamard matrices, and cyclic N-roots. Our goals are to describe these mathematical connections, to provide a unified exposition of the theory of CAZAC sequences integrating several diverse ideas, to introduce new techniques for constructing CAZAC sequences alongside established methods, and to give an exposition of the fascinating unpublished theorem of Uffe Haagerup (1949–2015), which proves that the number of CAZAC generating cyclic N-roots is finite. The role of the uncertainty principle in the proof is essential.

Mathematical Subject Classification: Primary 42Bxx · Secondary 42-06 · 42-02

1.1 Introduction

1.1.1 Background and Goal

In this subsection, we *define* a constant amplitude zero autocorrelation (CAZAC) sequence, *describe* some scenarios where CAZAC sequences play a role and state the *goal* of this paper.

Definition 1.1.1 (*CAZAC sequence*) Given a function, $x : \mathbb{Z}/N\mathbb{Z} \longrightarrow \mathbb{C}$.

a. The *autocorrelation*, $A_x : \mathbb{Z}/N\mathbb{Z} \longrightarrow \mathbb{C}$, of x is defined by

---

J. J. Benedetto (✉) · K. Cordwell · M. Magsino
Norbert Wiener Center, Department of Mathematics, University of Maryland,
College Park, MD 20742, USA
e-mail: jjb@math.umd.edu
URL: http://www.math.umd.edu/~jjb

A. Aldroubi et al. (eds.), *New Trends in Applied Harmonic Analysis, Volume 2*, Applied and Numerical Harmonic Analysis,
https://doi.org/10.1007/978-3-030-32353-0_1

$$\forall m \in \mathbb{Z}/N\mathbb{Z}, \quad A_x[m] = \frac{1}{N} \sum_{k=0}^{N-1} x[m+k]\overline{x[k]}.$$

b. The function (sequence), $x : \mathbb{Z}/N\mathbb{Z} \to \mathbb{C}$, is a *constant amplitude zero autocorrelation* (CAZAC) sequence if

$$\forall m \in \mathbb{Z}/N\mathbb{Z}, \quad |x[m]| = 1, \tag{CA}$$

and

$$\forall m \in \mathbb{Z}/N\mathbb{Z}\backslash\{0\}, \quad \frac{1}{N} \sum_{k=0}^{N-1} x[m+k]\overline{x[k]} = 0. \tag{ZAC}$$

Equation (CA) is the condition that x has constant amplitude 1. Equation (ZAC) is the condition that u has zero autocorrelation for $m \in (\mathbb{Z}/N\mathbb{Z})\backslash\{0\}$, i.e., off the dc component.

The construction of CAZAC sequences, or modifications where ZAC is replaced by low autocorrelation, is a central problem in the general area of waveform design; and it is particularly relevant in several applications in the areas of radar and communications.

In radar, CAZAC sequences can play a role in effective target recognition and other fundamental applications, see, e.g., [1, 21, 26, 28, 30, 35, 38, 39, 42, 43, 45, 46, 49, 51, 55, 58, 63, 65]. There has been a striking recent application of low correlation sequences to radar in terms of compressed sensing [36].

In communications, CAZAC sequences can be used to address synchronization issues in cellular access technologies, especially code-division multiple access (*CDMA*), e.g., [64, 66].

The radar and communication methods have combined in recent advanced multifunction RF systems (*AMRFS*).

In radar, there are two main reasons that the sequences x should have the constant amplitude property (CA). First, a transmitter can operate at peak power if x has constant peak amplitude—the system does not have to deal with the surprise of greater than expected amplitudes. Second, amplitude variations during transmission due to additive noise can be theoretically eliminated. The zero autocorrelation property (ZAC) ensures minimum interference between signals sharing the same channel.

The applications referenced above are part of a broad range of applications of the narrowband and wideband ambiguity function. The ZAC or low autocorrelation property can be viewed as the boundary value of an ambiguity function, which in the narrowband case is essentially the short-time Fourier transform (STFT), see [6, 9, 12, 29, 68, 69]. We shall not deal with the ambiguity function in this paper.

There are also purely mathematical roots for the construction of CAZAC sequences, e.g., [9]. One example, which inspired the role of probability theory in the subject, is due to Wiener, see [7]. Our interest in CAZAC sequences was inspired by the deep ideas and techniques of Björck and Saffari, e.g., [15, 18, 54], *and* by the good fortune of the first named author to benefit personally by discussions with both Björck and Saffari.

Our *goal* is simply stated: for various values of N, count and construct the CAZAC sequences of length N. This entails providing a unified exposition relating cyclic N-roots, complex circulant Hadamard matrices, and CAZAC sequences. We require a profound theorem due to Haagerup [32], see Sect. 1.1.3. His work builds on a brilliant counterexample by Björck, see Sects. 1.1.2 and 1.4.1, as well as explicit calculations by many others, e.g., [13, 19]. In pursuit of our goal, we give several new explicit calculations with the point of view of constructing new CAZAC sequences.

1.1.2 Gaussian and Non-Gaussian CAZAC Sequences

The beautiful story of this subsection was told expertly by Saffari in [54]. To begin, we define the discrete Fourier transform (DFT).

Definition 1.1.2 a. Given a finite sequence, $x = (x[0], x[1], \ldots, x[N-1]) \in \mathbb{C}^N$. The *discrete Fourier transform* (DFT), $\mathcal{F}_N(x) = \widehat{x} \in \mathbb{C}^N$, of x is defined by

$$\mathcal{F}_N(x)[n] = \widehat{x}[n] = \frac{1}{N^{1/2}} \sum_{m=0}^{N-1} x[m] e^{-2\pi i m n/N}, \quad n = 0, 1, \ldots, N-1.$$

Elementary calculations yield the *inversion formula*,

$$x[m] = \frac{1}{N^{1/2}} \sum_{n=0}^{N-1} \widehat{x}[n] e^{2\pi i m n/N}, \quad m = 0, 1, \ldots, N-1, \tag{1.1}$$

and *Parseval's formula*,

$$\sum_{m=0}^{N-1} |x[m]|^2 = \sum_{n=0}^{N-1} |\widehat{x}[n]|^2. \tag{1.2}$$

b. Notationally, for a given N, let $e_m = e^{-2\pi i m/N}$ and $W_N = e^{2\pi i/N} = e_1$. Also, for a given $x \in \mathbb{C}^N$, we denote translation by τ so that $\tau_m[k] = x[k-m]$. Clearly, W_N is an N-root of unity, and recall that it is *primitive* N-root of unity if it is not also an M-root of unity for some $M < N$. Thus, W_N^M is a primitive N-root of unity if and only if $\gcd(M, N) = 1$.

c. For a given N, the DFT matrix, $\mathcal{D}_N$, is defined as the $N \times N$ matrix,

$$\mathcal{D}_N = \left[\frac{1}{N^{1/2}} W_N^{-mn} \right]_{m,n=0}^{N-1},$$

and, for convenience, assume that W_N is a primitive N-root of unity. Using Eq. (1.2) we see that $\mathcal{D}_N$ is a unitary matrix, i.e., $\mathcal{D}_N^*\mathcal{D}_N = I$, the $N \times N$-identity matrix, where $\mathcal{D}_N^*$ is the complex conjugate of the transpose of $\mathcal{D}_N$. The trace of $\mathcal{D}_N$ is a sum of Gaussians, as defined in Example 1.1.3. The remarkable properties of these *Gauss sums* are stated and proved, with perspective, in [4, Chap. 3.9].

d. We have that

$$\forall x \in \mathbb{C}^N, \quad \mathcal{F}_N(x) = \widehat{x} = \mathcal{D}_N(x) \in \mathbb{C}^N,$$

see [4, 62] for much more on the DFT.

We shall say that a sequence, $x = (x[0], x[1], \dots, x[N-1]) \in \mathbb{C}^N$, is *unimodular* if each $|x[j]| = 1$, and it is *bi-unimodular* if each $|x[m]| = |\widehat{x}[n]| = 1$. In [15], Björck began his analysis of bi-equimodular sequence, i.e., $|x[m]| = A$ for all $m \in \mathbb{Z}/N\mathbb{Z}$ and $|\widehat{x}[n]| = B$ for all $n \in \mathbb{Z}/N\mathbb{Z}$, also see [18]. It is an interesting fact, and elementary to verify, that *a sequence, $x = (x[0], x[1], \dots, x[N-1]) \in \mathbb{C}^N$, is bi-unimodular if and only if it is a CAZAC sequence*, see Proposition 1.2.1.

Example 1.1.3 (Gaussian sequence) Given an integer $N \geq 2$, and define the *Gaussian sequence*, $g_{N,a,b}[m]$, $m = 0, \dots, N-1$, by the formula

$$g_{N,a,b}[m] = W_N^{am^2+bm}, \quad m = 0, \dots, N-1,$$

where $a, b \in \mathbb{Z}$ and $\gcd(a, N) = 1$, that is, a and N are relatively prime, see Definition 1.1.2, part *b*. We write $g_N = g_{N,1,0}$.

Björck and Saffari noted, by an elementary calculation, that if $N \geq 3$ is odd, then $\{g_N[m]\}_{m=0}^{N-1} = \{e^{2\pi i m^2/N}\}_{m=0}^{N-1}$ is a CAZAC sequence, and also noted that Gauss was aware of this fact, probably in terms of the bi-unimodular equivalence! In this regard, see Example 1.2.5.

At Stockholm University in 1983, Per Enflo asked the following question for a given odd prime p. Is it true that the modified Gaussian sequences, $\{g_p[m]\, W_p^{jm}\}_{m=0}^{p-1}$, $j \in \mathbb{Z}$, are the only bi-unimodular sequences of length p? Gaussian sequences are the special case when $j = 0$. The answer was known to be "yes" for $p = 3$ and $p = 5$. A positive answer generally would have helped Enflo with estimates he was making on exponential sums. Ultimately, he made these estimates independent of his question, but it led to deep mathematical questions in other directions.

The $p = 3$ case is elementary to resolve. It is much more involved for the $p = 5$ case, which was first *checked* and settled by L. Lovász in 1983 (private communication to Björck), and proved by Haagerup in 1996 [31], also see Remark 1.1.7 and Sect. 1.3.

Björck tried to answer Enflo's question positively by computer search for $p = 7$. However, the counterexample,

$$(1, 1, 1, e^{i\theta}, 1, e^{i\theta}, e^{i\theta}), \quad \theta = \arccos\left(-\frac{3}{4}\right), \tag{1.3}$$

"popped out" as Björck put it! see [54] and Sect. 1.4.

The rest is history, or, rather, the start of an important, and still unresolved and incomplete quest.

1.1.3 Haagerup's Theorem

We shall now state Haagerup's theorem mentioned in Sect. 1.1.1. In order to do this, we shall require several notions, which are equivalent to the CAZAC sequence property. To this end, we begin by defining a cyclic N-root, see [14].

Definition 1.1.4 A *cyclic N-root* is a solution $z = (z_0, z_1, \cdots z_{N-1}) \in \mathbb{C}^N$ to the following set of equations:

$$\begin{cases} z_0 + z_1 + \cdots + z_{N-1} = 0 \\ z_0 z_1 + z_1 z_2 + \cdots + z_{N-1} z_0 = 0 \\ \qquad \cdots \\ z_0 z_1 \cdots z_{N-2} + \cdots + z_{N-1} z_0 \cdots z_{N-3} = 0 \\ z_0 z_1 \cdots z_{N-1} = 1. \end{cases}$$

The second definition we shall need to state Haagerup's theorem, and to provide basic perspective, is that of a complex circulant Hadamard matrix.

Definition 1.1.5 a. A *complex $N \times N$ circulant matrix C_N* is a square $N \times N$ matrix, where each row vector is rotated one element to the right relative to the preceding row vector. Thus, a circulant matrix, C_N, is defined by one vector, $c \in \mathbb{C}^N$, which appears as the first row of C_N. The remaining rows of C_N are each cyclic permutation of the vector c with offset equal to the row index, see [40].

A complex $N \times N$ *permutation matrix P_N* is defined by the property that it has exactly one entry of 1 in each row and each column and 0s elsewhere.
A complex $N \times N$ *unitary matrix U_N* is defined by the property that $U_N U_N^* = Id$, where U_N^* is the conjugate transpose or adjoint of U_N and Id is the $N \times N$ identity matrix. Thus, the rows and columns of U_N form orthonormal bases for $\mathbb{C}^N$.

b. An important application of circulant matrices is that they are diagonalized by the DFT. Thus, a system of N linear equations, $C_N X = Y \in \mathbb{C}^N$, can be solved quickly using the fast Fourier transform (FFT), e.g., see [22].

c. A *complex* $N \times N$ *Hadamard matrix* H_N is a square $N \times N$ matrix with unimodular entries $c_{m,n} \in \mathbb{C}$ and mutually orthogonal rows, i.e., $H_N H_N^* = N\,Id$.
d. Let H_1, H_2 be two Hadamard matrices. As matrices they are equivalent if they can be transformed one into the other by elementary row and column operations. In the case of Hadamard matrices, this is the same as saying that H_1 and H_2 are *equivalent* if there exist diagonal unitary matrices D_1, D_2 and permutation matrices P_1, P_2 such that

$$H_2 = D_1 P_1 H_1 P_2 D_2. \tag{1.4}$$

Motivation for the definition of equivalence is spelled out for dephased Hadamard matrices in Sect. 1.2.5.
e. Bruzda et al. [13, 19] maintain a website that characterizes $N \times N$ Hadamard matrices for various, small values of N. Also, see [60].

There is a characterization of CAZAC sequences in terms of complex circulant Hadamard matrices [18]. In particular, the first row of any complex circulant Hadamard matrix is a CAZAC sequence. Moreover, if $x : \mathbb{Z}^N \to \mathbb{C}$ is a given function and if H_x is a circulant matrix with first row $x = (x[0], x[1], \ldots, x[N-1])$, then x is a CAZAC sequence if and only if H_x is a Hadamard matrix, see Proposition 1.2.2. Finally, there is a one-to-one correspondence between cyclic N-roots and CAZAC sequences of length N. This correspondence will be stated clearly in Proposition 1.2.4, and we shall prove Propositions 1.2.1, 1.2.2, and 1.2.4 in Sect. 1.2.1.

In [32], Haagerup proved the following deep and fundamental theorem: Theorem 1.1.6, cf. his earlier related work [31].

Theorem 1.1.6 *For every prime number p, the set of cyclic p-roots in $\mathbb{C}^p$ is finite. Moreover, the number of cyclic p-roots counted with multiplicity is equal to*

$$\binom{2p-2}{p-1} = \frac{(2p-2)!}{(p-1)!^2}.$$

In particular, the number of complex $p \times p$ circulant Hadamard matrices with diagonal entries equal to 1 is less than or equal to $(2p-2)!/(p-1)!^2$.

Remark 1.1.7 a. Before Haagerup's theorem, Theorem 1.1.6, it was not known whether there were finitely many or infinitely many cyclic p-roots for most primes p.
b. Although elementary for $N = 2, 3, 4$, it is generally difficult to compute the number of cyclic N-roots. In fact, prior to Theorem 1.1.6, computer algebra, as opposed to theoretical means, was the only available technology for such computation, and in this setting N was necessarily small, see Björck and Fröberg [16, 17] for the cases, $5 \leq N \leq 8$, as well as [2].

Table 1.1 $r(N)$ and $r_u(N)$ for $N = 2, \ldots, 9$

N	2	3	4	5	6	7	8	9
$r(N)$	2	6	∞	70	156	924	∞	∞
$r_u(N)$	2	6	∞	20	48	532	∞	∞

For a given N, let

$$r(N) = \binom{2N-2}{N-1},$$

resp., $r_u(N)$, be the number of cyclic N-roots, resp., unimodular cyclic N-roots, see Table 1.1 which is taken from [32]. Backelin and Fröberg [2] contain the proof that $r(7) = 924$.

With this information, Faugère conjectured that for a given prime p there are $\binom{2p-2}{p-1}$ cyclic p-roots. This is the content of Theorem 1.1.6 when the number of cyclic p-roots is counted with multiplicity. The multiplicity is 1 for $p = 2, 3, 5, 7$, but it is not known if this is true for all primes. In the case $p = 9$, Faugère [25] showed there are cyclic 9-roots with multiplicity 4.

c. In Sect. 1.5 we shall outline Haagerup's proof that the set of cyclic p-roots in $\mathbb{C}^p$ is finite. Although this part of Haagerup's proof is ingenious, the real depth is involved in his proof that the number of cyclic p-roots counted with multiplicity is equal to $\frac{(2p-2)!}{(p-1)!^2}$.

1.1.4 Outline

Besides the material in Sects. 1.1.1, 1.1.2, and 1.1.3, the outline of what we do is as follows.

Section 1.2 gives the basic theory of CAZAC sequences. In Sect. 1.2.1 we prove the various elementary characterizations of CAZAC sequences in terms of the DFT and bi-unimodular sequences, Hadamard matrices, and cyclic N-roots. Section 1.2.2 is devoted to analyzing equivalence classes of CAZAC sequences. Then, in Sects. 1.2.3 and 1.2.4, we study cyclic p-roots and CAZAC sequences of non-square-free length, respectively. Finally, Sect. 1.2.5 deals with technical but useful properties of dephased Hadamard matrices. These subsections contain new techniques for computation.

Then, in Sect. 1.3, we construct all CAZAC sequences of lengths 3 and 5 in several ways, e.g., in the case $p = 3$ we use cyclic 3-roots, Hadamard matrices, and equivalence classes. In fact, for lengths 3 and 5 all CAZAC sequences are Gaussian roots of unity, and so we do not have to generalize to other roots of unity. Although technical, these cases are straightforward and solved by elementary means. However, we provide careful detail to illustrate various computation methods, which may be generalized to CAZAC sequences of larger prime lengths where not all CAZAC sequences can be explicitly listed.

In order to deal with the length 7 case, new ideas arise and this is the content of Sect. 1.4. We construct two CAZAC sequences, which are not generated by roots of unity; and, as a result, when written in Hadamard matrix form, they are not equivalent to the Fourier matrix. One of these sequences can be generalized to other prime lengths and is known as the Björck sequence, see Sect. 1.4.1.

In Sect. 1.5, we present part of Haagerup's theorem on counting the number of CAZAC sequences of prime length p. We write out that part of his proof which shows that the number of CAZAC sequences of prime length must be finite. His original work goes on to count them as well, and we refer the reader to his paper [32], which is available on the Internet albeit unpublished. Even with the assertion of a finite number of CAZAC sequences and Haagerup's actual count, it is still not known how to construct all CAZAC sequences. One of the fascinating aspects of Haagerup's assertion of a finite number of CAZAC sequences of prime length is the natural *requirement* of Tchebotorev's theorem, re-discovered by Tao as an uncertainty principle inequality used in compressed sensing, and also re-discovered by Haagerup for this work on CAZAC sequences.

We close with Appendix 1.6 dealing mostly with real Hadamard matrices, but also with natural forays into topics as diverse as bent functions in coding theory, Walsh functions, and wavelet packets, and the solution of the Littlewood conjecture related to crest factors in antenna theory.

1.2 Characterizations and Properties of CAZAC Sequences

1.2.1 Characterizations of CAZAC Sequences

Proposition 1.2.1 *Given a finite sequence,* $x = (x[0], x[1], \ldots, x[N-1]) \in \mathbb{C}^N \backslash \{0\}$.

a. x is a CA sequence if and only if $\widehat{x}$ *is a ZAC sequence, and x is a ZAC sequence if and only if* $\widehat{x}$ *is a CA sequence, although the constant amplitude is not necessarily* 1.

b. $x = (x[0], x[1], \ldots, x[N-1]) \in \mathbb{C}^N$ *is a bi-unimodular sequence if and only if it is a CAZAC sequence.*

c. x is a CAZAC sequence if and only if $\widehat{x}$ *is a CAZAC sequence.*

Proof Parts *b* and ıc are immediate consequences of part *a*.

To prove part *a* we proceed as follows. Suppose $x \in \mathbb{C}^N$ is CA and let $n \neq 0$. Then, using the Parseval formula for the third equality, we have

$$
\begin{aligned}
NA_{\widehat{x}}[n] &= \sum_{k=0}^{N-1} \widehat{x}[n+k]\overline{\widehat{x}[k]} = \langle \tau_{-n}\widehat{x}, \widehat{x}\rangle = \langle e_n x, x\rangle \\
&= \sum_{k=0}^{N-1} |x[k]|^2 e^{2\pi ikn/N} = \sum_{k=0}^{N-1} e^{2\pi ikn/N} = 0,
\end{aligned}
$$

and so $\widehat{x}$ is ZAC.

Next, suppose that $x \in \mathbb{C}^N\backslash\{0\}$ is ZAC and let $m \neq 0$. Then,

$$
0 = NA_x[m] = \sum_{k=0}^{N-1} x[m+k]\overline{x[k]} = \langle \tau_{-m}x, x\rangle = \sum_{k=0}^{N-1} |\widehat{x}[k]|^2\, e^{2\pi imk/N}, \qquad (1.5)
$$

where the last step follows from the Parseval formula. Let $\widehat{y} = |\widehat{x}|^2$, so that by the inversion theorem, we have

$$
\forall m \in \mathbb{Z}/N\mathbb{Z}, \quad y[m] = \frac{1}{N^{1/2}} \sum_{k=0}^{N-1} \widehat{y}[k]\, e^{2\pi imk/N}.
$$

Thus, because of (1.5), we know that $y[m] = 0$ for $m \in \mathbb{Z}/N\mathbb{Z}\backslash\{0\}$, and so

$$
\forall n \in \mathbb{Z}/N\mathbb{Z}, \quad \widehat{y}[n] = \frac{1}{N^{1/2}}\, y[0]
$$

by the definition of the DFT. Hence, by the definition of $\widehat{y}$, $\widehat{x}$ is constant on $\mathbb{Z}/N\mathbb{Z}$, although not necessarily taking the value 1.

The converse directions in each case are proved by replacing x with $\widehat{x}$.

Proposition 1.2.2 *Given a sequence $x : \mathbb{Z}/N\mathbb{Z} \longrightarrow \mathbb{C}$, and let C_N be a complex circulant matrix with first row $x = (x[0], \cdots, x[N-1])$. Then, $x = \{x[0], \ldots, x[N-1]\}$ is a CAZAC sequence, where each $x[j] = x_j$, if and only if C_N is a Hadamard matrix.*

Proof First, we show that if $x = (x_0, \cdots, x_{N-1})$ is the first row of a complex $N \times N$ circulant Hadamard matrix H, then x is a CAZAC sequence, $\{x[0], \ldots, x[N-1]\}$, where each $x[m] = x_m$. Because H is a Hadamard matrix, each entry has norm 1, so x satisfies the CA condition defining CAZAC sequences. Next, because H is circulant, H has the form

$$
\begin{bmatrix}
x_0 & x_1 & \cdots & x_{N-1} \\
x_1 & x_2 & \cdots & x_0 \\
 & & \ddots & \\
x_{N-1} & x_0 & \cdots & x_{N-2}
\end{bmatrix}.
$$

Now, as noted in Definition 1.4.1, the orthogonality property of Hadamard matrices implies that $HH^* = NId$. In particular, this means that the inner product of x with column i of H is zero, where $2 \leq i \leq n$. Note that column i is of the form $x_i, x_{i+1}, \cdots, x_{i+(N-1)}$ where subscripts are taken modulo N. So the ith column is a rotation of x, and, taken together, columns $2, \ldots, N$ comprise all the nonidentity rotations of x. So, the inner product of x with any nonidentity rotation of x is 0, and thus x satisfies the zero autocorrelation property of CAZAC sequences. Hence, $x = \{x[0], \ldots, x[N-1]\}$ is a CAZAC sequence.

Conversely, we show that if $x = \{x[0], \ldots, x[N-1]\}$ is a CAZAC sequence, then $x = (x[0], \cdots, x[N-1])$ is the first row of a complex circulant Hadamard matrix $H = C_N$, of the form

$$\begin{bmatrix} x_0 & x_1 & \cdots & x_{N-1} \\ x_1 & x_2 & \cdots & x_0 \\ & & \ddots & \\ x_{N-1} & x_0 & \cdots & x_{n-2} \end{bmatrix}.$$

Now, because x is a CAZAC sequence, the absolute value of each x_i is 1. Thus, H satisfies the unimodular condition of complex Hadamard matrices.

Next, choose any row $(x_i, x_{i+1}, \cdots, x_{i+(N-1)})$ of H, where we consider subscripts mod N. Then, when we take the inner product of $(x_i, x_{i+1}, \cdots, x_{i+(N-1)})$ with itself, we obtain $x_i\overline{x_i} + x_{i+1}\overline{x_{i+1}} + \cdots + x_{N-1}\overline{x_{N-1}} = 1 + 1 + \cdots + 1 = N$, since the absolute value of each x_i is 1. If we take any other row, $(x_j, x_{j+1}, \ldots, x_{j+(N-1)})$, of H, where $i \neq j$, then the inner product $\langle (x_i, x_{i+1}, \ldots, x_{i+(N-1)}), (x_j, x_{j+1}, \ldots, x_{j+(N-1)}) \rangle$ is zero because x has zero autocorrelation. This implies that

$$HH^* = \begin{bmatrix} x_0 & x_1 & \cdots & x_{N-1} \\ x_1 & x_2 & \cdots & x_0 \\ & & \ddots & \\ x_{N-1} & x_0 & \cdots & x_{N-2} \end{bmatrix} \begin{bmatrix} \overline{x_0} & \overline{x_1} & \cdots & \overline{x_{N-1}} \\ \overline{x_1} & \overline{x_2} & \cdots & \overline{x_0} \\ & & \ddots & \\ \overline{x_{N-1}} & \overline{x_0} & \cdots & \overline{x_{N-2}} \end{bmatrix} = NId,$$

where Id is the identity matrix. Thus, H satisfies the orthogonality property of complex Hadamard matrices, and hence $H = C_N$ is a complex circulant Hadamard matrix.

First, the proofs of Propositions 1.2.1 and 1.2.2 should be compared with those in [8]. Further, due to the characterization of CAZAC sequences in Proposition 1.2.2, there is a basic relation between vector-valued CAZAC sequences and finite unit norm tight frames (FUNTFs) X for $\mathbb{C}^d$. In order to state this relation, we shall say that $x : \mathbb{Z}/N\mathbb{Z} \longrightarrow \mathbb{C}^d$ is a CAZAC sequence in $\mathbb{C}^d$ if each $\|x[k]\| = 1$ and

$$\forall k = 1, \ldots, N-1, \quad \frac{1}{N} \sum_{m=0}^{N-1} \langle x[m+k] x[m] \rangle = 0.$$

Here, $x[m] = (x_1[m], \dots, x_d[m])$, where $x_j[m] \in \mathbb{C}$, $m \in \mathbb{Z}/N\mathbb{Z}$, and $j = 1, \dots, d$; and the inner product is

$$\langle x[k]x[m]\rangle = \sum_{j=1}^{d} x_j[k]\overline{x_j[m]}.$$

Also, recall that $X = \{x_n\}_{n=0}^{N-1} \subseteq \mathbb{C}^d$ is a FUNTF if span$X = \mathbb{C}^d$ and each $\|x_n\| = 1$. This definition does not reflect the complexity of frames even in the finite FUNTF case, e.g., see [20], but it is sufficient to state Proposition 1.2.3, see [9] for its proof.

Proposition 1.2.3 *Let $x = \{x[n]\}_{n=0}^{N-1}$ be a CAZAC sequence in $\mathbb{C}$. Define*

$$\forall k = 0, \dots, N-1, \quad v[k] = \frac{1}{\sqrt{d}}\left(x[k], x[k+1], \dots, x[k+d-1]\right).$$

Then, $v = \{v[k]\}_{k=0}^{N-1}$ is a CAZAC sequence in $\mathbb{C}^d$ and $v = \{v[k]\}_{k=0}^{N-1}$ is a FUNTF for $\mathbb{C}^d$ with frame constant N/d.

The following fundamental result was proved by Björck in 1985 [14].

Proposition 1.2.4 *There is a one-to-one correspondence between unimodular cyclic N-roots and CAZAC sequences of length N and with first term $x[0] = 1$. In fact, given such a CAZAC sequence, x, we can obtain the corresponding cyclic N-root with the formula*

$$(z_0, z_1, \cdots, z_{N-1}) = \left(\frac{x[1]}{x[0]}, \frac{x[2]}{x[1]}, \cdots, \frac{x[N-1]}{x[N-2]}, \frac{x[0]}{x[N-1]}\right).$$

Proof First, we show that if $x = (x_0, \cdots, x_{N-1})$ is a CAZAC sequence with $x_0 = 1$, then $(z_0, \dots, z_{N-1}) = (x_1/x_0, x_2/x_1, \cdots, x_0/x_{N-1}) = (x_1, x_2/x_1, \cdots, x_0/x_{N-1})$ is a unimodular cyclic N-root. First note that $|z_i| = x_1/x_0 \cdot \overline{x_1/x_0} = x_1/x_0 \cdot x_0/x_1 = 1$ for all i, so $(z_0, \dots, z_{N-1})$ is unimodular.

Next, multiplying $z_0 \cdots z_{N-1}$ gives

$$\frac{x_1}{x_0} \cdot \frac{x_2}{x_1} \cdots \frac{x_0}{x_{N-1}} = \frac{x_1 \cdot \cdots x_{N-1} \cdot x_0}{x_0 \cdot x_1 \cdots x_{N-1}} = 1,$$

because all numerators and denominators cancel out.

Because each x_i is a unimodular complex number, we can write $x_i = e^{i\theta}$ for some θ. Then, $\overline{x_i} = e^{-i\theta} = 1/x_i$. Now, adding $z_0 + \cdots + z_{N-1}$ gives

$$\frac{x_1}{x_0} + \frac{x_2}{x_1} \cdots + \frac{x_0}{x_{N-1}} = x_1\overline{x_0} + x_2\overline{x_1} + \cdots + x_0\overline{x_{N-1}} = \langle (x_1, x_2, \cdots, x_0), (x_0, x_1, \dots, x_{N-1})\rangle = 0$$

since $(x_0, \dots, x_N)$ is a CAZAC sequence and thus satisfies the zero autocorrelation property.

Next, taking $z_0z_1 + z_1z_2 + \cdots + z_{N-1}x_0$ gives

$$\begin{aligned}&\frac{x_1}{x_0} \cdot \frac{x_2}{x_1} + \cdots + \frac{x_{N-1}}{x_{N-2}} \frac{x_0}{x_{N-1}} + \frac{x_0}{x_{N-1}} \frac{x_1}{x_0} \\ &= x_2\overline{x_0} + x_3\overline{x_1} + \cdots + x_0\overline{x_{N-2}} + x_1\overline{x_{N-1}} \\ &= \langle (x_2, x_3, \cdots, x_1), (x_0, x_1, \ldots, x_{N-1}) \rangle,\end{aligned}$$

which is 0 by the zero autocorrelation of $(x_0, \ldots, x_N)$.

In general, if we take $z_0z_1 \cdots z_i + z_1 \cdots z_{i+1} + \cdots + z_N \cdot z_0 \cdots z_{i-1}$, where $0 \leq i \leq N-1$, we get

$$\begin{aligned}&\frac{x_1}{x_0} \cdot \frac{x_2}{x_1} \cdots \frac{x_i}{x_{i-1}} + \cdots + \frac{x_{N-1}}{x_{N-2}} \frac{x_0}{x_{N-1}} \cdots \frac{x_{i-2}}{x_{i-3}} + \frac{x_0}{x_{N-1}} \frac{x_1}{x_0} \cdots \frac{x_{i-1}}{x_{i-2}} \\ &= x_i\overline{x_0} + x_{i+1}\overline{x_1} + \cdots + x_{i-2}\overline{x_{N-2}} + x_{i-1}\overline{x_{N-1}} \\ &= \langle (x_i, x_{i+1}, \cdots, x_{i-1}), (x_0, x_1, \ldots, x_{N-1}) \rangle,\end{aligned}$$

which is 0 by the zero autocorrelation of $(x_0, \ldots, x_N)$.

Thus, we see that $(z_0, \ldots, z_{N-1})$ is a cyclic unimodular N-root, as desired.

Now, we show that if $(z_0, \ldots, z_{N-1})$ is a cyclic N-root where $x_0 = 1$, then if we recursively define $x_0 = 1, x_k = x_{k-1}z_{k-1}$, then we get a CAZAC sequence (Note that this forces $z_{k-1} = x_k/x_{k-1}$, as before).

Certainly $|x_0| = 1$. Assume inductively that $|x_{k-1}| = 1$. Then $|x_k| = |x_{k-1}|$ $|z_{k-1}| = |z_{k-1}| = 1$ because $(z_0, \ldots, z_{N-1})$ is unimodular.

Also, we can compute

$$\langle (x_0, \ldots, x_{N-1}), (x_i, \ldots, x_{i+(N-1)}) \rangle = x_0\overline{x_i} + \cdots + x_{N-1}\overline{x_{i+(N-1)}} = \frac{x_0}{x_i} + \cdots + \frac{x_{N-1}}{x_{i+(N-1)}},$$

and we have already seen that this is $z_0z_1 \cdots z_i + z_1 \cdots z_{i+1} + \cdots + z_N \cdot z_0 \cdots z_{i-1}$, which we know to be 0 because $(z_0, \ldots, z_{N-1})$ is a cyclic N-root.

Example 1.2.5 We state the following modest extensions of the Gaussian CAZAC sequence example of Example 1.1.3.

a. Given an integer $N \geq 2$.

$$M = \begin{cases} N, & N \text{ odd}, \\ 2N, & N \text{ even}, \end{cases}$$

and let W_M be a primitive M-root of unity. Then, $\{W_M^{m^2}\}_{m=0}^{N-1}$ is a CAZAC sequence of length N. We refer to $\{W_M^{m^2}\}_{m=0}^{N-1}$ as the *Wiener sequence*, see [9] and Sect. 1.3.4.

b. Given an odd integer $N \geq 3$. Then, $\{g_{N,a,b}[m]\}_{m=0}^{N-1}$ is a CAZAC sequence.

c. Generally, for any CAZAC sequence of length N, we can construct a sequence of length N^2 in a systematic way. The construction is due to Milewski, see [9].

1.2.2 Equivalence Classes of CAZAC Sequences

There are several meaningful ways of defining equivalence classes on CAZAC sequences. We shall employ the following elementary definition. Two CAZAC sequences, x and y, on $\mathbb{Z}/N\mathbb{Z}$, are *equivalent* if $x = cy$ for some $|c| = 1$, e.g., see Haagerup [32]. Do there exist only finitely many non-equivalent CAZAC sequences in $\mathbb{Z}/N\mathbb{Z}$? The answer to this question is "yes" for N prime and "no" for $N = MK^2$, see, e.g., [9, 54]. For the case of non-prime square-free N, special cases are known, and there are published arguments asserting general results.

Another definition of equivalence, which was developed in [9], is the following. Two CAZAC sequences, x and y, on $\mathbb{Z}/N\mathbb{Z}$, are defined to be *5-operation equivalent* if they can be obtained from one another by means of compositions of the five operations: rotation, translation, decimation, linear frequency modulation, and conjugation. These *5-equivalence operations* for CAZAC sequences are defined as follows for all $k \in \mathbb{Z}/N\mathbb{Z}$:

1. (Rotation) $y[k] = cx[k]$, for some $|c| = 1$.
2. (Translation) $y[k] = x[k - m]$, for some $m \in \mathbb{Z}/N\mathbb{Z}$.
3. (Decimation) $y[k] = x[mk]$, for some $m \in \mathbb{Z}/N\mathbb{Z}$ for which $\gcd(m, N) = 1$.
4. (Linear Frequency Modulation) $y[k] = W_N^k x[k]$.
5. (Conjugation) $y[k] = \overline{x[k]}$.

Example 1.2.6 (Equivalence relations between CAZAC sequences and cyclic roots) Suppose two CAZAC sequences, x and y, defined on $\mathbb{Z}/N\mathbb{Z}$ have associated cyclic N-roots $\{z_k\}$ and $\{w_k\}$. It is straightforward to verify the following relations (stated for all $k \in \mathbb{Z}/N\mathbb{Z}$) between the 5-equivalence operations for CAZAC sequences, and how they become relations between cyclic N-roots.

1. $y[k] = cx[k] \Longrightarrow w_k = z_k$.
2. $y[k] = x[k - m] \Longrightarrow w_k = z_{k-m}$.
3. $y[k] = x[mk] \Longrightarrow w_k = \prod_{j=mk+1}^{mk+m} z_j$.
4. $y[k] = \underline{W_N^k} x[k] \Longrightarrow w_k = W_N z_k$.
5. $y[k] = \overline{x[k]} \Longrightarrow w_k = \overline{z_k}$.

In particular, the 5-equivalence operations for CAZAC sequences give rise to operations under which cyclic N-roots are closed.

Thus, CAZAC sequences that are equivalent are also 5-operation equivalent. However, generally, CAZAC sequences that are 5-operation equivalent are not equivalent. This is significant because the numbers in Table 1.1 refer to the number of equivalent CAZAC sequences.

In practice, two CAZAC sequences, x and y, are not equivalent if $x[0] = y[0] = 1$, but $x[k] \neq y[k]$ for some $k > 0$. It is clearly more difficult to check if two CAZAC sequences are 5-operation equivalent than if they are equivalent.

Additionally, there are questions about how these two notions of equivalence among CAZAC sequences relate to equivalence classes of the corresponding

Hadamard matrices. It is not true that CAZAC sequences from two equivalent Hadamard matrices will be equivalent in the sense of the 5-equivalence operations.

1.2.3 Cyclic p-roots

In order to address the problem of finding all cyclic p-roots computationally, where p is prime, we developed a Python script which checks every permutation of the p-roots of unity by brute force and tried to see if and when they are cyclic p-roots. Based on this script, we were led to formulate the following result. The result itself, along with a combinatorial argument, leads to all 20 cyclic 5-roots with modulus 1, see Sect. 1.3.5.

Proposition 1.2.7 *Let p be an odd prime, and recall that $W_p = e^{2\pi i/p}$. If $s \in \{1, \cdots, p-1\}$ and $r \in \{1, \cdots, p\}$, then $(W_p^r, W_p^{r+s}, W_p^{r+2s}, \cdots, W_p^{r+(p-1)s})$ is a cyclic p-root.*

Proof Given any cyclic p-root, we can obtain another cyclic p-root by multiplying by W_p^r. In particular, we can assume without loss of generality that $r = 0$. Fix $s \in \{0, \cdots, p-1\}$. The tth equation $(0 \leq t < p)$ in the cyclic p-root system can be written as

$$\sum_{k=0}^{p-1} \prod_{\ell=0}^{t-1} x_{k+\ell} = 0$$

so we would like to verify that

$$\sum_{k=0}^{p-1} \prod_{\ell=0}^{t-1} W_p^{s(k+\ell)} = 0. \tag{1.6}$$

To this end, we compute directly and obtain

$$\sum_{k=0}^{p-1} \prod_{\ell=0}^{t-1} W_p^{s(k+\ell)} = \sum_{k=0}^{p-1} W_p^{skt} \prod_{\ell=0}^{t-1} W_p^{s\ell} = \sum_{k=1}^{p-1} W_p^{skt} W_p^{s\sum_{\ell=0}^{t-1}\ell} = \sum_{k=1}^{p-1} W_p^{skt} W_p^{st(t-1)/2}$$
$$= W_p^{st(t-1)/2} \sum_{k=1}^{p-1} W_p^{skt} = 0,$$

since W_p^{st} is a primitive p-root of unity. For the last (pth) equation we want to show

$$\prod_{k=0}^{p-1} W_p^{sk} = 1. \tag{1.7}$$

To verify this, we once again compute directly, and obtain

$$\prod_{k=0}^{p-1} W_p^{sk} = W_p^{s\sum_{k=0}^{p-1} k} = W_p^{sp(p-1)/2} = e^{s(p-1)\pi i} = 1,$$

since p is odd. Combining Eqs. (1.6) and (1.7) gives us that $(1, W_p^s, \ldots, W_p^{(p-1)s})$ is a cyclic p-root.

Corollary 1.2.8 *The number of cyclic p-roots that are comprised of p-roots of unity is bounded below by $p(p-1)$.*

Proof Each cyclic p-root in Proposition 1.2.7 is comprised of roots of unity. There are two parameters: s and r. Note that s can take up to $p-1$ different values, and r can take up to p different values. Thus, the number of possible cyclic p-roots that can be formed by Proposition 1.2.7 is $p(p-1)$. This gives us the desired lower bound.

In particular, as a consequence of Corollary 1.2.8, all 20 cyclic 5-roots with modulus 1 are generated by Proposition 1.2.7. It is natural to speculate that all cyclic p-roots are given by Proposition 1.2.7, and that the lower bound $p(p-1)$ of Corollary 1.2.8 is the exact number of cyclic p-roots that are comprised of p-roots of unity.

1.2.4 CAZAC Sequences of Non-square-free Length

Much of the following material is found in [18] but has been recorded here for completeness.

Theorem 1.2.9 *Let $c \in \mathbb{C}^N$ be any constant amplitude sequence of length $N \geq 2$, and let σ be any permutation of the set $\{0, 1, \cdots, N-1\}$. Define a new sequence, $x \in \mathbb{C}^{N^2}$, by the formula,*

$$\forall a, b \in \{0, 1, \cdots, N-1\}, \quad x[aN+b] = c[b]e^{2\pi i a\sigma(b)/N}.$$

Then, x is a CAZAC sequence of length N^2.

Proof Without loss of generality, assume $|c[i]| = 1$ for all $i = 0, 1, \ldots, N-1$. The sequence, x, is CA by its definition. We shall prove that x is also ZAC by verifying that $\widehat{x}$ is CA. Let $W_{N^2} = e^{2\pi i/N^2}$. Then,

$$|\widehat{x}[j]| = \left| \sum_{k=0}^{N^2-1} x[k] W_{N^2}^{-kj} \right| = \left| \sum_{a=0}^{N-1} \sum_{b=0}^{N-1} x[aN+b] W_{N^2}^{-(aN+b)j} \right|$$

$$= \left| \sum_{a=0}^{N-1} \sum_{b=0}^{N-1} c[b] W_{N^2}^{Na\sigma(b)} W_{N^2}^{-aNj} W_{N^2}^{-bj} \right| = \left| \sum_{b=0}^{N-1} c[b] W_{N^2}^{-bj} \sum_{a=0}^{N-1} W_{N^2}^{Na\sigma(b)} W_{N^2}^{-aNj} \right|$$

$$= \left| \sum_{b=0}^{N-1} c[b] W_{N^2}^{-bj} \sum_{a=0}^{N-1} W_{N^2}^{N(\sigma(b)-j)a} \right|. \tag{1.8}$$

Note that the inner sum of (1.8) is 0 unless $\sigma(b) \equiv j \bmod N$, in which case the inner sum is N. Thus, we can rewrite (1.8), taking j modulo N if necessary, as

$$|\widehat{x}[j]| = \left| \sum_{b=0}^{N-1} c[b] W_{N^2}^{-bj} \sum_{a=0}^{N-1} W_{N^2}^{N(\sigma(b)-j)a} \right| = |Nc[\sigma^{-1}(j) W_{N^2}^{-\sigma^{-1}(j)j}| = N.$$

Corollary 1.2.10 *Given an integer $N \geq 2$. There are infinitely many non-equivalent CAZAC sequences of length N^2 whose first term is 1.*

We now wish to extend Theorem 1.2.9 to arbitrary sequences whose length is not square-free.

Theorem 1.2.11 *Let $Q \geq 2$ be an integer, let N^2 be the largest square dividing Q, let σ be any permutation of $\{0, 1, \cdots, N-1\}$, and consider the primitive M-root W_M, where $M = Q/N$. If $c \in \mathbb{C}^N$ is a constant amplitude sequence of length N, then define a new sequence, $x \in \mathbb{C}^Q$, by the formula*

$$\forall a \in \{0, \cdots, M-1\} \text{ and } \forall b \in \{0, \cdots, N-1\}, \quad x[aN+b] = c[b] W_M^{a\sigma(b)+Na(a-1)/2}.$$

If at least one of N and $M-1$ is even, then x is a CAZAC sequence of length Q.

Proof Without loss of generality, assume $|c[i]| = 1$ for every $i \in \{0, \cdots, N-1\}$. First, we note that x can be extended to an Q-periodic function on all of $\mathbb{Z}$. Indeed, if we let $k \in \mathbb{C}^Q$ be written as $k = aN + b$, then

$$\frac{x[Q+k]}{x[k]} = \frac{x[(M+a)N+b]}{x[aN+b]} = \frac{c[b] W_M^{(M+a)\sigma(b)+N(M+a)(M+a-1)/2}}{c[b] W_M^{a\sigma(b)+Na(a-1)/2}}$$

$$= \frac{W_M^{M\sigma(b)} W_M^{a\sigma(b)} W_M^{N(M^2+2Ma-M)} W_M^{Na(a-1)/2}}{W_M^{a\sigma(b)} W_M^{Na(a-1)/2}}$$

$$= W_M^{M\sigma(b)} W_M^{NM(2a+(M-1))/2} = 1,$$

since both terms are an Mth root of unity raised to a power which is an integer multiple of M.

Using this, we can directly compute the autocorrelation of x at $u = rN + s$, where $r \in \{0, \cdots, M-1\}$ and $s \in \{0, \cdots, N-1\}$ and at least one of r and s is nonzero, i.e., $u \neq 0$. Let $k = aN + b$ and $\theta = \lfloor \frac{b+s}{N} \rfloor$. Then,

$$\begin{aligned}A_x[u] &= \sum_{k=0}^{Q-1} x[k+u]\overline{x[k]} = \sum_{a=0}^{M-1}\sum_{b=0}^{N-1} x[(a+r+\theta)N+(b+s)]\overline{x[aN+b]} \\ &= \sum_{a=0}^{M-1}\sum_{b=0}^{N-1} c[b+s]W_M^{(a+r+\theta)\sigma(b+s)+N(a+r+\theta)(a+r+\theta-1)/2}\overline{c[b]W_M^{-a\sigma(b)-Na(a-1)/2}} \\ &= C_r\sum_{b=0}^{N-1} c[b+s]\overline{c[b]}W_M^{\frac{N\theta(2r+\theta-1)}{2}+(r+\theta)\sigma(b+s)}\sum_{a=0}^{M-1} W_M^{a(\sigma(b+s)-\sigma(b)+N(r+\theta))},\end{aligned} \tag{1.9}$$

where $C_r = W_M^{N(r^2-r)/2}$. If $s=0$, then $\theta = 0$ for every $b \in \{0,\cdots,N-1\}$, and we can write (1.9) as

$$C_r\sum_{b=0}^{N-1}|c[b]|^2 W_M^{r\sigma(b)}\sum_{a=0}^{M-1} W_M^{aNr} = C_r\sum_{a=0}^{M-1}\sum_{b=0}^{N-1} W_M^{r(\sigma(b)+aN)}. \tag{1.10}$$

Since σ is a permutation of $\{0,\cdots,N-1\}$, we can make a substitution $q=\sigma(b)$ and reorder as necessary to rewrite (1.10) as

$$C_r\sum_{a=0}^{M-1}\sum_{q=0}^{N-1} W_M^{r(aN+q)} = C_r\sum_{k=0}^{Q-1} W_M^{rk} = 0,$$

since $r \neq 0 \mod N$. If $s \neq 0$, then in the inner sum of (1.9) we observe that $0 < |\sigma(b+s)-\sigma(b)| < N$, and thus N does not divide $(\sigma(b+s)-\sigma(b)+N(r+\theta))$ for any fixed b. It then follows that M does not divide $(\sigma(b+s)-\sigma(b)+N(r+\theta))$ for any fixed b and the inner sum is 0 for every b. Thus, if $s \neq 0$ then (1.9) is 0 as well.

Corollary 1.2.12 *Given an integer $Q \geq 2$ that is not square-free. There are infinitely many non-equivalent CAZAC sequences of length Q whose first term is 1.*

Proof Take $c[0]=1$. Let N^2 be the largest square dividing Q and $M=Q/N$. If either N is even or M is odd, then Theorem 1.2.11 applies immediately and the sequence given in Theorem 1.2.11 gives us infinitely many CAZAC sequences. If N is odd and M is even, then Q has exactly one factor of 2. Thus, we can write $M=2M'$ with M' odd. In Theorem 1.2.11, replace Q by $Q/2$ and M with M', and let y be the resulting CAZAC sequence of length $Q/2$. We can then construct a CAZAC sequence of length Q by taking the Kronecker product $z \otimes y$ of $z=(1,i) \in \mathbb{C}^2$ by $y \in \mathbb{C}^{Q/2}$, so that

$$z \otimes y = \left(y[0], y[1], \ldots, y\left[\frac{Q}{2}-1\right], i\,y[0], i\,y[1], \ldots, i\,y\left[\frac{Q}{2}-1\right]\right).$$

1.2.5 Dephased Hadamard Matrices

One property that follows from the definition of equivalence classes of complex Hadamard matrices is that every complex Hadamard matrix is equivalent to a unique *dephased* Hadamard matrix, i.e., a Hadamard matrix with a first row and first column of $1s$.

The website [19] maintained by Bruzda et al. states the following construction of this dephased form.

Proposition 1.2.13 *[19] Given an $N \times N$ complex Hadamard matrix,*

$$H = \begin{bmatrix} h_{0,0} & h_{0,1} & \cdots & h_{0,N-1} \\ h_{1,0} & h_{1,1} & \cdots & h_{1,N-1} \\ \vdots & \vdots & \ddots & \vdots \\ h_{N-1,0} & h_{N-1,1} & \cdots & h_{N-1,N-1} \end{bmatrix}.$$

The equivalent dephased form is

$$D_1 H D_2 = \begin{bmatrix} \overline{h}_{0,0} & 0 & \cdots & 0 \\ 0 & \overline{h}_{1,0} & \cdots & 0 \\ \vdots & \vdots & \ddots & \vdots \\ 0 & 0 & \cdots & \overline{h}_{N-1,0} \end{bmatrix} \begin{bmatrix} h_{0,0} & h_{0,1} & \cdots & h_{0,N-1} \\ h_{1,0} & h_{1,1} & \cdots & h_{1,N-1} \\ \vdots & \vdots & \ddots & \vdots \\ h_{N-1,0} & h_{N-1,1} & \cdots & h_{N-1,N-1} \end{bmatrix} \begin{bmatrix} 1 & 0 & \cdots & 0 \\ 0 & h_{0,0}\overline{h}_{0,1} & \cdots & 0 \\ \vdots & \vdots & \ddots & \vdots \\ 0 & 0 & \cdots & h_{0,0}\overline{h}_{0,N-1} \end{bmatrix}. \tag{1.11}$$

Moreover, the equivalent dephased form is unique.

Proof We compute the matrix product in Eq. (1.11). In the first step, we have

$$D_1 H D_2 = \begin{bmatrix} \overline{h}_{0,0}h_{0,0} & \overline{h}_{0,0}h_{0,1} & \cdots & \overline{h}_{0,0}h_{0,N-1} \\ \overline{h}_{1,0}h_{1,0} & \overline{h}_{1,0}h_{1,1} & \cdots & \overline{h}_{1,0}h_{1,N-1} \\ \vdots & \vdots & \ddots & \vdots \\ \overline{h}_{N-1,0}h_{N-1,0} & \overline{h}_{N-1,0}h_{N-1,1} & \cdots & \overline{h}_{N-1,0}h_{N-1,N-1} \end{bmatrix} \begin{bmatrix} 1 & 0 & \cdots & 0 \\ 0 & h_{0,0}\overline{h}_{0,1} & \cdots & 0 \\ \vdots & \vdots & \ddots & \vdots \\ 0 & 0 & \cdots & h_{0,0}\overline{h}_{0,N-1} \end{bmatrix},$$

from which we obtain

$$D_1 H D_2 = \begin{bmatrix} \overline{h}_{0,0}h_{0,0} & \overline{h}_{0,0}h_{0,0}h_{0,1}\overline{h}_{0,1} & \cdots & h_{0,0}\overline{h}_{0,0}h_{0,N-1}\overline{h}_{0,N-1} \\ \overline{h}_{1,0}h_{1,0} & * & \cdots & * \\ \vdots & \vdots & \ddots & \vdots \\ \overline{h}_{N-1,0}h_{N-1,0} & * & \cdots & * \end{bmatrix}.$$

We now verify that $D_1 H D_2$ is a Hadamard matrix. First, because each h_{ij} has norm 1, and by norm multiplicativity, we see that each entry of $D_1 H D_2$ has norm 1. Next, note that $(D_1 H D_2)(D_1 H D_2)^* = D_1 H D_2 D_2^* H^* D_1^*$. Because D_1 and D_2 are unitary matrices, we have $D_2 D_2^* = I$ and $D_1 D_1^* = I$; and because H is a Hadamard matrix, $HH^* = N\,Id$. Thus, we obtain $D_1 H D_2 D_2^* H^* D_1^* = N D_1 I D_1^* = N\,Id$.

This matrix is dephased because the ith entry of the first column is of the form $\overline{h}_{i,0}h_{i,0} = 1$ for $0 \le i \le N-1$ and the ith entry of the first row, for $1 \le i \le N-1$, is of the form $h_{0,0}\overline{h}_{0,0}h_{0,i}\overline{h}_{0,i} = 1$.

Next, assume that there are $a_0, \ldots, a_{N-1}, b_0, \ldots, b_{N-1}$ such that $|a_i| = |b_i| = 1$ for $0 \le i \le N-1$ and

$$\begin{bmatrix} a_0 & 0 & \cdots & 0 \\ 0 & a_1 & \cdots & 0 \\ \vdots & \vdots & \ddots & \vdots \\ 0 & 0 & \cdots & a_{N-1} \end{bmatrix} \begin{bmatrix} 1 & 1 & \cdots & 1 \\ 1 & * & \cdots & * \\ \vdots & \vdots & \ddots & \vdots \\ 1 & * & \cdots & * \end{bmatrix} \begin{bmatrix} b_0 & 0 & \cdots & 0 \\ 0 & b_1 & \cdots & 0 \\ \vdots & \vdots & \ddots & \vdots \\ 0 & 0 & \cdots & b_{N-1} \end{bmatrix} = \begin{bmatrix} 1 & 1 & \cdots & 1 \\ 1 & * & \cdots & * \\ \vdots & \vdots & \ddots & \vdots \\ 1 & * & \cdots & * \end{bmatrix}. \tag{1.12}$$

Calculating the left side of (1.12), we have

$$\begin{bmatrix} a_0 & a_0 & \cdots & a_0 \\ a_1 & * & \cdots & * \\ \vdots & \vdots & \ddots & \vdots \\ a_{N-1} & * & \cdots & * \end{bmatrix} \begin{bmatrix} b_0 & 0 & \cdots & 0 \\ 0 & b_1 & \cdots & 0 \\ \vdots & \vdots & \ddots & \vdots \\ 0 & 0 & \cdots & b_{N-1} \end{bmatrix} = \begin{bmatrix} 1 & 1 & \cdots & 1 \\ 1 & * & \cdots & * \\ \vdots & \vdots & \ddots & \vdots \\ 1 & * & \cdots & * \end{bmatrix},$$

and so we obtain

$$\begin{bmatrix} a_0b_0 & a_0b_1 & \cdots & a_0b_{N-1} \\ a_1b_0 & * & \cdots & * \\ \vdots & \vdots & \ddots & \vdots \\ a_{N-1}b_0 & * & \cdots & * \end{bmatrix} = \begin{bmatrix} 1 & 1 & \cdots & 1 \\ 1 & * & \cdots & * \\ \vdots & \vdots & \ddots & \vdots \\ 1 & * & \cdots & * \end{bmatrix}.$$

From this, we find $a_0 = \cdots = a_{N-1} = b_0^{-1}$ and $b_0 = \cdots = b_{N-1} = a_0^{-1}$. Say $a_0 = x$ and $b_0 = x^{-1}$. Then, Eq. (1.12) becomes

$$\begin{bmatrix} x & 0 & \cdots & 0 \\ 0 & x & \cdots & 0 \\ \vdots & \vdots & \ddots & \vdots \\ 0 & 0 & \cdots & x \end{bmatrix} \begin{bmatrix} 1 & 1 & \cdots & 1 \\ 1 & * & \cdots & * \\ \vdots & \vdots & \ddots & \vdots \\ 1 & * & \cdots & * \end{bmatrix} \begin{bmatrix} x^{-1} & 0 & \cdots & 0 \\ 0 & x^{-1} & \cdots & 0 \\ \vdots & \vdots & \ddots & \vdots \\ 0 & 0 & \cdots & x^{-1} \end{bmatrix} = \begin{bmatrix} 1 & 1 & \cdots & 1 \\ 1 & * & \cdots & * \\ \vdots & \vdots & \ddots & \vdots \\ 1 & * & \cdots & * \end{bmatrix},$$

or

$$xx^{-1} \begin{bmatrix} 1 & 0 & \cdots & 0 \\ 0 & 1 & \cdots & 0 \\ \vdots & \vdots & \ddots & \vdots \\ 0 & 0 & \cdots & 1 \end{bmatrix} \begin{bmatrix} 1 & 1 & \cdots & 1 \\ 1 & * & \cdots & * \\ \vdots & \vdots & \ddots & \vdots \\ 1 & * & \cdots & * \end{bmatrix} \begin{bmatrix} 1 & 0 & \cdots & 0 \\ 0 & 1 & \cdots & 0 \\ \vdots & \vdots & \ddots & \vdots \\ 0 & 0 & \cdots & 1 \end{bmatrix} = \begin{bmatrix} 1 & 1 & \cdots & 1 \\ 1 & * & \cdots & * \\ \vdots & \vdots & \ddots & \vdots \\ 1 & * & \cdots & * \end{bmatrix},$$

and, thus, we have

$$\begin{bmatrix} 1 & 1 & \cdots & 1 \\ 1 & * & \cdots & * \\ \vdots & \vdots & \ddots & \vdots \\ 1 & * & \cdots & * \end{bmatrix} = \begin{bmatrix} 1 & 1 & \cdots & 1 \\ 1 & * & \cdots & * \\ \vdots & \vdots & \ddots & \vdots \\ 1 & * & \cdots & * \end{bmatrix}.$$

Therefore, the equivalent dephased form is unique.

1.3 Roots of Unity CAZAC Sequences of Prime Length

1.3.1 Introduction

All CAZAC sequences of lengths 3 and 5 are roots of unity, and they are known. We shall give the calculations, not only for exposition, but because they provide us with the ability to explore various methods for explicitly computing CAZAC sequences in different ways. For the case of $p = 3$, we shall give three different techniques. The first uses the correspondence between cyclic N-roots and CAZAC sequences, the second uses the correspondence between CAZAC sequences and Hadamard matrices, and the last capitalizes on the various notions of equivalence of CAZAC sequences. Then we proceed similarly for the case $p = 5$. There are 6 unimodular cyclic 3-roots and 20 unimodular cyclic 5-roots, see Table 1.1. We shall see that some of our calculations apply for arbitrary prime lengths and shall close the section by putting the material in the context of the 5-operation equivalence relations defined in Sect. 1.2.2.

1.3.2 Constructing CAZAC Sequences of Length 3 Using Cyclic 3-roots

We first would like to look at the specific case of cyclic 3-roots, which correspond to CAZAC sequences of length 3. In this case we are looking for solutions $(x, y, z) \in \mathbb{C}^3$ to the system of equations,

$$\begin{cases} x + y + z = 0 \\ xy + yz + zx = 0 \\ xyz = 1. \end{cases} \tag{1.13}$$

This system is easily solvable in the following way. First, multiply the second equation in (1.13) by z. This yields

$$xyz + yz^2 + xz^2 = 0.$$

By factoring z in the last two terms on the left-hand side and using the third equation in (1.13) we have

$$1 + z^2(x + y) = 0.$$

Rearranging the first equation in (1.13) gives us that $x + y = -z$. Substituting this into the above, we obtain

$$1 - z^3 = 0,$$

or, in other words, z must be a third root of unity. Note that the same computations can also be applied to x and y, and thus x and y must also be third roots of unity.

This leads to the conjecture that the six permutations of the third roots of unity $(1, e^{2\pi i/3}, e^{4\pi i/3})$ indeed generate all six CAZAC sequences of length 3. To this end, first let us write all six permutations of the third roots of unity and the corresponding candidate CAZAC sequences. Then, we verify that the sequences really are CAZAC sequences by observing that they are known CAZAC sequences or 5-operation equivalent.

The six permutations of the third roots of unity are

1. $(1, e^{2\pi i/3}, e^{4\pi i/3})$
2. $(1, e^{4\pi i/3}, e^{2\pi i/3})$
3. $(e^{2\pi i/3}, 1, e^{4\pi i/3})$
4. $(e^{2\pi i/3}, e^{4\pi i/3}, 1)$
5. $(e^{4\pi i/3}, 1, e^{2\pi i/3})$
6. $(e^{4\pi i/3}, e^{2\pi i/3}, 1)$.

Let (z_0, z_1, z_2) be a permutation of the third roots of unity. To convert (z_0, z_1, z_2) to the corresponding CAZAC sequence, we begin by letting $x[0] = 1$. Then, we define $x[1]$ and $x[2]$ as

$$x[1] = z_0$$

$$x[2] = z_0 z_1.$$

Using this, we can construct Table 1.2.

Table 1.2 Cyclic 3-roots and CAZAC sequences of length 3

Cyclic 3-root	CAZAC sequence
$(1, e^{2\pi i/3}, e^{4\pi i/3})$	$(1, 1, e^{2\pi i/3})$
$(1, e^{4\pi i/3}, e^{2\pi i/3})$	$(1, 1, e^{4\pi i/3})$
$(e^{2\pi i/3}, 1, e^{4\pi i/3})$	$(1, e^{2\pi i/3}, e^{2\pi i/3})$
$(e^{2\pi i/3}, e^{4\pi i/3}, 1)$	$(1, e^{2\pi i/3}, 1)$
$(e^{4\pi i/3}, 1, e^{2\pi i/3})$	$(1, e^{4\pi i/3}, e^{4\pi i/3})$
$(e^{4\pi i/3}, e^{2\pi i/3}, 1)$	$(1, e^{4\pi i/3}, 1)$

In particular, each of the six sequences generated by the six permutations of the roots of unity generates either a known CAZAC sequence or an aforementioned transformation of a known CAZAC sequence. Thus, Table 1.2 lists all six CAZAC sequences of length 3.

1.3.3 Constructing CAZAC Sequences of Length 3 Using Hadamard Matrices

In [13, 19], it is stated that all 3×3 Hadamard matrices are equivalent to the Fourier matrix. In fact, we shall obtain this result from computations in this subsection. The website [19] further characterizes the set of 3×3 Hadamard matrices into two types:

$$\left\{\begin{bmatrix} e^{ia} & 0 & 0 \\ 0 & e^{ib} & 0 \\ 0 & 0 & e^{ic} \end{bmatrix} \cdot \begin{bmatrix} 1 & 1 & 1 \\ 1 & W_3 & W_3^2 \\ 1 & W_3^2 & W_3 \end{bmatrix} \cdot \begin{bmatrix} 1 & 0 & 0 \\ 0 & e^{ix} & 0 \\ 0 & 0 & e^{iy} \end{bmatrix}\right\} \cup$$

$$\left\{\begin{bmatrix} e^{ia} & 0 & 0 \\ 0 & e^{ib} & 0 \\ 0 & 0 & e^{ic} \end{bmatrix} \cdot \begin{bmatrix} 1 & 1 & 1 \\ 1 & W_3^2 & W_3 \\ 1 & W_3 & W_3^2 \end{bmatrix} \cdot \begin{bmatrix} 1 & 0 & 0 \\ 0 & e^{ix} & 0 \\ 0 & 0 & e^{iy} \end{bmatrix}\right\},$$

where $a, b, c, x, y \in [0, 2\pi)$. We shall use these two forms to find all 3×3 circulant Hadamard matrices.

First, we consider the first form and compute the matrix product:

$$\begin{bmatrix} e^{ia} & 0 & 0 \\ 0 & e^{ib} & 0 \\ 0 & 0 & e^{ic} \end{bmatrix} \cdot \begin{bmatrix} 1 & 1 & 1 \\ 1 & W_3 & W_3^2 \\ 1 & W_3^2 & W_3 \end{bmatrix} \cdot \begin{bmatrix} 1 & 0 & 0 \\ 0 & e^{ix} & 0 \\ 0 & 0 & e^{iy} \end{bmatrix} = \begin{bmatrix} e^{ia} & e^{i(a+x)} & e^{i(a+y)} \\ e^{ib} & e^{i(b+x+\frac{2}{3}\pi)} & e^{i(b+y-\frac{2}{3}\pi)} \\ e^{ic} & e^{i(c+x-\frac{2}{3}\pi)} & e^{i(c+y+\frac{2}{3}\pi)} \end{bmatrix}. \tag{1.14}$$

In order for this matrix to be circulant, the following system of equations must hold mod 2π:

$$\begin{cases} a = b + x + \frac{2\pi}{3} = c + y + \frac{2\pi}{3} \\ c = a + x = b + y + \frac{4\pi}{3} \\ a + y = b = c + x + \frac{4\pi}{3}. \end{cases} \tag{1.15}$$

From the first equation in (1.15) we have

$$a = b + x + \frac{2}{3}\pi. \tag{1.16}$$

Using (1.16) in the second equation of (1.15), we calculate that

$$c = a + x = \left(b + x + \frac{2}{3}\pi\right) + x = b + 2x + \frac{2}{3}\pi. \tag{1.17}$$

From (1.16), (1.17), and the third equation of (1.15), we obtain

$$y = c + x + \frac{4}{3}\pi - a = \left(b + 2x + \frac{2}{3}\pi\right) + x + \frac{4}{3}\pi - \left(b + x + \frac{2}{3}\pi\right) = 2x + \frac{4}{3}\pi. \tag{1.18}$$

Finally, returning to the first equation of (1.15) and using (1.17) and (1.18), we have

$$b = c + y - x = \left(b + 2x + \frac{2}{3}\pi\right) + \left(2x + \frac{4}{3}\pi\right) - x = b + 3x + 2\pi. \tag{1.19}$$

In particular, (1.19) implies that $3x \equiv 0 \bmod 2\pi$, i.e., x is $\frac{2}{3}\pi, 0$, or $-\frac{2}{3}\pi$. Letting $x = \frac{2}{3}\pi$, we obtain as one solution: $x = \frac{2}{3}\pi, a = \frac{4}{3}\pi + b, c = b$, and $y = \frac{2}{3}\pi$, where b is left indeterminate.

As such, we return to (1.14) and use this solution to compute

$$\begin{bmatrix} e^{i(\frac{4}{3}\pi+b)} & 0 & 0 \\ 0 & e^{ib} & 0 \\ 0 & 0 & e^{ib} \end{bmatrix} \cdot \begin{bmatrix} 1 & 1 & 1 \\ 1 & W_3 & W_3^2 \\ 1 & W_3^2 & W_3 \end{bmatrix} \cdot \begin{bmatrix} 1 & 0 & 0 \\ 0 & e^{\frac{2}{3}\pi i} & 0 \\ 0 & 0 & e^{\frac{2}{3}\pi i} \end{bmatrix} = e^{ib} \begin{bmatrix} e^{-\frac{2}{3}\pi i} & 1 & 1 \\ 1 & e^{-\frac{2}{3}\pi i} & 1 \\ 1 & 1 & e^{-\frac{2}{3}\pi i} \end{bmatrix}.$$

The first row of the resulting circulant Hadamard matrix is $e^{ib}(e^{\frac{-2}{3}\pi i}, 1, 1)$. We let $b = \frac{2}{3}\pi$ and choose $(1, e^{\frac{2}{3}\pi i}, e^{\frac{2}{3}\pi i})$ as the representative for this class of CAZAC sequences and find our first CAZAC sequence.

As a second solution, we choose $x = 0$, which gives $a = \frac{2}{3}\pi + b, c = \frac{2}{3}\pi + b$, and $y = \frac{4}{3}\pi$, where b is again indeterminate. We return to (1.14) and compute

$$\begin{bmatrix} e^{i(\frac{2}{3}\pi+b)} & 0 & 0 \\ 0 & e^{ib} & 0 \\ 0 & 0 & e^{i(\frac{2}{3}\pi+b)} \end{bmatrix} \cdot \begin{bmatrix} 1 & 1 & 1 \\ 1 & W_3 & W_3^2 \\ 1 & W_3^2 & W_3 \end{bmatrix} \cdot \begin{bmatrix} 1 & 0 & 0 \\ 0 & 1 & 0 \\ 0 & 0 & e^{\frac{4}{3}\pi i} \end{bmatrix} = e^{ib} \begin{bmatrix} e^{\frac{2}{3}\pi i} & e^{\frac{2}{3}\pi i} & 1 \\ 1 & e^{\frac{2}{3}\pi i} & e^{\frac{2}{3}\pi i} \\ e^{\frac{2}{3}\pi i} & 1 & e^{\frac{2}{3}\pi i} \end{bmatrix}.$$

The first row of this circulant Hadamard matrix is $e^{ib}(e^{\frac{2}{3}\pi i}, e^{\frac{2}{3}\pi i}, 1)$. Letting $b = -\frac{2}{3}\pi$, we have $(1, 1, e^{\frac{4}{3}\pi i})$ as our second CAZAC sequence.

The final solution is $x = -\frac{2}{3}\pi, a = b, c = b - \frac{2}{3}\pi$, and $y = 0$, where b is indeterminate. Returning to (1.14), we take

$$\begin{bmatrix} e^{ib} & 0 & 0 \\ 0 & e^{ib} & 0 \\ 0 & 0 & e^{i(b-\frac{2}{3}\pi)} \end{bmatrix} \cdot \begin{bmatrix} 1 & 1 & 1 \\ 1 & W_3 & W_3^2 \\ 1 & W_3^2 & W_3 \end{bmatrix} \cdot \begin{bmatrix} 1 & 0 & 0 \\ 0 & e^{-\frac{2}{3}\pi i} & 0 \\ 0 & 0 & 1 \end{bmatrix} = e^{ib} \begin{bmatrix} 1 & e^{-\frac{2}{3}\pi i} & 1 \\ 1 & 1 & e^{-\frac{2}{3}\pi i} \\ e^{-\frac{2}{3}\pi i} & 1 & 1 \end{bmatrix}.$$

The first row of this circulant Hadamard matrix is $e^{ib}(1, e^{-\frac{2}{3}\pi i}, 1)$, and so letting $b = 0$, we have $(1, e^{\frac{4}{3}\pi i}, 1)$ as our third CAZAC sequence.

Now, we consider the second form of 3×3 Hadamard matrices in the union written at the beginning of this subsection, and take the product,

$$\begin{bmatrix} e^{ia} & 0 & 0 \\ 0 & e^{ib} & 0 \\ 0 & 0 & e^{ic} \end{bmatrix} \cdot \begin{bmatrix} 1 & 1 & 1 \\ 1 & W_3^2 & W_3 \\ 1 & W_3 & W_3^2 \end{bmatrix} \cdot \begin{bmatrix} 1 & 0 & 0 \\ 0 & e^{ix} & 0 \\ 0 & 0 & e^{iy} \end{bmatrix} = \begin{bmatrix} e^{ia} & e^{i(a+x)} & e^{i(a+y)} \\ e^{ib} & e^{i(b+x-\frac{2}{3}\pi)} & e^{i(b+y+\frac{2}{3}\pi)} \\ e^{ic} & e^{i(c+x+\frac{2}{3}\pi)} & e^{i(c+y-\frac{2}{3}\pi)} \end{bmatrix}. \tag{1.20}$$

In order for the right-hand side matrix to be circulant, the following equations must hold mod 2π:

$$\begin{cases} a = b + x + \frac{4\pi}{3} = c + y + \frac{4\pi}{3} \\ c = a + x = b + y + \frac{2\pi}{3} \\ a + y = b = c + x + \frac{2\pi}{3}. \end{cases} \tag{1.21}$$

Using the first equation in (1.21), we have

$$a = b + x + \frac{4}{3}\pi. \tag{1.22}$$

Next, using the second equation in (1.21) and as well as (1.22), we calculate that

$$c = a + x = \left(b + x + \frac{4}{3}\pi\right) + x = b + 2x + \frac{4}{3}\pi. \tag{1.23}$$

We now use the third equation in (1.21) along with (1.22) and (1.23), and obtain

$$y = c + x + \frac{2}{3}\pi - a = \left(b + 2x + \frac{4}{3}\pi\right) + x + \frac{2}{3}\pi - \left(b + x + \frac{4}{3}\pi\right) = 2x + \frac{2}{3}\pi \tag{1.24}$$

Finally, we return to the first equation of (1.21) and use (1.23) and (1.24) to compute

$$b = c + y - x = \left(b + 2x + \frac{4}{3}\pi\right) + \left(2x + \frac{2}{3}\pi\right) - x = b + 3x + 2\pi. \tag{1.25}$$

Similar to the previous calculations, (1.25) gives $3x \equiv 0 \bmod 2\pi$, or $x = \frac{2}{3}\pi, x = 0$, i.e., x is 0, $\frac{2}{3}\pi$ or $-\frac{2}{3}\pi$.

Our first solution is $x = \frac{2}{3}\pi$, $a = b$, $c = b + \frac{2}{3}\pi$, and $y = 0$, where b is arbitrary. As such, we return to (1.20) and compute

$$\begin{bmatrix} e^{ib} & 0 & 0 \\ 0 & e^{ib} & 0 \\ 0 & 0 & e^{i(b+\frac{2}{3}\pi)} \end{bmatrix} \cdot \begin{bmatrix} 1 & 1 & 1 \\ 1 & W_3 & W_3^2 \\ 1 & W_3^2 & W_3 \end{bmatrix} \cdot \begin{bmatrix} 1 & 0 & 0 \\ 0 & e^{\frac{2}{3}\pi} & 0 \\ 0 & 0 & 1 \end{bmatrix} = e^{ib} \begin{bmatrix} 1 & e^{\frac{2}{3}\pi i} & 1 \\ 1 & 1 & e^{\frac{2}{3}\pi i} \\ e^{\frac{2}{3}\pi i} & 1 & 1 \end{bmatrix}.$$

The first row of the resulting circulant Hadamard matrix is $e^{ib}(1, e^{\frac{2}{3}\pi i}, 1)$, and so letting $b = 0$ we obtain $(1, e^{\frac{2}{3}\pi i}, 1)$ as the fourth CAZAC sequence.

Our second solution is $x = 0$, $y = \frac{2}{3}\pi$, $a = b + \frac{4}{3}\pi$, and $c = b + \frac{4}{3}\pi$ where b is arbitrary. In this case, we take

$$\begin{bmatrix} e^{i(b+\frac{4}{3}\pi)} & 0 & 0 \\ 0 & e^{ib} & 0 \\ 0 & 0 & e^{i(b+\frac{4}{3}\pi)} \end{bmatrix} \cdot \begin{bmatrix} 1 & 1 & 1 \\ 1 & W_3 & W_3^2 \\ 1 & W_3^2 & W_3 \end{bmatrix} \cdot \begin{bmatrix} 1 & 0 & 0 \\ 0 & 1 & 0 \\ 0 & 0 & e^{\frac{2}{3}\pi} \end{bmatrix} = e^{ib} \begin{bmatrix} e^{-\frac{2}{3}\pi i} & e^{-\frac{2}{3}\pi i} & 1 \\ 1 & e^{-\frac{2}{3}\pi i} & e^{-\frac{2}{3}\pi i} \\ e^{-\frac{2}{3}\pi i} & 1 & e^{-\frac{2}{3}\pi i} \end{bmatrix}.$$

The first row of this circulant Hadamard matrix is $e^{ib}(e^{-\frac{2}{3}\pi i}, e^{-\frac{2}{3}\pi i}, 1)$, and so letting $b = \frac{2}{3}\pi$ we obtain $(1, 1, e^{\frac{2}{3}\pi i})$ as the fifth CAZAC sequence.

A third solution is $x = -\frac{2}{3}\pi$, $y = -\frac{2}{3}\pi$, $a = b + \frac{2}{3}\pi$, and $c = b$. We take

$$\begin{bmatrix} e^{i(b+\frac{2}{3}\pi)} & 0 & 0 \\ 0 & e^{ib} & 0 \\ 0 & 0 & e^{ib} \end{bmatrix} \cdot \begin{bmatrix} 1 & 1 & 1 \\ 1 & W_3 & W_3^2 \\ 1 & W_3^2 & W_3 \end{bmatrix} \cdot \begin{bmatrix} 1 & 0 & 0 \\ 0 & e^{-\frac{2}{3}\pi} & 0 \\ 0 & 0 & e^{-\frac{2}{3}\pi} \end{bmatrix} = e^{ib} \begin{bmatrix} e^{\frac{2}{3}\pi i} & 1 & 1 \\ 1 & e^{\frac{2}{3}\pi i} & 1 \\ 1 & 1 & e^{\frac{2}{3}\pi i} \end{bmatrix}.$$

The first row of this Hadamard matrix is $e^{ib}(e^{\frac{2}{3}\pi i}, 1, 1)$, and so letting $b = -\frac{2}{3}\pi$ we obtain $(1, e^{\frac{4}{3}\pi i}, e^{\frac{4}{3}\pi i})$ as the sixth and final CAZAC sequence.

To summarize, the six CAZAC sequences that we have obtained are

$$(1, e^{2\pi i/3}, e^{2\pi i/3})$$

$$(1, 1, e^{4\pi i/3})$$

$$(1, e^{4\pi i/3}, 1)$$

$$(1, e^{2\pi i/3}, 1)$$

$$(1, 1, e^{2\pi i/3})$$

$$(1, e^{4\pi i/3}, e^{4\pi i/3});$$

and they are associated with the following circulant Hadamard matrices:

$$\begin{bmatrix} 1 & e^{2\pi i/3} & e^{2\pi i/3} \\ e^{2\pi i/3} & 1 & e^{2\pi i/3} \\ e^{2\pi i/3} & e^{2\pi i/3} & 1 \end{bmatrix}, \begin{bmatrix} 1 & 1 & e^{4\pi i/3} \\ e^{4\pi i/3} & 1 & 1 \\ 1 & e^{4\pi i/3} & 1 \end{bmatrix}, \begin{bmatrix} 1 & e^{4\pi i/3} & 1 \\ 1 & 1 & e^{4\pi i/3} \\ e^{4\pi i/3} & 1 & 1 \end{bmatrix}$$

$$\begin{bmatrix} 1 & e^{2\pi i/3} & 1 \\ 1 & 1 & e^{2\pi i/3} \\ e^{2\pi i/3} & 1 & 1 \end{bmatrix}, \begin{bmatrix} 1 & 1 & e^{2\pi i/3} \\ e^{2\pi i/3} & 1 & 1 \\ 1 & e^{2\pi i/3} & 1 \end{bmatrix}, \begin{bmatrix} 1 & e^{4\pi i/3} & e^{4\pi i/3} \\ e^{4\pi i/3} & 1 & e^{4\pi i/3} \\ e^{4\pi i/3} & e^{4\pi i/3} & 1 \end{bmatrix}.$$

Note that these CAZAC sequences match the CAZAC sequences found in Sect. 1.3.2.

1.3.4 5-Operation Equivalence Relations

Let p be prime. In this subsection, we show that the 5-operation equivalence relation is an equivalence relation which is generated by a group G acting on $U_p^p \subseteq \mathbb{C}^p$, where U_p^p is the group of ordered p-tuples of p-roots of unity. We apply this to the specific cases of $p = 3$ and $p = 5$ in Sect. 1.3.5 to illustrate yet another way to generate all CAZAC sequences of length 3 and also to generate all sequences of length 5 in the process. We define the five operations again in the following way:

1. $c_0 x[n] = x[n]$ and $c_1 x[n] = \overline{x[n]}$;
2. $\tau_b x[n] = x[n-b]$, $b \in \mathbb{Z}/p\mathbb{Z}$;
3. $\pi_c x[n] = x[cn]$, $c \in \mathbb{Z}/p\mathbb{Z}$, $c \neq 0$;
4. $e_d x[n] = e^{2\pi i dn/p} x[n]$, $d \in \mathbb{Z}/p\mathbb{Z}$;
5. $\omega_f x[n] = e^{2\pi i f/p} x[n]$, $f \in \mathbb{Z}/p\mathbb{Z}$.

With this, we can define the set G as

$$G = \{(a, b, c, d, f) : a \in \{0, 1\},\ b, c, d, f \in \mathbb{Z}/p\mathbb{Z},\ c \neq 0\},$$

which has size $|G| = 2p^3(p-1)$. To each element $(a, b, c, d, f) \in G$ we associate the operator $\omega_f e_d \pi_c \tau_b c_a$. To motivate the group operation, we take (a, b, c, d, f), $(h, j, k, \ell, m) \in G$. One can show that the composition of the associated operators is

$$(\omega_m e_\ell \pi_k \tau_j c_h) \circ (\omega_f e_d \pi_c \tau_b c_a) = \omega_{m+(-1)^h(f-jc)} e_{\ell+(-1)^h kd} \pi_{ck} \tau_{cj+b} c_{a+h}.$$

As such we define the operation: $G \times G \to G$ by

$$(a, b, c, d, f) \cdot (h, j, k, \ell, m) = (a+h, cj+b, ck, \ell + (-1)^h kd, m + (-1)^h (f - jc)).$$

Theorem 1.3.1 *The operation $\cdot$ defines a group operation for G. In particular, $(G, \cdot)$ is a group.*

Proof We need to show that the operation is associative, has an identity element, and that each element has an inverse. It is easily verified that the identity element is $(0, 0, 1, 0, 0)$. Given an element $(a, b, c, d, f) \in G$, it is elementary to verify that

$$(a, b, c, d, f)^{-1} = (-a, -bc^{-1}, c^{-1}, (-1)^{-a+1} c^{-1} d, (-1)^{-a+1}(f + bc^{-1} d)).$$

Finally, for associativity we first compute

$$\begin{aligned}
&(v, w, x, y, z) \cdot ((a, b, c, d, f) \cdot (h, j, k, \ell, m)) \\
&= (v, w, x, y, z) \cdot (a+h, cj+b, ck, \ell + (-1)^h kd, m + (-1)^h (f - jc)) \\
&= (a+h+v, cjx + bx + w, ckx, \ell + (-1)^h kd + (-1)^{a+h} cky, \\
&\quad m + (-1)^h (f - jc) + (-1)^{a+h} (z - cjx - bx)).
\end{aligned}$$

Then, we compute

$$\begin{aligned}&((v, w, x, y, z) \cdot (a, b, c, d, f)) \cdot (h, j, k, \ell, m)\\&= (a + v, bx + w, cx, d + (-1)^a cy, f + (-1)^a (z - bx)) \cdot (h, j, k, \ell, m)\\&= (a + h + v, cxj + bx + w, ckx, \ell + (-1)^h kd + (-1)^{a+h} kcy,\\&\quad m + (-1)^h (f - jc) + (-1)^{a+h} (z - cjx - bx)).\end{aligned}$$

Consequently, $\cdot$ is an associative operation.

Since $(G, \cdot)$ is a group, it defines a proper group action on U_p^p. There are $p(p-1)$ many CAZAC sequences which start with 1 in U_p^p. If we construct all CAZAC sequences in U_p^p, including those whose first term is not 1, we see that there are $p^2(p-1)$ CAZAC sequences in U_p^p.

Theorem 1.3.2 *Let p be an odd prime and let $x \in U_p^p$ be the Wiener sequence $x[n] = e^{2\pi i s n^2/p}$, where $s \in \mathbb{Z}/p\mathbb{Z}$, see Example 1.2.5. Denote the stabilizer of x under the group $(G, \cdot)$ as G_x. If $p \equiv 1 \mod 4$, then $|G_x| = 4p$. If $p \equiv 3 \mod 4$, then $|G_x| = 2p$. In particular, the orbit of x has size $p^2(p-1)/2$ if $p \equiv 1 \mod 4$ and has size $p^2(p-1)$ if $p \equiv 3 \mod 4$.*

Proof First, let $(a, b, c, d, f) \in G$, and note that

$$(\omega_f e_d \pi_c \tau_b c_a)(x)[n] = W_p^{f+dn} c_a x[cn - b] = W_p^{f+dn+(-1)^a s(cn-b)^2}.$$

Setting $n = 0$ gives the condition that for $(a, b, c, d, f) \in G$,

$$f + (-1)^a s b^2 \equiv 0 \mod p, \tag{1.26}$$

from which we conclude that

$$f \equiv -(-1)^a s b^2 \mod p. \tag{1.27}$$

Setting $n = 1$ and substituting for f as in (1.27) gives us another condition, viz.,

$$(-1)^a s(c - b)^2 + d - (-1)^a s b^2 \equiv s \mod p. \tag{1.28}$$

From (1.28) we can solve for d to obtain

$$d \equiv s + (-1)^a s(2bc - c^2) \mod p. \tag{1.29}$$

Now, note that for any other $n > 1$, we can use (1.27) and (1.29) to obtain the equation

$$(-1)^a s(nc - b)^2 + n + (-1)^a s(2bc - c^2)n - (-1)^a s b^2 \equiv s n^2 \mod p. \tag{1.30}$$

After expanding and cancelling terms, we reduce (1.30) to

$$c^2 \equiv (-1)^a \mod p. \tag{1.31}$$

If $a = 0$, then (1.31) has two solutions, which we shall denote by c_0^+ and c_0^-. If $a = 1$, then by the law of quadratic reciprocity, (1.31) has two solutions, c_1^+ and c_1^- if $p \equiv 1 \mod 4$, but no solutions if $p \equiv 3 \mod 4$. Thus, if $p \equiv 3 \mod 4$, we obtain the following as stabilizers of x:

1. $(0, b, c_0^+, 1 + 2bc_0^+ - (c_0^+)^2, -b^2)$
2. $(0, b, c_0^-, 1 + 2bc_0^- - (c_0^-)^2, -b^2)$

which holds for any $b \in \mathbb{Z}/p\mathbb{Z}$. If $p \equiv 1 \mod 4$, then the following two sets of stabilizers also hold:

1. $(0, b, c_1^+, 1 + 2bc_1^+ - (c_1^+)^2, -b^2)$
2. $(0, b, c_1^-, 1 + 2bc_1^- - (c_1^-)^2, -b^2)$

for any $b \in \mathbb{Z}/p\mathbb{Z}$. Thus, if $p \equiv 1 \mod p$ there are $4p$ stabilizers for x, and if $p \equiv 1 \mod p$ there are $2p$ stabilizers for x.

Corollary 1.3.3 *If $p \equiv 3 \mod 4$, there is only one equivalence class of CAZAC sequences in U_p^p.*

Theorem 1.3.4 *Let $p \equiv 1 \mod 4$, and $x, y \in \mathbb{C}^N$. Let $x = e^{2\pi i n^2/p}$ and $y = e^{2\pi i s n^2/p}$, where s is not a quadratic residue modulo p. Then, x and y belong to different 5-operation equivalence classes.*

Proof Let $s = 1$ in the proof of Theorem 1.3.2, and for $(a, b, c, d, f) \in G$, we have

$$(\omega_f e_d \pi_c \tau_b c_a)(\varphi)[n] = W_p^{f+dn} c_a \varphi[cn - b] = W_p^{f+dn+(-1)^a(cn-b)^2}.$$

Emulating the proof of Theorem 1.3.2, we let $n = 0$ and obtain the condition,

$$f \equiv -(-1)^a b^2 \mod p.$$

Now letting $n = 1$ we have the condition,

$$d \equiv s + (-1)^a(2bc - c^2) \mod p.$$

For arbitrary $n > 1$, we obtain

$$(-1)^a(nc - b)^2 + sn + (-1)^a(2bc - c^2)n - (-1)^a b^2 \equiv sn^2 \mod p. \tag{1.32}$$

After expanding and cancelling terms, we calculate that

$$c^2 \equiv (-1)^a s \mod p.$$

Since $p \equiv 1 \mod 4$ and s is not a residue modulo p, (1.32) cannot be solved for either value of a. Thus, x and y must belong to different equivalence classes.

Corollary 1.3.5 *If $p \equiv 1 \mod 4$, then there are exactly two equivalence classes of CAZAC sequences in U_p^p both of which have size $p^2(p-1)/2$.*

1.3.5 5-Operation Equivalence for Lengths 3 and 5

We now apply the results from Sect. 1.3.4 to show there is only one 5-operation equivalence class for length 3 CAZAC sequences. Indeed, suppose that $x = (1, 1, e^{2\pi i/3})$. Then, the other five CAZAC sequences can be obtained from 5-operation equivalency as follows:

1. $c_1 x = (1, 1, e^{4\pi i/3})$
2. $e_1 c_1 x = (1, e^{2\pi i/3}, e^{2\pi i/3})$
3. $e_1 x = (1, e^{2\pi i/3}, 1)$
4. $e_2 x = (1, e^{4\pi i/3}, e^{4\pi i/3})$
5. $e_2 c_1 x = (1, e^{4\pi i/3}, 1)$.

Corollary 1.3.5 tells us that there are two 5-operation equivalence classes in the case $p = 5$. To write them explicitly, we start with the Wiener sequence,

$$x = (1, e^{2\pi i/5}, e^{8\pi i/5}, e^{8\pi i/5}, e^{2\pi i/5}).$$

We show that we can obtain 10 CAZAC sequences by applying 5-operation equivalencies to x:

1. $x = (1, e^{2\pi i/5}, e^{8\pi i/5}, e^{8\pi i/5}, e^{2\pi i/5})$
2. $c_1 x = (1, e^{8\pi i/5}, e^{2\pi i/5}, e^{2\pi i/5}, e^{8\pi i/5})$
3. $\omega_1 \tau_1 c_1 x = (1, e^{2\pi i/5}, 1, e^{4\pi i/5}, e^{4\pi i/5})$
4. $\omega_4 \tau_1 x = (1, e^{8\pi i/5}, 1, e^{6\pi i/5}, e^{6\pi i/5})$
5. $\omega_4 \tau_2 c_1 x = (1, e^{6\pi i/5}, e^{8\pi i/5}, e^{6\pi i/5}, 1)$
6. $\omega_1 \tau_2 x = (1, e^{4\pi i/5}, e^{2\pi i/5}, e^{4\pi i/5}, 1)$
7. $\omega_4 \tau_3 c_1 x = (1, 1, e^{6\pi i/5}, e^{8\pi i/5}, e^{6\pi i/5})$
8. $\omega_1 \tau_3 x = (1, 1, e^{4\pi i/5}, e^{2\pi i/5}, e^{4\pi i/5})$
9. $\omega_1 \tau_4 c_1 x = (1, e^{4\pi i/5}, e^{4\pi i/5}, 1, e^{2\pi i/5})$
10. $\omega_4 \tau_4 x = (1, e^{6\pi i/5}, e^{6\pi i/5}, 1, e^{8\pi i/5})$.

To find the other orbit, we use the fact that 3 is not a quadratic residue modulo 5 and apply Theorem 1.3.4. We then let x be the Wiener sequence,

$$x = (1, e^{6\pi i/5}, e^{4\pi i/5}, e^{4\pi i/5}, e^{6\pi i/5}),$$

and we compute

1. $x = (1, e^{6\pi i/5}, e^{4\pi i/5}, e^{4\pi i/5}, e^{6\pi i/5})$

2. $c_1 x = (1, e^{4\pi i/5}, e^{6\pi i/5}, e^{6\pi i/5}, e^{4\pi i/5})$
3. $\omega_3 \tau_1 c_1 x = (1, e^{6\pi i/5}, 1, e^{2\pi i/5}, e^{2\pi i/5})$
4. $\omega_2 \tau_1 x = (1, e^{4\pi i/5}, 1, e^{8\pi i/5}, e^{8\pi i/5})$
5. $\omega_2 \tau_2 c_1 x = (1, e^{8\pi i/5}, e^{4\pi i/5}, e^{8\pi i/5}, 1)$
6. $\omega_3 \tau_2 x = (1, e^{2\pi i/5}, e^{6\pi i/5}, e^{2\pi i/5}, 1)$
7. $\omega_2 \tau_3 c_1 x = (1, 1, e^{8\pi i/5}, e^{4\pi i/5}, e^{8\pi i/5})$
8. $\omega_3 \tau_3 x = (1, 1, e^{2\pi i/5}, e^{6\pi i/5}, e^{2\pi i/5})$
9. $\omega_3 \tau_4 x = (1, e^{2\pi i/5}, e^{2\pi i/5}, 1, e^{6\pi i/5})$
10. $\omega_2 \tau_4 x = (1, e^{8\pi i/5}, e^{8\pi i/5}, 1, e^{4\pi i/5})$.

In conclusion, we have explicitly shown that the $p = 3$ case has exactly one orbit and have shown which 5-operation transformations generate them starting with

$$x = (1, 1, e^{2\pi i/3}).$$

In the $p = 5$ case we have explicitly shown that there are two orbits under 5-operation equivalence. We generated both orbits using two different Wiener sequences and have written the 5-operation transformations that generate them.

1.4 Non-roots of Unity CAZAC Sequences of Prime Length

1.4.1 Björck sequences of prime length

In Sect. 1.1.2, we stated Björck's 1984 counterexample, Eq. (1.3), showing that not all CAZAC sequences of length 7 are Gaussian sequences or even roots of unity.

Let p be a prime number, and let $(\frac{k}{p})$ denote the *Legendre symbol* modulo p, defined as

$$\left(\frac{k}{p}\right) = \begin{cases} 0, & \text{if } k \equiv 0 \pmod{p}, \\ 1, & \text{if } k \equiv n^2 \pmod{p} \ \text{ for some } n \in \mathbb{Z}, \\ -1, & \text{if } k \not\equiv n^2 \pmod{p} \ \text{ for all } n \in \mathbb{Z}. \end{cases}$$

Thus, we can define the function $\Lambda : \mathbb{Z}/p\mathbb{Z} \longrightarrow \{+1, 0, -1\}$ as

$$\Lambda[k] = \left(\frac{k}{p}\right).$$

The pre-image of $+1$ under the function Λ is the set $\mathcal{Q}$ of nonzero *quadratic residues* modulo p; and the pre-image of -1 under the function Λ is the set $\mathcal{Q}^C$ of *quadratic non-residues* modulo p. Λ is a *character* of the multiplicative group $(\mathbb{Z}/p\mathbb{Z})^{\times}$. This means that Λ, when restricted to $(\mathbb{Z}/p\mathbb{Z})^{\times}$, is a group homomorphism into the multiplicative group $\mathbb{C}^{\times} = \mathbb{C}\backslash\{0\}$. See [34], Chaps. 5 and 6, for a classical treatment, and [6] for a critical application estimating values of the ambiguity function by means of estimates in terms of Weil's proof of the Riemann hypothesis for finite fields.

Definition 1.4.1 Let p be a prime number, and so $\mathbb{Z}/p\mathbb{Z}$ is a field.

If $p \equiv 1 \pmod 4$, the *Björck sequence*, $b_p : \mathbb{Z}/p\mathbb{Z} \to \mathbb{C}$, of length p, is defined as

$$\forall k = 0, 1, ..., p-1, \quad b_p[k] = e^{i\theta_p(k)},$$

where

$$\theta_p(k) = \left(\frac{k}{p}\right) \arccos\left(\frac{1}{1+\sqrt{p}}\right).$$

If $p \equiv 3 \pmod 4$, or, equivalently, for $p \equiv -1 \pmod 4$, the *Björck sequence*, $b_p : \mathbb{Z}/p\mathbb{Z} \to \mathbb{C}$, of length p, is defined as

$$\forall k = 0, 1, ..., p-1, \quad b_p[k] = \begin{cases} e^{i\theta_p(k)}, & \text{if } k \in \mathcal{Q}^C \subseteq (\mathbb{Z}/p\mathbb{Z})^\times, \\ 1, & \text{otherwise,} \end{cases}$$

where

$$\theta_p(k) = \arccos\left(\frac{1-p}{1+p}\right).$$

In [14] Björck proved that Björck sequences are CAZAC sequences, and elaborated on it in [15] by analyzing the structure of bi-equimodular functions. The structure is related to the subgroup of the multiplicative group $(\mathbb{Z}/p\mathbb{Z})^\times$, e.g., the group of quadratic residues. It was in this context that he used Proposition 1.2.4 in [15], and which he had originally proved in [14]. The following is Björck's main theorem on the topic. Because of the role of the Legendre symbol in the definition of Björck sequences, it is natural to expect a more computational proof of Theorem 1.4.2 in terms of the Legendre symbol. This was done by [5].

Theorem 1.4.2 *Let p be prime.*

a. *If* $p \equiv 1 \pmod 4$, *then* $b_p : \mathbb{Z}/p\mathbb{Z} \to \mathbb{C}$ *is a 3-valued CAZAC sequence of length p.*
b. *If* $p \equiv 3 \pmod 4$, *then* $b_p : \mathbb{Z}/p\mathbb{Z} \to \mathbb{C}$ *is a 2-valued CAZAC sequence of length p.*

Remark 1.4.3 a. Let $p \equiv 1 \pmod 4$. Note that the Legendre symbol sequence of length p has the form $\{0, 1, \ldots, -1, \ldots, 1\}$, i.e., $(\frac{p-1}{p}) = 1$, see Example 1.4.4. In this case of $p \equiv 1 \pmod 4$, Definition 1.4.1 is equivalent to the following sequence constructed by replacing elements of the Legendre sequence. We replace 0 by 1, every term 1 by

$$\eta = \exp\left(i \arccos\frac{\sqrt{p}-1}{p-1}\right) = \frac{1}{\sqrt{p}+1} + i\frac{\sqrt{p+2\sqrt{p}}}{\sqrt{p}+1},$$

and every term -1 by the complex conjugate $\overline{\eta}$ of η. Then, $1, \eta, \overline{\eta}$ are the three values of this Björck CAZAC sequence. See Saffari [54] for a generalization.

b. Let $p \equiv 3 \pmod 4$. Note that the Legendre symbol sequence of length p has the form $\{0, 1, \ldots, -1, \ldots, -1\}$, i.e., $(\frac{p-1}{p}) = -1$, see Example 1.4.4. In this case of $p \equiv 3 \pmod 4$, Definition 1.4.1 is equivalent to the following sequence constructed by replacing elements of the Legendre sequence. We replace 0 by 1, every term -1 by

$$\xi = \exp\left(i \arccos \frac{1-p}{1+p}\right) = \frac{1-p}{1+p} + i\frac{2\sqrt{p}}{1+p},$$

and leave the original $1s$ as they are. Then, $1, \xi$ are the two values of this Björck CAZAC sequence.

Example 1.4.4 a. As an example of the assertion in Remark 1.4.3 that if $p \equiv 1 \pmod 4$, then the Legendre symbol sequence of length p has the form $\{0, 1, \ldots, -1, \ldots, 1\}$, i.e., $(\frac{p-1}{p}) = 1$, let $p = 13$. Consequently, $12 \equiv 5^2 \pmod{13}$.

b. As an example of the assertion in Remark 1.4.3 that if $p \equiv 3 \pmod 4$, then the Legendre symbol sequence of length p has the form $\{0, 1, \ldots, -1, \ldots, -1\}$, i.e., $(\frac{p-1}{p}) = -1$, let $p = 19$. In this case, it is generally difficult to prove assertions of the form,

$$k \not\equiv n^2 \pmod p \quad \text{for all } n \in \mathbb{Z}.$$

Fortunately, we have Legendre's theorem, which asserts for $k \neq 0$ that

$$\left(\frac{k}{p}\right) \equiv k^{(p-1)/2} \pmod p,$$

and so

$$\left(\frac{p-1}{p}\right) = \left(\frac{-1}{p}\right) = \begin{cases} 1, & \textit{if } p \equiv 1 \pmod 4, \\ -1, & \textit{if } p \equiv 3 \pmod 4, \end{cases}$$

[34].

c. By straightforward calculations, we see that Björck sequences are Gaussian for $p = 3, 5$.

d. The theory of frames and CAZAC sequences are natural allies, especially in the case of non-Gaussian CAZAC sequences such as the Björck sequences, e.g., see [10, 44]. In fact, finite Gabor frames for $\mathbb{C}^d$ with CAZAC sequences as generating functions are a natural source of examples and direction for finding further examples, in order to deal with open questions in topics such as compressed sensing and Zauner's conjecture in quantum mechanics.

1.4.2 Circulant Hadamard Matrices not Equivalent to $\mathcal{D}_7$

If we consider $p \times p$ Hadamard matrices, where p is prime, we want to know if the Hadamard matrices generated by CAZAC sequences are always equivalent to $\mathcal{D}_p$, the $p \times p$ DFT matrix. If $p = 2, 3, 5$, then we have already noted that all Hadamard matrices are equivalent to $\mathcal{D}_p$, regardless of whether or not they are generated by a CAZAC sequence [19].

If $p = 7$, then Björck's example shows that there are Hadamard matrices not equivalent to $\mathcal{D}_7$ [19]. One such Hadamard matrix H_1 is defined as follows. Let $\theta = \arccos(-3/4)$ and let $d = \exp(i\theta)$, and set

$$H_1 = \begin{bmatrix} 1 & 1 & 1 & 1 & 1 & 1 & 1 \\ 1 & d^{-1} & 1 & d & d^{-1} & d & 1 \\ 1 & d^{-1} & d^{-1} & d & 1 & 1 & d \\ 1 & d^{-2} & d^{-2} & d^{-1} & d^{-1} & 1 & d^{-1} \\ 1 & 1 & d^{-1} & 1 & d^{-1} & d & d \\ 1 & d^{-2} & d^{-1} & d^{-1} & d^{-2} & d^{-1} & 1 \\ 1 & d^{-1} & d^{-2} & 1 & d^{-2} & d^{-1} & d^{-1} \end{bmatrix},$$

[19, 31]. To continue the process, let $P_1 = P_2 = Id_7$ and D_1, D_2 be the following matrices:

$$D_1 = \begin{bmatrix} 1&0&0&0&0&0&0 \\ 0&1&0&0&0&0&0 \\ 0&0&1&0&0&0&0 \\ 0&0&0&d&0&0&0 \\ 0&0&0&0&1&0&0 \\ 0&0&0&0&0&d&0 \\ 0&0&0&0&0&0&d \end{bmatrix}, \quad D_2 = \begin{bmatrix} 1&0&0&0&0&0&0 \\ 0&d&0&0&0&0&0 \\ 0&0&d&0&0&0&0 \\ 0&0&0&1&0&0&0 \\ 0&0&0&0&d&0&0 \\ 0&0&0&0&0&1&0 \\ 0&0&0&0&0&0&1 \end{bmatrix}.$$

Then, we can define an equivalent circulant Hadamard matrix H_2 by

$$H_2 = D_1 H_1 D_2 = \begin{bmatrix} 1&d&d&1&d&1&1 \\ 1&1&d&d&1&d&1 \\ 1&1&1&d&d&1&d \\ d&1&1&1&d&d&1 \\ 1&d&1&1&1&d&d \\ d&1&d&1&1&1&d \\ d&d&1&d&1&1&1 \end{bmatrix}.$$

In particular, the first column of H_2 is the length 7 Björck sequence and so H_2 is the Hadamard matrix associated with the length 7 Björck sequence.

Another matrix, which is equivalent to neither $\mathcal{D}_7$ nor H_1, is

$$J_1 = \begin{bmatrix} 1 & 1 & 1 & 1 & 1 & 1 & 1 \\ 1 & a^{-2} & a^{-1}b^{-1} & a^{-1}c^{-1} & a^{-1} & a^{-1}c & a^{-1}b \\ 1 & a^{-1}b^{-1} & a^{-2}b^{-2} & a^{-1}b^{-2}c^{-1} & a^{-1}b^{-1}c^{-1} & a^{-1}b^{-1}c & a^{-1}c \\ 1 & a^{-1}c^{-1} & a^{-1}b^{-2}c^{-1} & a^{-2}b^{-2}c^{-2} & a^{-1}b^{-2}c^{-2} & a^{-1}b^{-1}c^{-1} & a^{-1} \\ 1 & a^{-1} & a^{-1}b^{-1}c^{-1} & a^{-1}b^{-2}c^{-2} & a^{-2}b^{-2}c^{-2} & a^{-1}b^{-2}c^{-1} & a^{-1}c^{-1} \\ 1 & a^{-1}c & a^{-1}b^{-1}c & a^{-1}b^{-1}c^{-1} & a^{-1}b^{-2}c^{-1} & a^{-2}b^{-2} & a^{-1}b^{-1} \\ 1 & a^{-1}b & a^{-1}c & a^{-1} & a^{-1}c^{-1} & a^{-1}b^{-1} & a^{-2} \end{bmatrix},$$

where $a \approx \exp(4.312839i)$, $b \approx \exp(1.356228i)$, $c \approx \exp(1.900668i)$, see [16, 19]. The numbers, a, b, and c, are algebraic numbers whose explicit values can be found in [16]. We can put these two matrices in circulant form by multiplying J_1 on the left and right by the matrix,

$$D = \begin{bmatrix} 1 & 0 & 0 & 0 & 0 & 0 & 0 \\ 0 & a & 0 & 0 & 0 & 0 & 0 \\ 0 & 0 & ab & 0 & 0 & 0 & 0 \\ 0 & 0 & 0 & abc & 0 & 0 & 0 \\ 0 & 0 & 0 & 0 & abc & 0 & 0 \\ 0 & 0 & 0 & 0 & 0 & ab & 0 \\ 0 & 0 & 0 & 0 & 0 & 0 & a \end{bmatrix}.$$

Carrying out the multiplication, the circulant form of J_1, denoted as J_2, can be written as

$$J_2 = DJ_1D = \begin{bmatrix} 1 & a & ab & abc & abc & ab & a \\ a & 1 & a & ab & abc & abc & ab \\ ab & a & 1 & a & ab & abc & abc \\ abc & ab & a & 1 & a & ab & abc \\ abc & abc & ab & a & 1 & a & ab \\ ab & abc & abc & ab & a & 1 & a \\ a & ab & abc & abc & ab & a & 1 \end{bmatrix}.$$

1.5 Haagerup's Theorem

1.5.1 Introduction

We shall now outline that part of Haagerup's proof of his Theorem 1.1.6 [32] in which he proves that there are only finitely many cyclic p-roots. The complete proof in which the precise number of cyclic p-roots is computed requires sophisticated complex analysis that is beyond the scope of our theme.

At the risk of oversimplifying, the proof that there are only finitely many cyclic p-roots is divided into two parts: an ingenious algebraic manipulation using the DFT, coupled with an application of the uncertainty principle for $\mathbb{Z}/p\mathbb{Z}$.

1.5.2 Algebraic Manipulation

Recall that cyclic N-roots are solutions $z = (z_0, \ldots, z_{N-1}) \in \mathbb{C}^N$ of the system of equations:

$$\begin{cases} z_0 + z_1 + \cdots + z_{N-1} = 0 \\ z_0 z_1 + z_1 z_2 + \cdots + z_{N-1} z_0 = 0 \\ \qquad \cdots \\ z_0 z_1 \cdots z_{N-2} + \cdots + z_{N-1} z_0 \cdots z_{N-3} = 0 \\ z_0 z_1 \cdots z_{N-1} = 1, \end{cases} \tag{1.33}$$

see Definition 1.1.4. In particular, because of the last equation of (1.33), $z_j \in \mathbb{C}^\times = \mathbb{C}\backslash\{0\}$ for any cyclic N-root $z \in \mathbb{C}^N$. Haagerup makes several substitutions to transform (1.33).

First, assume $z \in \mathbb{C}^N$ is a cyclic N-root. Let $x_0 = 1$ and $x_j = z_0 z_1 \cdots z_{j-1}$ for all $j = 1, \ldots, N-1$. Thus, $x_{j+1}/x_j = z_j$ for $j = 0, \ldots, N-2$, where the last equation of (1.33) guarantees that x_{j+1}/x_j is well defined. Further, for $j = N-1$, we have

$$\frac{x_0}{x_{N-1}} = \frac{1}{z_0 z_1 \cdots z_{N-2}} = z_{N-1},$$

because $z_0 z_1 \cdots z_{N-1} = 1$ by the last equation of (1.33). Substituting these equations, which relate the x_j and z_i, into (1.33) we see that $x = (x_0, \ldots, x_{N-1})$ is a solution to the system,

$$\begin{cases} x_0 = 1 \\ \frac{x_1}{x_0} + \frac{x_2}{x_1} + \cdots + \frac{x_0}{x_{N-1}} = 0 \\ \frac{x_2}{x_0} + \frac{x_3}{x_1} + \cdots + \frac{x_1}{x_{N-1}} = 0 \\ \qquad \cdots \\ \frac{x_{N-1}}{x_0} + \frac{x_0}{x_1} + \cdots + \frac{x_{N-2}}{x_{N-1}} = 0. \end{cases} \tag{1.34}$$

Conversely, if $x = (x_0, \ldots, x_{N-1}) \in (\mathbb{C}^\times)^N$ is a solution to the system (1.34), then it is easy to check that

$$z = (z_0, \ldots, z_{N-1}) = \left(\frac{x_1}{x_0}, \frac{x_2}{x_1}, \ldots, \frac{x_0}{x_{N-1}}\right) \in (\mathbb{C}^\times)^N$$

is a solution to (1.33). Haagerup says that solutions x to (1.34) are *cyclic N-roots on the x-level.*

Second, assume $x = (x_0, \ldots, x_{N-1}) \in (\mathbb{C}^\times)^N$ is a cyclic N-root on the x-level. Let $y_j = 1/x_j$, for $j = 0, \ldots, N-1$. Then,

$$(x, y) = (x_0, \ldots, x_{N-1}, y_0, \ldots, y_{N-1}) \in (\mathbb{C}^\times)^N \times (\mathbb{C}^\times)^N$$

is a solution to the system,

$$\begin{cases} x_0 = y_0 = 1, \\ x_k y_k = 1, 1 \le k \le N-1, \\ \sum_{m=0}^{N-1} x_{k+m} y_m = 0, 1 \le k \le N-1. \end{cases} \tag{1.35}$$

Conversely, if $(x, y) \in \mathbb{C}^N \times \mathbb{C}^N$ is solution to (1.35), then, noting the condition $x_k y_k = 1$ of (1.35), it is easy to check that $x \in (\mathbb{C}^\times)^N$ and that x is a solution to (1.34). Haagerup says solutions $(x, y) \in \mathbb{C}^N \times \mathbb{C}^N$ to (1.35) are *cyclic N-roots on the (x, y)-level.*

Third, Haagerup introduces the DFT into the mix and proves that the system of equations (1.35) for the cyclic N-roots on the (x, y)-level are equivalent to the following system of equations for $(x, y) \in \mathbb{C}^N \times \mathbb{C}^N$:

$$\begin{cases} x_0 = y_0 = 1 \\ x_m y_m = 1, 1 \le m \le N-1 \\ \widehat{x}_n \widehat{y}_{-n} = 1, 1 \le n \le N-1. \end{cases} \tag{1.36}$$

Without providing the details, we can see how the third equation of (1.36) is deduced from (1.35) by writing out the product $\widehat{x}_n \widehat{y}_{-n}$.

Since we began recording these equivalences with cyclic N-roots $z = (z_0, \ldots, z_N)$ as defined in Sect. 1.1.3, we wrote $x_m, \widehat{x}_n$ in (1.36), but this is really $x[m], \widehat{x}[n]$ in the notation from Sect. 1.1.2.

None of the details in this subsection is difficult to prove, *but* Haagerup's strategy is dazzling! The transformations from the cyclic N-roots problem (1.33) to that of (1.36) preserve the number of distinct solutions, and so solving (1.36) is equivalent to solving (1.33), viz., if there are $0 \le M \le \infty$ solutions to one, then there are $0 \le M \le \infty$ solutions to the other. As such, Haagerup's proof that the set of cyclic p-roots is finite will be to solve (1.36).

1.5.3 The Uncertainty Principle for $\mathbb{Z}/p\mathbb{Z}$

In order to prove that the set of cyclic p-roots is finite (Theorem 1.5.6), Haagerup's strategy required Theorems 1.5.3 and 1.5.5. Theorem 1.5.3 is an uncertainty principle for the finite Abelian group $\mathbb{Z}/p\mathbb{Z}$, where p is prime. Its proof uses Chebotarëv's theorem, a fact known to Haagerup in 1996. We should point out that Gabidulin also understood the role of Chebotarëv's theorem if one wanted to prove Theorem 1.5.6.

Theorem 1.5.1 (Chebotarëv 1926) *Let p be prime and let $\mathcal{D}_p$ be the unitary Fourier matrix on $\mathbb{C}^p$, defined as*

$$\mathcal{D}_N = \left[\frac{1}{N^{1/2}} W_N^{-mn} \right]_{m,n=0}^{N-1},$$

see Definition 1.1.2. Then, all square submatrices of $\mathcal{D}_p$ have nonzero determinant.

Remark 1.5.2 There have been many different proofs of this theorem since Chebotarëv's original proof in 1926. A sampling of authors of published proofs is Danilevskii (1937), Reshetnyak (1955), Dieudonné (1970), M. Newman (1975), Evans and I. Stark (1977), Stevenhagen and Lenstra (1996), Goldstein, Guralnick, and Isaacs (c. 2004), and Tao (2005). There is also the proof by Frenkel (2004) that he first wrote down as a solution to a problem in the 1998 Schweitzer Competition! In fact, Chebotarëv's original proof provides much more information than Theorem 1.5.1 asserts, see [57], which is also a spectacular exposition of Chebotarëv's life and mathematical contributions, including his celebrated density theorem.

Independently, Tao [61] used Theorem 1.5.1 in order to prove Theorem 1.5.3. Further, he noted that the two results are equivalent, a fact discovered independently by András Biró. Theorem 1.5.3 itself is a refinement for the setting of $\mathbb{Z}/p\mathbb{Z}$ of the uncertainty principle inequality,

$$|\text{supp}(u)|\,|\text{supp}(\widehat{u})| \geq |G|, \tag{1.37}$$

where G is a finite Abelian group, $u : G \longrightarrow \mathbb{C}$ is a function, $\widehat{u}$ is the discrete Fourier transform of u, $|X|$ is the cardinality of X, and $\text{supp}(u) = \{x \in G : u(x) \neq 0\}$ is the *support* of u, see [62] for a systematic treatment of the discrete Fourier transform. The inequality, (1.37), is due to Donoho and Stark [24], cf., [56].

Theorem 1.5.3 *If $u \neq 0 \in \mathbb{C}^p$ and $\widehat{u} = \mathcal{F}_p u$ is the discrete Fourier transform of u, then $|\text{supp}(u)| + |\text{supp}(\widehat{u})| \geq p + 1$, where $|\text{supp}(u)|$, the support of u, denotes the number of nonzero coordinates of u.*

Algebraic varieties are a central object of study in algebraic geometry. Classically, and for us, an *algebraic variety* is defined as the set of solutions of a system of polynomial equations over the real or complex numbers. The following is a basic theorem.

Theorem 1.5.4 *A compact algebraic variety in $\mathbb{C}^N$ is a finite set, e.g., see [52], Theorem 13.3.*

Theorem 1.5.5 *If the number of solutions $(x, y) \in \mathbb{C}^N \times \mathbb{C}^N$ to (1.36) is infinite, then there are $u, v \in \mathbb{C}^N \backslash \{0\}$ such that $u_k v_k = 0$ and $\widehat{u}_k \widehat{v}_{-k} = 0$ for each $0 \leq k \leq N - 1$.*

Proof Let $W \subseteq \mathbb{C}^N \times \mathbb{C}^N$ denote the set of solutions to (1.36), and assume W is an infinite set. Since W is an algebraic variety, then, by Theorem 1.5.4 and the Heine–Borel theorem, W must be unbounded. Choose a sequence $\{(x^{(m)}, y^{(m)})\} \subseteq W$ for which

$$\lim_{m\to\infty} (||x^{(m)}||_2^2 + ||y^{(m)}||_2^2)^{1/2} = \infty.$$

Let $u^{(m)}$ and $v^{(m)}$ be the normalizations of $x^{(m)}$ and $y^{(m)}$, respectively, i.e., $u^{(m)} = x^{(m)}/||x^m||_2$. Therefore, the sequence, $\{(u^{(m)}, v^{(m)})\}$, is contained in a compact set.

Suppose that this sequence converges to (u, v), passing to a subsequence if necessary. Because each $(x^{(m)}, y^{(m)})$ is a solution to (1.36), $x_0^{(m)} = y_0^{(m)} = 1$ for all $m \in \mathbb{N}$. Thus, $||x^{(m)}||_2^2 = 1 + c_m$ and $||y^{(m)}||_2^2 = 1 + d_m$, where $c_m, d_m > 0$. It follows that

$$||x^{(m)}||_2^2||y^{(m)}||_2^2 = (1 + c_m)(1 + d_m) \geq 1 + c_m + d_m = ||x^{(m)}||_2^2 + ||y^{(m)}||_2^2 - 1.$$

Hence, by our choice of $\{(x^{(m)}, y^{(m)})\}$, we have

$$\lim_{m\to\infty} ||x^{(m)}||_2^2||y^{(m)}||_2^2 = \infty.$$

Now, from (1.36), we know that $x_k^{(m)} y_k^{(m)} = \widehat{x^{(m)}}_k \widehat{y^{(m)}}_{-k} = 1$ for each $m \geq 1$ and each $1 \leq k \leq N - 1$; and so

$$u_k v_k = \widehat{u}_k \widehat{v}_{-k} = \lim_{m\to\infty} (||x^{(m)}||_2||y^{(m)}||_2)^{-1}$$

for $1 \leq k \leq N - 1$. In addition, this equality is also true for $k = 0$, because $x_0^{(m)} = y_0^{(m)} = 1$. Therefore, since

$$\lim_{m\to\infty} ||x^{(m)}||_2||y^{(m)}||_2 = \infty,$$

we have that $u_k v_k = \widehat{u}_k \widehat{v}_{-k} = 0$.

Theorem 1.5.6 (Haagerup) *The set of cyclic p-roots is finite.*

Proof Let $N = p$ in (1.36). Assume for the sake of obtaining a contradiction that the set of solutions to (1.36) is infinite. Then, by Theorem 1.5.5, there are $u, v \in \mathbb{C}^p\backslash\{0\}$ with $u_k v_k = 0$ and $\widehat{u}_k \widehat{v}_{-k} = 0$, $k = 0, 1, \ldots, p - 1$.

This means that $\text{supp}(u) \cap \text{supp}(v) = \emptyset$ and $\text{supp}(\widehat{u}) \cap (-\text{supp}(\widehat{v})) = \emptyset$. In particular, we obtain $|\text{supp}(u)| + |\text{supp}(v)| \leq p$ and $|\text{supp}(\widehat{u})| + |\text{supp}(\hat{v})| \leq p$; and so,

$$|\text{supp}(u)| + |\text{supp}(v)| + |\text{supp}(\widehat{u})| + |\text{supp}(\widehat{v})| \leq 2p.$$

However, by Theorem 1.5.3, we have

$$|\text{supp}(u)| + |\text{supp}(v)| + |\text{supp}(\widehat{u})| + |\text{supp}(\widehat{v})| \geq 2(p + 1),$$

and this gives the desired contradiction.

1.6 Appendix—Real Hadamard Matrices

Definition 1.6.1 A *real Hadamard matrix* is a square matrix whose entries are either $+1$ or -1 and whose rows are mutually orthogonal.

Let H be a real Hadamard matrix of order n. Then, the matrix

$$\begin{bmatrix} H & H \\ H & -H \end{bmatrix}$$

is a real Hadamard matrix of order $2n$. This observation can be applied repeatedly, as Kronecker products, to obtain the following sequence of real Hadamard matrices:

$$H_1 = \begin{bmatrix} 1 \end{bmatrix},$$

$$H_2 = \begin{bmatrix} H_1 & H_1 \\ H_1 & -H_1 \end{bmatrix} = \begin{bmatrix} 1 & 1 \\ 1 & -1 \end{bmatrix},$$

$$H_4 = \begin{bmatrix} H_2 & H_2 \\ H_2 & -H_2 \end{bmatrix} = \begin{bmatrix} 1 & 1 & 1 & 1 \\ 1 & -1 & 1 & -1 \\ 1 & 1 & -1 & -1 \\ 1 & -1 & -1 & 1 \end{bmatrix}, \ldots .$$

Thus,

$$\begin{aligned} H_{2^k} &= \begin{bmatrix} H_{2^{k-1}} & H_{2^{k-1}} \\ H_{2^{k-1}} & -H_{2^{k-1}} \end{bmatrix} \\ &= \begin{bmatrix} H_{2^{k-2}} & H_{2^{k-2}} & H_{2^{k-2}} & H_{2^{k-2}} \\ H_{2^{k-2}} & -H_{2^{k-2}} & H_{2^{k-2}} & -H_{2^{k-2}} \\ H_{2^{k-2}} & H_{2^{k-2}} & -H_{2^{k-2}} & -H_{2^{k-2}} \\ H_{2^{k-2}} & H_{2^{k-2}} & -H_{2^{k-2}} & H_{2^{k-2}} \end{bmatrix}. \end{aligned} \tag{1.38}$$

This method of constructing real Hadamard matrices is due to Sylvester (1867) [59]. In this manner, he constructed real Hadamard matrices of order 2^k for every nonnegative integer k.

Hadamard conjecture 1. The *Hadamard conjecture 1* is that a real Hadamard matrix of order $4k$ exists for every positive integer k [37]. Real Hadamard matrices of orders 12 and 20 were constructed by Hadamard in 1893 [33]. He also proved that if U is a unimodular matrix of order n, then $|\det(U)| \leqslant n^{n/2}$, with equality in the case U is real if and only if U is a real Hadamard matrix [33]. In 1933, Paley discovered a construction that produces a real Hadamard matrix of order $q+1$ when q is a prime power that is congruent to 3 modulo 4, and that produces a real Hadamard matrix of order $2(q+1)$ when q is a prime power that is congruent to 1 modulo 4 [47]. In fact, the Hadamard conjecture 1 should probably be attributed to Paley. The smallest order that cannot be constructed by a combination of Sylvester's and Paley's methods is 92. A real Hadamard matrix of this order was found by computer by Baumert, Golomb, and Hall in 1962. They used a construction, due to Williamson, that has yielded many additional orders. In 2004, Hadi Kharaghani and Behruz Tayfeh-Rezaie constructed a real Hadamard matrix of order 428. As a result, the smallest order for which no real Hadamard matrix is presently known is 668.

Hadamard conjecture 2. If $x : \mathbb{Z}/N\mathbb{Z} \longrightarrow \mathbb{R}$ is CAZAC sequence for $N \geq 2$, then the *Hadamard conjecture 2* asserts that $N = 4$ and x is a translate of the 4-tuple $\pm[1, 1, 1, -1]$. The conjecture goes back to Ryser [53]. From definitions it is straightforward to show that $N = 4M^2$. The major progress has been made by Turyn (1965), B. Schmidt (1999 and 2000), Leung, Ma, and B. Schmidt (2004), see [41]. They proved that the Hadamard conjecture 2 is true if M is a power of a prime greater than 3 as well as it being true for all $N \leq 10^{11}$.

Remark 1.6.2 (Finite abelian groups) It is natural to pose the problems that we have considered about CAZAC sequences on $\mathbb{Z}/N\mathbb{Z}$ for the general case of finite Abelian groups, G. In fact, Gauss' theorem asserts that every such G can be written as

$$G = \mathbb{Z}/N_1\mathbb{Z} \times \cdots \mathbb{Z}/N_n\mathbb{Z},$$

where the N_j can be chosen as powers of primes. Beyond its purely mathematical interest, see [27, 62], this extension is important in coding theory, e.g., the analysis of bent functions and difference sets for the group $\mathbb{Z}/2\mathbb{Z}^n$ by Dillon (1975) and Rothaus (1976), independently, see, e.g., [23, 48, 54].

Remark 1.6.3 (Walsh functions and wavelet packets) Hadamard matrices are closely connected with Walsh functions [3]. The normalized Walsh functions [67] form an orthonormal basis for $L^2(\mathbb{T})$. Every Walsh function is constant over each of a finite number of subintervals of $(0, 1)$. A set of Walsh functions written down in appropriate order as rows of a matrix will give a real Hadamard matrix of order 2^n as obtained by Sylvester's method. When Walsh functions are transported to the real line in the correct way, they not only provide an orthonormal basis for $L^2(\mathbb{R})$ but are the primordial example of wavelet packets using multiresolution analysis in wavelet theory, e.g., see [11].

Remark 1.6.4 (The Littlewood flatness problem and antenna theory) Let $\mathcal{U}_N$ denote the class of unimodular trigonometric polynomials $U(\gamma) = \sum_{n=0}^{N} u_n \, e^{2\pi i n \gamma}$, i.e., $|u_n| = 1$ for $n = 0, \ldots, N$. The *Littlewood flatness problem* is to determine whether or not there are $U_N \in \mathcal{U}_N$ for which

$$\lim_{N \to \infty} \frac{\|U_N\|_\infty}{\|U_N\|_2} = 1. \tag{1.39}$$

It turns out that Gauss sums and their variants play a natural role in dealing with (1.39). There have been herculean efforts to prove (1.39), sometimes in concert with subtle failures only discovered by relatively herculean efforts. Finally, Kahane (1980) proved that such polynomials exist, but it still remains to construct them, see [50]. The ratio in (1.39) is the crest factor of U_N, and $\mathcal{U}_N$ combined with (1.39) play a role in antenna array signal processing where crest factors are analyzed, see [4] for details and references.

Acknowledgements The first named author gratefully acknowledges the support of DTRA Grant 1-13-1-0015 and ARO Grants W911NF 15-1-0112, 16-1-0008, and 17-1-0014. The second named author gratefully acknowledges the support of the Norbert Wiener Center (NWC) as a Daniel Sweet Undergraduate Research Fellow, as well as being a Banneker–Key Scholar. The third named author gratefully acknowledges the support of the NWC and the Department of Mathematics of the University of Maryland. The authors all want to give their thanks to Professor Enrico Au-Yeung of DePaul University for sharing his notes about some of this material from 2010–2012, when he was at the NWC. The first named author also gratefully acknowledges many number-theoretic insights on this material by Professor Robert L. Benedetto of Amherst College.

References

1. L. Auslander, P.E. Barbano, Communication codes and Bernoulli transformations. Appl. Comput. Harmon. Anal. **5**(2), 109–128 (1998)
2. J. Backelin, R. Fröberg, How we proved that there are exactly 924 cyclic 7-roots, in *Proceddings of ISSAC1991* (ACM Press, 1991)
3. D.A. Bell, Walsh functions and Hadamard matrices. Electr. Lett. **2**, 340–341 (1966)
4. J.J. Benedetto, *Harmonic Analysis and Applications*. Studies in Advanced Mathematics (CRC Press, Boca Raton, FL, 1997). MR MR1400886 (97m:42001)
5. J.J. Benedetto, R.L. Benedetto, J.T. Woodworth, *Björck CAZACs and the Discrete Ambiguity Function* (2012), 62 p. Technical Report
6. J.J. Benedetto, R.L. Benedetto, J.T. Woodworth, optimal ambiguity functions and Weil's exponential sum bound. J. Fourier Anal. Appl. **18**(3), 471–487 (2012)
7. J.J. Benedetto, S. Datta, *Construction of Infinite Unimodular Sequences with Zero Autocorrelation*. Advances in Computational Mathematics, vol. 32 (2010), pp. 191–207
8. J.J. Benedetto, S. Datta, *Constructions and a Generalization of Perfect Autocorrelation Sequences on Z*, ed. by X. Shen, A. Zayed. Invited Chapter 8 in volume dedicated to Gil Walter (2012)
9. J.J. Benedetto, J.J. Donatelli, Ambiguity function and frame theoretic properties of periodic zero autocorrelation waveforms. IEEE J. Spec. Top. Sig. Process. **1**, 6–20 (2007)
10. J.J. Benedetto, J.J. Donatelli, Frames and a vector-valued ambiguity function, in *Asilomar Conference on Signals, Systems, and Computers*, invited (October 2008)
11. J.J. Benedetto, M. Frazier (eds.), *Wavelets: Mathematics and Applications*. Studies in Advanced Mathematics (CRC Press, Boca Ratan, FL, 1994)
12. J.J. Benedetto, I. Konstantinidis, M. Rangaswamy, Phase-coded waveforms and their design. IEEE Sig. Process. Mag. invited **26**(1), 22–31 (2009)
13. I. Bengtsson, W. Bruzda, A. Ericsson, J.-A. Larsson, W. Tadej, K. Zyczkowski, Mutually unbiased bases and Hadamard matrices of order six, 1–34 (2007), arXiv:quan-ph/0610161v3
14. G. Björck, Functions of modulus one on Z_p whose Fourier transforms have constant modulus, in *A. Haar Memorial Conference*, vols. I, II (Budapest, 1985) (Colloquia Mathematica Societatis Janos Bolyai, Budapest, vol. 49. Amsterdam, North-Holland, 1987), pp. 193–197
15. G. Björck, Functions of modulus one on $\mathbb{Z}_n$ whose Fourier transforms have constant modulus, and cyclic n-roots, in *Proceedings of the NATO Advanced Study Institute on Fourier Analysis and Its Applications* (1990), pp. 131–140
16. G. Björck, R. Fröberg, A faster way to count the solutions of inhomogeneous systems of algebraic equations, with applications to cyclic n-roots. J. Symb. Comput. **12**(3), 329–336 (1991)
17. G. Björck, R. Fröberg, *Methods to "divide out" Certain Solutions from Systems of Algebraic Equations, Applied to Find All Cyclic 8 Roots* (Lulea 1992). Lecture Notes in Pure and Applied Mathematics, vol. 156 (Dekker, New York, 1994), pp. 57–70

18. G. Björck, B. Saffari, New classes of finite unimodular sequences with unimodular Fourier transforms. Circulant Hadamard matrices with complex entries. Comptes Rendus Mathematique Academie des Sciences, Paris **320**, 319–324 (1995)
19. W. Bruzda, W. Tadej, K. Zyczkowski, Complex Hadamard matrices—a catalogue (2006), http://chaos.if.uj.edu.pl/~karol/hadamard/index.phph=0301
20. O. Christensen, *An Introduction to Frames and Riesz Bases*, vol. 2016, 2nd edn. (Springer-Birkhäuser, New York, 2003)
21. D.C. Chu, Polyphase codes with good periodic correlation properties. IEEE Trans. Inf. Theory **18**, 531–532 (1972)
22. P.J. Davis, *Circulant Matrices* (Wiley, New York, 1970)
23. J.F. Dillon, Elementary Hadamard difference sets, in *Proceedings of the sixth Southeastern Conference on Combinatorics, Graph Theory, and Computing* (Boca Ratan, FL, 1975), pp. 237–249
24. D.L. Donoho, P.B. Stark, Uncertainty principles and signal recovery. SIAM J. Appl. Math. **49**(3), 906–931 (1989)
25. J.C. Faugère, *Finding all the Ssolutions of cycli-9 using Gröbner Basis Techniques*. Computer Mathematics, Lecture Notes in Computer Science, vol. 9 (World Science Publishing, Matsuyama, 2001), pp. 1–12
26. R.L. Frank, S.A. Zadoff, Phase shift pulse codes with good periodic correlation properties. IRE Trans. Inf. Theory **8**, 381–382 (1962)
27. J. Gilbert, Z. Rzeszotnik, The norm of the Fourier transform on finite Abelian groups. Ann. Inst. Fourier, Grenoble **60**(4), 1317–1346 (2010)
28. S.W. Golomb, G. Gong, *Signal Design for Good Correlation* (Cambridge University Press, Cambridge, 2005)
29. K.H. Gröchenig, *Foundations of Time-Frequency Analysis* (Springer - Birkhäuser, New York, 2001)
30. J.-C. Guey, M.R. Bell, Diversity waveform sets for delay-doppler imaging. IEEE Trans. Inf. Theory **44**(4), 1504–1522 (1998)
31. U. Haagerup, Orthogonal maximal abelian *-subalgebras of the $n \times n$ matrices and cyclic n-roots, in *Operator Algebras and Quantum Field Theory* (Rome, 1996) (International Press, Cambridge, MA, 1997), pp. 296–322
32. U. Haagerup, Cyclic p-roots of prime length p and related complex Hadamard matrices, 1–29 (2008), arXiv:0803.2629v1
33. J. Hadamard, Résolution d'une question relative aux déterminants. Bulletin des Sciences Mathématiques **17**, 240–246 (1893)
34. G.H. Hardy, E.M. Wright, *An Introduction to the Theory of Numbers*, 4th edn. (Oxford University, Oxford, 1965)
35. T. Helleseth, P. Vijay Kumar, *Sequences with Low Correlation*. Handbook of Coding Theory, vols. I, II, ed. by V.S. Pless, W. Cary Huffman (North-Holland, Amsterdam, 1998), pp. 1765–1853
36. M.A. Herman, T. Strohmer, High-resolution radar via compressed sensing. IEEE Trans. Sig. Process. **57**, 2275–2284 (2009)
37. K.J. Horadam, *Hadamard Matrices and Their Applications* (Princeton University Press, Princeton, NJ, 2007)
38. J.R. Klauder, The design of radar signals having both high range resolution and high velocity resolution. Bell Syst. Tech. J. **39**, 809–820 (1960)
39. J.R. Klauder, A.C., Price, S. Darlington, W.J. Albersheim, The theory and design of chirp radars. Bell Syst. Tech. J. **39**, 745–808 (1960)
40. I. Kra, S.R. Simanca, On circulant matrices. Notices Amer. Math. Soc. **59**(3), 368–377 (2012)
41. K.H. Leung, S.L. Ma, B. Schmidt, Non-existence of abelian difference sets: Lander's conjecture for prime power orders. Trans. Amer. Math. Soc. **356**(11), 4343–4358 (2004)
42. N. Levanon, E. Mozeson, *Radar Signals* (Wiley Interscience, IEEE Press, Hoboken, NJ, 2004)
43. M.L. Long, *Radar Reflectivity of Land and Sea* (Artech House, 2001)

44. M. Magsino, Constructing tight Gabor frames using CAZAC sequences. Sampl. Theory Signal Image Process. **16**, 73–99 (2017)
45. W.H. Mow, A new unified construction of perfect root-of-unity sequences, in *Proceedings of IEEE 4th International Symposium on Spread Spectrum Techniques and Applications*, Germany, September 1996, pp. 955–959
46. F.E. Nathanson, *Radar Design Principles—Signal Processing and the Environment* (SciTech Publishing Inc., Mendham, NJ, 1999)
47. R.E.A.C. Paley, On orthogonal matrices. J. Math. Phys. **12**, 311–320 (1933)
48. V.S. Pless, W.C. Huffman, R.A. Brualdi (eds.), *Handbook of Coding Theory*, vol. I (North-Holland, Amsterdam, II, 1998)
49. B.M. Popovic, Generalized chirp-like polyphase sequences with optimum correlation properties. IEEE Trans. Inf. Theory **38**(4), 1406–1409 (1992)
50. H. Queffelec, B. Saffari, On bernstein's inequality and Kahane's ultraflat polynomials. J. Fourier Anal. Appl. **2**, 519–582 (1996)
51. M.A. Richards, J.A. Scheer, W.A. Holm (eds.), *Principles of Modern Radar* (SciTech Publishing Inc., Raleigh, NC, 2010)
52. W. Rudin, *Function Theory in the Unit Ball*. Gründlehren Series, vol. 241 (Springer, New York, 1980)
53. H.J. Ryser, *Combinatorial Mathematics* (Wiley, New York, 1963)
54. B. Saffari, Some polynomial extremal problems which emerged in the twentieth century, in *Twentieth Century Harmonic Analysis–A Celebration* (Il Ciocco, 2000). NATO Science Series II: Mathematics, Physics and Chemistry, vol. 33 (Kluwer Academic Publishers, Dordrecht, 2001), pp. 201–233
55. M.I. Skolnik, *Introduction to Radar Systems* (McGraw-Hill Book Company, New York, 1980)
56. K.T. Smith, The uncertainty principle on groups. SIAM J. Applied. Math. **50**, 876–882 (1990)
57. P. Stevenhagen, H.W. Lenstra, Chebotarëv and his density theorem. Math. Intell. **18**(2), 26–37 (1996)
58. G.W. Stimson, *Airborne Radar* (SciTech Publishing Inc., Mendham, NJ, 1998)
59. J.J. Sylvester, Thoughts on inverse orthogonal matrices, simultaneous sign successions, and tessellated pavements in two or more colours, with applications to newton's rule, ornamental tile-work, and the theory of numbers. Philos. Mag. **34**, 461–475 (1867)
60. W. Tadej, K. Zyczkowski, A concise guide to complex Hadamard matrices. Open Syst. Inf. Dyn. **13**, 133–177 (2006)
61. T. Tao, An uncertainty principle for cyclic groups of prime order. Math. Res. Lett. **12**, 121–127 (2005)
62. A. Terras, *Fourier Analysis on Finite Groups and Applications*, no. 43 (Cambridge University Press, 1999)
63. R.J. Turyn, Sequences with small correlation, Error Correcting Codes, in *Proceedings of a Symposium*, Mathematics Research Center, Madison, Wisconsin (Wiley, New York, 1968), pp. 195–228
64. S. Ulukus, R.D. Yates, Iterative construction of optimum signature sequence sets in synchronous CDMA systems. IEEE Trans. Inform. Theory **47**(5), 1989–1998 (2001)
65. D.E. Vakman, *Sophisticated Signals and the Uncertainty Principle in Radar* (Springer, New York, 1969)
66. S. Verdú, *Multiuser Detection* (Cambridge University Press, Cambridge, UK, 1998)
67. J.L. Walsh, A closed set of normal orthogonal functions. Am. J. Math. **45**, 5–24 (1923)
68. P.M. Woodward, *Probability and Iinformation Theory, with Applications to Radar* (Pergamon Press, Oxford, 1953)
69. P.M. Woodward, Theory Theory of radar information. IEEE Trans. Inf. **1**(1), 108–113 (1953)

Chapter 2
Hardy Spaces with Variable Exponents

Víctor Almeida, Jorge J. Betancor, Estefanía Dalmasso and Lourdes Rodríguez-Mesa

Abstract In this paper, we make a survey on some recent developments of the theory of Hardy spaces with variable exponents in different settings.

2.1 Introduction

In this paper, we gather the essential contents of the talk "Hardy spaces with variable exponents" given by the second author at the CIMPA2017 Research School-IX Escuela Santaló Harmonic Analysis, Geometric Measure Theory and Applications, which took place in Buenos Aires, Argentina, from July 31 to August 11, 2017. We give an overview of the latest advances about Hardy and local Hardy spaces with variable exponents in different contexts. We cannot be exhaustive and we recommend the interested reader to consult the references at the end of this paper and others that can be found in them.

V. Almeida · J. J. Betancor (✉) · L. Rodríguez-Mesa
Departamento de Análisis Matemático, Universidad de La Laguna, Campus de Anchieta, Avda. Astrofísico Sánchez, s/n, 38721 La Laguna (Santa Cruz de Tenerife), Spain
e-mail: jbetanco@ull.es

V. Almeida
e-mail: valmeida@ull.es

L. Rodríguez-Mesa
e-mail: lrguez@ull.es

E. Dalmasso
Instituto de Matemática Aplicada del Litoral, UNL, CONICET, FIQ, Colectora Ruta Nac. NÂř 168, Paraje El Pozo, S3007ABA Santa Fe, Argentina
e-mail: edalmasso@santafe-conicet.gov.ar

A. Aldroubi et al. (eds.), *New Trends in Applied Harmonic Analysis, Volume 2*, Applied and Numerical Harmonic Analysis,
https://doi.org/10.1007/978-3-030-32353-0_2

2.2 Lebesgue Spaces with Variable Exponents

Lebesgue spaces with variable exponents appeared for the first time in a paper of Orlicz [51] in 1931. He considered the discrete case $\ell^{(p_i)}$, where $(p_i)_{i=1}^{\infty}$ is a sequence in $[1,\infty)$, and the continuous one $L^{p(\cdot)}(\mathbb{R})$, where p is a real $[1,\infty)$-valued function. Orlicz proved variable exponents versions of Hölder's inequality. Nakano [49] in 1950 studied the so-called modular spaces which constituted a generalization of the variable exponent Lebesgue spaces, that is, they are a special case of the modular one.

Several groups of researchers have developed the theory of modular spaces, in particular, of variable exponents Lebesgue spaces during the second half of the twentieth century. In [23, pp. 2 and 3], some highlights in the evolution of the theory are mentioned. We point out that many of the basic properties of variable exponents Lebesgue and Sobolev spaces in $\mathbb{R}^n$ were established by Kováčik and Rákosník [42].

With the arrival of the new century, the study of variable exponent function spaces has gained a new impetus. This has been due to inter alia to applications where these spaces play an important role. For instance, in recent years increasing attention has been paid to the study of the so-called electrorheological fluids that have high technological interest. They have an important feature; their viscosity and other of their mechanical characteristics can change drastically when they are in the presence of an electromagnetic field. To model electrorheological fluids, variable exponent spaces are the natural setting [22, 54, 55, 67]. Also, these spaces appear in the modeling of non-Newtonian fluids with thermo-corrective effects [5]. Variable exponent spaces are connected with variational integrals with nonstandard growth and coercitivity conditions [1, 74]. Variable exponent models for image restoration have been studied in [14, 35, 45].

On the other hand, from a theoretical point of view, a fundamental fact that justifies the new interest in the study of variable exponent spaces was the discovery of the correct regularity condition, the so-called log-Hölder continuity condition, for the variable exponents. Diening [24] uses this condition to establish the boundedness of the Hardy–Littlewood maximal function on $L^{p(\cdot)}(\Omega)$ when Ω is bounded. We will come back to this question later.

The theory of Lebesgue spaces with variable exponents has been collected in an exhaustive way in the monographs of Diening et al. [23] and Cruz-Uribe and Fiorenza [16]. We recall here the definitions and some properties of $L^{p(\cdot)}(\Omega)$ that also serve to fix our notation.

Let Ω be an open subset of $\mathbb{R}^n$ and let $p:\Omega\to(0,\infty)$ be a measurable function. We consider $p_-(\Omega)=\operatorname*{ess\,inf}_{x\in\mathbb{R}^n} p(x)$, $\quad p_+(\Omega)=\operatorname*{ess\,sup}_{x\in\mathbb{R}^n} p(x)$ and assume that $0<p_-(\Omega)\le p_+(\Omega)<\infty$ except other thing is said. When $\Omega=\mathbb{R}^n$, we just write p_- and p_+, respectively.

If f is a measurable complex-valued function defined on Ω, we define the modular $\varrho_{p(\cdot)}(f)$ of f by

$$\varrho_{p(\cdot)}(f):=\int_{\Omega}|f(x)|^{p(x)}dx,$$

and we say that $f \in L^{p(\cdot)}(\Omega)$ when $\varrho_{p(\cdot)}(\lambda f) < \infty$, for some $\lambda > 0$.

The Luxemburg norm $\|\cdot\|_{p(\cdot)}$ on $L^{p(\cdot)}(\Omega)$ is defined by

$$\|f\|_{p(\cdot)} = \inf\left\{\lambda > 0 : \varrho_{p(\cdot)}\left(\frac{f}{\lambda}\right) \leq 1\right\}, \quad f \in L^{p(\cdot)}(\Omega).$$

A first definition of $L^{p(\cdot)}(\Omega)$ including the case $p_+(\Omega) = \infty$ was introduced by Sharpudinov [57] in one dimension and then, in higher dimension, by Kováčik and Rákosník [42]. $(L^{p(\cdot)}(\Omega), \|\cdot\|_{p(\cdot)})$ is a Banach space provided that $p_-(\Omega) \geq 1$ and it is reflexive when $p_-(\Omega) > 1$.

The Hardy–Littlewood maximal function $\mathcal{M}$ defined by

$$\mathcal{M}(f)(x) = \sup_{B \ni x} \frac{1}{|B|} \int_{B \cap \Omega} |f(y)| dy, \quad x \in \mathbb{R}^n,$$

where the supremum is taken over all those balls B in $\mathbb{R}^n$ such that $x \in B$ is a very useful tool in harmonic analysis. Many authors have studied the behavior of the maximal function $\mathcal{M}$ on variable exponent Lebesgue spaces.

Lerner proved in [44, Theorem 1.1] that if $1 < p_- \leq p_+ < \infty$ there exists $C > 0$ for which

$$\int_{\mathbb{R}^n} (\mathcal{M}(f)(x))^{p(x)} dx \leq C \int_{\mathbb{R}^n} |f(x)|^{p(x)} dx$$

if and only if $p(x) = p_0$, $x \in \mathbb{R}^n$, for some $1 < p_0 < \infty$.

The following conditions for the exponents are usual in variable exponent settings. We say that an exponent p is locally log-Hölder continuous on Ω when there exists $C > 0$ for which

$$|p(x) - p(y)| \leq \frac{C}{\log(e + 1/|x - y|)}, \quad x, y \in \Omega.$$

It is said that p satisfies the log-Hölder decay condition if there exists $\alpha_\infty \in \mathbb{R}$ and $C > 0$ such that

$$|p(x) - \alpha_\infty| \leq \frac{C}{\log(e + |x|)}, \quad x \in \Omega.$$

When p is locally log-Hölder continuous and satisfies the log-Hölder decay condition in Ω, we say that p is globally log-Hölder continuous in Ω.

Diening [24] proved that the maximal function defines a bounded operator from $L^{p(\cdot)}(\mathbb{R}^n)$ into itself provided that p is locally log-Hölder continuous on $\mathbb{R}^n$ and p is constant outside a ball. This result was improved by Cruz-Uribe et al. [18] relaxing the second property which is replaced by the log-Hölder decay condition for p in $\mathbb{R}^n$. Nekvinda [50] proved the boundedness of $\mathcal{M}$ in $L^{p(\cdot)}(\mathbb{R}^n)$ by replacing the assumption that p is constant outside a ball in $\mathbb{R}^n$ by a weaker decay condition at infinity involving an integral, that is, there exist $\beta_\infty > 1$ and $C > 0$ such that

$$\int_{\mathbb{R}^n\setminus\{p(x)=\beta_\infty\}} |p(x)-\beta_\infty| C^{1/|p(x)-\beta_\infty|} dx < \infty.$$

In [18, 24, 50], $1 < p_- \le p_+ < \infty$ is assumed. On the other hand, Pick and Ružička [53] established that the local log-Hölder condition is the optimal continuity modulus. The condition $p_- > 1$ is necessary for the maximal operator to be bounded from $L^{p(\cdot)}(\mathbb{R}^n)$ into itself (see [18]). The $L^{p(\cdot)}$-boundedness of $\mathcal{M}$ including the case $p_+ = \infty$ was shown by Diening et al. [25].

When Ω is a bounded open subset of $\mathbb{R}^n$, Diening [24] proved that if p is locally log-Hölder continuous in Ω, $\mathcal{M}$ defines a bounded operator in $L^{p(\cdot)}(\Omega)$.

Many others have investigated $L^{p(\cdot)}$-boundedness properties for the maximal operator in different directions: weak-type inequalities, weighted inequalities, metric spaces, homogeneous type spaces, ... Our previous comments are only concerned about strong-type results.

It is well known that if $q \in (0, \infty)$ and $\{\Omega_i\}_{i=0}^\infty$ is a partition of $\mathbb{R}^n$, then

$$\|f\|_q^q = \sum_{i=0}^{\infty} \|f\chi_{\Omega_i}\|_q^q. \tag{2.1}$$

This property allows us to pass from local to global results. It is clear that it is not possible to get a property like the last one in variable exponent settings.

Gogatishvili et al. [32] and Hästö [36] have proved partial versions of (2.1) in $L^{p(\cdot)}$ contexts. Suppose that $\{Q_j\}_{j=1}^\infty$ is a partition of $\mathbb{R}^n$ into cubes with equal size and ordered so that $i > j$ if $dist(0, Q_i) > dist(0, Q_j)$, and $\alpha > 0$. We define a partition norm $\|\cdot\|_{p(\cdot),\{Q_j\},\alpha}$ by

$$\|f\|_{p(\cdot),\{Q_j\},\alpha} = \left(\sum_{j=1}^{\infty} \|f\chi_{Q_j}\|_{p(\cdot)}^\alpha\right)^{1/\alpha}.$$

Note that if $p(x) = q = \alpha$, $x \in \mathbb{R}^n$, (2.1) says that $\|f\|_{p(\cdot),\{Q_j\},\alpha} = \|f\|_q$. In [36, Theorem 2.4] it was proved that if $p(\cdot)$ is globally log-Hölder continuous in $\mathbb{R}^n$, $\|f\|_{p(\cdot),\{Q_j\},\alpha_\infty} \sim \|f\|_{p(\cdot)}$, $f \in L^{p(\cdot)}(\mathbb{R}^n)$. This result allows us to upgrade properties proved on bounded sets, to global results, valid in all of $\mathbb{R}^n$ [32, 36].

2.3 Hardy Spaces with Variable Exponents

The study of Hardy spaces started at the beginning of the twentieth century in the context of Fourier series and complex analysis of one variable. The theory of classical real Hardy spaces $H^p(\mathbb{R}^n)$ was originated by the paper of Stein and Weiss [61] in the early 1960s and it was initially tied closely to the theory of harmonic functions. Real variable methods were incorporated into this topic in the celebrated paper of Fefferman and Stein [31]. Hardy spaces $H^p(\mathbb{R}^n)$ reduce to Lebesgue spaces $L^p(\mathbb{R}^n)$

when $1 < p < \infty$, and they are suitable substitutes of $L^p(\mathbb{R}^n)$ when $0 < p \leq 1$, especially with respect to the boundedness of operators, playing an important role in various fields of analysis and partial differential equations. Since the appearance of [31] Hardy spaces have been the object of much research. In [60, Chaps. III and IV], a systematic presentation of the main properties of Hardy spaces in $\mathbb{R}^n$ can be encountered.

Hardy spaces with variable exponent have been introduced in recent years by Nakai and Sawano [48] and, independently, by Cruz-Uribe and Wang [19]. In this exposition we take [19] as a starting point.

By $S(\mathbb{R}^n)$ we denote the Schwartz space and $S'(\mathbb{R}^n)$ represents the dual space of $S(\mathbb{R}^n)$. The first definitions of Hardy space $H^{p(\cdot)}(\mathbb{R}^n)$ are given by using maximal functions. Suppose that $f \in S'(\mathbb{R}^n)$. For every $\phi \in S(\mathbb{R}^n)$ we define the radial maximal function $M_\phi(f)$ by

$$M_\phi(f) = \sup_{t>0} |(f * \phi_t)|,$$

where $\phi_t(x) = \frac{1}{t^n}\phi(\frac{x}{t})$, $x \in \mathbb{R}^n$ and $t > 0$.

For every $N \in \mathbb{N}$ we denote by S_N the following subset of $S(\mathbb{R}^n)$

$$S_N = \left\{\phi \in S(\mathbb{R}^n) : \max_{|\alpha| \leq N} \sup_{x \in \mathbb{R}^n} (1 + |x|)^N |D^\alpha \phi(x)| \leq 1\right\},$$

and the grand maximal functions $M_N(f)$ of f defined by

$$M_N(f) = \sup_{\phi \in S_N} M_\phi(f).$$

Also the corresponding nontangential Poisson maximal function $\mathcal{N}(f)$, of f, is given by

$$\mathcal{N}(f)(x) = \sup_{t>0} \sup_{|x-y|<t} |P_t(f)(y)|, \quad x \in \mathbb{R}^n,$$

where P represents the Poisson kernel in $\mathbb{R}^n$:

$$P(x) = \frac{\Gamma(\frac{n+1}{2})}{\pi^{(n+1)/2}} \frac{1}{(1 + |x|^2)^{(n+1)/2}}, \quad x \in \mathbb{R}^n.$$

In order to the definition of $P_t(f)$, $t > 0$ makes sense, in the corresponding one of $\mathcal{N}(f)$, we need to restrict the action of $\mathcal{N}$ to the set of bounded tempered distributions f. That is, those $f \in S'(\mathbb{R}^n)$ such that $f * \phi \in L^\infty(\mathbb{R}^n)$, for every $\phi \in S(\mathbb{R}^n)$.

As in [19] we say that a measurable function $p : \mathbb{R}^n \to (0, \infty)$ is in $\mathcal{M}P_0$ when $0 < p_- \leq p_+ < \infty$ and there exists $0 < p_0 < p_-$ such that the Hardy–Littlewood maximal operator $\mathcal{M}$ is bounded from $L^{p(\cdot)/p_0}(\mathbb{R}^n)$ into itself.

A variable exponent version of the classical result in [60, Theorem 1, p. 91] is the following.

Theorem 2.3.1 ([19, Theorem 3.1]) *Given $p(\cdot) \in \mathcal{M}P_0$, for every $f \in S'(\mathbb{R}^n)$, the following assertions are equivalent.*

(a) There exists $\phi \in S(\mathbb{R}^n)$ being $\int_{\mathbb{R}^n} \phi(x)dx \neq 0$, such that $M_\phi(f) \in L^{p(\cdot)}(\mathbb{R}^n)$.
(b) For all $N > \frac{n}{p_0} + n + 1$, $M_N(f) \in L^{p(\cdot)}(\mathbb{R}^n)$.
(c) f is a bounded distribution and $\mathcal{N}(f) \in L^{p(\cdot)}(\mathbb{R}^n)$.

Furthermore, the quantities $\|M_\phi(f)\|_{p(\cdot)}$, $\|M_N(f)\|_{p(\cdot)}$, and $\|\mathcal{N}(f)\|_{p(\cdot)}$ are comparable with constants that depend only on $p(\cdot)$ and n but not on f.

Let $p(\cdot) \in \mathcal{M}P_0$. A tempered distribution $f \in S'(\mathbb{R}^n)$ is said to be in $H^{p(\cdot)}(\mathbb{R}^n)$ when (a) (equivalently, (b) or (c)) in Theorem 2.3.1 is satisfied. For every $f \in H^{p(\cdot)}(\mathbb{R}^n)$ we define

$$\|f\|_{H^{p(\cdot)}(\mathbb{R}^n)} = \|M_N(f)\|_{p(\cdot)},$$

being $N \in \mathbb{N}$, $N > \frac{n}{p_0} + n + 1$.

Latter [43] and Coifman [15] obtained atomic characterizations for classical Hardy spaces. Stromberg and Torchinsky [62] described the distributions in weighted Hardy spaces by using atoms. Inspired by the last two mentioned atom decompositions, in [19, 48] (see also [56]) the variable exponent Hardy space $H^{p(\cdot)}(\mathbb{R}^n)$ is characterized by using atoms.

Let $p(\cdot) \in \mathcal{M}P_0$. Assume that $1 < q \leq \infty$. We say that a function $\mathfrak{a}$ is a $(p(\cdot), q)$-atom associated with the ball B when

(i) supp $a \subset B$;
(ii) $\|\mathfrak{a}\|_q \leq |B|^{1/q}\|\chi_B\|_{p(\cdot)}^{-1}$;
(iii) $\int_{\mathbb{R}^n} \mathfrak{a}(x)x^\alpha dx = 0$, for every $\alpha \in \mathbb{N}^n$, $|\alpha| \leq [n(1/p_0 - 1)]$.

Here $|\alpha| = \alpha_1 + \alpha_2 + ... + \alpha_n$, when $\alpha = (\alpha_1, \alpha_2, ..., \alpha_n) \in \mathbb{N}^n$, and, for every $\beta > 0$, $[\beta]$ denotes the unique $k \in \mathbb{N}$ such that $k \leq \beta < k + 1$.

Theorem 2.3.2 ([19, Theorem 7.1]) *Suppose $p(\cdot) \in \mathcal{M}P_0$. Then, a distribution $f \in H^{p(\cdot)}(\mathbb{R}^n)$ if, and only if, for every $q > 1$, there exist, for each $j \in \mathbb{N}$, $\lambda_j > 0$ and a $(p(\cdot), q)$-atom $\mathfrak{a}_j$ associated with the ball B_j such that $f = \sum_{j\in\mathbb{N}}^\infty \lambda_j \mathfrak{a}_j$, where the series converges in $H^{p(\cdot)}(\mathbb{R}^n)$, and $\displaystyle\sum_{j\in\mathbb{N}}^\infty \frac{\lambda_j}{\|\chi_{B_j}\|_{p(\cdot)}} \chi_{B_j} \in L^{p(\cdot)}(\mathbb{R}^n)$. Furthermore, if $q > 1$ and we define, for every $f \in H^{p(\cdot)}(\mathbb{R}^n)$,*

$$\|f\|_{H^{p(\cdot)}_{at}(\mathbb{R}^n)} = \inf \left\| \sum_{j\in\mathbb{N}}^\infty \frac{\lambda_j}{\|\chi_{B_j}\|_{p(\cdot)}} \chi_{B_j} \right\|_{p(\cdot)},$$

where the infimum is taken over all the pair of sequences $(\{\lambda_j\}_{j=1}^\infty, \{B_j\}_{j=1}^\infty)$ such that, for every $j \in \mathbb{N}$, $\lambda_j > 0$ and there exists a $(p(\cdot), q)$-atom $\mathfrak{a}_j$ associated with the ball B_j, and that $f = \sum_{j=1}^\infty \lambda_j \mathfrak{a}_j$, in $H^{p(\cdot)}(\mathbb{R}^n)$, then $\|f\|_{H^{p(\cdot)}(\mathbb{R}^n)} \sim \|f\|_{H^{p(\cdot)}_{at}(\mathbb{R}^n)}$.

We now define, for every $f \in S'(\mathbb{R}^n)$, $\phi \in S(\mathbb{R}^n)$ and $j \in \mathbb{Z}$,

$$\phi_{(j)} = \phi(2^{-j}x), \quad x \in \mathbb{R}^n,$$

and

$$\phi_{(j)}(D)f = (\phi_{(j)}\hat{f})\check{}.$$

If $\phi \in S(\mathbb{R}^n)$ the Fourier transform $\hat{\phi}$ of ϕ is given by

$$\hat{\phi}(y) = \int_{\mathbb{R}^n} e^{-2\pi i x\cdot y}\phi(x)dx, \quad y \in \mathbb{R}^n,$$

and the inverse Fourier transform $\check{\phi}$ of ϕ is defined by $\check{\phi}(y) = \hat{\phi}(-y)$, $y \in \mathbb{R}^n$. The Fourier transform $\hat{f}$ and the inverse Fourier transform $\check{f}$ of $f \in S'(\mathbb{R}^n)$ are defined by duality.

A Triebel–Lizorkin-type characterization of $H^{p(\cdot)}(\mathbb{R}^n)$ was established in [48].

Theorem 2.3.3 ([48, Theorem 5.7]) *Assume that $p(\cdot)$ is globally log-Hölder continuous in $\mathbb{R}^n$. Let $\varphi \in S(\mathbb{R}^n)$ such that* $\operatorname{supp}\varphi \subset B(0,4)\setminus B(0,1/4)$ *verifying that* $\sum_{j=-\infty}^{\infty} |\varphi_{(j)}(y)|^2 > 0$, $y \in \mathbb{R}^n\setminus\{0\}$. *Then, there exists $C > 0$ such that, for every $f \in H^{p(\cdot)}(\mathbb{R}^n)$,*

$$\frac{1}{C}\|f\|_{H^{p(\cdot)}(\mathbb{R}^n)} \leq \left\| \left(\sum_{j\in\mathbb{Z}} |\varphi_{(j)}(D)f|^2 \right)^{1/2} \right\|_{p(\cdot)} \leq C\|f\|_{H^{p(\cdot)}(\mathbb{R}^n)}. \tag{2.2}$$

Variable exponent Hardy spaces $H^{p(\cdot)}(\mathbb{R}^n)$ also can be characterized by using other Littlewood–Paley functions (see [77]). We consider the following square-type functions. By $S_\infty(\mathbb{R}^n)$ we denote the subspace of $S(\mathbb{R}^n)$ constituted by all those $\phi \in S(\mathbb{R}^n)$ such that $0 \notin \operatorname{supp}\hat{\phi}$. $S'_\infty(\mathbb{R}^n)$ represent the dual space of $S_\infty(\mathbb{R}^n)$. Suppose that $\phi \in S(\mathbb{R}^n)$ satisfying that

$$\operatorname{supp}\hat{\phi} \subset \{y \in \mathbb{R}^n : 1/2 \leq |y| \leq 2\}$$

and, for a certain $C > 0$,

$$|\hat{\phi}(y)| \geq C, \quad 3/5 \leq |y| \leq 5/3.$$

It is clear that $\phi \in S_\infty(\mathbb{R}^n)$. For every $f \in S'_\infty(\mathbb{R}^n)$, we define

$$g(f)(x) = \left\{ \int_0^\infty |f * \phi_t(x)|^2 \frac{dt}{t} \right\}^{1/2}, \quad x \in \mathbb{R}^n,$$

that can be seen as a continuous version of the square function being in the center of the chain of inequalities (2.2),

$$S(f)(x)=\left\{\int_0^\infty\int_{|x-y|<t}|f*\phi_t(x)|^2\frac{dydt}{t^{n+1}}\right\}^{1/2},\quad x\in\mathbb{R}^n,$$

and, for every $\lambda>0$,

$$g_\lambda^*(f)(x)=\left\{\int_0^\infty\int_{\mathbb{R}^n}\left(\frac{t}{t+|x-y|}\right)^{\lambda n}|f*\phi_t(x)|^2\frac{dydt}{t^{n+1}}\right\}^{1/2},\quad x\in\mathbb{R}^n.$$

Theorem 2.3.4 ([77, Theorem 1.4 and Corollary 1.5]) *Suppose that $p(\cdot)$ is globally log-Hölder continuous in $\mathbb{R}^n$. Then, $f\in H^{p(\cdot)}(\mathbb{R}^n)$ if, and only if, $f\in S'_\infty(\mathbb{R}^n)$ and $S(f)\in L^{p(\cdot)}(\mathbb{R}^n)$. Furthermore, there exists $C>0$ such that*

$$\frac{1}{C}\|f\|_{H^{p(\cdot)}(\mathbb{R}^n)}\leq\|S(f)\|_{p(\cdot)}\leq C\|f\|_{H^{p(\cdot)}(\mathbb{R}^n)}\quad f\in H^{p(\cdot)}(\mathbb{R}^n).$$

The same assertion is true if $S(f)$ is replaced by $g(f)$ or $g_\lambda^*(f)$ when $\lambda\in(1+\frac{2}{\min\{2,p_-\}},\infty)$.

The property in Theorem 2.3.4 must be understood in the following way: if $f\in S'_\infty(\mathbb{R}^n)$ and $S(f)\in L^{p(\cdot)}(\mathbb{R}^n)$, then there exists a unique $g\in S'(\mathbb{R}^n)$ such that $<g,\psi>=<f,\psi>$, $\psi\in S_\infty(\mathbb{R}^n)$, $g\in H^{p(\cdot)}(\mathbb{R}^n)$, and $\|g\|_{H^{p(\cdot)}(\mathbb{R}^n)}\leq\|S(f)\|_{p(\cdot)}$.

Hardy spaces $H^{p(\cdot)}(\mathbb{R}^n)$ also can be characterized by using the corresponding intrinsic square function in the sense of Wilson [66]. This was proved in [77, Theorem 1.8].

For every $j=1,2,...,n$ and $f\in L^p(\mathbb{R}^n)$, $1\leq p<\infty$, the jth Riesz transform R_jf of f is defined by

$$R_j(f)(x)=\lim_{\epsilon\to 0^+}\frac{\Gamma((n+1)/2)}{\pi^{(n+1)/2}}\int_{|y|>\epsilon}\frac{y_j}{|y|^{n+1}}f(x-y)dy,\quad a.e.\ x\in\mathbb{R}^n.$$

It is well known that the Riesz transforms define bounded operators from $L^p(\mathbb{R}^n)$ into itself, for every $1<p<\infty$, and from $L^1(\mathbb{R}^n)$ into $L^{1,\infty}(\mathbb{R}^n)$.

We say that a distribution $f\in S'(\mathbb{R}^n)$ is restricted at infinity when there exists $r_0>1$ such that $f*\phi\in L^r(\mathbb{R}^n)$, for every $r\geq r_0$ and $\phi\in S(\mathbb{R}^n)$. The definition of Riesz transforms $R_j(f)$, $j=1,2,...,n$, for every restricted at infinity distribution $f\in S'(\mathbb{R}^n)$ can be encountered in [60, p. 123].

In [72, Theorems 1.5 and 1.6], Yang, Zhuo and Nakai characterized $H^{p(\cdot)}(\mathbb{R}^n)$ by Riesz transforms.

Theorem 2.3.5 ([72, Theorem 1.6]) *Assume that $m\in\mathbb{N}$, $m\geq 2$, and $p(\cdot)$ is a globally log-Hölder continuous function in $\mathbb{R}^n$ such that $p_-\in(\frac{n-1}{n+m-1},\infty)$. Let*

$f \in S'(\mathbb{R}^n)$ and $\phi \in S(\mathbb{R}^n)$ such that $\int_{\mathbb{R}^n} \phi = 1$. The following assertions are equivalent:

(a) $f \in H^{p(\cdot)}(\mathbb{R}^n)$,
(b) *$f \in S'(\mathbb{R}^n)$ is restricted at infinity and*

$$\sup_{t>0} \left(\|f * \phi_t\|_{p(\cdot)} + \sum_{k=1}^{m} \sum_{j_1,\dots,j_k=1}^{n} \|R_{j_1}...R_{j_k}(f) * \phi_t\|_{p(\cdot)} \right) < \infty.$$

Furthermore, there exists $C > 0$ such that

$$\frac{1}{C}\|f\|_{H^{p(\cdot)}(\mathbb{R}^n)} \leq \sup_{t>0} \left(\|f * \phi_t\|_{p(\cdot)} + \sum_{k=1}^{m} \sum_{j_1,\dots,j_k=1}^{n} \|R_{j_1}...R_{j_k}(f) * \phi_t\|_{p(\cdot)} \right)$$
$$\leq C\|f\|_{H^{p(\cdot)}(\mathbb{R}^n)}, \quad f \in H^{p(\cdot)}(\mathbb{R}^n).$$

Riesz transforms are a typical example of Calderón–Zygmund singular integrals. $L^{p(\cdot)}(\mathbb{R}^n)$ and $H^{p(\cdot)}(\mathbb{R}^n)$-boundedness properties of singular integrals have been studied, for instance, in [17, 19, 20, 27].

Fefferman and Stein [31] proved that the dual space of $H^1(\mathbb{R}^n)$ is the space $BMO(\mathbb{R}^n)$ of functions with bounded mean oscillation in $\mathbb{R}^n$. The dual of the space $H^p(\mathbb{R}^n)$ when $0 < p < 1$ was characterized as a Lipschitz space of exponent $1/p - 1$ [29]. The dual space of $H^{p(\cdot)}(\mathbb{R}^n)$ was described in [48].

By $\mathcal{P}_d(\mathbb{R}^n)$ we denote the set of polynomials in $\mathbb{R}^n$ having degree less or equal than d, being $d \in \mathbb{N}$. Suppose that $f \in L^1_{loc}(\mathbb{R}^n)$, $d \in \mathbb{N}$ and Q is a cube in $\mathbb{R}^n$. There exists a unique polynomial $P \in \mathcal{P}_d(\mathbb{R}^n)$ such that, for every $\mathfrak{q} \in \mathcal{P}_d(\mathbb{R}^n)$,

$$\int_Q (f(x) - P(x))\mathfrak{q}(x)dx = 0.$$

We denote this unique polynomial by $P^d_Q(f)$.

If $1 \leq q \leq \infty$, $\Psi : \mathcal{Q} \to (0, \infty)$ is a function defined on the set $\mathcal{Q}$ of cubes in $\mathbb{R}^n$ with sides parallel to the coordinates axis, and $f \in L^q_{loc}(\mathbb{R}^n)$, we define

$$\|f\|_{\mathcal{L}_{q,\Psi,d}} = \sup_{Q\in\mathcal{Q}} \frac{1}{\Psi(Q)} \left(\frac{1}{|Q|} \int_Q |f(x) - P^d_Q(f)(x)|^q dx \right)^{1/q},$$

when $q < \infty$, and

$$\|f\|_{\mathcal{L}_{q,\Psi,d}} = \sup_{Q\in\mathcal{Q}} \frac{1}{\Psi(Q)} \|f(x) - P^d_Q(f)\|_{L^\infty(Q)},$$

when $q = \infty$. The Campanato space $\mathcal{L}_{q,\Psi,d}(\mathbb{R}^n)$ is that one constituted by those $f \in L^q_{loc}(\mathbb{R}^n)$ such that $\|f\|_{\mathcal{L}_{q,\Psi,d}} < \infty$. By identifying $f \in \mathcal{L}_{q,\Psi,d}(\mathbb{R}^n)$ with $g = f + P$, where $P \in \mathcal{P}_d(\mathbb{R}^n)$, in $\mathcal{L}_{q,\Psi,d}(\mathbb{R}^n)$, the Campanato space is a Banach space.

Note that when $\Psi(Q) = 1$, $Q \in \mathcal{Q}$, $\mathcal{L}_{q,\Psi,d}(\mathbb{R}^n)$ coincides with $BMO(\mathbb{R}^n)$, and if $\alpha \in (0, 1)$ and $\Psi(Q) = |Q|^{1/\alpha-1}$, $Q \in \mathcal{Q}$, $\mathcal{L}_{q,\Psi,d}(\mathbb{R}^n)$ reduces to the Lipschitz space of exponent α in $\mathbb{R}^n$.

Theorem 2.3.6 ([48, Theorem 7.5]) *Assume that $p(\cdot)$ is a global log-Hölder continuous function such that $0 < p_- \leq p_+ \leq 1$, $p_+ < q \leq \infty$, and $d_{p(\cdot)} = \{d \in \mathbb{N} \cup \{0\} : p_-(n + d + 1) > n\}$. We consider the function $\Psi : \mathcal{Q} \to (0, \infty)$ defined by $\Psi(Q) = \|\chi_Q\|_{p(\cdot)}|Q|^{-1}$, $Q \in \mathcal{Q}$. Then, the dual space $(H^{p(\cdot)}(\mathbb{R}^n))'$ can be identified with the Campanato space $\mathcal{L}_{q',\Psi,d_{p(\cdot)}}(\mathbb{R}^n)$, where $1/q' + 1/q = 1$.*

Variable exponent weak Hardy spaces in $\mathbb{R}^n$ were considered in [73].

Bownik [9] introduced Hardy spaces associated with anisotropies defined by an expansive matrix in $\mathbb{R}^n$. Recently, Hardy–Lorentz spaces with variable exponents in anisotropic settings have been studied in [4, 46].

Zhuo et al. [75] defined variable exponent Hardy spaces $H^{p(\cdot)}(X)$ where X is a RD-homogeneous-type space.

2.4 Hardy Spaces with Variable Exponents Associated with Operators

Classical Hardy spaces $H^p(\mathbb{R}^n)$ are adapted, in some sense, to the Laplacian operator. However, these spaces are not useful in important problems involving other operators different from the Laplacian. It is necessary to introduce Hardy spaces adapted to the linear operators in the same way as the classical Hardy spaces are related to the Laplacian. The theory of Hardy spaces associated with operators has been developed by many authors in the last years (see, for instance, [6, 8, 28, 30, 37–39, 58, 59, 68]).

Motivated by the above papers and those ones mentioned in Sect. 2.3, Dachun Yang and his collaborators have introduced and studied variable exponent Hardy spaces associated to operators [39, 69, 71, 76].

We now precise the class of operators that we consider here. By L we denote a nonnegative selfadjoint operator on $L^2(\mathbb{R}^n)$. Then, $-L$ generates a bounded analytic semigroup $\{e^{-tL}\}_{t>0}$. It is usual to assume on $\{e^{-tL}\}_{t>0}$ some kind of exponential bound properties as the following.

(A1) (Gaussian estimates [71, 76]) For every $t > 0$ the kernel of the operator e^{-tL} is a measurable function K_t on $\mathbb{R}^n \times \mathbb{R}^n$ satisfying that there exist $C, c > 0$ such that

$$|K_t(x, y)| \leq \frac{C}{t^{n/2}} \exp\left(-c\frac{|x - y|^2}{t}\right), \quad t > 0, \ x, y \in \mathbb{R}^n.$$

(A2) (Reinforced off-diagonal estimates Davies–Gaffney estimates [37]) There exist $C, c > 0$ such that

$$| < e^{-tL}(f_1), f_2 > | \leq C \exp\left(-\frac{[dist(U_1, U_2)]^2}{ct}\right) \|f_1\|_2 \|f_2\|_2, \quad t > 0,$$

for every $f_1, f_2 \in L^2(\mathbb{R}^n)$ such that $\mathrm{supp}(f_i) \subset U_i \subset \mathbb{R}^n$, $i = 1, 2$.

(A3) ([69]) There exist $p_L \in [1, 2)$ and $q_L \in (2, \infty]$ such that, for every $k \in \mathbb{N}$, $\{(tL)^k e^{-tL}\}_{t>0}$ satisfy the reinforced (p_L, q_L) off-diagonal estimates on balls [7].

Many interesting operators satisfy these assumptions. Some of them are the following ones:

(a) Schrödinger operator $L = -\Delta + V$, where as usual, Δ represents the Laplacian operator and $0 \leq V \in L^1_{loc}(\mathbb{R}^n)$.
(b) Second-order divergence forms elliptic operators $L = -div(A\nabla)$, where $A = (a_{i,j})_{i,j=1}^n$ is such that $a_{i,j} = a_{j,i}$, $i, j = 1, 2, ..., n$ and for a certain $\lambda > 0$,

$$\frac{1}{\lambda}|y|^2 \leq \sum_{i,j=1}^{n} a_{i,j}(x) y_i y_j \leq \lambda |y|^2, \quad x, y \in \mathbb{R}^n.$$

(c) Magnetic operators: $L = -(\nabla + ia)(\nabla - ia)$, where $a = (a_j)_{j=1}^n \in L^2_{loc}(\mathbb{R}^n, \mathbb{R}^n)$.

In order to define variable exponent Hardy spaces associated to L, we consider the following Littlewood–Paley functions:

- The area square function defined by the heat semigroup $\{e^{-tL}\}_{t>0}$

$$S_{L,h}(f)(x) = \left(\int_0^\infty \int_{|x-y|<t} |t^2 L e^{-t^2 L}(f)(y)|^2 \frac{dydt}{t^{n+1}}\right)^{1/2}, \quad x \in \mathbb{R}^n.$$

- The area square function defined by the Poisson semigroup $\{e^{-t\sqrt{L}}\}_{t>0}$

$$S_{L,P}(f)(x) = \left(\int_0^\infty \int_{|x-y|<t} |tLe^{-t\sqrt{L}}(f)(y)|^2 \frac{dydt}{t^{n+1}}\right)^{1/2}, \quad x \in \mathbb{R}^n.$$

Let Ψ be an even function in $S(\mathbb{R})$ such that $\int_{\mathbb{R}} \Psi(z)dz \neq 0$. We define $\Psi(t\sqrt{L})$, $t > 0$, by using functional calculus. Note that if $\Psi(z) = e^{-z^2}$, $z \in \mathbb{R}$, then $\Psi(t\sqrt{L}) = e^{-t^2 L}$, $t > 0$. We consider the following maximal functions:

– Radial maximal function

$$\Psi_L^+(f)(x) = \sup_{t>0} |\Psi(t\sqrt{L})(f)(x)|, \quad x \in \mathbb{R}^n.$$

– Non tangential maximal function: with $\alpha > 0$,

$$\Psi^*_{L,\alpha}(f)(x) = \sup_{t>0} \sup_{|x-y|<\alpha t} |\Psi(t\sqrt{L})(f)(y)|, \quad x \in \mathbb{R}^n.$$

– Grand maximal function

$$\ominus^*_{L,\alpha}(f)(x) = \sup_{\Psi \in S_{N,even}(\mathbb{R})} \Psi^*_L(f)(x), \quad x \in \mathbb{R}^n,$$

where $\Psi^*_L = \Psi^*_{L,1}$ and $S_{N,even}(\mathbb{R}) = \{\phi \in S(\mathbb{R}) : \phi \text{ is even and } \max_{0\le m\le N} \sup_{z\in\mathbb{R}}(1+|z|)^N|D^m\Psi(z)| \le 1\}$.

The operators $S_{L,h}$, $S_{L,P}$, Ψ^+_L, $\Psi^*_{L,\alpha}$, and $\ominus^*_{L,\alpha}$ are sublinear operators and are bounded from $L^2(\mathbb{R}^n)$ into itself.

Suppose that T is a sublinear and bounded operator from $L^2(\mathbb{R}^n)$ into itself and $p(\cdot)$ is a global log-Hölder continuous in $\mathbb{R}^n$.

We define the space

$$\mathbb{H}^{p(\cdot)}_T(\mathbb{R}^n) = \{f \in L^2(\mathbb{R}^n) : T(f) \in L^{p(\cdot)}(\mathbb{R}^n)\}.$$

The $(p(\cdot), T)$-Hardy space $H^{p(\cdot)}_T(\mathbb{R}^n)$ is defined as the completion of $\mathbb{H}^{p(\cdot)}_T(\mathbb{R}^n)$ with respect to the quasinorm $\|\cdot\|_{H^{p(\cdot)}_T(\mathbb{R}^n)}$ given by

$$\|f\|_{H^{p(\cdot)}_T(\mathbb{R}^n)} = \|T(f)\|_{p(\cdot)}, \quad f \in \mathbb{H}^{p(\cdot)}_T(\mathbb{R}^n).$$

According to this definition, we can consider variable exponent Hardy spaces associated with the operator L by using the sublinear operators $S_{L,h}$, $S_{L,P}$, Ψ^+_L, $\Psi^*_{L,\alpha}$, and $\ominus^*_{L,\alpha}$.

Let $1 < q \le \infty$ and $M \in \mathbb{N}$. A function $\mathfrak{a} \in L^q(\mathbb{R}^n)$ is called a $(p(\cdot), q, M)_L$-atom associated with the ball B in $\mathbb{R}^n$ when there exists $\mathfrak{b}$ in the domains of L^M such that

(i) $\mathfrak{a} = L^M\mathfrak{b}$,
(ii) $\operatorname{supp}(L^k\mathfrak{b}) \subset B, \quad k = 0, 1, 2, ..., M$,
(iii) $\|(r_B^2 L)^k\mathfrak{b}\|_q \le r_B^{2M}|B|^{1/q}\|\chi_B\|^{-1}_{p(\cdot)}, \quad k = 0, 1, 2, ..., M$. Here r_B represents the radius of B.

We say that $f \in L^2(\mathbb{R}^n)$ has a $(p(\cdot), q, M)_L$-atomic representation when $f = \sum_{j\in\mathbb{N}} \lambda_j\mathfrak{a}_j$, where the series converges in $L^2(\mathbb{R}^n)$, and, for every $j \in \mathbb{N}$, $\lambda_j > 0$ and $\mathfrak{a}_j$ is a $(p(\cdot), q, M)_L$-atom associated with the ball B_j, satisfying that

$$\left(\sum_{j\in\mathbb{N}} \left(\frac{\lambda_j}{\|\chi_{B_j}\|_{p(\cdot)}}\right)^{\underline{p}} \chi_{B_j}\right)^{1/\underline{p}} \in L^{p(\cdot)}(\mathbb{R}^n).$$

Here and in the sequel $\underline{p} = \min\{1, p_-\}$.

The space $\mathbb{H}^{p(\cdot)}_{L,at,q,M}(\mathbb{R}^n)$ consists of all those $f \in L^2(\mathbb{R}^n)$ having a $(p(\cdot), q, M)_L$-atomic representation. For every $f \in \mathbb{H}^{p(\cdot)}_{L,at,q,M}(\mathbb{R}^n)$, we define

$$\|f\|_{H^{p(\cdot)}_{L,at,q,M}(\mathbb{R}^n)} = \inf \left\| \left(\sum_{j\in\mathbb{N}} \left(\frac{\lambda_j}{\|\chi_{B_j}\|_{p(\cdot)}} \right)^{\underline{p}} \chi_{B_j} \right)^{1/\underline{p}} \right\|_{p(\cdot)},$$

where the infimum is taken over all those pair of sequences $\{\lambda_j\}_{j\in\mathbb{N}}$ and $\{B_j\}_{j\in\mathbb{N}}$ such that, for every $j \in \mathbb{N}$, $\lambda_j > 0$ and there exists a $(p(\cdot), q, M)_L$-atom $\mathfrak{a}_j$ associated with the ball B_j satisfying that $f = \sum_{j\in\mathbb{N}} \lambda_j \mathfrak{a}_j$ in $L^2(\mathbb{R}^n)$ and

$$\left(\sum_{j\in\mathbb{N}} \left(\frac{\lambda_j}{\|\chi_{B_j}\|_{p(\cdot)}} \right)^{\underline{p}} \chi_{B_j} \right)^{1/\underline{p}} \in L^{p(\cdot)}(\mathbb{R}^n).$$

The atomic Hardy space $H^{p(\cdot)}_{L,at,q,M}(\mathbb{R}^n)$ is defined as the completion of $\mathbb{H}^{p(\cdot)}_{L,at,q,M}(\mathbb{R}^n)$ with respect to the quasinorm $\|\cdot\|_{H^{p(\cdot)}_{L,at,q,M}(\mathbb{R}^n)}$.

Let $1 < q \leq \infty$, $\epsilon > 0$ and $M \in \mathbb{N}$. A function $\mathfrak{a} \in L^q(\mathbb{R}^n)$ is called a $(p(\cdot), q, \epsilon, M)_L$-molecule associated with the ball B in $\mathbb{R}^n$ when there exists $\mathfrak{b}$ in the domains of L^M such that

(i) $\mathfrak{a} = L^M \mathfrak{b}$,
(ii) $\|\chi_{S_i(B)} L^k \mathfrak{b}\|_q \leq 2^{-i\epsilon} r_B^{2(M-k)} |B|^{1/q} \|\chi_B\|_{p(\cdot)}^{-1}$, $k = 0, 1, 2, ..., M$, and $i \in \mathbb{N}$, where r_B is the radius of B, $S_0(B) = B$ and $S_i(B) = 2^i B \backslash 2^{i-1} B$, $i \in \mathbb{N}$, $i \geq 1$.

We say that $f \in L^2(\mathbb{R}^n)$ has a $(p(\cdot), q, \epsilon, M)_L$-molecular representation when $f = \sum_{j\in\mathbb{N}} \lambda_j \mathfrak{a}_j$, where the series converges in $L^2(\mathbb{R}^n)$, and, for every $j \in \mathbb{N}$, $\lambda_j > 0$ and $\mathfrak{a}_j$ is $(p(\cdot), q, \epsilon, M)_L$-molecule associated with the ball B_j, satisfying that

$$\left(\sum_{j\in\mathbb{N}} \left(\frac{\lambda_j}{\|\chi_{B_j}\|_{p(\cdot)}} \right)^{\underline{p}} \chi_{B_j} \right)^{1/\underline{p}} \in L^{p(\cdot)}(\mathbb{R}^n).$$

By $\mathbb{H}^{p(\cdot)}_{L,mol,q,\epsilon,M}(\mathbb{R}^n)$, we denote the space constituted by all those $f \in L^2(\mathbb{R}^n)$ having a $(p(\cdot), q, \epsilon, M)_L$-molecular representation. We define, for every $f \in \mathbb{H}^{p(\cdot)}_{L,mol,q,\epsilon,M}(\mathbb{R}^n)$,

$$\|f\|_{H^{p(\cdot)}_{L,mol,q,\epsilon,M}(\mathbb{R}^n)} = \inf \left\| \left(\sum_{j\in\mathbb{N}} \left(\frac{\lambda_j}{\|\chi_{B_j}\|_{p(\cdot)}} \right)^{\underline{p}} \chi_{B_j} \right)^{1/\underline{p}} \right\|_{p(\cdot)},$$

where the infimum is taken over all those pair of sequences $\{\lambda_j\}_{j\in\mathbb{N}}$ and $\{B_j\}_{j\in\mathbb{N}}$ such that, for every $j \in \mathbb{N}$, $\lambda_j > 0$ and there exists a $(p(\cdot), q, \epsilon, M)_L$-molecule $\mathfrak{a}_j$

associated with the ball B_j satisfying that $f = \sum_{j\in\mathbb{N}} \lambda_j \mathfrak{a}_j$ in $L^2(\mathbb{R}^n)$ and

$$\left(\sum_{j\in\mathbb{N}} \left(\frac{\lambda_j}{\|\chi_{B_j}\|_{p(\cdot)}}\right)^{\underline{p}} \chi_{B_j}\right)^{1/\underline{p}} \in L^{p(\cdot)}(\mathbb{R}^n).$$

The molecular Hardy space $H^{p(\cdot)}_{L,mol,q,\epsilon,M}(\mathbb{R}^n)$ is defined as the completion of $\mathbb{H}^{p(\cdot)}_{L,mol,q,\epsilon,M}(\mathbb{R}^n)$ with respect to the quasinorm $\| \; \|_{H^{p(\cdot)}_{L,mol,q,\epsilon,M}(\mathbb{R}^n)}$.

Atomic and molecular representations are very useful to study the boundedness properties of operators in Hardy spaces.

In the following results, the coincidence of the above-defined Hardy spaces is established.

Theorem 2.4.1 ([71, Theorem 3.13] and [76, Theorems 1.8 and 1.11]) *Assume that L is a nonnegative and selfadjoint operator in $L^2(\mathbb{R}^n)$ and it satisfies (A1). Let $p(\cdot)$ be a globally log-Hölder continuous function in $\mathbb{R}^n$ such that $0 < p_+ \leq 1$, $q \in (1,\infty)$ and $M \in \mathbb{N} \cap (n/2[1/p_- - 1], \infty)$. Then, for every $\alpha \in (0,\infty)$ and $\Psi \in S(\mathbb{R}^n)$ such that Ψ is even and $\int_{\mathbb{R}^n} \Psi(x)dx \neq 0$, the spaces $H^{p(\cdot)}_{L,at,q,M}(\mathbb{R}^n)$, $H^{p(\cdot)}_{S_L,h}(\mathbb{R}^n)$, $H^{p(\cdot)}_{\Theta^*_{L,\alpha}}(\mathbb{R}^n)$, $H^{p(\cdot)}_{\Psi^*_{L,\alpha}}(\mathbb{R}^n)$ and $H^{p(\cdot)}_{L,mol,q,\epsilon,M}(\mathbb{R}^n)$ coincide, and the corresponding quasi-norms are equivalent provided that $p_- > n/(n+\epsilon)$ with $\epsilon > 0$ and $N \in \mathbb{N}$ large enough.*

Theorem 2.4.2 ([76, Theorems 1.17]) *Suppose that L is a nonnegative and selfadjoint operator in $L^2(\mathbb{R}^n)$ and it satisfies (A1). Assume also that if, for every $t > 0$, the function K_t denotes the kernel of the integral operator e^{-tL}, there exist $C > 0$ and $\mu \in (0,1]$ such that*

$$|K_t(y_1,x) - K_t(y_2,x)| \leq \frac{C}{t^{n/2}} \frac{|y_1 - y_2|^k}{t^{\mu/2}}, \quad t > 0 \text{ and } x, y_1, y_2 \in \mathbb{R}^n.$$

Let $p(\cdot)$ be a globally log-Hölder continuous function in $\mathbb{R}^n$ such that $0 < p_+ \leq 1$, $q \in (1,\infty)$, and $M \in \mathbb{N} \cap (n/2[1/p_- - 1], \infty)$. Then, the spaces $H^{p(\cdot)}_{L,at,q,M}(\mathbb{R}^n)$, and $H^{p(\cdot)}_{\Psi^+_L}(\mathbb{R}^n)$ coincide and the corresponding quasinorms are equivalent provided that $\Psi \in S(\mathbb{R}^n)$ is even and $\int_{\mathbb{R}^n} \Psi(x)dx \neq 0$.

In order to prove $H^{p(\cdot)}_{L,at,q,M}(\mathbb{R}^n) = H^{p(\cdot)}_{\Psi^+_{L,\alpha}}(\mathbb{R}^n)$ in Theorem 2.4.2, a Lipschitz condition for the kernel functions is needed. We are going to present a particular case where this extra property is not necessary in order to that the equality of radial and atomic Hardy spaces holds.

We consider the magnetic Schrödinger operator

$$L = (\nabla - ia)(\nabla - ia) + V,$$

where $a \in L^2_{loc}(\mathbb{R}^n, \mathbb{R}^n)$ and $0 \leq V \in L^1_{loc}(\mathbb{R}^n)$. This operator satisfies all the assumptions in Theorem 2.4.2 except, in general, the Lipschitz one.

We consider $\Psi(z) = e^{-z^2}$, $z \in \mathbb{R}$, and $\alpha > 0$. It is clear that $\Psi_L^+(f) \leq \Psi_{L,\alpha}^*(f)$, for every $f \in L^2(\mathbb{R}^n)$. Hence, $H^{p(\cdot)}_{\Psi^*_{L,\alpha}}(\mathbb{R}^n)$ is continuously contained in $H^{p(\cdot)}_{\Psi^+_L}(\mathbb{R}^n)$. By proceeding as in [39, p. 483] we can see that, for every $\omega \in A_\infty(\mathbb{R}^n)$ (see [60] for definitions) and $0 < p \leq 1$, there exists $C > 0$ such that

$$\|\Psi^*_{L,\alpha}(f)\|_{L^p(\mathbb{R}^n,\omega)} \leq C\|\Psi^+_L(f)\|_{L^p(\mathbb{R}^n,\omega)}, \quad f \in L^2(\mathbb{R}^n) \cap L^p(\mathbb{R}^n, \omega).$$

According to the extension of Rubio de Francia extrapolation theorem proved in [17, Corollary 1.10], if $p(\cdot)$ is a globally log-Hölder continuous function in $\mathbb{R}^n$, then

$$\|\Psi^*_{L,\alpha}(f)\|_{p(\cdot)} \leq C\|\Psi^+_L(f)\|_{p(\cdot)}, \quad f \in H^{p(\cdot)}_{\Psi^+_L}(\mathbb{R}^n),$$

and $H^{p(\cdot)}_{\Psi^+_L}(\mathbb{R}^n)$ is continuously contained in $H^{p(\cdot)}_{\Psi^*_{L,\alpha}}(\mathbb{R}^n)$.

Thus, it is proved that $H^{p(\cdot)}_{\Psi^*_{L,\alpha}}(\mathbb{R}^n) = H^{p(\cdot)}_{\Psi^+_L}(\mathbb{R}^n)$ with equivalent quasinorms provided that $p(\cdot)$ is globally log-Hölder continuous in $\mathbb{R}^n$.

The dual space of $H^{p(\cdot)}_{\Theta^*_{L,\alpha}}(\mathbb{R}^n)$ is characterized as a Campanato-type space in [71, Theorem 4.3] provided that the operator L satisfies the following properties:

(a) L is one to one and it has dense range in $L^2(\mathbb{R}^n)$ and a bounded H^∞ functional calculus in $L^2(\mathbb{R}^n)$.
(b) For every $t > 0$ the kernel of the integral operator e^{-tL} is a measurable bounded function K_t in $\mathbb{R}^n \times \mathbb{R}^n$ such that

$$|K_t(x, y)| \leq t^{-n/m} g\left(\frac{|x-y|}{t^{1/m}}\right), \quad t > 0 \text{ and } x, y \in \mathbb{R}^n,$$

where $m > 0$ and g is a positive, bounded, and decreasing function on $(0, \infty)$ satisfying that, for every $\epsilon > 0$,

$$\lim_{r\to\infty} r^{n+\epsilon} g(r) = 0.$$

In [2] are defined variable exponent Hardy spaces on graphs associated with weighted discrete Laplacians. It is remarkable that in this discrete setting the unitary sets have positive measure. This fact implies that the arguments developed by Uchiyama [65] or Grafakos et al. [34] do not work for the variable exponent Hardy spaces studied in [2], so the authors have not been able to characterize it by maximal functions.

2.5 Local Hardy Spaces with Variable Exponents

Local Hardy spaces $h^p(\mathbb{R}^n)$, $0 < p \leq 1$, were introduced by Goldberg [33]. They are spaces of tempered distributions that are much better suited to problems associated with partial differential equations than the classical Hardy spaces. In particular, some pseudo-differential operators are bounded in local Hardy spaces $h^p(\mathbb{R}^n)$ but they are not bounded on Hardy spaces $H^p(\mathbb{R}^n)$ [33].

If $\phi \in S(\mathbb{R}^n)$ we define the local maximal function

$$m_\phi(f) = \sup_{0<t\leq 1} |f * \phi_t|, \quad f \in S'(\mathbb{R}^n),$$

and, for every $N \in \mathbb{N}$, the local grand maximal function

$$m_N(f) = \sup_{\phi \in S_N} |m_\phi(f)|, \quad f \in S'(\mathbb{R}^n).$$

We consider the strip $S = \{(x,t) : x \in \mathbb{R}^n \text{ and } t \in (0,1)\}$. By $P_t^0(x)$, $x \in \mathbb{R}^n$, $t \in (0,1)$, we denote the function whose Fourier transform $\hat{P}_t^0$ with respect to x is given by

$$\hat{P}_t^0(y) = \frac{\sinh((1-t)2\pi|y|)}{\sinh(2\pi|y|)}, \quad y \in \mathbb{R}^n \text{ and } t \in (0,1).$$

The Poisson kernel for the strip S is

$$P_t(x) = P_t^0(x) + P_{1-t}^0(x), \quad x \in \mathbb{R}^n \text{ and } t \in (0,1).$$

Note that, for every $t \in (0,1)$, $P_t \in S(\mathbb{R}^n)$. For every $f \in S'(\mathbb{R}^n)$ and $\phi \in S(\mathbb{R}^n)$, we define the nontangetial maximal function m^* by

$$m^*(f)(x) = \sup_{|x-y|\leq t<1/2} |(f * \phi_t)(y)|, \quad x \in \mathbb{R}^n.$$

Goldberg [33] proved the following fundamental result.

Theorem 2.5.1 ([33]) *For every $f \in S'(\mathbb{R}^n)$ and $0 < p \leq \infty$ the following assertions are equivalent.*

(i) There exists $\phi \in S(\mathbb{R}^n)$ such that $\int_{\mathbb{R}^n} \phi(x)dx \neq 0$ and $m_\phi(f) \in L^p(\mathbb{R}^n)$.
(ii) For every N large enough $m_N(f) \in L^p(\mathbb{R}^n)$.
(iii) $m^*(f) \in L^p(\mathbb{R}^n)$.

Furthermore, for every $f \in S'(\mathbb{R}^n)$ and $0 < p \leq \infty$ the quantities $\|m^*(f)\|_p$, $\|m_\phi(f)\|_p$ and $\|m_N(f)\|_p$ are equivalent being the equivalence constant dependent of p but independent of f.

We say that a tempered distribution f is in $h^p(\mathbb{R}^n)$, $0 < p \leq \infty$, when properties (i), (ii), or (iii) in Theorem 2.5.1 are satisfied. For every $0 < p \leq \infty$, $h^p(\mathbb{R}^n)$ is endowed with the quasinorm $\|\cdot\|_{h^p(\mathbb{R}^n)}$ defined by

$$\|f\|_{h^p(\mathbb{R}^n)} = \|m_N(f)\|_p, \quad f \in h^p(\mathbb{R}^n).$$

The classical Hardy space $H^p(\mathbb{R}^n)$ is contained properly in $h^p(\mathbb{R}^n)$, for every $0 < p \leq \infty$, and $h^p(\mathbb{R}^n)$ can be identified with $L^p(\mathbb{R}^n)$ when $1 < p \leq \infty$.

The following result connects $H^p(\mathbb{R}^n)$ and $h^p(\mathbb{R}^n)$.

Theorem 2.5.2 ([33]) *Assume that $\phi \in S(\mathbb{R}^n)$ such that $\int_{\mathbb{R}^n} x^\alpha \phi(x)dx = 0$, when $\alpha \in \mathbb{N}^n$ and $|\alpha| \leq [n(1/p - 1)]$. Then, there exists $C > 0$ such that, for every $f \in h^p(\mathbb{R}^n)$, $f - f * \phi \in H^p(\mathbb{R}^n)$ and $\|f - f * \phi\|_{H^p(\mathbb{R}^n)} \leq C\|f\|_{h^p(\mathbb{R}^n)}$.*

As a consequence of Theorem 2.5.2 atomic characterizations of the distribution in $h^p(\mathbb{R}^n)$ can be obtained.

Let $0 < p \leq 1$. We say that a measurable function $\mathfrak{a}$ in $\mathbb{R}^n$ is a local p-atom associated with the ball B when

(i) $\operatorname{supp} \mathfrak{a} \subset B$;
(ii) $\|\mathfrak{a}\|_\infty \leq |B|^{-1/p}$;
(iii) $\int_{\mathbb{R}^n} \mathfrak{a}(x)x^\alpha dx = 0$, for every $\alpha \in \mathbb{N}^n$, $|\alpha| \leq 1 + [n(1/p - 1)]$, when the radius of B, $r_B \geq 1$.

Note that any null moment is required when $r_B \geq 1$ in contrast with the global $H^p(\mathbb{R}^n)$ spaces.

Theorem 2.5.3 ([33]) *Let $0 < p \leq 1$. A distribution $f \in S'(\mathbb{R}^n)$ is in $h^p(\mathbb{R}^n)$ if and only if $f = \sum_{j\in\mathbb{N}} \lambda_j \mathfrak{a}_j$ in $S'(\mathbb{R}^n)$, where $\{\lambda_j\}_{j\in\mathbb{N}} \subset (0, \infty)$ such that $\sum_{j\in\mathbb{N}} \lambda_j^p < \infty$ and, for every $j \in \mathbb{N}$, $\mathfrak{a}_j$ is a local p-atom. Furthermore, the quantities $\|f\|_{h^p(\mathbb{R}^n)}$ and $(\sum_{j\in\mathbb{N}} \lambda_j^p)^{1/p}$ are equivalent.*

In [52] the local Hardy spaces $h^p(\mathbb{R}^n)$ are characterized by using local Riesz transforms. Bui [10], see also [64], proved that $h^p(\mathbb{R}^n)$ coincides with the Triebel–Lizorkin spaces $F_p^{0,2}(\mathbb{R}^n)$. Local Hardy spaces in different contexts can be encountered in [11, 21, 63, 70].

Local Hardy spaces associated with operators have been studied in [12, 13, 34, 40].

Diening et al. [26] defined the variable exponent local Hardy space $h^{p(\cdot)}(\mathbb{R}^n)$ as the Triebel–Lizorkin space $F_{p(\cdot)}^{0,2}(\mathbb{R}^n)$ with variable integrability. Nakai and Sawano [48] characterized $h^{p(\cdot)}(\mathbb{R}^n)$ by using local maximal functions.

Assume that $p(\cdot)$ is globally log-Hölder continuous in $\mathbb{R}^n$. A distribution $f \in S'(\mathbb{R}^n)$ is in $h^{p(\cdot)}(\mathbb{R}^n)$ when $m_N(f) \in L^{p(\cdot)}(\mathbb{R}^n)$. The quasinorm $\|\cdot\|_{h^{p(\cdot)}(\mathbb{R}^n)}$ is defined by

$$\|f\|_{h^{p(\cdot)}(\mathbb{R}^n)} = \|m_N(f)\|_{p(\cdot)}, \quad f \in h^{p(\cdot)}(\mathbb{R}^n).$$

We have that (see [48, Theorem 3.3])

$$\|f\|_{h^{p(\cdot)}(\mathbb{R}^n)} \sim \|\sup_{j\in\mathbb{N}} \Psi_{(j)}(D)f\|_{p(\cdot)}, \quad f \in h^{p(\cdot)}(\mathbb{R}^n),$$

where $\Psi \in S(\mathbb{R}^n)$ and $\int_{\mathbb{R}^n} \Psi(x)dx \neq 0$.

A variable exponent version of Theorem 2.5.2 was established in [48, Lemma 9.1]. Using this result the authors proved in [48, Theorem 9.2] proved that $h^{p(\cdot)}(\mathbb{R}^n)$ and $F^{0,2}_{p(\cdot)}(\mathbb{R}^n)$ are isomorphic.

In [3] we define local Hardy spaces with variable exponents associated to operators. We consider an operator L that is nonnegative and selfadjoint in $L^2(\mathbb{R}^n)$ which satisfies estimation (A1) in Sect. 2.4.

Our definition of variable exponent local Hardy space associated with L is motivated by those ones due to Carbonaro, McIntosh and Morris [13], who defined the local Hardy space h^1 of differential forms on Riemannian manifolds, and Cao et al. [12] who studied local Hardy spaces associated with inhomogeneous higher order elliptic operators.

We consider now the localized area square integral function S_L^{loc} defined by

$$S_L^{loc}(f)(x) = \left(\int_0^1 \int_{B(x,t)} |t^2 L e^{-t^2 L}(f)(y)|^2 \frac{dydt}{t^{n+1}} \right)^{1/2}, \quad x \in \mathbb{R}^n,$$

for every $f \in S'(\mathbb{R}^n)$.

We say a function $f \in L^2(\mathbb{R}^n)$ is in $\mathfrak{h}_L^{p(\cdot)}(\mathbb{R}^n)$ when $S_L^{loc}(f) \in L^{p(\cdot)}(\mathbb{R}^n)$ and $S_I(f) \in L^{p(\cdot)}(\mathbb{R}^n)$. Here S_I denotes the area square integral function associated with the identity operator, that is,

$$S_I(f)(x) = \left(\int_0^\infty \int_{|x-y|<t} |t^2 e^{-t^2}(f)(y)|^2 \frac{dydt}{t^{n+1}} \right)^{1/2}, \quad x \in \mathbb{R}^n.$$

The quasinorm $\|\cdot\|_{h_L^{p(\cdot)}(\mathbb{R}^n)}$ is defined by

$$\|f\|_{h_L^{p(\cdot)}(\mathbb{R}^n)} = \|S_L^{loc}(f)\|_{p(\cdot)} + \|S_I(f)\|_{p(\cdot)}, \quad f \in \mathfrak{h}_L^{p(\cdot)}(\mathbb{R}^n).$$

We defined the local Hardy space $h_L^{p(\cdot)}(\mathbb{R}^n)$ as the completion of $\mathfrak{h}_L^{p(\cdot)}(\mathbb{R}^n)$ with respect to $\|\cdot\|_{h_L^{p(\cdot)}(\mathbb{R}^n)}$.

We now establish molecular characterizations of $h_L^{p(\cdot)}(\mathbb{R}^n)$. Let $M \in \mathbb{N}$ and $\epsilon > 0$. A function $\mathfrak{m} \in L^2(\mathbb{R}^n)$ is said to be a $(p(\cdot), 2, M, \epsilon)_{L,loc}$-molecule associated with the ball B with radius r_B when

(i) $r_B \geq 1$ and $\|\chi_{S_i(B)}\mathfrak{m}\|_2 \leq 2^{-i\epsilon}|2^i B|^{1/2}\|\chi_{2^i B}\|^{-1}_{p(\cdot)}$ $i \in \mathbb{N}$;
(ii) $r_B \in (0, 1)$ and there exists $\mathfrak{b}$ in the domain of L^M such that $\mathfrak{m} = L^M \mathfrak{b}$ and, for every $k \in \{0, 1, ..., M\}$,

$$\|\chi_{S_i(B)} L^k \mathfrak{b}\|_2 \leq 2^{-i\epsilon} r_B^{2(M-k)} |2^i B|^{1/2} \|\chi_{2^i B}\|^{-1}_{p(\cdot)}, \quad i \in \mathbb{N}.$$

We say that $f \in L^2(\mathbb{R}^n)$ is in $\mathfrak{h}^{p(\cdot)}_{L,mol,M,\epsilon}(\mathbb{R}^n)$ when $f = \sum_{j \in \mathbb{N}} \lambda_j \mathfrak{m}_j$, where the series converges in $L^2(\mathbb{R}^n)$, and, for every $j \in \mathbb{N}$, $\lambda_j > 0$ and $\mathfrak{m}_j$ is a $(p(\cdot), 2,$

$M, \epsilon)_{L,loc}$-molecule associated with the ball B_j, satisfying that

$$\left(\sum_{j\in\mathbb{N}}\left(\frac{\lambda_j}{\|\chi_{B_j}\|_{p(\cdot)}}\right)^{\underline{p}}\chi_{B_j}\right)^{1/\underline{p}} \in L^{p(\cdot)}(\mathbb{R}^n).$$

The quasinorm $\|\cdot\|_{h^{p(\cdot)}_{L,mol,M,\epsilon}(\mathbb{R}^n)}$ is defined as follows, for every $f \in \mathfrak{h}^{p(\cdot)}_{L,mol,M,\epsilon}(\mathbb{R}^n)$,

$$\|f\|_{H^{p(\cdot)}_{L,mol,M,\epsilon}(\mathbb{R}^n)} = \inf\left\|\left(\sum_{j\in\mathbb{N}}\left(\frac{\lambda_j}{\|\chi_{B_j}\|_{p(\cdot)}}\right)^{\underline{p}}\chi_{B_j}\right)^{1/\underline{p}}\right\|_{p(\cdot)},$$

where the infimum is taken over all the pairs of sequences $\{\lambda_j\}_{j\in\mathbb{N}}$ and $\{B_j\}_{j\in\mathbb{N}}$ being, for every $j \in \mathbb{N}$, $\lambda_j > 0$ and there exists a $(p(\cdot), 2, M, \epsilon)_{L,loc}$-molecule $\mathfrak{m}_j$ associated with the ball B_j such that $f = \sum_{j\in\mathbb{N}} \lambda_j \mathfrak{m}_j$ in $L^2(\mathbb{R}^n)$ and

$$\left(\sum_{j\in\mathbb{N}}\left(\frac{\lambda_j}{\|\chi_{B_j}\|_{p(\cdot)}}\right)^{\underline{p}}\chi_{B_j}\right)^{1/\underline{p}} \in L^{p(\cdot)}(\mathbb{R}^n).$$

By $h^{p(\cdot)}_{L,mol,2,M,\epsilon}(\mathbb{R}^n)$ we denote the completion of $\mathfrak{h}^{p(\cdot)}_{L,mol,2,M,\epsilon}(\mathbb{R}^n)$ with respect to $\|\cdot\|_{h^{p(\cdot)}_{L,mol,2,M,\epsilon}(\mathbb{R}^n)}$.

Theorem 2.5.4 ([3, Theorem 1.1]) *Suppose that $p(\cdot)$ is a global log-Hölder continuous function in $\mathbb{R}^n$ such that $p_+ < 2$.*

(i) If $\epsilon > n(1/p_- - 1/p_+)$ and $M \in \mathbb{N}$ such that $2M > n(2/p_- - 1/2 - 1/p_+)$, then

$$h^{p(\cdot)}_{L,mol,2,M,\epsilon}(\mathbb{R}^n) \subset h^{p(\cdot)}_L(\mathbb{R}^n).$$

(ii) If $\epsilon > 0$ and $M \in \mathbb{N}$, $h^{p(\cdot)}_L(\mathbb{R}^n) \subset h^{p(\cdot)}_{L,mol,2,M,\epsilon}(\mathbb{R}^n)$.

The embeddings in (i) and (ii) are algebraic and topological.

By Theorem 2.5.4, we can see that $H^{p(\cdot)}_L(\mathbb{R}^n) \subset h^{p(\cdot)}_L(\mathbb{R}^n)$ and that $h^{p(\cdot)}_L(\mathbb{R}^n) = h^q_L(\mathbb{R}^n)$, when $p(x) = q$, $x \in \mathbb{R}^n$, with $0 < q \leq 1$. Also we obtain the next property.

Theorem 2.5.5 ([3, Corollary 1.4]) *Let $p(\cdot)$ be a globally log-Hölder continuous function in $\mathbb{R}^n$ such that $p_+ < 2$. Then, $h^{p(\cdot)}_L(\mathbb{R}^n) \subset L^{p(\cdot)}(\mathbb{R}^n)$. Furthermore, $h^{p(\cdot)}_L(\mathbb{R}^n) = H^{p(\cdot)}_L(\mathbb{R}^n) = L^{p(\cdot)}(\mathbb{R}^n)$, provided that $p_- > 1$.*

The following result is a generalization of another one obtained by Kemppainen when $p(x) = 1$, $x \in \mathbb{R}^n$ [41, Theorem 7].

Theorem 2.5.6 ([3, Theorem 1.5]) *Assume that $p(\cdot)$ is a globally log-Hölder continuous function in $\mathbb{R}^n$ such that $p_+ < 2$. If $\inf \sigma(L) > 0$, where $\sigma(L)$ denotes the spectrum of L in $L^2(\mathbb{R}^n)$, then $h_L^{p(\cdot)}(\mathbb{R}^n) = H_L^{p(\cdot)}(\mathbb{R}^n)$ and their quasinorms are equivalent.*

The Hermite operator (also called harmonic oscillator) is the Schrödinger operator with potential $V(x) = |x|^2$, that is, $H = -\Delta + |x|^2$. The spectrum of H in $L^p(\mathbb{R}^n)$, $1 < p < \infty$, is $\sigma(H) = \{2k + n : k \in \mathbb{N}\}$. The twisted Laplacian operator $\mathcal{L}$ is defined by

$$\mathcal{L} = \frac{1}{2}\sum_{j=1}^{n}((\partial x_j + iy_j)^2 + (\partial y_j - ix_j))^2, \quad (x = (x_1, ..., x_n), y = (y_1, ..., y_n)) \in \mathbb{R}^n \times \mathbb{R}^n.$$

The spectrum of $\mathcal{L}$ in $L^2(\mathbb{R}^n)$ is $\sigma(\mathcal{L}) = \{2k + n : k \in \mathbb{N}\}$. Our Theorem 2.5.6 applies for H and $\mathcal{L}$. We note that $\mathcal{L}$ is a magnetic Laplacian operator. Mauceri et al. [47] defined the Hardy space $H^1_{\mathcal{L}}(\mathbb{R}^n)$. Our result in Theorem 2.5.6 extends other ones obtained in [47].

Molecular characterizations established in Theorem 2.5.4 allow us to get the following generalization to variable exponents settings for the results given in [40].

Theorem 2.5.7 ([3, Theorem 1.6]) *Let $p(\cdot)$ be a globally log-Hölder continuous function in $\mathbb{R}^n$ such that $p_+ < 2$. Then, $h_L^{p(\cdot)}(\mathbb{R}^n) = H_{L+I}^{p(\cdot)}(\mathbb{R}^n)$, and the quasinorms are equivalent.*

References

1. E. Acerbi, G. Mingione, Regularity results for a class of functionals with non-standard growth. Arch. Ration. Mech. Anal. **156**(2), 121–140 (2001)
2. V. Almeida, J.J. Betancor, A.J. Castro, L. Rodríguez-Mesa, Variable exponent Hardy spaces associated with discrete Laplacians on graphs. Sci. China Math. **62**(1), 73–124 (2019)
3. V. Almeida, J.J. Betancor, E. Dalmasso, L. Rodríguez-Mesa, Local Hardy spaces with variable exponents associated to non-negative self-adjoint operators satisfying Gaussian estimates. J. Geom. Anal. (2019). https://doi.org/10.1007/s12220-019-00199-y
4. V. Almeida, J.J. Betancor, L. Rodríguez-Mesa, Anisotropic Hardy-Lorentz spaces with variable exponents. Canad. J. Math. **69**(6), 1219–1273 (2017)
5. S.N. Antontsev, J.F. Rodrigues, On stationary thermo-rheological viscous flows. Ann. Univ. Ferrara Sez. VII Sci. Mat. **52**(1), 19–36 (2006)
6. P. Auscher, X.T. Duong, A. McIntosh, Boundedness of banach space valued singular integral and Hardy spaces. Unplublished Manuscript (2005)
7. P. Auscher, J.M.A. Martell, Weighted norm inequalities, off-diagonal estimates and elliptic operators. II. Off-diagonal estimates on spaces of homogeneous type. J. Evol. Equ. 7, **2**, 265–316 (2007)
8. P. Auscher, A. McIntosh, E. Russ, Hardy spaces of differential forms on Riemannian manifolds. J. Geom. Anal. **18**(1), 192–248 (2008)
9. M. Bownik, Anisotropic Hardy spaces and wavelets. Mem. Amer. Math. Soc. 164, **781**, vi+122 (2003)

10. H.Q. Bui, Weighted Hardy spaces. Math. Nachr. **103**, 45–62 (1981)
11. J. Cao, D.-C. Chang, D. Yang, S. Yang, Weighted local Orlicz-Hardy spaces on domains and their applications in inhomogeneous Dirichlet and Neumann problems. Trans. Amer. Math. Soc. **365**(9), 4729–4809 (2013)
12. J. Cao, S. Mayboroda, D. Yang, Local Hardy spaces associated with inhomogeneous higher order elliptic operators. Anal. Appl. (Singap.) **15**(2), 137–224 (2017)
13. A. Carbonaro, A. McIntosh, A.J. Morris, Local Hardy spaces of differential forms on Riemannian manifolds. J. Geom. Anal. **23**(1), 106–169 (2013)
14. Y. Chen, S. Levine, M. Rao, Variable exponent, linear growth functionals in image restoration. SIAM J. Appl. Math. **66**(4), 1383–1406 (2006)
15. R.R. Coifman, A real variable characterization of H^p. Studia Math. **51**, 269–274 (1974)
16. D.V. Cruz-Uribe, A. Fiorenza, *Variable Lebesgue Spaces: Foundations and Harmonic Analysis*. Applied and Numerical Harmonic Analysis (Birkhäuser/Springer, Heidelberg, 2013)
17. D. Cruz-Uribe, A. Fiorenza, J.M. Martell, C. Pérez, The boundedness of classical operators on variable L^p spaces. Ann. Acad. Sci. Fenn. Math. **31**(1), 239–264 (2006)
18. D. Cruz-Uribe, A. Fiorenza, C.J. Neugebauer, The maximal function on variable L^p spaces. Ann. Acad. Sci. Fenn. Math. **28**(1), 223–238 (2003)
19. D. Cruz-Uribe, L.-A.D. Wang, Variable Hardy spaces. Indiana Univ. Math. J. **63**(2), 447–493 (2014)
20. D. Cruz-Uribe, L.-A.D. Wang, Extrapolation and weighted norm inequalities in the variable Lebesgue spaces. Trans. Amer. Math. Soc. **369**(2), 1205–1235 (2017)
21. G. Dafni, H. Yue, Some characterizations of local bmo and h^1 on metric measure spaces. Anal. Math. Phys. **2**(3), 285–318 (2012)
22. L. Diening, Theoretical and numerical results for electrorheological fluids. Ph.D. thesis, Univ. Freiburg im Breisgau, Mathematische Fakultät, 156 p. (2002)
23. L. Diening, P. Harjulehto, P. Hästö, M. Ružička, *Lebesgue and Sobolev Spaces with Variable Exponents*. Lecture Notes in Mathematics, vol. 2017 (Springer, Heidelberg, 2011)
24. L. Diening, Maximal function on generalized Lebesgue spaces $L^{p(\cdot)}$. Math. Inequal. Appl. **7**(2), 245–253 (2004)
25. L. Diening, P. Harjulehto, P. Hästö, Y. Mizuta, T. Shimomura, Maximal functions in variable exponent spaces: limiting cases of the exponent. Ann. Acad. Sci. Fenn. Math. **34**(2), 503–522 (2009)
26. L. Diening, P. Hästö, S. Roudenko, Function spaces of variable smoothness and integrability. J. Funct. Anal. **256**(6), 1731–1768 (2009)
27. L. Diening, M. Ružička, Calderón-Zygmund operators on generalized Lebesgue spaces $L^{p(\cdot)}$ and problems related to fluid dynamics. J. Reine Angew. Math. **563**, 197–220 (2003)
28. X.T. Duong, L. Yan, Duality of Hardy and BMO spaces associated with operators with heat kernel bounds. J. Amer. Math. Soc. **18**(4), 943–973 (2005) (electronic)
29. P.L. Duren, B.W. Romberg, A.L. Shields, Linear functionals on H^p spaces with $0<p<1$. J. Reine Angew. Math. **238**, 32–60 (1969)
30. J. Dziubański, M. Preisner, Hardy spaces for semigroups with Gaussian bounds. Ann. Mat. Pura Appl. **197**(3), 965–987 (2018)
31. C. Fefferman, E.M. Stein, H^p spaces of several variables. Acta Math. **129**(3–4), 137–193 (1972)
32. A. Gogatishvili, A. Danelia, T. Kopaliani, Local Hardy-Littlewood maximal operator in variable Lebesgue spaces. Banach J. Math. Anal. **8**(2), 229–244 (2014)
33. D. Goldberg, A local version of real Hardy spaces. Duke Math. J. **46**(1), 27–42 (1979)
34. L. Grafakos, L. Liu, D. Yang, Radial maximal function characterizations for Hardy spaces on RD-spaces. Bull. Soc. Math. France **137**(2), 225–251 (2009)
35. P. Harjulehto, P. Hästö, V. Latvala, O. Toivanen, Critical variable exponent functionals in image restoration. Appl. Math. Lett. **26**(1), 56–60 (2013)
36. P.A. Hästö, Local-to-global results in variable exponent spaces. Math. Res. Lett. **16**(2), 263–278 (2009)

37. S. Hofmann, G. Lu, D. Mitrea, M. Mitrea, L. Yan, Hardy spaces associated to non-negative self-adjoint operators satisfying Davies-Gaffney estimates. Memoirs of the Amer. Math. Soc. **214** (2011)
38. S. Hofmann, S. Mayboroda, Hardy and BMO spaces associated to divergence form elliptic operators. Math. Ann. **344**(1), 37–116 (2009)
39. R. Jiang, D. Yang, D. Yang, Maximal function characterizations of Hardy spaces associated with magnetic Schrödinger operators. Forum Math. **24**(3), 471–494 (2012)
40. R. Jiang, D. Yang, Y. Zhou, Localized Hardy spaces associated with operators. Appl. Anal. **88**(9), 1409–1427 (2009)
41. M. Kemppainen, A note on local Hardy spaces. Forum Math. **29**(4), 941–949 (2017)
42. O. Kováčik, J. Rákosník, On spaces $L^{p(x)}$ and $W^{k,p(x)}$. Czechoslovak Math. J. **41**(4)(116), 592–618 (1991)
43. R.H. Latter, A characterization of $H^p(R^n)$ in terms of atoms. Studia Math. **62**(1), 93–101 (1978)
44. A.K. Lerner, On modular inequalities in variable L^p spaces. Arch. Math. (Basel) **85**(6), 538–543 (2005)
45. F. Li, Z. Li, L. Pi, Variable exponent functionals in image restoration. Appl. Math. Comput. **216**(3), 870–882 (2010)
46. J. Liu, D. Yang, W. Yuan, Anisotropic variable Hardy-Lorentz spaces and their real interpolation. J. Math. Anal. Appl. **456**(1), 356–393 (2017)
47. G. Mauceri, M.A. Picardello, F. Ricci, A Hardy space associated with twisted convolution. Adv. in Math. **39**(3), 270–288 (1981)
48. E. Nakai, Y. Sawano, Hardy spaces with variable exponents and generalized Campanato spaces. J. Funct. Anal. **262**(9), 3665–3748 (2012)
49. H. Nakano, *Modulared Semi-ordered Linear Spaces* (Maruzen Co. Ltd., Tokyo, 1950)
50. A.S. Nekvinda, Hardy-Littlewood maximal operator on $L^{p(x)}(\mathbb{R})$. Math. Inequal. Appl. **7**(2), 255–265 (2004)
51. W. Orlicz, Über Konjugierte Exponentenfolgen. Studia Math. **3**(1), 200–211 (1931)
52. M.M. Peloso, S. Secco, Local Riesz transforms characterization of local Hardy spaces. Collect. Math. **59**(3), 299–320 (2008)
53. L.S. Pick, M. Ružička, An example of a space $L^{p(x)}$ on which the Hardy-Littlewood maximal operator is not bounded. Expo. Math. **19**(4), 369–371 (2001)
54. K.R. Rajagopal, M. Ružička, On the modelling of electrorheological materials. Mech. Res. Commun. **23**(4), 401–407 (1996)
55. M. Ružička, *Electrorheological Fluids: Modeling and Mathematical Theory*. Lecture Notes in Mathematics, vol. 1748 (Springer, Berlin, 2000)
56. Y. Sawano, Atomic decompositions of Hardy spaces with variable exponents and its application to bounded linear operators. Int. Equ. Oper. Theory **77**(1), 123–148 (2013)
57. I.I. Sharapudinov, The topology of the space $\mathcal{L}^{p(t)}([0,\ 1])$. Mat. Zametki **26**(4), 613–632, 655 (1979)
58. L. Song, L. Yan, Maximal function characterization for Hardy spaces associated with nonnegative self-adjoint operators on spaces of homogeneous type. J. Evol. Equations. **18**(1), 221–243 (2018)
59. L. Song, L. Yan, A maximal function characterization for Hardy spaces associated to nonnegative self-adjoint operators satisfying Gaussian estimates. Adv. Math. **287**, 463–484 (2016)
60. E.M. Stein, *Harmonic Analysis: Real-variable Methods, Orthogonality, and Oscillatory Integrals*. Princeton Mathematical Series, vol. 43 (Princeton University Press, Princeton, NJ, 1993)
61. E.M. Stein, G. Weiss, On the theory of harmonic functions of several variables. I. The theory of H^p-spaces. Acta Math. **103**, 25–62 (1960)
62. J.-O. Strömberg, A. Torchinsky, *Weighted Hardy Spaces*. Lecture Notes in Mathematics, vol. 1381 (Springer, Berlin, 1989)
63. L. Tang, Weighted local Hardy spaces and their applications. Illinois J. Math. **56**(2), 453–495 (2012)

64. H. Triebel, *Theory of Function Spaces*. Monographs in Mathematics, vol. 78 (Birkhäuser Verlag, Basel, 1983)
65. A. Uchiyama, A maximal function characterization of H^p on the space of homogeneous type. Trans. Amer. Math. Soc. **262**(2), 579–592 (1980)
66. M. Wilson, The intrinsic square function. Rev. Mat. Iberoam. **23**(3), 771–791 (2007)
67. A.S. Wineman, K.R. Rajagopal, On a constitutive theory for materials undergoing microstructural changes. Arch. Mech. (Arch. Mech. Stos.) **42**(1), 53–75 (1990)
68. L. Yan, Classes of Hardy spaces associated with operators, duality theorem and applications. Trans. Amer. Math. Soc. **360**(8), 4383–4408 (2008)
69. D. Yang, J. Zhang, C. Zhuo, Variable Hardy spaces associated with operators satisfying Davies-Gaffney estimates. Proc. Edinburgh Math. Soc. **61**(3), 759–810 (2018)
70. D. Yang, S. Yang, Local Hardy spaces of Musielak-Orlicz type and their applications. Sci. China Math. **55**(8), 1677–1720 (2012)
71. D. Yang, C. Zhuo, Molecular characterizations and dualities of variable exponent Hardy spaces associated with operators. Ann. Acad. Sci. Fenn. Math. **41**(1), 357–398 (2016)
72. D. Yang, C. Zhuo, E. Nakai, Characterizations of variable exponent Hardy spaces via Riesz transforms. Rev. Mat. Complut. **29**(2), 245–270 (2016)
73. X. Yan, D. Yang, W. Yuan, C. Zhuo, Variable weak Hardy spaces and their applications. J. Funct. Anal. **271**(10), 2822–2887 (2016)
74. V.V. Zhikov, Meyer-type estimates for solving the nonlinear Stokes system. Differ. Uravn. **33**(1), 107–114, 143 (1997)
75. C. Zhuo, Y. Sawano, D. Yang, Hardy spaces with variable exponents on RD-spaces and applications. Dissertationes Math. (Rozprawy Mat.) **520**, 74 (2016)
76. C. Zhuo, D. Yang, Maximal function characterizations of variable Hardy spaces associated with non-negative self-adjoint operators satisfying Gaussian estimates. Nonlinear Anal. **141**, 16–42 (2016)
77. C. Zhuo, D. Yang, Y. Liang, Intrinsic square function characterizations of Hardy spaces with variable exponents. Bull. Malays. Math. Sci. Soc. **39**(4), 1541–1577 (2016)

Chapter 3
Regularity of Maximal Operators: Recent Progress and Some Open Problems

Emanuel Carneiro

Abstract This is an expository paper on the regularity theory of maximal operators, when these act on Sobolev and BV functions, with a special focus on some of the current open problems in the topic. Overall, a list of fifteen research problems is presented. It summarizes the contents of a talk delivered by the author in the CIMPA 2017 Research School—Harmonic Analysis, Geometric Measure Theory, and Applications, in Buenos Aires, Argentina.

Subject Classification: 42B25 · 26A45 · 46E35 · 46E39

3.1 Introduction

Maximal operators are classical objects in analysis. They usually arise as important tools to prove different sorts of pointwise convergence results, e.g., Lebesgue's differentiation theorem, Carleson's theorem on the pointwise convergence of Fourier series, pointwise convergence of solutions of PDEs to the initial datum, and so on. Despite being extensively studied for decades, maximal operators still conceal some of their secrets, and understanding the intrinsic mapping properties of these operators in different function spaces still remains an active topic of research.

Throughout this paper, we focus on the most classical of these objects, the Hardy–Littlewood maximal operator, and some of its variants. As we shall see, it will be important for our discussion to consider the centered and uncentered versions of this operator, discrete analogues, fractional analogues, and convolution-type analogues. For $f \in L^1_{loc}(\mathbb{R}^d)$, we define the *centered Hardy–Littlewood maximal function* Mf

E. Carneiro (✉)
IMPA—Instituto Nacional de Matemática Pura e Aplicada, Estrada Dona Castorina 110, Rio de Janeiro 22460-320, Brazil
e-mail: carneiro@impa.br; carneiro@ictp.it

ICTP - The Abdus Salam International Centre for Theoretical Physics, Strada Costiera, 11, 34151 Trieste, Italy

A. Aldroubi et al. (eds.), *New Trends in Applied Harmonic Analysis, Volume 2*, Applied and Numerical Harmonic Analysis,
https://doi.org/10.1007/978-3-030-32353-0_3

by

$$Mf(x) = \sup_{r>0} \frac{1}{m(B(x,r))} \int_{B(x,r)} |f(y)|\, dy\,, \tag{3.1.1}$$

where $B(x, r)$ is the open ball of center x and radius r, and $m(B(x, r))$ denotes its d-dimensional Lebesgue measure. The *uncentered maximal function* $\widetilde{M}f$ at a point x is defined analogously, taking the supremum of averages over open balls that contain the point x, *but that are not necessarily centered at* x.

One of the fundamental results in harmonic analysis is the theorem of Hardy and Littlewood that states that $M : L^1(\mathbb{R}^d) \to L^{1,\infty}(\mathbb{R}^d)$ is a bounded operator. By interpolation with the trivial L^∞-estimate, this yields the boundedness of $M : L^p(\mathbb{R}^d) \to L^p(\mathbb{R}^d)$ for $1 < p \le \infty$. Another consequence of the weak-$(1, 1)$ bound for M is the Lebesgue's differentiation theorem. The L^p-mapping properties of the uncentered maximal operator $\widetilde{M}$ are exactly the same.

One may consider the action of the Hardy–Littlewood maximal operator in other function spaces and investigate whether it improves, preserves, or destroys the a priori regularity of an initial datum f. This type of question is essentially the main driver of what we refer to here as *regularity theory for maximal operators*. Let us denote by $W^{1,p}(\mathbb{R}^d)$ the Sobolev space of functions $f \in L^p(\mathbb{R}^d)$ that have a weak gradient $\nabla f \in L^p(\mathbb{R}^d)$, with norm given by

$$\|f\|_{W^{1,p}(\mathbb{R}^d)} = \|f\|_{L^p(\mathbb{R}^d)} + \|\nabla f\|_{L^p(\mathbb{R}^d)}.$$

In 1997, J. Kinnunen wrote an enlightening paper [17], establishing the boundedness of the operator $M : W^{1,p}(\mathbb{R}^d) \to W^{1,p}(\mathbb{R}^d)$ for $1 < p \le \infty$. This marks the beginning of our story. After that, a number of interesting works have devoted their attention to the investigation of the action of maximal operators on Sobolev spaces and on the closely related space of functions of bounded variation. This survey paper is brief account of some of the developments in this topic over the last 20 years, with a special focus on a list of 15 open problems that may guide new endeavors.

The choice of topics and problems presented here is obviously biased by the personal preferences of the author and is by no means exhaustive. We shall present just a couple of brief proofs of some of the earlier results to give a flavor to the reader of what is going on, for the main purpose of this expository paper is to provide a light and inviting reading on the topic, especially to newcomers. We refer the reader to the original papers for the proofs of the results mentioned here.

For simplicity, all functions considered in this paper are real-valued functions.

3.2 Kinnunen's Seminal Work

Let us start by revisiting the main result of [17] and its elegant proof.

Theorem 3.2.1 (Kinnunen, 1997—cf. [17]) *Let* $1 < p < \infty$ *and let* $f \in W^{1,p}(\mathbb{R}^d)$. *Then* Mf *is weakly differentiable and*

$$\left|\partial_i Mf(x)\right| \leq M\big(\partial_i f\big)(x) \tag{3.2.1}$$

for almost every $x \in \mathbb{R}^d$. Therefore, $M : W^{1,p}(\mathbb{R}^d) \to W^{1,p}(\mathbb{R}^d)$ is bounded.

Proof Let $B = B(0,1) \subset \mathbb{R}^d$ be the unit ball and define

$$\varphi(x) := \frac{\chi_B(x)}{m(B)}, \tag{3.2.2}$$

where χ_B is the characteristic function of B. For $r > 0$ let us define

$$\varphi_r(x) := r^{-d}\,\varphi(x/r).$$

With this notation we plainly have

$$Mf(x) = \sup_{r>0}\,(\varphi_r * |f|)(x).$$

Fix $1 \leq i \leq d$. Recall that if $f \in W^{1,p}(\mathbb{R}^d)$ then $|f| \in W^{1,p}(\mathbb{R}^d)$ and $|\partial_i |f|| = |\partial_i f|$ almost everywhere (see, for instance, [22, Theorem 6.17]). Let us enumerate the positive rational numbers as $\{r_1, r_2, r_3, \ldots\}$ and define $h_j := \varphi_{r_j} * |f|$. Then $h_j \in W^{1,p}(\mathbb{R}^d)$ and $\partial_i h_j = \varphi_{r_j} * \partial_i |f|$.

Let $N \geq 1$ be a natural number and define $g_N(x) := \max_{1\leq j\leq N} h_j(x)$. Note that $g_N \in W^{1,p}(\mathbb{R}^d)$ with

$$g_N(x) \leq Mf(x)$$

and (see [22, Theorem 6.18])

$$|\partial_i g_N(x)| \leq \max_{1\leq j\leq N} |\partial_i h_j(x)| \leq M(\partial_i f)(x) \tag{3.2.3}$$

for almost every $x \in \mathbb{R}^d$. Then $\{g_N\}_{N\geq 1}$ is a bounded sequence in $W^{1,p}(\mathbb{R}^d)$ with the property that $g_N(x) \to Mf(x)$ pointwise as $N \to \infty$. Since $W^{1,p}(\mathbb{R}^d)$ is a reflexive Banach space, by passing to a subsequence if necessary, we may assume that g_N converges weakly to a function $g \in W^{1,p}(\mathbb{R}^d)$ (a crucial point in this argument is that this weak limit is already born in $W^{1,p}(\mathbb{R}^d)$). Standard functional analysis tools (for instance, using Mazur's lemma [6, Corollary 3.8 and Exercise 3.4]) lead to the conclusion that $Mf = g \in W^{1,p}(\mathbb{R}^d)$ and that the upper bound (3.2.3) is preserved almost everywhere up to the weak limit. The latter assertion leads to (3.2.1). □

We call the attention of the reader for the use of the reflexivity of the space $W^{1,p}(\mathbb{R}^d)$, for $1 < p < \infty$, in the conclusion of the proof above. This is one of the obstacles when one considers the endpoint case $p = 1$, as we shall see in the next section. The case $p = \infty$ can be dealt with directly. In fact, if $f \in W^{1,\infty}(\mathbb{R}^d)$ then f can be modified on a set of measure zero to become Lipschitz continuous, with Lipschitz constant $L \leq \|\nabla f\|_\infty$. With the notation of the proof above, for a fixed r, each average $\varphi_r * |f|$ is Lipschitz with constant at most L. The pointwise supremum of uniformly Lipschitz functions is still Lipschitz with (at most) the same constant. This shows that $M : W^{1,\infty}(\mathbb{R}^d) \to W^{1,\infty}(\mathbb{R}^d)$ is a bounded operator.

If one is not necessarily interested in pointwise estimates for the derivative of the maximal function, there is a simpler argument using the characterization of Sobolev spaces via difference quotients [16, Theorem 1]. This covers the general situation of sublinear operators that commute with translations. Recall that an operator $A : X \to Y$, acting between linear function spaces X and Y, is said to be sublinear if $Af \geq 0$ a.e. for $f \in X$, and $A(f+g) \leq Af + Ag$ a.e. for $f, g \in X$. In what follows we let $f_y(x) := f(x+y)$ for $x, y \in \mathbb{R}^d$.

Theorem 3.2.2 (Hajłasz and Onninen, 2004—cf. [16]) *Assume that the operator $A : L^p(\mathbb{R}^d) \to L^p(\mathbb{R}^d)$, $1 < p < \infty$, is bounded and sublinear. If $A(f_y) = (Af)_y$ for all $f \in L^p(\mathbb{R}^d)$ and all $y \in \mathbb{R}^d$, then $A : W^{1,p}(\mathbb{R}^d) \to W^{1,p}(\mathbb{R}^d)$ is also bounded.*

Proof Let e_i be the unit coordinate vector in the x_i direction. For $t > 0$ we have

$$\begin{aligned}\|(Af)_{te_i} - Af\|_{L^p(\mathbb{R}^d)} &= \|A(f_{te_i}) - Af\|_{L^p(\mathbb{R}^d)}\\ &\leq \|A(f_{te_i} - f)\|_{L^p(\mathbb{R}^d)} + \|A(f - f_{te_i})\|_{L^p(\mathbb{R}^d)}\\ &\leq C\,\|f_{te_i} - f\|_{L^p(\mathbb{R}^d)}\\ &\leq C\,t\,\|\partial_i f\|_{L^p(\mathbb{R}^d)}.\end{aligned}$$

The last inequality above follows from [14, Lemma 7.23]. Since the difference quotients $\big\{((Af)_{te_i} - Af)/t\big\}_{t>0}$ are uniformly bounded in $L^p(\mathbb{R}^d)$, an application of [14, Lemma 7.24] guarantees that $Af \in W^{1,p}(\mathbb{R}^d)$ and

$$\|\partial_i Af\|_{L^p(\mathbb{R}^d)} \leq C\,\|\partial_i f\|_{L^p(\mathbb{R}^d)}.$$

This concludes the proof. □

In the scope of Theorem 3.2.2, one may consider the spherical maximal operator. Letting $S^{d-1}(x,r) \subset \mathbb{R}^d$ be the $(d-1)$-dimensional sphere of center x and radius r, this operator is defined as

$$M_S f(x) = \sup_{r>0} \frac{1}{\omega_{d-1} r^{d-1}} \int_{S^{n-1}(x,r)} |f(z)|\, \mathrm{d}\sigma(z), \tag{3.2.4}$$

where σ is the canonical surface measure on $S^{d-1}(x,r)$ and $\omega_{d-1} = \sigma\big(S^{d-1}(0,1)\big)$. A remarkable result of Stein [37] in dimension $d \geq 3$, and Bourgain [5] in dimension $d = 2$, establishes that $M_S : L^p(\mathbb{R}^d) \to L^p(\mathbb{R}^d)$ is a bounded operator for $p > d/(d-1)$. It plainly follows from Theorem 3.2.2 that $M_S : W^{1,p}(\mathbb{R}^d) \to W^{1,p}(\mathbb{R}^d)$ is also bounded for $p > d/(d-1)$ (the case $p = \infty$ is treated directly).

Theorem 3.2.1 has been extended in many different ways over the last years, and we now mention a few of such related results. Kinnunen and Lindqvist [18] extended Theorem 3.2.1 to a local version of the maximal operator. In this setting, one considers a domain $\Omega \subset \mathbb{R}^d$, functions $f \in W^{1,p}(\Omega)$, and the maximal operator is taken over balls entirely contained in the domain Ω. Extensions of Theorem 3.2.1 to a multilinear setting are considered in the work of the author and Moreira [11] and by Liu and Wu [24], and a similar result in fractional Sobolev spaces is the subject of the work of Korry [20]. A fractional version of the Hardy–Littlewood maximal operator is considered in the paper [19] by Kinnunen and Saksman (we will return to this particular operator later on). An interesting variant of this result on Hardy–Sobolev spaces is considered in the recent work of Pérez et al. [33].

3.3 The Endpoint Sobolev Space

With the philosophy that averaging is a smoothing process, we would like to understand if certain smoothing features are still preserved when we take a pointwise supremum over averages. Understanding the situation described in Theorem 3.2.1 at the endpoint case $p = 1$ is a subtle issue. Of course, if $f \in L^1(\mathbb{R}^d)$ is nonidentically zero, we already know that $f \notin L^1(\mathbb{R}^d)$, and the interesting question is whether one can control the behavior of the derivative of the maximal function. The following question was raised in the work of Hajłasz and Onninen [16, Question 1] and remains one of the main open problems in the subject.

Question 1 (Hajłasz and Onninen, 2004—cf. [16]) *Is the operator $f \mapsto |\nabla Mf|$ bounded from $W^{1,1}(\mathbb{R}^d)$ to $L^1(\mathbb{R}^d)$? Same question for the uncentered operator $\widetilde{M}$.*

Naturally, this involves proving that Mf is weakly differentiable, and establishing the bound

$$\|\nabla Mf\|_{L^1(\mathbb{R}^d)} \leq C\big(\|f\|_{L^1(\mathbb{R}^d)} + \|\nabla f\|_{L^1(\mathbb{R}^d)}\big), \tag{3.3.1}$$

for some universal constant $C = C(d)$. If the global estimate (3.3.1) holds for every $f \in W^{1,1}(\mathbb{R}^d)$, a simple dilation argument implies that one should actually have

$$\|\nabla Mf\|_{L^1(\mathbb{R}^d)} \leq C\, \|\nabla f\|_{L^1(\mathbb{R}^d)},$$

which reveals the true nature of the problem: if one can control the variation of the maximal function by the variation of the original function (the term *variation* here is used as the L^1-norm of the gradient).

Several interesting papers addressed Question 1, which has been answered affirmatively in dimension $d = 1$, but remains vastly open in dimensions $d \geq 2$. We now comment a bit on these results.

3.3.1 One-Dimensional Results

The achievements in dimension $d = 1$ started with the work of Tanaka [38], for the uncentered maximal operator $\widetilde{M}$. In this particular work, Tanaka showed that if $f \in W^{1,1}(\mathbb{R})$ then $\widetilde{M}f$ is weakly differentiable and

$$\|(\widetilde{M} f)'\|_{L^1(\mathbb{R})} \leq 2\,\|f'\|_{L^1(\mathbb{R})} \tag{3.3.2}$$

(see also [23]). This result was later refined by Aldaz and Pérez Lázaro in [1, Theorem 2.5]. Letting Var (f) denote the total variation of a function $f : \mathbb{R} \to \mathbb{R}$, they proved the following very interesting result.

Theorem 3.3.1 (Aldaz and Pérez Lázaro, 2007—cf. [1]) *Let $f : \mathbb{R} \to \mathbb{R}$ be a function of bounded variation. Then $\widetilde{M} f$ is an absolutely continuous function and we have the inequality*

$$Var\big(\widetilde{M} f\big) \leq Var\,(f). \tag{3.3.3}$$

We comment on the two main features of this theorem. First, the regularizing effect of the operator $\widetilde{M}$ takes a mere function of bounded variation into an absolutely continuous function. The proof of this fact relies on the classical Banach–Zarecki theorem. This regularizing effect is not shared by the centered maximal operator M, as it can be seen by simply taking f to be the characteristic function of an interval. In this sense, the uncentered operator is more regular than the centered one, and in many instances in this theory it is a more tractable object. Second, the inequality (3.3.3) with constant $C = 1$ is sharp, as it can be seen again by taking f to be the characteristic function of an interval. Note, in particular, that (3.3.3) indeed refines (3.3.2), since any function $f \in W^{1,1}(\mathbb{R})$ can be modified on a set of measure zero to become absolutely continuous. The core of this argument comes from the fact that the maximal function does not have points of local maxima in the set where it disconnects from the original function (sometimes referred to here as *detachment set*).

Proving an inequality of the same spirit as (3.3.3) for the one-dimensional centered Hardy–Littlewood maximal operator is a harder task. In this situation, there may be local maxima of Mf in the detachment set (one may see this, for instance, by considering $f = \delta_{-1} + \delta_1$, where δ_{x_0} denotes the Dirac delta function at the point x_0; in this case, the point $x = 0$ is a local maximum for Mf; of course, technically one would have to smooth out this example to view the Dirac deltas as actual functions) and the previous argument of Aldaz and Pérez Lázaro cannot directly be adapted. In the work [21], O. Kurka proved the following remarkable result.

Theorem 3.3.2 (Kurka, 2015—cf. [21]) *Let $f : \mathbb{R} \to \mathbb{R}$ be a function of bounded variation. Then*

$$Var\,(Mf) \leq 240004\; Var\,(f). \tag{3.3.4}$$

The proof of this theorem relies on a beautiful, yet rather intricate, argument of induction on scales (from which one arrives at the particular constant $C = 240004$). Things seem to be tailor-made to the case of the Hardy–Littlewood maximal function, and it would be interesting to see if the argument can be adapted to treat other convolution kernels (discussed in the next section). The constant $C = 240004$ is certainly intriguing, but there seems to be no philosophical reason to justify this order of magnitude. This leaves the natural open question.

Question 2 *Let $f:\mathbb{R}\to\mathbb{R}$ be a function of bounded variation. Do we have*

$$Var\,(Mf) \leq Var\,(f)?$$

Or at least, can one substantially improve on Kurka's constant $C = 240004$?

Despite the innocence of the statement of Question 2, the reader should not underestimate its difficulty. As a matter of fact, the reader is invited to think a little bit about this question to get acquainted with some of its obstacles. This is a beautiful example of an open question in this research topic. These usually have relatively simple statements and their solutions might only require "elementary" tools, but the difficulty lies in *how to properly combine these tools.*

Recently, Ramos [34] considered a hybrid version between M and $\widetilde{M}$ in dimension $d = 1$. For $\alpha \geq 0$, we may define the non-tangential maximal operator M^{α} by

$$M^{\alpha} f(x) := \sup_{(y,t)\,:\,|x-y|\leq \alpha t} \frac{1}{2t}\int_{y-t}^{y+t} |f(s)|\,ds.$$

In this setting, we notice that $M^0 = M$ and $M^1 = \widetilde{M}$. Ramos shows that

$$\text{Var}\,(M^{\alpha} f) \leq \text{Var}\,(M^{\beta} f) \tag{3.3.5}$$

if $\alpha \geq \beta$, and from Theorem 3.3.2 one readily sees that

$$\text{Var}\,(M^{\alpha} f) \leq C\,\text{Var}\,(f)$$

for all $\alpha \geq 0$. From Theorem 3.3.1 we may take $C = 1$ if $\alpha \geq 1$. Ramos [34, Theorem 1] goes further and establishes the following result.

Theorem 3.3.3 (Ramos, 2017—cf. [34]) *Let $\alpha \in [\frac{1}{3}, \infty)$ and let $f:\mathbb{R}\to\mathbb{R}$ be a function of bounded variation. Then*

$$Var(M^{\alpha} f) \leq Var(f). \tag{3.3.6}$$

The constant $C = 1$ in inequality (3.3.6) is sharp as it can be easily seen by taking f to be the characteristic function of an interval. The proof of Ramos for Theorem 3.3.3 extends the argument of Aldaz and Pérez Lázaro [1], in particular establishing the crucial property that $M^{\alpha} f$ has no local maxima in the detachment set for $\alpha > 1/3$. The case $\alpha = 1/3$ in (3.3.6) is obtained by a limiting argument. The interesting thing here is that $\alpha = 1/3$ is the threshold for this property. Indeed, if $\alpha < 1/3$, by taking $f = \delta_{-1} + \delta_1$ we see that $x = 0$ is a local maximum of $M^{\alpha} f$, see [34, Theorem 2] (again, one must smooth out this example, since the Dirac deltas are actually singular measures and not exactly functions of bounded variation—but this can be done with no harm). We conclude the one-dimensional discussion with the following question (which, by (3.3.5), would follow from an affirmative answer to Question 2).

Question 3 *Let $f : \mathbb{R} \to \mathbb{R}$ be a function of bounded variation. Do we have*

$$Var(M^{\alpha} f) \leq Var(f) \tag{3.3.7}$$

for $0 \leq \alpha < \frac{1}{3}$? Alternatively, what is the smallest value of α for which (3.3.7) *holds?*

3.3.2 Multidimensional Results

Question 1 remains open, in general, for dimensions $d \geq 2$. There have been a few particular works that made interesting partial progress and we now comment on three of them, namely [15, 28, 35].

In the paper [15], Hajłasz and Malý consider a slightly weaker notion of differentiability. A function $f : \mathbb{R}^d \to \mathbb{R}$ is said to be *approximately differentiable* at the point $x_0 \in \mathbb{R}^d$ if there exists a vector $L = (L_1, L_2, \ldots, L_d)$ such that for any $\varepsilon > 0$ "…the set

$$A_{\varepsilon} := \left\{ x \in \mathbb{R}^d \ : \ \frac{|f(x) - f(x_0) - L(x - x_0)|}{|x - x_0|} < \varepsilon \right\}$$

has x_0 as a density point." If this is the case, the vector L is unique determined and it is called the *approximate differential* of f at x_0. This is a weaker notion than that of classical differentiability or weak differentiability. In fact, if a function f is differentiable at a point x_0 then it is approximately differentiable at x_0 and $L = \nabla f(x_0)$, and similarly, if f is weakly differentiable then it is approximately differentiable and its approximate differential is equal to the weak derivative a.e., see, for instance, [13, Sect. 6.1.3, Theorem 4]. The main result of [15] reads as follows.

Theorem 3.3.4 (Hajłasz and Malý, 2010—cf. [15]) *If $f \in L^1(\mathbb{R}^d)$ is approximately differentiable a.e. then the maximal function Mf is approximately differentiable a.e.*

The recent interesting work of Luiro [28] answers Question 1 affirmatively in the case of the uncentered maximal function $\widetilde{M}$ and *restricted to radial functions f*.

Theorem 3.3.5 (Luiro, 2017—cf. [28]) *If $f \in W^{1,1}(\mathbb{R}^d)$ is radial, then $\widetilde{M}f$ is weakly differentiable and*

$$\left\| \nabla \widetilde{M} f \right\|_{L^1(\mathbb{R}^d)} \leq C \, \|\nabla f\|_{L^1(\mathbb{R}^d)},$$

where $C = C(d)$ is a universal constant.

This raises a natural question, another interesting particular case of Question 1.

Question 4 *Does Theorem 3.3.5 hold for the centered maximal operator M acting on radial functions f?*

In [35], Saari studies the regularity of maximal operators via generalized Poincaré inequalities. An interesting corollary [35, Corollary 4.1] of the main result of this paper establishes that, if $f \in W^{1,1}(\mathbb{R}^d)$, then the distributional partial derivatives $\partial_i \widetilde{M} f$ (or $\partial_i M f$) can be represented as functions $h_i \in L^{1,\infty}(\mathbb{R}^d)$ when they act on smooth functions with compact support not meeting a certain singularity set.

Finally, let us briefly return to the spherical maximal operator M_S defined in (3.2.4). Recall that we have shown in the previous section that M_S is bounded in $W^{1,p}(\mathbb{R}^d)$ for $d \geq 2$ and $p > d/(d-1)$. We conclude this section with the following question, originally proposed in [16, Question 2].

Question 5 (Hajłasz and Onninen, 2004—cf. [16]) *Let $d \geq 2$. Is $M_S : W^{1,p}(\mathbb{R}^d) \to W^{1,p}(\mathbb{R}^d)$ bounded for $1 < p \leq d/(d-1)$?*

Note that $p = 1$ is not actually part of Question 5. In this case, the operator $f \mapsto |\nabla M_S f|$ is not bounded from $W^{1,1}(\mathbb{R}^d)$ to $L^1(\mathbb{R}^d)$. A counterexample is given in [16].

3.4 Maximal Operators of Convolution Type

Let $\varphi : \mathbb{R}^d \to \mathbb{R}$ be a nonnegative and integrable function with

$$\int_{\mathbb{R}^d} \varphi(x)\,\mathrm{d}x = 1.$$

As before, for $t > 0$, we let $\varphi_t = t^{-d}\,\varphi(x/t)$. We define here the *maximal operator of convolution type* associated to φ by

$$M_\varphi f(x) = \sup_{t>0} \big(\varphi_t * |f|\big)(x).$$

Recall that the centered Hardy–Littlewood maximal operator arises when the kernel φ is given by (3.2.2). When φ admits a radial decreasing majorant in $L^1(\mathbb{R}^d)$, a classical result of Stein [36, Chap. III, Theorem 2] establishes that $M_\varphi : L^p(\mathbb{R}^d) \to L^p(\mathbb{R}^d)$ is a bounded operator for $1 < p \leq \infty$, and at $p = 1$ we have a weak-$(1, 1)$ estimate. Theorem 3.2.2 plainly implies that $M_\varphi : W^{1,p}(\mathbb{R}^d) \to W^{1,p}(\mathbb{R}^d)$ is bounded for $p > 1$ (again, the case $p = \infty$ can be dealt with directly), and we may ask ourselves the same sort of questions as in the previous section, with respect to the regularity of this operator at the endpoint Sobolev space $W^{1,1}(\mathbb{R}^d)$.

As it turns out, we may have some advantages in considering certain smooth kernels. This additional leverage may come, for instance, from partial differential equations naturally associated to the kernel φ. This is well exemplified in the work [12], where two special kernels are considered: the Poisson kernel

$$\varphi(x) = \frac{\Gamma\left(\frac{d+1}{2}\right)}{\pi^{(d+1)/2}}\,\frac{1}{(|x|^2+1)^{(d+1)/2}}, \tag{3.4.1}$$

and the Gauss kernel

$$\varphi(x) = \frac{1}{(4\pi)^{d/2}}\, e^{-|x|^2/4}. \tag{3.4.2}$$

For the Poisson kernel (3.4.1), the function $u(x,t) = \varphi_t(x)$ solves Laplace's equation $u_{tt} + \Delta_x u = 0$ on the upper half-space $(x,t) \in \mathbb{R}^d \times (0,\infty)$. For the Gauss kernel (3.4.2), the function $u(x,t) = \varphi_{\sqrt{t}}(x)$ solves the heat equation $u_t - \Delta_x u = 0$ on the upper half-space $(x,t) \in \mathbb{R}^d \times (0,\infty)$. The qualitative properties of these two partial differential equations (namely, the corresponding maximum principles and the semigroup property) can be used to establish a positive answer for the convolution-type analogue of Question 2 in these cases [12, Theorems 1 and 2].

Theorem 3.4.1 (Carneiro and Svaiter, 2013—cf. [12]) *Let φ be given by* (3.4.1) *or* (3.4.2)*, and let $f : \mathbb{R} \to \mathbb{R}$ be a function of bounded variation. Then*

$$Var\,(M_\varphi f) \leq Var\,(f).$$

Remark In [12, Theorems 1 and 2] it is also proved that, for every dimension $d \geq 1$, if $f \in W^{1,2}(\mathbb{R}^d)$, then we have (for φ given by (3.4.1) or (3.4.2))

$$\big\|\nabla M_\varphi f\big\|_{L^2(\mathbb{R}^d)} \leq \|\nabla f\|_{L^2(\mathbb{R}^d)}. \tag{3.4.3}$$

Additionally, if $d = 1$, an analogous inequality to (3.4.3) holds on $L^p(\mathbb{R})$ for all $p > 1$.

Theorem 3.4.1 has been extended to a larger family of kernels in the work [7]. The general version of the result in Theorem 3.4.1 is the theme of the following question.

Question 6 *Let φ be a convolution kernel as described above, and let $f : \mathbb{R} \to \mathbb{R}$ be a function of bounded variation. Can we show that*

$$Var\,(M_\varphi f) \leq C\; Var\,(f) \tag{3.4.4}$$

with $C = C(\varphi)$? For which φ can we actually show (3.4.4) *with $C = 1$?*

3.5 Fractional Maximal Operators

For $0 \leq \beta < d$ and $f \in L^1_{loc}(\mathbb{R}^d)$, we define the centered Hardy–Littlewood fractional maximal function $M_\beta f$ by

$$M_\beta f(x) = \sup_{r>0} \frac{1}{m(B(x,r))^{1-\frac{\beta}{d}}} \int_{B(x,r)} |f(y)|\, dy.$$

when $\beta = 0$ we plainly recover (3.1.1). The uncentered fractional maximal function $\widetilde{M}_\beta f$ is defined analogously, with the supremum of the fractional averages being taken over balls that simply contain the point x but are not necessarily centered at x. Such fractional maximal operators have connections to potential theory and partial differential equations. By comparison with an appropriate Riesz potential, one can show that if $1 < p < \infty$, $0 < \beta < d/p$ and $q = dp/(d - \beta p)$, then $M_\beta : L^p(\mathbb{R}^d) \to L^q(\mathbb{R}^d)$ is bounded. When $p = 1$ we have again a weak-type bound (for details, see [36, Chap. V, Theorem 1]).

In [19], Kinnunen and Saksman studied the regularity properties of such fractional maximal operators, arriving at the following interesting conclusions [19, Theorems 2.1 and 3.1].

Theorem 3.5.1 (Kinnunen and Saksman, 2003—cf. [19]) *Let $1 < p < \infty$.*

(i) For $0 \leq \beta < d/p$ and $q = dp/(d - \beta p)$ the operator $M_\beta : W^{1,p}(\mathbb{R}^d) \to W^{1,q}(\mathbb{R}^d)$ is bounded.
(ii) Assume that $1 \leq \beta < d/p$ and that $f \in L^p(\mathbb{R}^d)$. Then $M_\beta f$ is weakly differentiable and there exists a constant $C = C(d, \beta)$ such that

$$|\nabla M_\beta f(x)| \leq C\, M_{\beta-1} f(x)$$

for almost every $x \in \mathbb{R}^d$.

Part (i) of Theorem 3.5.1 extends the original result of Kinnunen (Theorem 3.2.1) to this fractional setting. One can prove it by using the characterization of the Sobolev spaces via the difference quotients as in the proof of Theorem 3.2.2. Part (ii) of Theorem 3.5.1 presents a beautiful regularization effect of this operator when the fractional parameter β is greater than or equal to 1.

In light of Theorem 3.5.1, it is then natural to ask ourselves what happens in the endpoint situation $p = 1$ and $q = d/(d - \beta)$. Let us first consider the case $1 \leq \beta < d$. If $f \in W^{1,1}(\mathbb{R}^d)$, by the Sobolev embedding we have $f \in L^{p^*}(\mathbb{R}^d)$, where $p^* = d/(d-1)$, and hence $f \in L^r(\mathbb{R}^d)$ for any $1 \leq r \leq p^*$. We may choose r with $1 < r < d$ such that $1 \leq \beta < d/r$. Using part (ii) of Theorem 3.5.1 we have that M_β is weakly differentiable and

$$\left\|\nabla M_\beta f\right\|_{L^q(\mathbb{R}^d)} \leq C \left\|M_{\beta-1} f\right\|_{L^q(\mathbb{R}^d)} \leq C' \,\|f\|_{L^{p^*}(\mathbb{R}^d)} \leq C''\|\nabla f\|_{L^1(\mathbb{R}^d)}.$$

This shows that the map $f \to |\nabla M_\beta f|$ is bounded from $W^{1,1}(\mathbb{R}^d)$ to $L^q(\mathbb{R}^d)$ in this case. We are thus left with the following endpoint question, first posed in [9].

Question 7 *Let $0 \leq \beta < 1$ and $q = d/(d - \beta)$. Is the map $f \to |\nabla M_\beta f|$ bounded from $W^{1,1}(\mathbb{R}^d)$ to $L^q(\mathbb{R}^d)$? Same question for the uncentered version $\widetilde{M}_\beta$.*

A complete answer to Question 7 was achieved in dimension $d = 1$ for the uncentered fractional maximal operator $\widetilde{M}_\beta$ in [9, Theorem 1]. To state this result we need to introduce a generalized version of the concept of variation of a function. For a function $f : \mathbb{R} \to \mathbb{R}$ and $1 \leq q < \infty$, we define its q-variation as

$$\operatorname{Var}_q(f) := \sup_{\mathcal{P}} \left(\sum_{n=1}^{N-1} \frac{|f(x_{n+1}) - f(x_n)|^q}{|x_{n+1} - x_n|^{q-1}} \right)^{1/q}, \tag{3.5.1}$$

where the supremum is taken over all finite partitions $\mathcal{P} = \{x_1 < x_2 < \ldots < x_N\}$. This is also known as the *Riesz q-variation* of f (see, for instance, the discussion in [2] for this object and its generalizations). Naturally, when $q = 1$, this is the usual total variation of the function. A classical result of F. Riesz (see [32, Chap. IX Sect. 4, Theorem 7]) states that, if $1 < q < \infty$, then $\operatorname{Var}_q(f) < \infty$ if and only if f is absolutely continuous and its derivative f' belongs to $L^q(\mathbb{R})$. Moreover, in this case, we have that

$$\|f'\|_{L^q(\mathbb{R})} = \operatorname{Var}_q(f).$$

In [9, Theorem 1], the author and J. Madrid proved the following regularizing effect.

Theorem 3.5.2 (Carneiro and Madrid, 2017—cf. [9]) *Let $0 \leq \beta < 1$ and $q = 1/(1-\beta)$. Let $f : \mathbb{R} \to \mathbb{R}$ be a function of bounded variation such that $\widetilde{M}_\beta f \not\equiv \infty$. Then $\widetilde{M}_\beta f$ is absolutely continuous and its derivative satisfies*

$$\left\| \left(\widetilde{M}_\beta f\right)' \right\|_{L^q(\mathbb{R})} = Var_q\left(\widetilde{M}_\beta f\right) \leq 8^{1/q}\, Var(f). \tag{3.5.2}$$

The constant $C = 8^{1/q}$ in (3.5.2) arises naturally with the methods employed in [9] and it is not necessarily sharp (in fact, we have seen that, when $\beta = 0$ and $q = 1$, this inequality holds with constant $C = 1$). The problem of finding the sharp constant in this inequality is certainly an interesting one. The strategy of [9] to prove Theorem 3.5.2 in the pure fractional case $\beta > 0$ is very different from that of the proof of Theorem 3.3.1. While in the proof of Theorem 3.3.1 the essential idea is to prove that the maximal function does not have any local maxima in the set where it disconnects from the original function, in the fractional case $\beta > 0$, the mere notion of the disconnecting set is ill-posed, since one does not necessarily have $\widetilde{M}_\beta(f)(x) \geq |f(x)|$ a.e. anymore. To overcome this challenge, the author and Madrid in [9] adopt a suitable bootstrapping procedure to bound the q-variation of $\widetilde{M}_\beta f$ on certain intervals by the variation of f in larger (but still somewhat comparable) intervals.

In the higher dimensional case, partial progress on Question 7 was obtained by Luiro and Madrid in the recent work [29]. They considered the uncentered fractional maximal operator $\widetilde{M}_\beta$ *acting on radial functions*. The following result is therefore the fractional analogue of Theorem 3.3.5.

Theorem 3.5.3 (Luiro and Madrid, 2017—cf. [29]) *Given $0 < \beta < 1$ and $q = d/(d-\beta)$, there is a constant $C = C(d, \beta)$ such that for every radial function $f \in W^{1,1}(\mathbb{R}^d)$ we have that $\widetilde{M}_\beta f$ is weakly differentiable and*

$$\left\| \nabla \widetilde{M}_\beta f \right\|_{L^q(\mathbb{R}^d)} \leq C\, \|\nabla f\|_{L^1(\mathbb{R}^d)}.$$

We can hence ask ourselves the follow-up question, which is a particular case of Question 7.

Question 8 *Does Theorem 3.5.3 hold for the centered fractional maximal operator M_β acting on radial functions f?*

Also in the higher dimensional case, the very interesting recent work of Beltran et al. [3] establishes endpoint bounds for derivatives of fractional maximal functions in the spirit of the ones proposed in Question 7. They consider the slightly different settings of maximal operators either associated to smooth convolution kernels or to a lacunary set of radii in dimensions $d \geq 2$ (see [3, Theorem 1]). In this work, they also show that the spherical maximal operator maps L^p into first-order Sobolev spaces in dimensions $d \geq 5$. One of the novelties in the approach of [3] is the use of Fourier analysis techniques.

3.6 Discrete Analogues

The problems we have discussed so far can also be considered in a discrete setup. A point $n \in \mathbb{Z}^d$ is a d-uple $n = (n_1, n_2, \ldots, n_d)$ with each $n_i \in \mathbb{Z}$. For a function $f : \mathbb{Z}^d \to \mathbb{R}$ (or, in general, for a vector-valued function $f : \mathbb{Z}^d \to \mathbb{R}^m$), we define its ℓ^p-norm as usual:

$$\|f\|_{\ell^p(\mathbb{Z}^d)} = \left(\sum_{n\in\mathbb{Z}^d} |f(n)|^p\right)^{1/p}, \tag{3.6.1}$$

if $1 \leq p < \infty$, and

$$\|f\|_{\ell^\infty(\mathbb{Z}^d)} = \sup_{n\in\mathbb{Z}^d} |f(n)|.$$

The gradient ∇f of a discrete function $f : \mathbb{Z}^d \to \mathbb{R}$ is the vector

$$\nabla f(n) = \big(\partial_1 f(n), \partial_2 f(n), \ldots, \partial_d f(n)\big),$$

where

$$\partial_i f(n) = f(n + e_i) - f(n),$$

and $e_i = (0, 0, \ldots, 1, \ldots, 0)$ is the canonical ith base vector. If $f : \mathbb{Z} \to \mathbb{R}$ is a given function, we define its total variation as

$$\mathrm{Var}\,(f) = \|f'\|_{\ell^1(\mathbb{Z})} = \sum_{n\in\mathbb{Z}} |f(n+1) - f(n)|.$$

If $f \in \ell^p(\mathbb{Z}^d)$, observe by the triangle inequality that we have $\nabla f \in \ell^p(\mathbb{Z}^d)$ as well. Therefore, if we were to copy and paste the definition of the Sobolev space

$W^{1,p}(\mathbb{R}^d)$ to the discrete setting, we would simply find the space $\ell^p(\mathbb{Z}^d)$ with a norm equivalent to (3.6.1). Hence, in what follows, some of the questions that were formulated using the Sobolev spaces $W^{1,p}(\mathbb{R}^d)$ in the continuous setting will now be formulated within $\ell^p(\mathbb{Z}^d)$.

3.6.1 One-Dimensional Results

We may start by defining the discrete analogue of (3.1.1) in the one-dimensional case. For $f : \mathbb{Z} \to \mathbb{R}$ we define the discrete centered one-dimensional Hardy–Littlewood maximal function $\mathcal{M}f : \mathbb{Z} \to \mathbb{R}$ by

$$\mathcal{M}f(n) = \sup_{r\geq 0} \frac{1}{(2r+1)} \sum_{k=-r}^{r} |f(n+k)|,$$

where the supremum is taken over nonnegative and integer values of r. Analogously, we define the uncentered version of this operator by

$$\widetilde{\mathcal{M}}f(n) = \sup_{r,s\geq 0} \frac{1}{(r+s+1)} \sum_{k=-r}^{s} |f(n+k)|,$$

where the supremum is taken over nonnegative and integer values of r and s. As in the continuous case, the uncentered version is more friendly for the sort of questions we investigate here. For instance, the analogue of Theorem 3.3.1 was established in [4, Theorem 1].

Theorem 3.6.1 (Bober, Carneiro, Pierce and Hughes, 2012—cf. [4]) *If $f : \mathbb{Z} \to \mathbb{R}$ is a function of bounded variation, then*

$$Var\left(\widetilde{\mathcal{M}}f\right) \leq Var(f).$$

This inequality is sharp as one can see by the "delta" example $f(0) = 1$ and $f(n) = 0$ for $n \neq 0$. The same sort of inequality in the centered case is subtler. Assume for a moment that we have

$$\mathrm{Var}(\mathcal{M}f) \leq \mathrm{Var}(f) \tag{3.6.2}$$

for any $f : \mathbb{Z} \to \mathbb{R}$ of bounded variation. Then, by (3.6.2) and an application of the triangle inequality, we would have the weaker inequality

$$\mathrm{Var}(\mathcal{M}f) \leq 2\|f\|_{\ell^1(\mathbb{Z})}. \tag{3.6.3}$$

Inequality (3.6.3) was proved in [4, Theorem 1] with constant $C = 2 + \frac{146}{315}$ replacing the constant $C = 2$, and it was proved with the sharp constant $C = 2$ in the recent work of Madrid [30, Theorem 1.1]. The fact that $C = 2$ is sharp in (3.6.3) is again seen by taking the delta example.

The interesting part of the story is that (3.6.2) is still not known. The BV-boundedness in the discrete centered case was proved by Temur [39], adapting the circle of ideas developed by Kurka [21] for the continuous case.

Theorem 3.6.2 (Temur, 2013—cf. [39]) *If $f : \mathbb{Z} \to \mathbb{R}$ is a function of bounded variation, then*

$$Var\,(\mathcal{M}f) \leq C\,Var\,(f)$$

with $C = (72000)2^{12} + 4 = 294912004$.

We record here the open inequality (3.6.2).

Question 9 *Let $f : \mathbb{Z} \to \mathbb{R}$ be a function of bounded variation. Do we have*

$$Var\,(\mathcal{M}f) \leq Var\,(f)?$$

Or at least, can one substantially improve on Temur's constant $C = 294912004$?

Having discussed the classical case, we may now consider the *discrete fractional* case. For $0 \leq \beta < 1$ and $f : \mathbb{Z} \to \mathbb{R}$, we define the one-dimensional discrete centered fractional maximal operator by

$$\mathcal{M}_\beta f(n) = \sup_{r \geq 0} \frac{1}{(2r+1)^{1-\beta}} \sum_{k=-r}^{r} |f(n+k)|$$

and its uncentered version by

$$\widetilde{\mathcal{M}}_\beta f(n) = \sup_{r,s \geq 0} \frac{1}{(r+s+1)^{1-\beta}} \sum_{k=-r}^{s} |f(n+k)|.$$

For $f : \mathbb{Z} \to \mathbb{R}$ and $1 \leq q < \infty$, the discrete analogue of (3.5.1) is the q-variation defined by

$$\mathrm{Var}_q(f) = \left(\sum_{n \in \mathbb{Z}} |f(n+1) - f(n)|^q \right)^{1/q} = \|f'\|_{\ell^q(\mathbb{Z})}.$$

The discrete analogue of Theorem 3.5.2 was also established in [9].

Theorem 3.6.3 (Carneiro and Madrid, 2017—cf. [9]) *Let $0 \leq \beta < 1$ and $q = 1/(1-\beta)$. Let $f : \mathbb{Z} \to \mathbb{R}$ be a function of bounded variation such that $\widetilde{\mathcal{M}}_\beta f \not\equiv \infty$. Then*

$$\|(\widetilde{\mathcal{M}}_\beta f)'\|_{\ell^q(\mathbb{Z})} = Var_q(\widetilde{\mathcal{M}}_\beta f) \leq 4^{1/q}\,Var\,(f).$$

As in the continuous case, we remark that the constant $C = 4^{1/q}$ above is not necessarily sharp. The same inequality for the centered fractional case is currently an open problem.

Question 10 *Let $0 < \beta < 1$ and $q = 1/(1-\beta)$. Let $f : \mathbb{Z} \to \mathbb{R}$ be a function of bounded variation such that $\mathcal{M}_\beta f \not\equiv \infty$. Do we have*

$$Var_q(\mathcal{M}_\beta f) \leq C\, Var(f)$$

for some universal constant C?

3.6.2 Multidimensional Results

In discussing the multidimensional discrete setting, we allow ourselves a more general formulation, in which we consider maximal operators associated to general convex sets. Let $\Omega \subset \mathbb{R}^d$ be a bounded open convex set with Lipschitz boundary. Let us assume that $0 \in \text{int}(\Omega)$. For $r > 0$ we write

$$\overline{\Omega}(x,r) = \left\{y \in \mathbb{R}^d;\ r^{-1}(y-x) \in \overline{\Omega}\right\},$$

and for $r = 0$ we consider

$$\overline{\Omega}(x,0) = \{x\}.$$

This object is the "Ω-ball of center x and radius r" in our maximal operators below. For instance, to work with regular ℓ^p-balls, one should consider $\Omega = \{x \in \mathbb{R}^d;\ \|x\|_p < 1\}$, where $\|x\|_p = (|x_1|^p + |x_2|^p + \cdots + |x_d|^p)^{\frac{1}{p}}$ for $x = (x_1, x_2, \ldots, x_d) \in \mathbb{R}^d$. These convex Ω-balls have roughly the same behavior as the regular Euclidean balls from the geometric and arithmetic points of view. For instance, we have the following asymptotics [25, Chap. VI , Sect. 2, Theorem 2], for the number of lattice points $N(x,r)$ of $\overline{\Omega}(x,r)$,

$$N(x,r) = C_\Omega\, r^d + O\big(r^{d-1}\big)$$

as $r \to \infty$, where $C_\Omega = m(\Omega)$ is the d-dimensional volume of Ω, and the constant implicit in the big O notation depends only on the dimension d and on the set Ω.

Given $0 \leq \beta < d$ and $f : \mathbb{Z}^d \to \mathbb{R}$, we denote by $\mathcal{M}_{\Omega,\beta}$ the discrete centered fractional maximal operator associated to Ω, i.e.,

$$\mathcal{M}_{\Omega,\beta} f(n) = \sup_{r \geq 0} \frac{1}{N(0,r)^{1-\frac{\beta}{d}}} \sum_{m \in \overline{\Omega}(0,r)} |f(n+m)|,$$

and we denote by $\widetilde{\mathcal{M}}_{\Omega,\beta}$ its uncentered version

$$\widetilde{\mathcal{M}}_{\Omega,\beta} f(n) = \sup_{\overline{\Omega}(x,r)\ni n} \frac{1}{N(x,r)^{1-\frac{\beta}{d}}} \sum_{m\in\overline{\Omega}(x,r)} |f(m)|.$$

Here r is a nonnegative real parameter.

The $\ell^p - \ell^q$ boundedness numerology for these discrete operators is the very same as the continuous fractional Hardy–Littlewood maximal operator (see [9] for a discussion), that is, if $1 < p < \infty$, $0 < \beta < d/p$ and $q = dp/(d - \beta p)$, then $\mathcal{M}_{\Omega,\beta} : \ell^p(\mathbb{Z}^d) \to \ell^q(\mathbb{Z}^d)$ is bounded (same for $\widetilde{\mathcal{M}}_{\Omega,\beta}$). Motivated by the endpoint philosophy in the continuous setting, a typical question here should be: let $0 \leq \beta < d$ and $q = d/(d - \beta)$; for a discrete function $f : \mathbb{Z}^d \to \mathbb{R}$ do we have $\|\nabla \mathcal{M}_{\Omega,\beta} f\|_{\ell^q(\mathbb{Z}^d)} \leq C(d, \Omega, \beta)\, \|\nabla f\|_{\ell^1(\mathbb{Z}^d)}$? (same question for $\widetilde{\mathcal{M}}_{\Omega,\beta}$). As in the continuous case, this question admits a positive answer if $1 \leq \beta < d$. In the harder case $0 \leq \beta < 1$, the current state of affairs is that one has a family of estimates that *approximate* the conjectured bounds (but unfortunately blow up when one tries to get exactly there). This was established in [9, Theorem 3] and we quote below.

Theorem 3.6.4 (Carneiro and Madrid, 2017—cf. [9]).

(i) Let $0 \leq \beta < d$ and $0 \leq \alpha \leq 1$. Let $q \geq 1$ be such that

$$q > \frac{d}{d - \beta + \alpha}.$$

Then there exists a constant $C = C(d, \Omega, \alpha, \beta, q) > 0$ such that

$$\|\nabla \mathcal{M}_{\Omega,\beta} f\|_{\ell^q(\mathbb{Z}^d)} \leq C\, \|\nabla f\|_{\ell^1(\mathbb{Z}^d)}^{1-\alpha}\, \|f\|_{\ell^1(\mathbb{Z}^d)}^{\alpha} \quad \forall f \in \ell^1(\mathbb{Z}^d). \tag{3.6.4}$$

Moreover, the operator $f \mapsto \nabla \mathcal{M}_{\Omega,\beta} f$ is continuous from $\ell^1(\mathbb{Z}^d)$ to $\ell^q(\mathbb{Z}^d)$.

(ii) Let $1 \leq \beta < d$ and $0 \leq \alpha < 1$. Let

$$q = \frac{d}{d - \beta + \alpha}.$$

Then there exists a constant $C = C(d, \Omega, \alpha, \beta) > 0$ such that

$$\|\nabla \mathcal{M}_{\Omega,\beta} f\|_{\ell^q(\mathbb{Z}^d)} \leq C\, \|\nabla f\|_{\ell^1(\mathbb{Z}^d)}^{1-\alpha}\, \|f\|_{\ell^1(\mathbb{Z}^d)}^{\alpha} \quad \forall f \in \ell^1(\mathbb{Z}^d).$$

Moreover, the operator $f \mapsto \nabla \mathcal{M}_{\Omega,\beta} f$ is continuous from $\ell^1(\mathbb{Z}^d)$ to $\ell^q(\mathbb{Z}^d)$.

The same results hold for the discrete uncentered fractional maximal operator $\widetilde{\mathcal{M}}_{\Omega,\beta}$.

Remark Theorem 3.6.4 already brings some continuity statements. These shall be further discussed in the next section.

By a suitable dilation argument, in [9] it is shown that inequality (3.6.4) can only hold if

$$q \geq \frac{d}{d - \beta + \alpha}. \tag{3.6.5}$$

The argument to show (3.6.5) goes roughly as follows. Consider, for instance, the uncentered case where $\Omega = (-1, 1)^d$ is the unit open cube. Let $k \in \mathbb{N}$ and consider the cube $Q_k = [-k, k]^d$ and its characteristic function $f_k := \chi_{Q_k}$. One has $\|f_k\|_{\ell^1(\mathbb{Z}^d)} \sim_d k^d$, $\|\nabla f_k\|_{\ell^1(\mathbb{Z}^d)} \sim_d k^{d-1}$, and $\|\nabla \widetilde{\mathcal{M}}_{\Omega,\beta} f_k\|_{\ell^q(\mathbb{Z}^d)} \gg_{\Omega,\beta,d} k^{\frac{d}{q}-1+\beta}$. One can see this last estimate by considering the region $H = \{n = (n_1, n_2, \ldots, n_d) \in \mathbb{Z}^d;\ n_1 \geq 4dk\,;\ |n_i| \leq k, \text{ for } i = 2, 3, \ldots, d\}$ and showing that the maximal function at $n \in H$ is realized by the cube of side $n_1 + k$ that contains the cube Q_k. Then we sum $|\widetilde{\mathcal{M}}_{\Omega,\beta} f_k(n + e_1) - \widetilde{\mathcal{M}}_{\Omega,\beta} f_k(n)|^q$ from $n_1 = 4dk$ to ∞, and then sum these contributions over the $\sim k^{d-1}$ possibilities for $(n_2, \ldots, n_d)$. Letting $k \to \infty$ we obtain the necessary condition (3.6.5).

This leaves us the following open question.

Question 11 *Let* $0 \leq \beta < 1$ *and* $q = d/(d - \beta)$. *For a discrete function* $f : \mathbb{Z}^d \to \mathbb{R}$ *do we have* $\|\nabla \mathcal{M}_{\Omega,\beta} f\|_{\ell^q(\mathbb{Z}^d)} \leq C(d, \Omega, \beta)\, \|\nabla f\|_{\ell^1(\mathbb{Z}^d)}$? *More generally, does the inequality* (3.6.4) *hold for all* $\alpha \leq \beta$ *and* $q = d/(d - \beta + \alpha)$? *(Analogous questions for* $\widetilde{\mathcal{M}}_{\Omega,\beta}$*).*

3.7 Continuity

We now turn to the final chapter of our discussion, in which we consider the continuity properties of the mappings we have addressed so far. The classical Hardy–Littlewood maximal operator M is a sublinear operator, i.e., $M(f + g)(x) \leq Mf(x) + Mg(x)$ pointwise. Having this property at hand, it is easy to see that the fact that $M : L^p(\mathbb{R}^d) \to L^p(\mathbb{R}^d)$ is bounded (for $1 < p \leq \infty$) implies that this map is also continuous. In fact, if $f_j \to f$ in $L^p(\mathbb{R}^d)$, then

$$\|Mf_j - Mf\|_{L^p(\mathbb{R}^d)} \leq \|M(f_j - f)\|_{L^p(\mathbb{R}^d)} \leq C\, \|f_j - f\|_{L^p(\mathbb{R}^d)} \to 0.$$

Same reasoning applies to its uncentered, fractional, or discrete versions (all being sublinear operators).

At the level of the (weak) derivatives, these operators, in principle, are not necessarily sublinear anymore. In light of the boundedness of the operator $M : W^{1,p}(\mathbb{R}^d) \to W^{1,p}(\mathbb{R}^d)$, for $1 < p \leq \infty$, established by Kinnunen, it is then a natural and nontrivial question to ask whether this operator is also continuous. This question is attributed to T. Iwaniec and was first explicitly posed in the work of Hajłasz and Onninen [16, Question 3], in the case $1 < p < \infty$. It was settled affirmatively by Luiro in [26, Theorem 4.1].

Theorem 3.7.1 (Luiro, 2007—cf. [26]) *The operator* $M : W^{1,p}(\mathbb{R}^d) \to W^{1,p}(\mathbb{R}^d)$ *is continuous for* $1 < p < \infty$.

Remark In the case $p = \infty$, the continuity of $M : W^{1,\infty}(\mathbb{R}^d) \to W^{1,\infty}(\mathbb{R}^d)$ does not hold, as pointed out to the author by H. Luiro. A counterexample may be constructed along the following lines (in dimension $d = 1$, say). Take a smooth f with compact support such that $(Mf)'$ has a point of discontinuity. Letting $f_h(x) = f(x + h)$, one sees that $f_h \to f$ in $W^{1,\infty}(\mathbb{R})$ as $h \to 0$, but $(M(f_h))' = (Mf)'_h \nrightarrow (Mf)'$ in $L^\infty(\mathbb{R})$ as $h \to 0$.

The proof of Luiro for Theorem 3.7.1 is very elegant. It provides an important qualitative study of the convergence properties of the sets of "good radii" (i.e., the radii that realize the supremum in the definition of the maximal function) and establishes an explicit formula for the derivative of the maximal function (in which one is able to move the derivative inside the integral over a ball of good radius). It also uses crucially the L^p-boundedness of the maximal operator. A similar study of the continuity properties of the local maximal operator on subdomains of $\mathbb{R}^d$ was also carried out by Luiro in [27].

3.7.1 Endpoint Study

As we have done many times before in this paper, we now turn our attention to the endpoint $p = 1$. So far, we have established several boundedness results at $p = 1$, and we now want to ask ourselves if such maps are continuous. For instance, the very first one of such boundedness results is Tanaka's inequality (3.3.2) that establishes that $f \mapsto (\widetilde{M} f)'$ is bounded from $W^{1,1}(\mathbb{R})$ to $L^1(\mathbb{R})$. The corresponding continuity question is: if $f_j \to f \in W^{1,1}(\mathbb{R})$ as $j \to \infty$, do we have $(\widetilde{M} f_j)' \to (\widetilde{M} f)'$ in $L^1(\mathbb{R})$ as $j \to \infty$? This was settled affirmatively in [10, Theorem 1].

Theorem 3.7.2 (Carneiro, Madrid and Pierce, 2017—cf. [10]) *The map* $f \mapsto \big(\widetilde{M} f\big)'$ *is continuous from* $W^{1,1}(\mathbb{R})$ *to* $L^1(\mathbb{R})$.

The proof of this result is quite subtle and different from Luiro's approach to Theorem 3.7.1 since one does not have the L^1-boundedness of the maximal operator. The authors in [10] develop a fine analysis toward the required convergence using the qualitative description of the uncentered maximal function (and the one-sided maximal functions) on the disconnecting set. The corresponding question for the one-dimensional centered maximal function M is even more challenging and it is currently open.

Question 12 *Is the map* $f \mapsto (Mf)'$ *is continuous from* $W^{1,1}(\mathbb{R})$ *to* $L^1(\mathbb{R})$*?*

In light of inequalities (3.3.3) and (3.3.4), one may ask similar (and harder) continuity questions on the Banach space of normalized functions of bounded variation.

Throughout the rest of this section, let us denote by $BV(\mathbb{R})$ the (Banach) space of functions $f : \mathbb{R} \to \mathbb{R}$ of bounded total variation with norm

$$\|f\|_{BV(\mathbb{R})} = |f(-\infty)| + \mathrm{Var}\,(f),$$

where $f(-\infty) = \lim_{x\to-\infty} f(x)$. Since

$$\|\widetilde{M} f\|_{L^\infty(\mathbb{R})} \leq \|f\|_{L^\infty(\mathbb{R})} \leq \|f\|_{BV(\mathbb{R})},$$

together with (3.3.3) we see that $\widetilde{M} : BV(\mathbb{R}) \to BV(\mathbb{R})$ is bounded. The same holds for $M : BV(\mathbb{R}) \to BV(\mathbb{R})$. The corresponding continuity statements arise as interesting open problems that would be qualitatively stronger than Theorem 3.7.2 or Question 12, if confirmed.

Question 13 *Is the map* $\widetilde{M} : BV(\mathbb{R}) \to BV(\mathbb{R})$ *continuous?*

Question 14 *Is the map* $M : BV(\mathbb{R}) \to BV(\mathbb{R})$ *continuous?*

3.7.2 Fractional Setting

We now move the endpoint discussion to the fractional setting as considered in Sect. 3.5. As in the classical setting considered above, we may think of the endpoint continuity questions assuming a (stronger) $W^{1,1}(\mathbb{R})$ convergence or a (weaker) $BV(\mathbb{R})$ convergence on the source space. With respect to the first type, the corresponding continuity statement to Theorem 3.5.2 was established in [31].

Theorem 3.7.3 (Madrid, 2018—cf. [31]) *Let* $0 < \beta < 1$ *and* $q = 1/(1-\beta)$. *The map* $f \mapsto \left(\widetilde{M}_\beta f\right)'$ *is continuous from* $W^{1,1}(\mathbb{R})$ *to* $L^q(\mathbb{R})$.

The analogous continuity question for the centered one-dimensional fractional maximal operator, for which the boundedness is not yet known (see Question 7), is an interesting open problem.

Question 15 *Let* $0 < \beta < 1$ *and* $q = 1/(1-\beta)$. *Is the map* $f \mapsto (M_\beta f)'$ *bounded and continuous from* $W^{1,1}(\mathbb{R})$ *to* $L^q(\mathbb{R})$*?*

With respect to the second type of continuity statement, in which one assumes the $BV(\mathbb{R})$-convergence on the source space, the interesting fact is that the fractional endpoint maps *are not continuous.* This was shown in [10, Theorems 3 and 4] and we quote the results below.

Theorem 3.7.4 (Carneiro, Madrid and Pierce, 2017—cf. [10]) *Let* $0 < \beta < 1$ *and* $q = 1/(1-\beta)$.

(i) *(Uncentered case) The map* $f \mapsto \big(\widetilde{M}_\beta f\big)'$ *is not continuous from* $BV(\mathbb{R})$ *to* $L^q(\mathbb{R})$, *i.e., there is a sequence* $\{f_j\}_{j\geq 1} \subset BV(\mathbb{R})$ *and a function* $f \in BV(\mathbb{R})$ *such that* $\|f_j - f\|_{BV(\mathbb{R})} \to 0$ *as* $j \to \infty$ *but*

$$\|\big(\widetilde{M}_\beta f_j\big)' - \big(\widetilde{M}_\beta f\big)'\|_{L^q(\mathbb{R})} = Var_q\big(\widetilde{M}_\beta f_j - \widetilde{M}_\beta f\big) \nrightarrow 0$$

as $j \to \infty$.

(ii) *(Centered case) There is a sequence* $\{f_j\}_{j\geq 1} \subset BV(\mathbb{R})$ *and a function* $f \in BV(\mathbb{R})$ *such that* $\|f_j - f\|_{BV(\mathbb{R})} \to 0$ *as* $j \to \infty$ *but* $Var_q\big(M_\beta f_j - M_\beta f\big) \nrightarrow 0$ *as* $j \to \infty$.

Notice the slightly different wordings in the items (i) and (ii) of the theorem above. The reason is that in the centered case we do not know yet if the analogue of Theorem 3.5.2 holds. Theorem 3.7.4 (ii) says that, regardless of the map $f \mapsto (M_\beta f)'$ being bounded from $BV(\mathbb{R})$ to $L^q(\mathbb{R})$ or not, it is not continuous.

3.7.3 Discrete Setting

To consider similar continuity issues in the discrete setting, we define the Banach space $BV(\mathbb{Z})$ as the space of functions $f : \mathbb{Z} \to \mathbb{R}$ of bounded total variation with norm

$$\|f\|_{\mathrm{BV}(\mathbb{Z})} = \big|f(-\infty)\big| + \mathrm{Var}\,(f),$$

where $f(-\infty) := \lim_{n\to-\infty} f(n)$. Recall the discussion on the beginning of Sect. 3.6 in which we said that there is no actual space $W^{1,1}(\mathbb{Z})$, as this is simple $\ell^1(\mathbb{Z})$ with a different norm. Then, in the instances where we assumed a $W^{1,1}(\mathbb{R})$-convergence in the continuous setting, we will be assuming an $\ell^1(\mathbb{Z})$-convergence in the discrete setting. As a particular case of the general framework of Theorem 3.6.4 (which is [9, Theorem 3]) we see that the maps $f \mapsto (\mathcal{M}_\beta f)'$ and $f \mapsto (\widetilde{\mathcal{M}}_\beta f)'$ are continuous from $\ell^1(\mathbb{Z})$ to $\ell^q(\mathbb{Z})$ for $0 \leq \beta < 1$ and $q = 1/(1-\beta)$ (the case $\beta = 0$ of these results had previously been obtained in [8]). Therefore, we have an affirmative answer for the discrete analogues of Theorems 3.7.2 and 3.7.3 and Questions 12 and 15.

The $BV(\mathbb{Z})$-continuity is a much more interesting issue. For the classical discrete Hardy–Littlewood maximal operators, we can affirmatively answer the analogues of Questions 13 and 14. This was accomplished in [10, Theorem 2] for the uncentered case and in [31, Theorem 1.2] for the centered case. We collect these results below.

Theorem 3.7.5 (Carneiro, Madrid and Pierce, 2017—cf. [10]) *The map* $\widetilde{\mathcal{M}} : BV(\mathbb{Z}) \to BV(\mathbb{Z})$ *is continuous.*

Theorem 3.7.6 (Madrid, 2018—cf. [31]) *The map* $\mathcal{M} : BV(\mathbb{Z}) \to BV(\mathbb{Z})$ *is continuous.*

As in the continuous cases, the fractional discrete maximal operators are not continuous on $BV(\mathbb{Z})$, as observed in [10, Theorems 5 and 6].

Theorem 3.7.7 (Carneiro, Madrid and Pierce, 2017—cf. [10]) *Let* $0 < \beta < 1$ *and* $q = 1/(1-\beta)$.

(i) (Uncentered case) The map $f \mapsto \big(\widetilde{\mathcal{M}}_\beta f\big)'$ *is not continuous from* $BV(\mathbb{Z})$ *to* $\ell^q(\mathbb{Z})$, *i.e., there is a sequence* $\{f_j\}_{j\geq 1} \subset BV(\mathbb{Z})$ *and a function* $f \in BV(\mathbb{Z})$ *such that* $\|f_j - f\|_{BV(\mathbb{Z})} \to 0$ *as* $j \to \infty$ *but*

$$\big\|(\widetilde{\mathcal{M}}_\beta f_j)' - (\widetilde{\mathcal{M}}_\beta f)'\big\|_{\ell^q(\mathbb{Z})} = Var_q\big(\widetilde{\mathcal{M}}_\beta f_j - \widetilde{\mathcal{M}}_\beta f\big) \nrightarrow 0$$

as $j \to \infty$.

(ii) (Centered case) There is a sequence $\{f_j\}_{j\geq 1} \subset BV(\mathbb{Z})$ *and a function* $f \in BV(\mathbb{Z})$ *such that* $\|f_j - f\|_{BV(\mathbb{Z})} \to 0$ *as* $j \to \infty$ *but* $Var_q\big(\mathcal{M}_\beta f_j - \mathcal{M}_\beta f\big) \nrightarrow 0$ *as* $j \to \infty$.

Note again the slight difference in the wording between parts (i) and (ii) of the statement above. This is due to the fact that the map $f \mapsto (\mathcal{M}_\beta f)'$ is not yet known to be bounded from $BV(\mathbb{Z})$ to $\ell^q(\mathbb{Z})$. Nevertheless, it is not continuous.

3.7.4 Summary

Table 3.1 collects the 16 different situations in which we analyzed the endpoint continuity (all of them one-dimensional). These arise from the following pairs of possibilities: (i) centered versus uncentered maximal operator; (ii) classical versus fractional

Table 3.1 One-dimensional endpoint continuity program

	$W^{1,1}$—continuity; continuous setting	BV—continuity; continuous setting	ℓ^1—continuity; discrete setting	BV—continuity; discrete setting
Centered classical maximal operator	OPEN: Question 12	OPEN: Question 14	YES[b]: Theorem 3.6.4	YES: Theorem 3.7.6
Uncentered classical maximal operator	YES: Theorem 3.7.2	OPEN: Question 13	YES[b]: Theorem 3.6.4	YES: Theorem 3.7.5
Centered fractional maximal operator	OPEN[a]: Question 15	NO[a]: Theorem 3.7.4	YES: Theorem 3.6.4	NO[a]: Theorem 3.7.7
Uncentered fractional maximal operator	YES: Theorem 3.7.3	NO: Theorem 3.7.4	YES: Theorem 3.6.4	NO: Theorem 3.7.7

[a]Corresponding boundedness result not yet known
[b]Result previously obtained in [8, Theorem 1]

maximal operator; (iii) continuous versus discrete setting; and (iv) $W^{1,1}$ (or ℓ^1) versus BV continuity. The word YES in a box below means that the continuity of the corresponding map has been established, whereas the word NO means that the continuity fails. The remaining boxes are marked as OPEN problems.

Acknowledgements The author acknowledges support from CNPq–Brazil grants 305612/2014-0 and 477218/2013-0, and FAPERJ grant E-26/103.010/2012. The author is thankful to José Madrid and João Pedro Ramos for reviewing the manuscript and providing valuable feedback. The author is also thankful to all of the members of the Scientific and Organizing Committees of the *CIMPA 2017 Research School—Harmonic Analysis, Geometric Measure Theory and Applications*, in Buenos Aires, Argentina, for the wonderful event.

References

1. J.M. Aldaz, J. Pérez Lázaro, Functions of bounded variation, the derivative of the one dimensional maximal function, and applications to inequalities. Trans. Amer. Math. Soc. **359**(5), 2443–2461 (2007)
2. S. Barza, M. Lind, A new variational characterization of Sobolev spaces. J. Geom. Anal. **25**(4), 2185–2195 (2015)
3. D. Beltran, J.P. Ramos, O. Saari, Regularity of fractional maximal functions through Fourier multipliers. J. Funct. Anal. **276** (6), 1875–1892 (2019)
4. J. Bober, E. Carneiro, K. Hughes, L.B. Pierce, On a discrete version of Tanaka's theorem for maximal functions. Proc. Am. Math. Soc. **140**, 1669–1680 (2012)
5. J. Bourgain, Averages in the plane over convex curves and maximal operators. J. Anal. Math. **47**, 69–85 (1986)
6. H. Brezis, *Functional Analysis, Sobolev Spaces and Partial Differential Equations* (Universitext. Springer, New York, 2011)
7. E. Carneiro, R. Finder, M. Sousa, On the variation of maximal operators of convolution type II, Rev. Mat. Iberoam. **34**(2), 739–766 (2018)
8. E. Carneiro, K. Hughes, On the endpoint regularity of discrete maximal operators. Math. Res. Lett. **19**(6), 1245–1262 (2012)
9. E. Carneiro, J. Madrid, Derivative bounds for fractional maximal functions. Trans. Am. Math. Soc. **369**(6), 4063–4092 (2017)
10. E. Carneiro, J. Madrid, L.B. Pierce, Endpoint Sobolev and BV continuity for maximal operators. J. Funct. Anal. **273**, 3262–3294 (2017)
11. E. Carneiro, D. Moreira, On the regularity of maximal operators. Proc. Am. Math. Soc. **136**(12), 4395–4404 (2008)
12. E. Carneiro, B.F. Svaiter, On the variation of maximal operators of convolution type. J. Funct. Anal. **265**, 837–865 (2013)
13. L.C. Evans, R.F. Gariepy, *Measure theory and fine properties of functions*, Studies in Advanced Mathematics (CRC Press, Boca Raton, FL, 1992)
14. D. Gilbarg, N.S. Trudinger, *Elliptic Partial Differential Equations of Second Order* (Springer, Berlin, 1983)
15. P. Hajłasz, J. Malý, On approximate differentiability of the maximal function. Proc. Am. Math. Soc. **138**, 165–174 (2010)
16. P. Hajłasz, J. Onninen, On boundedness of maximal functions in Sobolev spaces. Ann. Acad. Sci. Fenn. Math. **29**(1), 167–176 (2004)
17. J. Kinnunen, The Hardy-Littlewood maximal function of a Sobolev function. Israel J. Math. **100**, 117–124 (1997)

18. J. Kinnunen, P. Lindqvist, The derivative of the maximal function. J. Reine Angew. Math. **503**, 161–167 (1998)
19. J. Kinnunen, E. Saksman, Regularity of the fractional maximal function. Bull. Lond. Math. Soc. **35**(4), 529–535 (2003)
20. S. Korry, A class of bounded operators on Sobolev spaces. Arch. Math. (Basel) **82**(1), 40–50 (2004)
21. O. Kurka, On the variation of the Hardy-Littlewood maximal function. Ann. Acad. Sci. Fenn. Math. **40**, 109–133 (2015)
22. E. Lieb, M. Loss, *Analysis*, Graduate Studies in Mathematics, vol. 14, 2nd edn. (American Mathematical Society, Providence, RI, 2001)
23. F. Liu, T. Chen, H. Wu, A note on the endpoint regularity of the Hardy-Littlewood maximal functions. Bull. Aust. Math. Soc. **94**(1), 121–130 (2016)
24. F. Liu, H. Wu, On the regularity of the multisublinear maximal functions. Canad. Math. Bull. **58**(4), 808–817 (2015)
25. S. Lang, *Algebraic Number Theory* (Addison-Wesley Publishing Co., Inc., 1970)
26. H. Luiro, Continuity of the maximal operator in Sobolev spaces. Proc. Am. Math. Soc. **135**(1), 243–251 (2007)
27. H. Luiro, On the regularity of the Hardy-Littlewood maximal operator on subdomains of $\mathbb{R}^n$, Proc. Edinb. Math. Soc. **53**(1), 211–237 (2010). 2nd edn
28. H. Luiro, The variation of the maximal function of a radial function. Ark. Mat. **56**(1), 147–161 (2018)
29. H. Luiro, J. Madrid, The variation of the fractional maximal function of a radial function. Int. Math. Res. Not. IMRN **17** 5284–5298 (2019)
30. J. Madrid, Sharp inequalities for the variation of the discrete maximal function. Bull. Aust. Math. Soc. **95**(1), 94–107 (2017)
31. J. Madrid, Endpoint Sobolev and BV continuity for maximal operators, II. to appear in Rev. Mat. Iberoam
32. I.P. Natanson, *Theory of Functions of a Real Variable* (Frederick Ungar Publishing Co., New York, 1950)
33. C. Pérez, T. Picon, O. Saari, M. Sousa, Regularity of maximal functions on Hardy-Sobolev spaces. Bull. Lond. Math. Soc. **50**(6), 1007–1015 (2018)
34. J.P. Ramos, Sharp total variation results for maximal functions. Ann. Acad. Sci. Fenn. Math. **44**(1), 41–64 (2019)
35. O. Saari, Poincaré inequalities for the maximal function. Ann. Sc. Norm. Super. Pisa Cl. Sci. **5**(3), 1065–1083 (2019)
36. E.M. Stein, *Singular Integrals and Differentiability Properties of Functions* (Princeton University Press, Princeton, 1970)
37. E.M. Stein, Maximal functions. I. Spherical means, Proc. Natl. Acad. Sci. USA **73** 2174–2175 (1976)
38. H. Tanaka, A remark on the derivative of the one-dimensional Hardy-Littlewood maximal function. Bull. Austral. Math. Soc. **65**(2), 253–258 (2002)
39. F. Temur, On regularity of the discrete Hardy-Littlewood maximal function. Preprint abs/1303.3993

Chapter 4
Gabor Frames: Characterizations and Coarse Structure

Karlheinz Gröchenig and Sarah Koppensteiner

Abstract This chapter offers a systematic and streamlined exposition of the most important characterizations of Gabor frames over a lattice.

4.1 Introduction

Given a point $z = (x, \xi) \in \mathbb{R}^{2d}$ in time–frequency space (phase space), we define the corresponding time–frequency shift $\pi(z)$ acting on a function $f \in L^2(\mathbb{R}^d)$ by

$$\pi(z) f(t) = e^{2\pi i \xi \cdot t} f(t - x) .$$

Gabor analysis deals with the spanning properties of sets of time–frequency shifts. Specifically, for a window function $g \in L^2(\mathbb{R}^d)$ and a discrete set $\Lambda \subseteq \mathbb{R}^{2d}$, which we will always assume to be a lattice, we would like to understand when the set

$$\mathcal{G}(g, \Lambda) = \{\pi(\lambda) g : \lambda \in \Lambda\}$$

is a frame. This means that there exist positive constants $A, B > 0$ such that

$$A \|f\|_{L^2}^2 \le \sum_{\lambda \in \Lambda} |\langle f, \pi(\lambda) g \rangle|^2 \le B \|f\|_{L^2}^2 \qquad \forall f \in L^2(\mathbb{R}^d). \tag{4.1.1}$$

For historical reasons a frame with this structure is called a *Gabor frame*, or sometimes a Weyl–Heisenberg frame.

The motivation for studying sets of time–frequency shifts is in the foundations of quantum mechanics by Neumann [35] and in information theory by Gabor [17].

K. Gröchenig (✉) · S. Koppensteiner
Faculty of Mathematics, University of Vienna, Oskar-Morgenstern-Platz 1, 1090 Vienna, Austria
e-mail: karlheinz.groechenig@univie.ac.at

S. Koppensteiner
e-mail: sarah.koppensteiner@univie.ac.at

A. Aldroubi et al. (eds.), *New Trends in Applied Harmonic Analysis, Volume 2*, Applied and Numerical Harmonic Analysis,
https://doi.org/10.1007/978-3-030-32353-0_4

Since 1980 the investigation of Gabor frames has stimulated the interest of many mathematicians in harmonic, complex, and numerical analysis and engineers in signal processing and wireless communications.

Whereas (4.1.1) expresses a strong form of completeness (with stability built in the definition), a complementary concept is the linear independence of time–frequency shifts. Specifically, we ask for constants $A, B > 0$ such that

$$A\|c\|_{\ell^2}^2 \leq \Big\|\sum_{\lambda\in\Lambda} c_\lambda \pi(\lambda) g\Big\|_{L^2}^2 \leq B\|c\|_{\ell^2}^2 \qquad \forall c \in \ell^2(\Lambda), \tag{4.1.2}$$

and in this case $\mathcal{G}(g, \Lambda)$ is called a Riesz sequence in $L^2(\mathbb{R}^d)$. Riesz sequences are important in wireless communications: a data set $(c_\lambda)_{\lambda\in\Lambda}$ is transformed into an analog signal $f = \sum_{\lambda\in\Lambda} c_\lambda \pi(\lambda) g$ and then transmitted. The task at the receiver's end is to decode the data (c_λ). In this context (4.1.2) expresses the fact that the coefficients c_λ are uniquely determined by f and that their recovery is feasible in a robust way.

In this chapter, we restrict our attention to sets of time–frequency shifts over a lattice Λ, i.e., $\Lambda = A\mathbb{Z}^{2d}$ for an invertible, real-valued $2d \times 2d$ matrix A. The lattice structure implies the translation invariance $\pi(\lambda)\mathcal{G}(g, \Lambda) = \mathcal{G}(g, \Lambda)$ (up to phase factors) and is at the basis of a beautiful and deep structure theory and of many characterizations of (4.1.1) and (4.1.2).

After three decades, we have a clear understanding of the structures governing Gabor systems. Our goal is to collect the most important characterizations of Gabor frames and offer a systematic exposition of these structures. In the center of these characterizations is the duality theorem for Gabor frames. To our knowledge, *all* other characterizations within the L^2-theory follow directly from this fundamental duality. In particular, the celebrated characterizations of Janssen and Ron–Shen are consequences of the duality theorem, and the characterization of Zeevi and Zibulski for rational lattices also becomes a corollary.

Even with this impressive list of different criteria at our disposal, it remains very difficult to determine whether a given window function and lattice generate a Gabor frame. Ultimately, each criterion (within the L^2-theory) is formulated by means of the invertibility of some operator, and proving invertibility is always difficult. This fact explains perhaps why there are so many general results about Gabor frames, but so few explicit results about concrete Gabor frames.

Yet, there are some success stories due to Lyubarski [34], Seip [38], Janssen [30, 31], and some recent progress for totally positive windows [23, 24]. All these results have applied some of the characterizations presented here, or even invented some new ones. On the other hand, most questions about concrete Gabor systems remain unanswered, and so far every explicit conjecture about Gabor frames (with one exception) has been disproved by counterexamples.

To document some of the many white spots on the map of Gabor frames, let us mention two specific examples. (i) Let $g_1(t) = (1 - |t|)_+$ be the hat function (or B-spline of order 1). It is known that for all $\alpha > 0$ the Gabor system $\mathcal{G}(g_1, \alpha\mathbb{Z} \times 2\mathbb{Z})$

is *not* a Gabor frame. But it is not known whether $\mathcal{G}(g_1, 0.33\mathbb{Z} \times 2.001\mathbb{Z})$ is a frame. (ii) Let $h_1(t) = te^{-\pi t^2}$ be the first Hermite function. It is known that $\mathcal{G}(h_1, \alpha\mathbb{Z} \times \beta\mathbb{Z})$ is *not* a Gabor frame whenever $\alpha\beta = 2/3$ [33]. But it is not known whether $\mathcal{G}(h_1, \mathbb{Z} \times 0.66666\mathbb{Z})$ is a frame. In both cases, there is numerical evidence that these Gabor systems are frames, but so far there is no proof despite an abundance of precise criteria to check.

The novelty of our approach is the streamlined sequence of proofs, so that most of the structure theory of Gabor frames fits into a single, short chapter. In view of dozens of contributions to every aspect of Gabor analysis, we hope that this survey will be useful and inspire work on concrete open questions. The only prerequisite is the thorough mastery of the Poisson summation formula and some basic facts about frames and Riesz sequences.

The chapter is organized as follows: Sect. 4.2 covers the main objects of Gabor analysis. Section 4.3 is devoted to the interplay between the short-time Fourier transform, the Poisson summation formula, and commutativity of time–frequency shifts. The central Sect. 4.4 offers a complete proof of the duality theorem for Gabor frames. Section 4.5 sketches the main theorems about the coarse structure of Gabor frames. A list of criteria that are tailored to rectangular frames is discussed and proved in Sect. 4.6. In Sect. 4.7, we derive the criterion of Zeevi and Zibulski for rational lattices, and Sect. 4.8 presents a number of (technically more advanced) criteria some of which have recently become useful. Except for the last section, we fully prove all statements.

4.2 The Objects of Gabor Analysis

Let $g \in L^2(\mathbb{R}^d)$ be a nonzero window function and $\Lambda \subseteq \mathbb{R}^{2d}$ a lattice. The set $\mathcal{G}(g, \Lambda)$ is called a *Gabor frame* if there exist positive constants $A, B > 0$ such that

$$A\|f\|_{L^2}^2 \le \sum_{\lambda\in\Lambda} |\langle f, \pi(\lambda)g\rangle|^2 \le B\|f\|_{L^2}^2 \qquad \forall f \in L^2(\mathbb{R}^d). \tag{4.2.1}$$

The frame inequality (4.2.1) can be recast by means of functional analytic properties of certain operators associated to a Gabor system $\mathcal{G}(g, \Lambda)$. We will use the *frame operator* $S = S_{g,\Lambda}$ defined by

$$S_{g,\Lambda} f = \sum_{\lambda\in\Lambda} \langle f, \pi(\lambda)g\rangle \pi(\lambda)g.$$

Then $\mathcal{G}(g, \Lambda)$ is a frame if and only if $S_{g,\Lambda}$ is bounded and invertible on $L^2(\mathbb{R}^d)$. The extremal spectral values A, B are called the frame bounds. If they can be chosen to be equal $A = B$, then the frame operator is a multiple of the identity, and $\mathcal{G}(g, \Lambda)$ is called a tight frame.

We will also use the *Gramian operator* $G = G_{g,\Lambda}$ defined by its entries

$$G_{\lambda\mu} = \langle \pi(\mu)g, \pi(\lambda)g \rangle.$$

In this notation, $\mathcal{G}(g, \Lambda)$ is a Riesz sequence if and only if $G_{g,\Lambda}$ is bounded and invertible on $\ell^2(\Lambda)$.

If the upper inequality in (4.2.1) is satisfied, then the frame operator is well-defined and bounded on $L^2(\mathbb{R}^d)$ and the Gramian operator is bounded on $\ell^2(\Lambda)$. In this case, we call $\mathcal{G}(g, \Lambda)$ a *Bessel sequence*.

The underlying object of this definition is the *short-time Fourier transform* of f with respect to the window function $g \in L^2(\mathbb{R}^d)$, which is defined by

$$V_g f(z) = V_g f(x, \xi) = \int_{\mathbb{R}^d} f(t)\bar{g}(t - x)e^{-2\pi i \xi \cdot t}\, dt\,.$$

We will need the following properties of the short-time Fourier transform.

Lemma 4.2.1 (Covariance property) *Let $f, g \in L^2(\mathbb{R}^d)$ and $w, z \in \mathbb{R}^{2d}$. Then*

$$V_g(\pi(w)f)(z) = e^{-2\pi i(z_2 - w_2)\cdot w_1} V_g f(z - w) \qquad \text{and} \tag{4.2.2}$$

$$V_{\pi(w)g}(\pi(w)f)(z) = e^{2\pi i z\cdot \mathcal{I} w} V_g f(z), \tag{4.2.3}$$

where $\mathcal{I} = \begin{pmatrix} 0 & I_d \\ -I_d & 0 \end{pmatrix}$ denotes the standard symplectic matrix and I_d is the d-dimensional identity matrix.

The covariance property follows by a straightforward computation.

Proposition 4.2.2 (Orthogonality relations) *Let $f, g, h, \gamma \in L^2(\mathbb{R}^d)$.*

(i) Then $V_g f, V_\gamma h \in L^2(\mathbb{R}^{2d})$ and

$$\langle V_g f, V_\gamma h \rangle_{L^2(\mathbb{R}^{2d})} = \langle f, h \rangle_{L^2(\mathbb{R}^d)} \overline{\langle g, \gamma \rangle}_{L^2(\mathbb{R}^d)}. \tag{4.2.4}$$

In particular, if $\|g\|_z = 1$, then V_g is an isometry from $L^2(\mathbb{R}^d)$ to $L^2(\mathbb{R}^{2d})$.

(ii) Furthermore, for all $z \in \mathbb{R}^{2d}$,

$$\big(V_g f\, \overline{V_\gamma h}\big)\hat{}\,(z) = \big(V_g \gamma\, \overline{V_f h}\big)(\mathcal{I} z). \tag{4.2.5}$$

Proof (i) We first write the short-time Fourier transform as

$$V_g f(x, \xi) = \int_{\mathbb{R}^d} f(t)\overline{g(t - x)}\, e^{-2\pi i \xi\cdot t}\, dt = \mathcal{F}_2 \mathcal{T}(f \otimes \bar{g})(x, \xi)\,,$$

with the coordinate transform $\mathcal{T} F(x, t) = F(t, t - x)$ and the partial Fourier transform $\mathcal{F}_2 F(x, \xi) = \int_{\mathbb{R}^d} F(x, t)e^{-2\pi i \xi\cdot t}\, dt$. Since $\mathcal{F}_2$ is unitary by Plancherel's theorem and $\mathcal{T}$ is unitary by the transformation formula for integrals, we obtain

$$\begin{aligned}\langle V_g f, V_\gamma h\rangle_{L^2(\mathbb{R}^{2d})} &= \langle \mathcal{F}_2\mathcal{T}(f\otimes\bar{g}), \mathcal{F}_2\mathcal{T}(h\otimes\bar{\gamma})\rangle_{L^2(\mathbb{R}^{2d})}\\ &= \langle f\otimes\bar{g}, h\otimes\bar{\gamma}\rangle_{L^2(\mathbb{R}^{2d})} = \langle f,h\rangle_{L^2(\mathbb{R}^d)}\overline{\langle g,\gamma\rangle}_{L^2(\mathbb{R}^d)}.\end{aligned}$$

(ii) By the Cauchy–Schwarz inequality the product $V_g f\,\overline{V_\gamma h}$ is in $L^1(\mathbb{R}^{2d})$, therefore the Fourier transform is defined pointwise, and we obtain

$$\begin{aligned}\big(V_g f\,\overline{V_\gamma h}\big)\hat{}\,(z) &= \int_{\mathbb{R}^{2d}} V_g f(w)\,\overline{V_\gamma h(w)}e^{-2\pi i w\cdot z}\mathrm{d}w\\ &= \int_{\mathbb{R}^{2d}} V_{\pi(\mathcal{I}z)g}\big(\pi(\mathcal{I}z)f\big)(w)\,\overline{V_\gamma h(w)}\mathrm{d}w\\ &= \langle\gamma,\pi(\mathcal{I}z)g\rangle\overline{\langle h,\pi(\mathcal{I}z)f\rangle},\end{aligned}$$

where we first used $\mathcal{I}^2 = -I_{2d}$ and the covariance property (4.2.3), then the orthogonality relations (4.2.4) to separate the integral into two inner products.

4.3 Commutation Rules and the Poisson Summation Formula in Gabor Analysis

In this section, we exploit the invariance properties of a Gabor system for the structural interplay between the short-time Fourier transform and time–frequency lattices.

4.3.1 *Poisson Summation Formula*

If Λ is a lattice, then the function $\Phi(z) = \sum_{\lambda\in\Lambda}|\langle f,\pi(z+\lambda)g\rangle|^2$ satisfies $\Phi(z+\nu) = \Phi(z)$ for $\nu\in\Lambda$ and thus is periodic with respect to Λ. It is therefore natural to study the Fourier series of Φ. The mathematical tool is the Poisson summation formula, and this is in fact the mathematical core of all existing characterizations of Gabor frames over a lattice.[1]

We formulate the Poisson summation formula explicitly for an arbitrary lattice $\Lambda = A\mathbb{Z}^{2d}$ where A denotes an invertible, real-valued $2d\times 2d$ matrix. We write $\Lambda^\perp = (A^T)^{-1}\mathbb{Z}^{2d}$ for the dual lattice and $\Lambda^\circ = \mathcal{I}\Lambda^\perp$ for the adjoint lattice with $\mathcal{I} = \left(\begin{smallmatrix}0 & I_d\\ -I_d & 0\end{smallmatrix}\right)$.

The *volume* of the lattice is $\mathrm{vol}\Lambda = |\det(A)|$, and the reciprocal value $D(\Lambda) = \mathrm{vol}(\Lambda)^{-1}$ is the *density* or *redundancy* of Λ.

We first formulate a sufficiently general version of the Poisson summation formula [39].

[1] The terminology is often a bit different, e.g., [37] uses a "fiberization technique."

Lemma 4.3.1 *Assume that* $\Lambda = A\mathbb{Z}^{2d}$ *and* $F \in L^1(\mathbb{R}^{2d})$. *Then the periodization* $\Phi(x) = \sum_{\lambda\in\Lambda} F(x-\lambda)$ *is in* $L^1(\mathbb{R}^{2d}/\Lambda)$.

(i) The Fourier coefficients of Φ *are given by* $\hat{\Phi}(\nu) = \hat{F}(\nu)$ *for all* $\nu \in \Lambda^\perp$.
(ii) Poisson summation formula—general version: Let K_n *be a summability kernel*[2] *then*

$$\sum_{\lambda\in\Lambda} F(z+\lambda) = \mathrm{vol}(\Lambda)^{-1} \lim_{n\to\infty} \sum_{\nu\in\Lambda^\perp} K_n(\nu)\hat{F}(\nu)\, e^{2\pi i\nu\cdot z}.$$

with convergence in $L^1(\mathbb{R}^{2d}/\Lambda)$.
(iii) If $(\hat{F}(\nu))_{\nu\in\Lambda^\perp} \in \ell^1(\Lambda^\perp)$, *then the Fourier series converges absolutely and* Φ *coincides almost everywhere with a continuous function.*

By applying the Poisson summation formula to the function $V_g f\, \overline{V_\gamma h}$ and a lattice Λ, we obtain an important identity for the analysis of Gabor frames. This technique is so ubiquitous in Gabor analysis that Janssen [29] and later Feichtinger and Luef [16] called it *the* "Fundamental Identity of Gabor Analysis."

Theorem 4.3.2 *Let* $f, g, h, \gamma \in L^2(\mathbb{R}^d)$, *and* $\Lambda = A\mathbb{Z}^{2d}$ *be a lattice.*

(i) Then

$$\sum_{\lambda\in\Lambda} V_g f(z+\lambda)\overline{V_\gamma h(z+\lambda)} = \mathrm{vol}(\Lambda)^{-1} \lim_{n\to\infty} \sum_{\mu\in\Lambda^\circ} K_n(-\mathcal{I}\mu) V_g\gamma(\mu)\overline{V_f h(\mu)} e^{2\pi i\mu\cdot\mathcal{I}z} \tag{4.3.1}$$

holds almost everywhere with convergence in $L^1(\mathbb{R}^{2d}/\Lambda)$.
(ii) Assume in addition that both $\mathcal{G}(g,\Lambda)$ *and* $\mathcal{G}(\gamma,\Lambda)$ *are Bessel sequences and that* $\sum_{\mu\in\Lambda^\circ} |V_g\gamma(\mu)| < \infty$. *Then*

$$\sum_{\lambda\in\Lambda} V_g f(z+\lambda)\overline{V_\gamma h(z+\lambda)} = \mathrm{vol}(\Lambda)^{-1} \sum_{\mu\in\Lambda^\circ} V_g\gamma(\mu)\overline{V_f h(\mu)} e^{2\pi i\mu\cdot\mathcal{I}z} \quad \forall z \in \mathbb{R}^{2d}. \tag{4.3.2}$$

Proof (i) We apply the Poisson summation formula to the product $V_g f\, \overline{V_\gamma h}$ and the lattice Λ and obtain

$$\begin{aligned}\sum_{\lambda\in\Lambda} V_g f(z+\lambda)\overline{V_\gamma h(z+\lambda)} &= \mathrm{vol}(\Lambda)^{-1} \lim_{n\to\infty} \sum_{\nu\in\Lambda^\perp} K_n(\nu) \big(V_g f\, \overline{V_\gamma h}\big)\widehat{\ }(\nu)\, e^{2\pi i\nu\cdot z} \\ &= \mathrm{vol}(\Lambda)^{-1} \lim_{n\to\infty} \sum_{\nu\in\Lambda^\perp} K_n(\nu) \big(V_g\gamma\, \overline{V_f h}\big)(\mathcal{I}\nu)\, e^{2\pi i\nu\cdot z} \\ &= \mathrm{vol}(\Lambda)^{-1} \lim_{n\to\infty} \sum_{\mu\in\Lambda^\circ} K_n(-\mathcal{I}\mu) V_g\gamma(\mu)\overline{V_f h(\mu)}\, e^{2\pi i\mu\cdot\mathcal{I}z},\end{aligned}$$

where we used Proposition 4.2.2 to rewrite the Fourier transform in the first line.

---

[2] It suffices to take the Fourier coefficients of the multivariate Fejer kernel $\hat{F}_n(k) = \prod_{j=1}^d \big(1 - \frac{|k_j|}{n+1}\big)_+$ and set $K_n(\nu) = \hat{F}_n(A^T\nu) = \hat{F}_n(k)$ for $\nu = (A^T)^{-1}k \in \Lambda^\perp$.

(ii) If $\sum_{\mu \in \Lambda^\circ} |V_g \gamma(\mu)| < \infty$, then the right-hand side of (4.3.2) converges absolutely to a continuous function, and we do not need the summability kernel. Next we rewrite the left-hand side with the help of identity (4.2.2) as

$$\Phi(z) = \sum_{\lambda \in \Lambda} V_g(\pi(-z)f)(\lambda) \overline{V_\gamma(\pi(-z)h)(\lambda)},$$

where, as so often, the phase factors cancel. Since $\mathcal{G}(g, \Lambda)$ is a Bessel sequence with Bessel bound B_g, we know that

$$\|V_g(\pi(-z)f - f)|_\Lambda\|_{\ell^2} \le B_g^{1/2} \|\pi(-z)f - f\|_{L^2}.$$

This means that the map $z \mapsto V_g(\pi(-z)f)|_\Lambda$ is continuous from $\mathbb{R}$ to $\ell^2(\Lambda)$ for all $f \in L^2(\mathbb{R}^d)$. Likewise, the map $z \mapsto V_\gamma(\pi(-z)h)|_\Lambda$ is continuous for all $h \in L^2(\mathbb{R}^d)$.

This observation implies that the left-hand side is also a continuous function. Thus both sides of (4.3.2) are continuous and coincide almost everywhere, therefore (4.3.2) must hold everywhere.

4.3.2 Commutation Rules

In the fundamental identity (4.3.1) the adjoint lattice Λ° appears as a consequence of the Poisson summation formula. We now present a more structural property of the adjoint lattice.

Lemma 4.3.3 *Let $\Lambda \subseteq \mathbb{R}^{2d}$ be a lattice. Then its adjoint lattice is characterized by the property*

$$\Lambda^\circ = \{\mu \in \mathbb{R}^{2d} : \pi(\lambda)\pi(\mu) = \pi(\mu)\pi(\lambda) \quad \forall \lambda \in \Lambda\}.$$

Proof Let $z \in \mathbb{R}^{2d}$ and $\lambda = Ak \in \Lambda$ for some $k \in \mathbb{Z}^{2d}$. A straightforward computation yields $\pi(z)\pi(\lambda) = e^{2\pi i(\lambda_1 \cdot z_2 - \lambda_2 \cdot z_1)}\pi(\lambda)\pi(z)$. Consequently, the time–frequency shifts commute if and only if

$$1 = e^{2\pi i(\lambda_1 \cdot z_2 - \lambda_2 \cdot z_1)} = e^{2\pi i \lambda \cdot \mathcal{I} z} = e^{2\pi i Ak \cdot \mathcal{I} z}.$$

This holds for all $k \in \mathbb{Z}^{2d}$ if and only if $Ak \cdot \mathcal{I}z = k \cdot A^T \mathcal{I} z \in \mathbb{Z}$ for all $k \in \mathbb{Z}^{2d}$, which is precisely the case when $A^T \mathcal{I} z \in \mathbb{Z}^{2d}$, or equivalently when $z \in \mathcal{I}^{-1}(A^T)^{-1}\mathbb{Z}^{2d} = \Lambda^\circ$.

This interpretation of the adjoint lattice is crucial for an important technical point.

Lemma 4.3.4 (Bessel duality) *Let $g \in L^2(\mathbb{R}^d)$ and $\Lambda \subseteq \mathbb{R}^{2d}$ be a lattice. Then $\mathcal{G}(g, \Lambda)$ is a Bessel sequence if and only if $\mathcal{G}(g, \Lambda^\circ)$ is a Bessel sequence.*

Proof The proof is inspired by [10].

Fix $h \in \mathcal{S}(\mathbb{R}^d)$ with $\|h\|_{L^2} = 1$. Then $\mathcal{G}(h, M)$ is a Bessel sequence for every lattice $M \subseteq \mathbb{R}^{2d}$. Next let $Q = A[0,1)^{2d}$ be a fundamental domain of $\Lambda = A\mathbb{Z}^{2d}$, i.e., $\mathbb{R}^{2d} = \bigcup_{\lambda\in\Lambda} \lambda + Q$ as a disjoint union. As a consequence we may write $\int_{\mathbb{R}^{2d}} f(z)\,dz = \int_Q \sum_{\lambda\in\Lambda} f(z-\lambda)\,dz$ for $f \in L^1(\mathbb{R}^{2d})$.

Now assume that $\mathcal{G}(g, \Lambda)$ is a Bessel sequence. Let $c = (c_\mu)_{\mu\in\Lambda^\circ} \in \ell^2(\Lambda^\circ)$ be a finite sequence and $f = \sum_{\mu\in\Lambda^\circ} c_\mu \pi(\mu) g$. Since $V_h : L^2(\mathbb{R}^d) \to L^2(\mathbb{R}^{2d})$ is an isometry by Proposition 4.2.2, we obtain

$$
\begin{aligned}
\|f\|^2_{L^2(\mathbb{R}^d)} &= \|V_h f\|^2_{L^2(\mathbb{R}^{2d})} \\
&= \int_Q \sum_{\lambda\in\Lambda} \Big| \langle \sum_{\mu\in\Lambda^\circ} c_\mu \pi(\mu) g, \pi(-\lambda+z)h\rangle \Big|^2 \,dz := \int_Q I(z)\,dz\,.
\end{aligned} \tag{4.3.3}
$$

We now reorganize the sum over μ. First we use $\pi(-\lambda+z) = \gamma_{z,\lambda}\pi(\lambda)^*\pi(z)$ for some phase factor $|\gamma_{z,\lambda}| = 1$. Then we use the commutativity $\pi(\lambda)\pi(\mu) = \pi(\mu)\pi(\lambda)$ for all $\lambda \in \Lambda$, $\mu \in \Lambda^\circ$ (Lemma 4.3.3). This is the heart of the proof, and the reader should convince herself that the proof does not work without this property. We obtain

$$
\begin{aligned}
\langle \sum_{\mu\in\Lambda^\circ} c_\mu \pi(\mu) g, \pi(-\lambda+z)h\rangle &= \bar{\gamma}_{\lambda,z} \langle \pi(\lambda) g, \sum_{\mu\in\Lambda^\circ} \bar{c}_\mu \pi(\mu)^* \pi(z) h\rangle \\
&= \bar{\gamma}_{\lambda,z} \langle \pi(\lambda) g, \pi(z) \sum_{\mu\in\Lambda^\circ} \bar{c}_{-\mu}\gamma_{\mu,z}\pi(\mu)h\rangle\,.
\end{aligned}
$$

Since both Gabor families $\mathcal{G}(g, \Lambda)$ and $\mathcal{G}(h, \Lambda^\circ)$ are Bessel sequences by assumption (with constants B_g and B_h), we obtain a pointwise estimate for the integrand $I(z)$ in (4.3.3):

$$
\begin{aligned}
I(z) &= \sum_{\lambda\in\Lambda} \Big| \bar{\gamma}_{\lambda,z} \langle \pi(z) \sum_{\mu\in\Lambda^\circ} \bar{c}_{-\mu}\gamma_{\mu,z}\pi(\mu)h, \pi(\lambda) g\rangle \Big|^2 \\
&\le B_g \Big\| \pi(z) \sum_{\mu\in\Lambda^\circ} \bar{c}_{-\mu}\gamma_{\mu,z}\pi(\mu)h \Big\|^2_{L^2(\mathbb{R}^d)} \\
&\le B_g B_h \sum_{\mu\in\Lambda^\circ} |\bar{c}_{-\mu}\gamma_{\mu,z}|^2 = B_g B_h \|c\|^2_{\ell^2}\,.
\end{aligned}
$$

Integration over z now yields

$$
\Big\| \sum_{\mu\in\Lambda^\circ} c_\mu \pi(\mu) g \Big\|^2_{L^2} = \int_Q I(z)\,dz \le B_g B_h |\det A| \|c\|^2_{\ell^2}\,,
$$

and thus $\mathcal{G}(g, \Lambda^\circ)$ is a Bessel sequence.

Since $\Lambda = (\Lambda^\circ)^\circ$, the proof of the converse is the same.

4.4 Duality Theory

The duality theory relates the spanning properties of a Gabor family $\mathcal{G}(g, \Lambda)$ on a lattice Λ to the spanning properties of $\mathcal{G}(g, \Lambda^\circ)$ over the adjoint lattice. The following duality theorem is the central result of the theory of Gabor frames. We will see that most structural results about Gabor frames can be derived easily from it.

Theorem 4.4.1 (Duality theorem) *Let $g \in L^2(\mathbb{R}^d)$ and $\Lambda \subseteq \mathbb{R}^{2d}$ be a lattice. Then the following are equivalent:*

(i) $\mathcal{G}(g, \Lambda)$ is a frame for $L^2(\mathbb{R}^d)$.
(ii) $\mathcal{G}(g, \Lambda^\circ)$ is a Bessel sequence and there exists a dual window $\gamma \in L^2(\mathbb{R}^d)$ such that $\mathcal{G}(\gamma, \Lambda^\circ)$ is a Bessel sequence satisfying

$$\langle \gamma, \pi(\mu)g \rangle = \mathrm{vol}(\Lambda)\delta_{\mu,0} \qquad \forall \mu \in \Lambda^\circ. \tag{4.4.1}$$

(iii) $\mathcal{G}(g, \Lambda^\circ)$ is a Riesz sequence for $L^2(\mathbb{R}^d)$.

We follow the proof sketch given in the survey article [20].

Proof (i) $\Rightarrow$ (ii): Since $\mathcal{G}(g, \Lambda)$ is a Gabor frame, there exists a dual window γ in $L^2(\mathbb{R}^d)$ such that $\mathcal{G}(\gamma, \Lambda)$ is a frame and the reconstruction formula

$$f = \sum_{\lambda \in \Lambda} \langle f, \pi(\lambda)g \rangle \pi(\lambda)\gamma$$

holds for all $f \in L^2(\mathbb{R}^d)$ with unconditional L^2-convergence. We apply the reconstruction formula to $\pi(z)^* f$ and take the inner product with $\pi(z)^* h$ for $z \in \mathbb{R}^{2d}$ and $h \in L^2(\mathbb{R}^d)$. Then we have

$$\begin{aligned} \langle f, h \rangle = \langle \pi(z)^* f, \pi(z)^* h \rangle &= \sum_{\lambda \in \Lambda} \langle \pi(z)^* f, \pi(\lambda)g \rangle \langle \pi(\lambda)\gamma, \pi(z)^* h \rangle \\ &= \sum_{\lambda \in \Lambda} V_g f(z+\lambda) \overline{V_\gamma h(z+\lambda)} := \Phi(z) \end{aligned}$$

for all $f, h \in L^2(\mathbb{R}^d)$ and all $z \in \mathbb{R}^{2d}$. This means that the Λ-periodic function Φ on the right-hand side is constant.

By Proposition 4.2.2 (ii) the Fourier coefficients of the right-hand side are given by

$$\hat{\Phi}(\nu) = \big(V_g f \, \overline{V_\gamma h}\big)\widehat{\ }(\nu) = V_g \gamma(\mu) \overline{V_f h(\mu)},$$

where $\nu \in \Lambda^\perp$ and $\mu = \mathcal{I}\nu \in \Lambda^\circ$. Since these are the Fourier coefficients of a constant function, they must satisfy

$$\mathrm{vol}(\Lambda)^{-1} V_g \gamma(\mu) \overline{V_f h(\mu)} = \langle f, h \rangle \delta_{\mu,0} \qquad \forall \mu \in \Lambda^\circ.$$

As this identity holds for all $f, h \in L^2(\mathbb{R}^d)$, we obtain the biorthogonality relation

$$\mathrm{vol}(\Lambda)^{-1}\langle \gamma, \pi(\mu)g\rangle = \delta_{\mu,0} \qquad \forall \mu \in \Lambda^\circ .$$

By assumption both $\mathcal{G}(g, \Lambda)$ and $\mathcal{G}(\gamma, \Lambda)$ are frames and thus Bessel sequences; therefore, Lemma 4.3.4 implies that both $\mathcal{G}(g, \Lambda^\circ)$ and $\mathcal{G}(\gamma, \Lambda^\circ)$ are Bessel sequences.

(ii) ⇒ (i): We use the biorthogonality and read the fundamental identity (4.3.2) backward:

$$\mathrm{vol}\,(\Lambda)^{-1} \sum_{\mu\in\Lambda^\circ} V_g\gamma(\mu)\overline{V_f h(\mu)} e^{2\pi i \mu\cdot \mathcal{I}z} = \sum_{\lambda\in\Lambda} V_g f(z+\lambda)\, V_\gamma h(z+\lambda)\,.$$

Since both $\mathcal{G}(g, \Lambda^\circ)$ and $\mathcal{G}(\gamma, \Lambda^\circ)$ are Bessel sequences, Lemma 4.3.4 implies that the Gabor systems $\mathcal{G}(g, \Lambda)$ and $\mathcal{G}(\gamma, \Lambda)$ are also Bessel sequences. Furthermore $\sum_{\mu\in\Lambda^\circ} |V_g\gamma(\mu)| < \infty$ by the biorthogonality relation (4.4.1); hence, all assumptions of Theorem 4.3.2 are satisfied and guarantee that (4.3.2) holds pointwise. For $z = 0$ and $f = h$ we thus obtain

$$\|f\|_{L^2}^2 = \sum_{\lambda\in\Lambda} \langle f, \pi(\lambda)g\rangle\, \langle \pi(\lambda)\gamma, f\rangle\,.$$

Since both sets $\mathcal{G}(g, \Lambda)$ and $\mathcal{G}(\gamma, \Lambda)$ are Bessel sequences with Bessel bounds B_g and B_γ, respectively, the frame inequality for $\mathcal{G}(g, \Lambda)$ is obtained as follows:

$$\begin{aligned}\|f\|_{L^2}^4 &\le \Big(\sum_{\lambda\in\Lambda} |\langle f, \pi(\lambda)g\rangle|^2\Big)\Big(\sum_{\lambda\in\Lambda} |\langle f, \pi(\lambda)\gamma\rangle|^2\Big)\\ &\le B_\gamma \|f\|_{L^2}^2 \sum_{\lambda\in\Lambda} |\langle f, \pi(\lambda)g\rangle|^2 \le B_g B_\gamma \|f\|_{L^2}^4\,.\end{aligned}$$

(ii) ⇒ (iii): By assumption, the Bessel sequences $\mathcal{G}(g, \Lambda^\circ)$ and $\mathcal{G}(\gamma, \Lambda^\circ)$ satisfy the biorthogonal condition (4.4.1), thus

$$\langle \pi(\nu)\gamma, \pi(\mu)g\rangle = e^{-2\pi i(\mu_2-\nu_2)\cdot\nu_1}\,\langle \gamma, \pi(\mu-\nu)g\rangle = \mathrm{vol}(\Lambda)\delta_{\mu-\nu,0} \quad \forall \mu, \nu \in \Lambda^\circ\,. \tag{4.4.2}$$

Define $\tilde\gamma := \mathrm{vol}(\Lambda)^{-1}\gamma$, then (4.4.2) implies that $\mathcal{G}(\tilde\gamma, \Lambda^\circ)$ is a biorthogonal Bessel sequence for $\mathcal{G}(g, \Lambda^\circ)$. This means that $\mathcal{G}(g, \Lambda^\circ)$ is a Riesz sequence.

(iii) ⇒ (ii): By assumption $\mathcal{G}(g, \Lambda^\circ)$ is a Riesz sequence, i.e., a Riesz basis for its closed linear span, which we denote by $\mathcal{K} := \overline{\mathrm{span}}\{\mathcal{G}(g, \Lambda^\circ)\}$. By the general properties of Riesz bases [5], there exists a Bessel sequence $\{e_\nu : \nu \in \Lambda^\circ\}$ in $\mathcal{K}$ such that

$$\langle e_\nu, \pi(\mu)g\rangle = \delta_{\nu,\mu} \qquad \forall \mu, \nu \in \Lambda^\circ .$$

On the other hand, since $\mathcal{K}$ is invariant with respect to $\pi(\nu)$ for all $\nu \in \Lambda^\circ$, we have that $\pi(\nu)e_0$ is also in $\mathcal{K}$ and satisfies the biorthogonality

$$\langle \pi(\nu)e_0, \pi(\mu)g\rangle = e^{-2\pi i(\mu_2-\nu_2)\cdot\nu_1}\langle e_0, \pi(\mu-\nu)g\rangle = \delta_{0,\mu-\nu}\,.$$

This implies that $e_\nu - \pi(\nu)e_0 \in \mathcal{K}\cap\mathcal{K}^\perp = \{0\}$.

After the normalization $\gamma := \mathrm{vol}(\Lambda)^{-1}e_0$, the set $\mathcal{G}(\gamma, \Lambda^\circ) = \{\mathrm{vol}(\Lambda)^{-1}e_\nu : \nu \in \Lambda^\circ\}$ satisfies the biorthogonality relations (4.4.1) and is a Bessel sequence by the properties of Riesz bases.

The duality theory was foreshadowed by Rieffel's abstract work on noncommutative tori [36]. The biorthogonality relations (4.4.1) were discovered by the engineers Wexler and Raz [41] and characterize all possible dual windows (see Corollary 4.4.4). Janssen [27, 28], Daubechies et al. [10], and Ron–Shen [37] made the results of Wexler and Raz rigorous and further expanded upon them which became the duality theory for separable lattices. The theory for general lattices is due to Feichtinger and Kozek [15]. Recently, Jakobsen and Lemvig [26] formulated density and duality theorems for Gabor frames along a closed subgroup of the time–frequency plane. We remark that the duality theory also holds verbatim for general locally compact Abelian groups admitting a lattice.

Remark 4.4.2 (Frame bounds and an alternative proof) By rewriting Janssen's proof of the duality theory in [28] for general lattices, one can show that

$$AI_{L^2} \le S_{g,\Lambda} \le BI_{L^2} \quad\Longleftrightarrow\quad AI_{\ell^2} \le \mathrm{vol}(\Lambda)^{-1}G_{g,\Lambda^\circ} \le BI_{\ell^2}.$$

Hence, the family $\mathcal{G}(g, \Lambda)$ is a frame with frame bounds $A, B > 0$ if and only if $\mathcal{G}(g, \Lambda^\circ)$ is a Riesz sequence with bounds $\mathrm{vol}(\Lambda)A$, $\mathrm{vol}(\Lambda)B > 0$, respectively.

Definition 4.4.3 Let $g \in L^2(\mathbb{R}^d)$ and $\mathcal{G}(g, \Lambda)$ be a Bessel sequence. We call $\gamma \in L^2(\mathbb{R}^d)$ a *dual window* for $\mathcal{G}(g, \Lambda)$ if $\mathcal{G}(\gamma, \Lambda)$ is a Bessel sequence and the reconstruction property

$$f = \sum_{\lambda\in\Lambda}\langle f, \pi(\lambda)g\rangle\pi(\lambda)\gamma = \sum_{\lambda\in\Lambda}\langle f, \pi(\lambda)\gamma\rangle\pi(\lambda)g$$

holds for all $f \in L^2(\mathbb{R}^d)$.

The duality theorem now yields the following characterization of all dual windows.

Corollary 4.4.4 *Suppose $g, \gamma \in L^2(\mathbb{R}^d)$ and $\Lambda \subseteq \mathbb{R}^{2d}$ such that $\mathcal{G}(g, \Lambda)$ and $\mathcal{G}(\gamma, \Lambda)$ are Bessel sequences. Then γ is a dual window for $\mathcal{G}(g, \Lambda)$ if and only if the Wexler–Raz biorthogonality relations* (4.4.1) *are satisfied.*

Proof This is simply the equivalence (i) $\Leftrightarrow$ (ii) of Theorem 4.4.1.

We conclude this section with a characterization of tight Gabor frames.

Corollary 4.4.5 *A Gabor system $\mathcal{G}(g, \Lambda)$ is a tight frame if and only if $\mathcal{G}(g, \Lambda^\circ)$ is an orthogonal system. In this case the frame bound satisfies $A = \mathrm{vol}(\Lambda)^{-1}\|g\|_{L^2}^2$.*

Proof If $\mathcal{G}(g, \Lambda)$ is a tight frame, then the frame operator is just a multiple of the identity, i.e., $S = AI_{L^2}$. Hence, the canonical dual window is of the form $\gamma = S^{-1}g = \frac{1}{A}g$ and the biorthogonality relations (4.4.1) yield

$$\langle g, \pi(\mu)g\rangle = A\langle \gamma, \pi(\mu)g\rangle = A\mathrm{vol}(\Lambda)\delta_{\mu,0} \qquad \forall \mu \in \Lambda^\circ.$$

Therefore, $\mathcal{G}(g, \Lambda^\circ)$ is an orthogonal system and in particular $A = \mathrm{vol}(\Lambda)^{-1}\|g\|_{L^2}^2$.

Conversely, let $\mathcal{G}(g, \Lambda^\circ)$ be an orthogonal system, i.e.,

$$\langle g, \pi(\mu)g\rangle = \|g\|_{L^2}^2\delta_{\mu,0} \qquad \forall \mu \in \Lambda^\circ.$$

Then by Theorem 4.3.2 with $\gamma = g$, $h = f$ and $z = 0$, we obtain

$$\mathrm{vol}(\Lambda)^{-1}\|g\|_{L^2}^2\|f\|_{L^2}^2 = \sum_{\lambda\in\Lambda} |\langle f, \pi(\lambda)g\rangle|^2$$

and thus $\mathcal{G}(g, \Lambda)$ is a tight frame.

4.5 The Coarse Structure of Gabor Frames

Many of the fundamental properties of Gabor frames can be derived with little effort from the duality theorem. In the following, we deal with the density theorem, the Balian–Low theorem, and the existence of Gabor frames.

4.5.1 Density Theorem

To recover f from the inner products $\langle f, \pi(\lambda)g\rangle$, we need enough information. The density theorem quantifies this statement. The density theorem has a long history and has been proved many times. We refer to Heil's comprehensive survey [25]. Our point is that it follows immediately from the duality theory.

Theorem 4.5.1 (Density theorem) *Let $g \in L^2(\mathbb{R}^d)$ and $\Lambda \subseteq \mathbb{R}^{2d}$ be a lattice. Then the following holds:*

(i) If $\mathcal{G}(g, \Lambda)$ is a frame for $L^2(\mathbb{R}^d)$, then $0 < \mathrm{vol}(\Lambda) \le 1$.
(ii) If $\mathcal{G}(g, \Lambda)$ is a Riesz sequence in $L^2(\mathbb{R}^d)$, then $\mathrm{vol}(\Lambda) \ge 1$.

(iii) $\mathcal{G}(g, \Lambda)$ *is a Riesz basis for* $L^2(\mathbb{R}^d)$ *if and only if it is a frame and* $\operatorname{vol}(\Lambda) = 1$.

Proof (i) Let $\gamma = S^{-1}g$ be the canonical dual window of $\mathcal{G}(g, \Lambda)$. Then g possesses the following two distinguished representations with respect to the frame $\mathcal{G}(g, \Lambda)$:

$$g = 1 \cdot g = \sum_{\lambda \in \Lambda} \langle g, \pi(\lambda)\gamma \rangle \pi(\lambda) g \,.$$

By the general properties of the dual frame [11], the latter expansion has the coefficients with the minimum ℓ^2-norm, therefore

$$\sum_{\lambda \in \Lambda} |\langle g, \pi(\lambda)\gamma \rangle|^2 \le 1 + \sum_{\lambda \ne 0} 0 = 1 \,.$$

Consequently, with the biorthogonality (4.4.1) (in fact, we only need the condition for $\mu = 0$) we obtain

$$\operatorname{vol}(\Lambda)^2 = \langle g, \gamma \rangle^2 \le \sum_{\lambda \in \Lambda} |\langle g, \pi(\lambda)\gamma \rangle|^2 \le 1 \,, \tag{4.5.1}$$

which is the density theorem.

(ii) The volume of the adjoint lattice is $\operatorname{vol}(\Lambda^\circ) = \operatorname{vol}(\Lambda)^{-1}$. Therefore, the claim is equivalent to (i) by Theorem 4.4.1.

(iii) A Riesz basis is a Riesz sequence that is complete in the Hilbert space, and therefore is also a frame. Consequently, both (i) and (ii) apply and thus $\operatorname{vol}(\Lambda) = 1$.

Conversely, if $\mathcal{G}(g, \Lambda)$ is a frame with $\operatorname{vol}(\Lambda) = 1$, then we have equality in (4.5.1) and thus $\langle g, \pi(\lambda)\gamma \rangle = \delta_{\lambda,0}$ for $\lambda \in \Lambda$. Since the Gabor system $\mathcal{G}(\gamma, \Lambda)$ is a Bessel sequence and biorthogonal to $\mathcal{G}(g, \Lambda)$, we deduce that $\mathcal{G}(g, \Lambda)$ is a Riesz sequence, and by the assumed completeness it is a Riesz basis for $L^2(\mathbb{R}^d)$.

The above proof of the density theorem is due to Janssen [27].

4.5.2 *Existence of Gabor Frames for Sufficiently Dense Lattices*

In the early treatments of Gabor frames, one finds many qualitative statements that assert the existence of Gabor frames. Typically, they claim that for a window function $g \in L^2(\mathbb{R}^d)$ with "sufficient" decay and smoothness and for a "sufficiently dense" lattice Λ the Gabor system $\mathcal{G}(g, \Lambda)$ is a frame for $L^2(\mathbb{R}^d)$. For a sample of results, we refer to [8, 12, 40]. In this section, we derive such a qualitative result as a consequence of the duality theorem.

We will measure decay and smoothness by means of time–frequency concentration as follows: we say that g belongs to the modulation space $M^\infty_{v_s}(\mathbb{R}^d)$ if

$$|V_g g(z)| \leq C(1+|z|)^{-s} \qquad \forall z \in \mathbb{R}^{2d} .$$

This is not the standard definition of the modulation space, but it is the most convenient definition for our purpose. A systematic exposition of modulation spaces is contained in [18].

To quantify the density of a lattice $\Lambda = A\mathbb{Z}^{2d}$, we set simply

$$\|\Lambda\| = \|A\|_{op} ,$$

with the understanding that this definition is highly ambiguous and depends more on the choice of a basis A for the lattice than on the lattice itself.

Theorem 4.5.2 *Assume that $g \in M^{\infty}_{v_s}(\mathbb{R}^d)$ for some $s > 2d$. Then there exists a τ_0 depending on g such that for every lattice $\Lambda = A\mathbb{Z}^{2d}$ with $\|A\|_{op} < \tau_0$ the Gabor system $\mathcal{G}(g, \Lambda)$ is a frame for $L^2(\mathbb{R}^d)$.*

In other words, there exists a sufficiently small neighborhood V of the zero matrix such that $\mathcal{G}(g, A\mathbb{Z}^{2d})$ is a frame for every $A \in V$.

Proof Invariably, qualitative existence theorems in Gabor analysis (and more generally in sampling theory) use the fact that an operator that is close enough to the identity operator is invertible. For this proof, we use the duality theorem and show that on the adjoint lattice, which is sufficiently sparse, the Gramian matrix is diagonally dominant and therefore invertible.

Without loss of generality we assume that $\|g\|_{L^2} = 1$, then the Gramian can be written as $G = \mathrm{I} + R$, where $G_{\mu,\nu} = \langle \pi(\nu)g, \pi(\mu)g\rangle$ and R is the off-diagonal part of G.

We now make the following observations about R:

(i) If $\|A\|_{op} = \delta$ and $\mu = \mathcal{I}(A^T)^{-1}k \in \Lambda^\circ$, then $|k| = |A^T\mathcal{I}^{-1}\mathcal{I}(A^T)^{-1}k| \leq \|A\|_{op}|\mu| = \delta|\mu|$ and therefore

$$(1+|\mu|)^{-s} \leq (1+\delta^{-1}|k|)^{-s} .$$

(ii) By applying a simplified version of Schur's test to the self-adjoint operator R, the operator norm of R can estimated by

$$\begin{aligned}
\|R\|_{op} &\leq \sup_{\mu\in\Lambda^\circ} \sum_{\nu\neq\mu} |\langle \pi(\nu)g, \pi(\mu)g\rangle| \\
&= \sup_{\mu\in\Lambda^\circ} \sum_{\nu\neq\mu} |\langle g, \pi(\mu-\nu)g\rangle| \\
&\leq \sum_{\mu\neq 0} (1+|\mu|)^{-s} \qquad (4.5.2) \\
&\leq \sum_{\substack{k\in\mathbb{Z}^d \\ k\neq 0}} (1+\delta^{-1}|k|)^{-s} := \varphi(\delta) .
\end{aligned}$$

(iii) Since $s > 2d$, $\varphi(\delta)$ is finite for all $\delta > 0$, and φ is a continuous, increasing function that satisfies

$$\lim_{\delta \to 0+} \varphi(\delta) = 0 \text{ and } \lim_{\delta \to \infty} \varphi(\delta) = \infty .$$

Consequently, there is a τ_0 such that $\varphi(\tau_0) = 1$. For $\delta < \tau_0$ we then obtain that

$$\|R\|_{op} \le \varphi(\delta) < 1 ,$$

therefore G is invertible on $\ell^2(\Lambda^\circ)$. This means that $\mathcal{G}(g, \Lambda^\circ)$ is a Riesz sequence, and by duality $\mathcal{G}(g, \Lambda)$ is a frame, whenever the matrix A defining $\Lambda = A\mathbb{Z}^{2d}$ satisfies $\|A\|_{op} < \tau_0$.

The above proof highlights the role of the duality theorem in the qualitative existence proof. By emphasizing some technicalities about modulation spaces, one may prove a slightly more general version of the existence theorem. We say that g belongs to the modulation space $M^1(\mathbb{R}^d)$ if

$$\int_{\mathbb{R}^{2d}} |\langle g, \pi(z)g\rangle| \, dz < \infty .$$

The proof of Theorem 4.5.2 can be extended to yield the following result.

Theorem 4.5.3 ([12, 13]) *Assume that $g \in M^1(\mathbb{R}^d)$. Then there exists a τ_0 depending on g such that for every lattice $\Lambda = A\mathbb{Z}^{2d}$ with $\|A\|_{op} < \tau_0$ the Gabor system $\mathcal{G}(g, \Lambda)$ is a frame for $L^2(\mathbb{R}^d)$.*

These existence results are complemented by an important theorem of Bekka [3]: *For every lattice Λ with* vol $(\Lambda) \le 1$, *there exists a window $g \in L^2(\mathbb{R}^d)$ such that $\mathcal{G}(g, \Lambda)$ is a frame.*

4.5.3 Balian–Low Theorem

The Balian–Low theorem (BLT) states that for a window with a mild decay in time–frequency the necessary density condition must be strict. In the standard formulation, the window a Gabor frame $\mathcal{G}(g, \Lambda)$ at the critical density vol $(\Lambda) = 1$ lacks time–frequency localization. We refer to the surveys [4, 7] for a detailed discussion of the Balian–Low phenomenon in dimension 1. For higher dimensions and arbitrary lattices, the BLT follows from an important deformation result of Feichtinger and Kaiblinger [14] with useful subsequent improvements in [1, 22].

Theorem 4.5.4 *Assume that $g \in M^\infty_{v_s}(\mathbb{R}^d)$ for some $s > 2d$ and that $\mathcal{G}(g, \Lambda)$ is a frame for $L^2(\mathbb{R}^d)$. Then there exists an $\varepsilon_0 > 0$ such that $\mathcal{G}(g, (1+\tau)\Lambda)$ is a frame for every τ with $|\tau| < \varepsilon_0$.*

Proof We only give the proof idea and indicate where the duality theorem enters. Let $\tilde{\Lambda} = (1+\tau)\Lambda$, then its adjoint lattice is $\tilde{\Lambda}^\circ = (1+\tau)^{-1}\Lambda^\circ$. If $\mathcal{G}(g,\Lambda)$ is a frame, then the Gramian operator $G = G_{g,\Lambda^\circ}$ is invertible on $\ell^2(\Lambda^\circ)$. Set $\rho = (1+\tau)^{-1}$ and we consider the cross-Gramian operator $\tilde{G}^\rho$ with entries

$$\tilde{G}^\rho_{\mu\nu} = \langle \pi(\rho\nu)g, \pi(\mu)g\rangle \qquad \mu,\nu \in \Lambda^\circ .$$

We argue that

$$\lim_{\rho\to 1} \|\tilde{G}^\rho - G\|_{op} = 0 . \tag{4.5.3}$$

This implies that for $|\rho - 1| < \varepsilon_0$ for some ε_0 the cross-Gramian operator $\tilde{G}^\rho$ is invertible on $\ell^2(\Lambda^\circ)$. Now a perturbation result for Riesz bases that goes back to Paley–Wiener (see, e.g., [5]) implies that $\mathcal{G}(g,(1+\tau)^{-1}\Lambda)$ is a Riesz sequence, and by the duality theorem $\mathcal{G}(g,(1+\tau)\Lambda)$ is a frame.

The proof of (4.5.3) is similar to the proof of Theorem 4.5.2. We apply Schur's test to estimate the operator norm of $\tilde{G}^\rho - G$. Given $\delta > 0$, we may choose $R > 0$ such that

$$\sum_{\mu:|\mu-\nu|>R} |\tilde{G}^\rho_{\mu\nu} - G_{\mu\nu}| < \delta/2 \qquad \text{for all } \nu \in \Lambda^\circ \text{ and } 1/2 < \rho < 2 ,$$

and likewise $\sum_{\nu:|\mu-\nu|>R} |\tilde{G}^\rho_{\mu\nu} - G_{\mu\nu}| < \delta/2$ for all $\mu \in \Lambda^\circ$. As in (4.5.2), this is possible because $g \in M^\infty_{v_s}(\mathbb{R}^d)$ guarantees the off-diagonal decay of G and $\tilde{G}^\rho$.

Next, we choose $\varepsilon_0 > 0$ such that for $|\rho - 1| < \varepsilon_0$

$$\sum_{\mu:|\mu-\nu|\le R} |\tilde{G}^\rho_{\mu\nu} - G_{\mu\nu}| = \sum_{\mu:|\mu-\nu|\le R} |\langle \pi(\rho\nu)g - \pi(\nu)g, \pi(\mu)g\rangle| < \delta/2$$

and $\sum_{\nu:|\mu-\nu|\le R} |\tilde{G}^\rho_{\mu\nu} - G_{\mu\nu}| < \delta/2$. Combining both estimates yields $\|\tilde{G}^\rho - G\|_{op} < \delta$.

Again, the optimal assumption on g in Theorem 4.5.4 is that it belongs to $M^1(\mathbb{R}^d)$.

Corollary 4.5.5 *Assume that* $\mathrm{vol}(\Lambda) = 1$ *and* $\mathcal{G}(g,\Lambda)$ *is a frame for* $L^2(\mathbb{R}^d)$. *Then* $g \notin M^\infty_{v_s}(\mathbb{R}^d)$ *for all* $s > 2d$.

Proof If $g \in M^\infty_{v_s}$ and $\mathcal{G}(g,\Lambda)$ were a frame for some lattice Λ with $\mathrm{vol}(\Lambda) = 1$, then by Theorem 4.5.4 the Gabor system $\mathcal{G}(g,(1+\tau)\Lambda)$ would also be a frame for some $\tau > 0$. But $\mathrm{vol}\big((1+\tau)\Lambda\big) = (1+\tau)^{2d}\mathrm{vol}(\Lambda) > 1$, and this contradicts the density theorem.

4.5.4 The Coarse Structure of Gabor Frames and Gabor Riesz Sequences

One of the principal questions of Gabor analysis is the question under which conditions on a window g and a lattice Λ the Gabor system $\mathcal{G}(g, \Lambda)$ is a frame for $L^2(\mathbb{R}^d)$ or a Riesz sequence in $L^2(\mathbb{R}^d)$. To formalize this, we define the full frame set $\mathcal{F}_{\text{full}}(g)$ of g to be the set of all lattices Λ such that $\mathcal{G}(g, \Lambda)$ is a frame and the reduced frame set $\mathcal{F}(g)$ to be the set of all rectangular lattices $\alpha\mathbb{Z}^d \times \beta\mathbb{Z}^d$ such that $\mathcal{G}(g, \alpha\mathbb{Z}^d \times \beta\mathbb{Z}^d)$ is a frame. Formally,

$$\begin{aligned}\mathcal{F}_{\text{full}}(g) &= \{\Lambda \subseteq \mathbb{R}^{2d} \text{ lattice } : \mathcal{G}(g, \Lambda) \text{ is a frame }\}\\ \mathcal{F}(g) &= \{(\alpha, \beta) \subseteq \mathbb{R}^2_+ : \mathcal{G}(g, \alpha\mathbb{Z}^d \times \beta\mathbb{Z}^d) \text{ is a frame }\}.\end{aligned}$$

We summarize the results of the previous sections in the main result about the coarse structure of Gabor frames, i.e., results that hold for arbitrary Gabor systems over a lattice.

Theorem 4.5.6 *If $g \in M^\infty_{v_s}(\mathbb{R}^d)$ for some $s > 2d$ or in $M^1(\mathbb{R}^d)$, then $\mathcal{F}_{\text{full}}(g)$ is an open subset of $\{\Lambda : \operatorname{vol}(\Lambda) < 1\}$ and contains a neighborhood of 0.*

Likewise, $\mathcal{F}(g)$ is open in $\{(\alpha, \beta) \in \mathbb{R}^2_+ : \alpha\beta < 1\}$ and contains a neighborhood of $(0, 0)$ in $\mathbb{R}^2_+$.

Theorem 4.5.6 should not be underestimated. It compresses the efforts of dozens of articles into a single statement. It contains the existence of Gabor frames, the density theorem, and the Balian–Low theorem. For each result, there are now several different proofs (with subtle differences in the hypotheses) and many ramifications. What is perhaps new in our presentation is the close connection of the coarse structure of Gabor frames to the duality theory.

4.6 The Criterion of Janssen, Ron, and Shen for Rectangular Lattices

In this and the following section, we consider rectangular lattices of the form $\Lambda = \alpha\mathbb{Z}^d \times \beta\mathbb{Z}^d$ for $\alpha, \beta > 0$. Observe that the adjoint of such a lattice is

$$\Lambda^\circ = \mathcal{I}\begin{pmatrix} \frac{1}{\alpha} I_d & 0 \\ 0 & \frac{1}{\beta} I_d \end{pmatrix} \mathbb{Z}^{2d} = \tfrac{1}{\beta}\mathbb{Z}^d \times \tfrac{1}{\alpha}\mathbb{Z}^d,$$

and is again a rectangular lattice.

For convenience, we denote $\mathcal{G}(g, \alpha, \beta) := \mathcal{G}(g, \alpha\mathbb{Z}^d \times \beta\mathbb{Z}^d)$.

Definition 4.6.1 Let $g \in L^2(\mathbb{R}^d)$ and $\Lambda = \alpha\mathbb{Z}^d \times \beta\mathbb{Z}^d$ with $\alpha, \beta > 0$. The *pre-Gramian matrix* $P(x)$ is defined by

$$P(x)_{j,k} = \overline{g}\big(x + \alpha j - \tfrac{k}{\beta}\big) \qquad \forall j, k \in \mathbb{Z}^d ,$$

and the *Ron–Shen matrix* $R(x) := P(x)^* P(x)$ has the entries

$$R(x)_{k,l} = \sum_{j\in\mathbb{Z}^d} g\big(x + \alpha j - \tfrac{k}{\beta}\big)\overline{g}\big(x + \alpha j - \tfrac{l}{\beta}\big) \qquad \forall k, l \in \mathbb{Z}^d .$$

Theorem 4.6.2 *Let $g \in L^2(\mathbb{R}^d)$ and $\Lambda = \alpha\mathbb{Z}^d \times \beta\mathbb{Z}^d$ with $\alpha, \beta > 0$ be a rectangular lattice. Then the following are equivalent:*

(i) $\mathcal{G}(g, \alpha, \beta)$ is a frame for $L^2(\mathbb{R}^d)$.
(ii) $\mathcal{G}(g, \alpha, \beta)$ is a Bessel sequence and there exists a dual window $\gamma \in L^2(\mathbb{R}^d)$ such that $\mathcal{G}(\gamma, \alpha, \beta)$ is a Bessel sequence satisfying

$$\sum_{j\in\mathbb{Z}^d} \gamma(x + \alpha j)\overline{g}\big(x + \alpha j - \tfrac{k}{\beta}\big) = \beta^d \delta_{k,0} \qquad \forall k \in \mathbb{Z}^d \text{ and a.e. } x \in \mathbb{R}^d . \tag{4.6.1}$$

(iii) $\mathcal{G}(g, \frac{1}{\beta}, \frac{1}{\alpha})$ is a Riesz sequence for $L^2(\mathbb{R}^d)$.
(iv) There exist positive constants $A, B > 0$ such that for all $c \in \ell^2(\mathbb{Z}^d)$ and almost all $x \in \mathbb{R}^d$

$$A\|c\|_{\ell^2}^2 \le \sum_{j\in\mathbb{Z}^d} \Big| \sum_{k\in\mathbb{Z}^d} c_k \overline{g}\big(x + \alpha j - \tfrac{k}{\beta}\big)\Big|^2 \le B\|c\|_{\ell^2}^2. \tag{4.6.2}$$

(v) There exist positive constants $A, B > 0$ such that the spectrum of almost every Ron–Shen matrix is contained in the interval $[A, B]$. This means

$$\sigma(R(x)) \subseteq [A, B] \quad \text{for a.e. } x \in \mathbb{R}^d.$$

(vi) The set of pre-Gramians $\{P(x)\}$ is uniformly bounded on $\ell^2(\mathbb{Z}^d)$ and has a set of uniformly bounded left-inverses. This means that there exist $\Gamma(x) : \ell^2(\mathbb{Z}^d) \to \ell^2(\mathbb{Z}^d)$ such that

$$\Gamma(x)P(x) = I_{\ell^2(\mathbb{Z}^d)} \quad \text{for a.e. } x \in \mathbb{R}^d,$$
$$\|\Gamma(x)\| \le C \quad \text{for a.e. } x \in \mathbb{R}^d.$$

Proof The equivalence of (i) and (iii) is Theorem 4.4.1. The equivalence of conditions (iv), (v), and (vi) is mainly of linguistic nature, and the mathematical content is in the equivalence (iii) $\Leftrightarrow$ (iv).

(iv) $\Leftrightarrow$ (v): For all sequences $c \in \ell^2(\mathbb{Z}^d)$, we have

$$\sum_{j\in\mathbb{Z}^d}\Big|\sum_{k\in\mathbb{Z}^d} c_k\overline{g}\big(x+\alpha j-\tfrac{k}{\beta}\big)\Big|^2 = \langle P(x)c, P(x)c\rangle = \langle R(x)c, c\rangle.$$

Hence, inequality (4.6.2) becomes

$$A\|c\|_{\ell^2}^2 \le \langle R(x)c, c\rangle \le B\|c\|_{\ell^2}^2 \qquad \forall c\in\ell^2(\mathbb{Z}^d),$$

for almost all $x\in\mathbb{R}^d$, which is equivalent to $\sigma(R(x))\subseteq[A,B]$ for almost all $x\in\mathbb{R}^d$.

(iv) $\Rightarrow$ (iii): For this proof we write time-frequency shifts with the translation operator $T_x f(t)=f(t-x)$ and the modulation operator $M_\xi f(t)=e^{2\pi i\xi\cdot t}f(t)$. Let $c\in\ell^2(\mathbb{Z}^{2d})$ be a finite sequence and $f=\sum_{k,l\in\mathbb{Z}^d} c_{k,l}M_{\frac{l}{\alpha}}T_{\frac{k}{\beta}}g$. For fixed k, the sum over l is a trigonometric polynomial

$$p_k(x) := \sum_{l\in\mathbb{Z}^d} c_{k,l}e^{2\pi i\frac{l}{\alpha}\cdot x}$$

with period α in each coordinate, and its L^2-norm over a period $Q_\alpha := [0,\alpha]^d$ given by

$$\int_{Q_\alpha}|p_k(x)|^2\mathrm{d}x = \alpha^d\sum_{l\in\mathbb{Z}^d}|c_{k,l}|^2.$$

To calculate the L^2-norm of f, we use the periodization trick and obtain

$$\begin{aligned}\|f\|_{L^2}^2 &= \Big\|\sum_{k\in\mathbb{Z}^d} p_k\cdot T_{\frac{k}{\beta}}g\Big\|_{L^2}^2\\ &= \int_{\mathbb{R}^d}\Big|\sum_{k\in\mathbb{Z}^d} p_k(x)g\big(x-\tfrac{k}{\beta}\big)\Big|^2\mathrm{d}x\\ &= \int_{Q_\alpha}\sum_{j\in\mathbb{Z}^d}\Big|\sum_{k\in\mathbb{Z}^d} p_k(x)g\big(x+\alpha j-\tfrac{k}{\beta}\big)\Big|^2\mathrm{d}x.\end{aligned}$$

Next, for every $x\in Q_\alpha$ we apply assumption (4.6.2) to the integrand and obtain

$$\begin{aligned}\|f\|_{L^2}^2 &\ge \int_{Q_\alpha} A\sum_{k\in\mathbb{Z}^d}|p_k(x)|^2\mathrm{d}x\\ &= \alpha^d A\sum_{k,l\in\mathbb{Z}^d}|c_{k,l}|^2 = \alpha^d A\|c\|_{\ell^2}^2\end{aligned}$$

for all finite sequences $c\in\ell^2(\mathbb{Z}^{2d})$. The upper bound follows analogously, and thus $\mathcal{G}(g,\frac{1}{\beta},\frac{1}{\alpha})$ is a Riesz sequence.

(iii) $\Rightarrow$ (iv): By assumption,

$$A\|c\|_{\ell^2}^2 \le \Big\| \sum_{k,l\in\mathbb{Z}^d} c_{k,l} M_{\frac{l}{\alpha}} T_{\frac{k}{\beta}} g \Big\|_{L^2}^2 \le B\|c\|_{\ell^2}^2$$

for all $c \in \ell^2(\mathbb{Z}^{2d})$. We apply this fact to sequences c of the form $c_{k,l} := a_k b_l$ for $a, b \in \ell^2(\mathbb{Z}^d)$. Then $\|c\|_{\ell^2(\mathbb{Z}^{2d})}^2 = \|a\|_{\ell^2(\mathbb{Z}^d)}^2 \|b\|_{\ell^2(\mathbb{Z}^d)}^2$.

Every $p \in L^2(Q_\alpha)$ can be written as Fourier series $p(x) = \sum_{l\in\mathbb{Z}^d} b_l e^{2\pi i l\cdot\frac{x}{\alpha}}$ with coefficients $b \in \ell^2(\mathbb{Z}^d)$. Hence, we obtain for arbitrary $a \in \ell^2(\mathbb{Z}^d)$ and $p \in L^2(Q_\alpha)$

$$\begin{aligned}
\frac{A}{\alpha^d}\|a\|_{\ell^2(\mathbb{Z}^d)}^2 \int_{Q_\alpha} |p(x)|^2 \mathrm{d}x &= A\|a\|_{\ell^2(\mathbb{Z}^d)}^2 \|b\|_{\ell^2(\mathbb{Z}^d)}^2 = A\|c\|_{\ell^2(\mathbb{Z}^{2d})}^2 \\
&\le \Big\| \sum_{k,l\in\mathbb{Z}^d} a_k b_l M_{\frac{l}{\alpha}} T_{\frac{k}{\beta}} g \Big\|_{L^2}^2 \\
&= \int_{\mathbb{R}^d} |p(x)|^2 \Big| \sum_{k\in\mathbb{Z}^d} a_k g\big(x - \tfrac{k}{\beta}\big)\Big|^2 \mathrm{d}x \\
&= \int_{Q_\alpha} \sum_{j\in\mathbb{Z}^d} |p(x+\alpha j)|^2 \Big| \sum_{k\in\mathbb{Z}^d} a_k g\big(x + \alpha j - \tfrac{k}{\beta}\big)\Big|^2 \mathrm{d}x \\
&= \int_{Q_\alpha} |p(x)|^2 \sum_{j\in\mathbb{Z}^d} \Big| \sum_{k\in\mathbb{Z}^d} a_k g\big(x + \alpha j - \tfrac{k}{\beta}\big)\Big|^2 \mathrm{d}x.
\end{aligned} \tag{4.6.3}$$

Since $L^2(Q_\alpha)$ contains all characteristic functions of measurable subsets in Q_α, (4.6.3) implies

$$\frac{A}{\alpha^d}\|a\|_{\ell^2}^2 \le \sum_{j\in\mathbb{Z}^d} \Big| \sum_{k\in\mathbb{Z}^d} a_k g\big(x + \alpha j - \tfrac{k}{\beta}\big)\Big|^2 \qquad \text{for a.e. } x \in \mathbb{R}^d$$

for all $a \in \ell^2(\mathbb{Z}^d)$. The upper bound follows analogously.

(v) $\Rightarrow$ (vi): Suppose that the spectrum of almost all $R(x)$ is contained in the interval $[A, B]$ for some positive constants $A, B > 0$. Then the set of pre-Gramians is uniformly bounded by $B^{1/2}$ since $R(x) = P(x)^* P(x)$.

As $R(x)$ is invertible, we may define the pseudo-inverse $\Gamma(x) := R(x)^{-1} P(x)^*$. Then

$$\Gamma(x)P(x) = I_{\ell^2(\mathbb{Z}^d)}$$

and

$$\|\Gamma(x)\| \le \|R(x)^{-1}\| \|P(x)\| \le A^{-1} B^{1/2}.$$

(vi) $\Rightarrow$ (v): By assumption, every $P(x)$ possesses a left inverse $\Gamma(x)$ with control of the operator norm. This implies

$$\begin{aligned}\|c\|_{\ell^2}^2 &= \|\Gamma(x)P(x)c\|_{\ell^2}^2 \le \|\Gamma(x)\|^2\|P(x)c\|_{\ell^2}^2 \\ &\le C^2\langle R(x)c, c\rangle \le C^2\|P(x)\|^2\|c\|_{\ell^2}^2 \le C^2D^2\|c\|_{\ell^2}^2 .\end{aligned}$$

for all $c \in \ell^2(\mathbb{Z}^{2d})$ and almost all $x \in \mathbb{R}^d$. This is (v).

(ii) $\Leftrightarrow$ (iii) and (ii) $\Leftrightarrow$ (vi): Condition (ii) can be understood as explicit version of (vi). Alternatively, it is a slight reformulation of the biorthogonality condition (4.4.1), once again with the Poisson summation formula:

$$\sum_{j\in\mathbb{Z}^d} \gamma(x+\alpha j)\bar{g}\big(x+\alpha j - \tfrac{k}{\beta}\big) = \frac{1}{\alpha^d}\sum_{j\in\mathbb{Z}^d}\langle \gamma, M_{\frac{j}{\alpha}}T_{\frac{k}{\beta}}g\rangle e^{2\pi i \frac{j}{\beta}\cdot x} = \beta^d\delta_{k,0} .$$

Formulation (4.6.1) of the biorthogonality is due to Janssen [29]. Conditions (iv) and (v) were discovered by Ron and Shen [37]. The criterion (vi) is from [24].

The results of Ron and Shen are more general and hold for *separable lattices* of the form $P\mathbb{Z}^d \times Q\mathbb{Z}^d$ with invertible, real-valued $d \times d$ matrices P, Q. In this setting, Theorem 4.6.2 holds with the appropriate modifications (just replace the scalar multiplication with $\alpha, \beta, 1/\alpha, 1/\beta$ by the matrix–vector multiplication with P, Q, P^{-1}, Q^{-1} and use appropriate fundamental domains).

Condition (iv) has been the master tool of Janssen in his work on exponential windows [30] and "Zak transforms with few zeros" [31]. The construction of a dual window was used by Janssen [27] to give a signal-analytic proof of the theorem of Lyubarski and Seip. Recently, the biorthogonality condition for the dual window was used successfully in the analysis of totally positive windows of finite type [24]. Christensen et al. [6] have used (4.6.1) to compute explicit formulas for dual windows.

Condition (iv) also lends itself to proving qualitative sufficient conditions for the existence of Gabor frames. By imposing the diagonal dominance of $R(x)$, one can derive some conditions on g to guarantee that $\mathcal{G}(g, \alpha, \beta)$ is a frame. The easiest case is $R(x)$ being a family of diagonal matrices. In this way, one obtains the "painless non-orthogonal expansions" of Daubechies et al. [9]. This fundamental result precedes the era of wavelets and Gabor analysis, and yields all Gabor frames that are used for real applications, e.g., in signal analysis or speech processing.

Theorem 4.6.3 (Painless non-orthogonal expansions) *Suppose $g \in L^\infty(\mathbb{R}^d)$ with $\operatorname{supp} g \subseteq [0, L]^d$. If $\alpha \le L$ and $\beta \le \frac{1}{L}$, then $\mathcal{G}(g, \alpha, \beta)$ is a frame if and only if*

$$0 < \operatorname*{ess\,inf}_{x\in\mathbb{R}^d} \sum_{k\in\mathbb{Z}^d} |g(x-\alpha k)|^2.$$

Proof By assumption, we have $\frac{1}{\beta} \ge L$. If $\frac{1}{\beta} > L$, then the supports of $T_{\frac{k}{\beta}}g$ and $T_{\frac{l}{\beta}}g$ are disjoint for $k \ne l$; if $\frac{1}{\beta} = L$, then the supports of $T_{\frac{k}{\beta}}g$ and $T_{\frac{l}{\beta}}g$ overlap on a set of measure zero and we may modify g so that $T_{\frac{k}{\beta}}g\, T_{\frac{l}{\beta}}g = 0$ everywhere for $k \ne l$.

Consequently,

$$
\begin{aligned}
R(x)_{k,l} &= \sum_{j\in\mathbb{Z}^d} g\big(x + j\alpha - \tfrac{k}{\beta}\big)\overline{g}\big(x + j\alpha - \tfrac{l}{\beta}\big) \\
&= \sum_{j\in\mathbb{Z}^d} \big|g\big(x + j\alpha - \tfrac{k}{\beta}\big)\big|^2 \delta_{k,l},
\end{aligned}
$$

and thus $R(x)$ is a diagonal matrix for almost all x. Clearly, a diagonal matrix is bounded and invertible if and only if its diagonal is bounded above and away from zero; therefore, the assertion of Theorem 4.6.3 follows immediately.

Theorem 4.6.2 can also be reformulated in terms of frames for $L^2(\mathbb{T}^d)$. For this we recall that the *Zak transform* with respect to the parameter $\alpha > 0$ is defined by

$$
Z_\alpha f(x, \xi) := \sum_{k\in\mathbb{Z}^d} f(x - \alpha k)e^{2\pi i\alpha k\cdot\xi}.
$$

Most characterizations of a Gabor frame over a rectangular lattice can be formulated by means of the Zak transform. Here is the general version attached to Theorem 4.6.2.

Theorem 4.6.4 *Let $g \in L^2(\mathbb{R}^d)$ and $\Lambda = \alpha\mathbb{Z}^d \times \beta\mathbb{Z}^d$ with $\alpha, \beta > 0$. Then the following are equivalent:*

(i) $\mathcal{G}(g, \alpha, \beta)$ is a frame for $L^2(\mathbb{R}^d)$.
(ii) $\{Z_{\frac{1}{\beta}} g(x + \alpha j, \beta\,.) : j \in \mathbb{Z}^d\}$ is a frame for $L^2(\mathbb{T}^d)$ for almost all $x \in \mathbb{R}^d$ with frame bounds independent of x.

Proof By Theorem 4.6.2, $\mathcal{G}(g, \alpha, \beta)$ is a Gabor frame if and only if there exist positive constants $A, B > 0$ such that

$$
A\|c\|_{\ell^2}^2 \le \sum_{j\in\mathbb{Z}^d} \Big| \sum_{k\in\mathbb{Z}^d} c_k\overline{g}\big(x + \alpha j - \tfrac{k}{\beta}\big)\Big|^2 \le B\|c\|_{\ell^2}^2 \tag{4.6.4}
$$

for all $c \in \ell^2(\mathbb{Z}^d)$ and almost all $x \in \mathbb{R}^d$.

Using Parseval's identity for Fourier series, we interpret the inner sum over k as an inner product of periodic L^2-functions. The Fourier series of c is $\hat{c}(\xi) = \sum_{k\in\mathbb{Z}^d} c_k e^{2\pi i k\cdot\xi}$, and the Fourier series of the sequence $\big(g(x + \alpha j - k/\beta)\big)_{k\in\mathbb{Z}^d}$ (for fixed x) is precisely the Zak transform

$$
Z_{\frac{1}{\beta}} g(x + \alpha j, \beta\xi) = \sum_{k\in\mathbb{Z}^d} g\big(x + \alpha j - \tfrac{k}{\beta}\big)e^{2\pi i k\cdot\xi}.
$$

Consequently,

$$
\sum_{k\in\mathbb{Z}^d} c_k\overline{g}\big(x + \alpha j - \tfrac{k}{\beta}\big) = \int_{\mathbb{T}^d} \hat{c}(\xi)\overline{Z_{\frac{1}{\beta}} g(x + \alpha j, \beta\xi)}\,\mathrm{d}\xi
$$

and (4.6.4) just says that the set $\{Z_{\frac{1}{\beta}}g(x+\alpha j, \beta\cdot) : j \in \mathbb{Z}^d\}$ is a frame for $L^2(\mathbb{T}^d)$ for almost all $x \in \mathbb{R}^d$. Furthermore, the frame bounds can be chosen to be A and B independent of x.

4.7 Zak Transform Criteria for Rational Lattices—The Criteria of Zeevi and Zibulski

All criteria formulated so far are expressed by the invertibility of an infinite matrix or of an operator on an infinite-dimensional space. For rectangular lattices $\Lambda = \alpha\mathbb{Z}^d \times \beta\mathbb{Z}^d$ with $\alpha\beta \in \mathbb{Q}$, one may further reduce the effort and study the invertibility of a family of finite-dimensional matrices.

Assume that $\alpha\beta = p/q \leq 1$ for $p, q \in \mathbb{N}$. In order to simplify the labeling of vectors and matrices, we define $E_q := \{0, 1, \ldots, q-1\}^d$ and $E_p := \{0, 1, \ldots, p-1\}^d$. We then write $j \in \mathbb{Z}^d$ uniquely as $j = ql + r$ for $l \in \mathbb{Z}^d$ and $r \in E_q$. Using the quasi-periodicity of the Zak transform, we obtain

$$Z_{\frac{1}{\beta}}g(x+\alpha j, \beta\xi) = Z_{\frac{1}{\beta}}g\big(x + \tfrac{p}{q\beta}(ql+r), \beta\xi\big) = e^{2\pi i p l\cdot\xi} Z_{\frac{1}{\beta}}g\big(x + \tfrac{p}{q\beta}r, \beta\xi\big)\,.$$

Thus for rational values of $\alpha\beta$, we obtain a function system which factors into certain complex exponentials and some functions. The frame property of such a system is characterized in the following lemma.

Lemma 4.7.1 *Let $\{h_r : r \in F\} \subseteq L^2(\mathbb{T}^d)$ be a finite set and $p \in \mathbb{N}$ such that $\operatorname{card} F \geq \operatorname{card} E_p = p^d$. Furthermore, let $\mathcal{A}(\xi)$ be the matrix with entries $\mathcal{A}(\xi)_{r,s} = \overline{h_r}(\xi + \frac{s}{p})$ for $r \in F$, $s \in E_p$. Then the following are equivalent:*

(i) The set $\{e^{2\pi i p l\cdot\xi} h_r(\xi) : l \in \mathbb{Z}^d, r \in F\}$ is a frame for $L^2(\mathbb{T}^d)$.
(ii) There exist $A, B > 0$ such that the singular values of $\mathcal{A}(\xi)$ are contained in $[A^{1/2}, B^{1/2}]$ for almost all $\xi \in \mathbb{T}^d$.
(iii) There exist $A, B > 0$ such that $\sigma(\mathcal{A}^*(\xi)\mathcal{A}(\xi)) \subseteq [A, B]$ for almost all $\xi \in \mathbb{T}^d$.

The condition $\operatorname{card} F \geq p^d$ is essential; otherwise, the matrix $\mathcal{A}(\xi)$ cannot be injective and $\mathcal{A}^*(\xi)\mathcal{A}(\xi)$ cannot be invertible.

Proof For $f \in L^2(\mathbb{T}^d)$ and $\xi \in Q_{1/p} = [0, \frac{1}{p}]^d$, we write the vector $y(\xi) = \big(f(\xi + \frac{s}{p})\big)_{s\in E_p}$. Then the inner product of f with the frame functions $h_r(\xi)e^{2\pi i p l\cdot\xi}$ can be written as

$$\begin{aligned}\langle f, h_r e^{2\pi i p l\cdot\xi}\rangle &= \int_{[0,\frac{1}{p}]^d} \sum_{s\in E_p} f(\xi + \tfrac{s}{p})\overline{h_r}(\xi + \tfrac{s}{p}) e^{-2\pi i p l\cdot\xi}\, \mathrm{d}\xi \\ &= \int_{[0,\frac{1}{p}]^d} \big(\mathcal{A}(\xi)y(\xi)\big)_r\, e^{-2\pi i p l\cdot\xi}\, \mathrm{d}\xi\,.\end{aligned}$$

Since $\{p^{d/2}e^{2\pi i p l\cdot\xi} : l \in \mathbb{Z}^d\}$ is an orthonormal basis for $L^2(Q_{1/p})$, we now obtain

$$\sum_{l\in\mathbb{Z}^d}\sum_{r\in F} |\langle f, h_r e^{2\pi i p l\cdot\xi}\rangle|^2 = \sum_{r\in F}\sum_{l\in\mathbb{Z}^d} \left| \int_{Q_{1/p}} (\mathcal{A}(\xi)y(\xi))_r e^{-2\pi i p l\cdot\xi}\, d\xi \right|^2$$
$$= \frac{1}{p^d}\sum_{r\in F}\int_{Q_{1/p}} |(\mathcal{A}(\xi)y(\xi))_r|^2\, d\xi$$
$$= \frac{1}{p^d}\int_{Q_{1/p}} |\mathcal{A}(\xi)y(\xi)|^2\, d\xi\,.$$

If the singular values of $\mathcal{A}$ are all in an interval $[A^{1/2}, B^{1/2}]$, then $|\mathcal{A}(\xi)y(\xi)|^2 = \langle \mathcal{A}(\xi)^*\mathcal{A}(\xi)y(\xi), y(\xi)\rangle \geq A|y(\xi)|^2$. Therefore

$$\sum_{l\in\mathbb{Z}^d}\sum_{r\in F} |\langle f, h_r e^{2\pi i p l\cdot\xi}\rangle|^2 = \frac{1}{p^d}\int_{Q_{1/p}} |\mathcal{A}(\xi)y(\xi)|^2\, d\xi$$
$$\geq \frac{A}{p^d}\int_{Q_{1/p}} |y(\xi)|^2\, d\xi \tag{4.7.1}$$
$$= \frac{A}{p^d}\int_{Q_{1/p}} \sum_{s\in E_p} |f(\xi + \tfrac{s}{p})|^2\, d\xi = \frac{A}{p^d}\|f\|^2_{L^2(\mathbb{T}^d)}\,.$$

Similarly, the upper frame bound follows. Thus the set $\{h_r(\xi)e^{2\pi i p l\cdot\xi} : l \in \mathbb{Z}^d, r \in F\}$ is a frame for $L^2(\mathbb{T}^d)$.

Conversely, assume that $\{h_r(\xi)e^{2\pi i p l\cdot\xi} : l \in \mathbb{Z}^d, r \in F\}$ is a frame for $L^2(\mathbb{T}^d)$. Then (4.7.1) says that for all $f \in L^2(\mathbb{T}^d)$ with associated vector-valued function $y(\xi) = \big(f(\xi + \frac{s}{p})\big)_{s\in E_p}$ we must have

$$\frac{A}{p^d}\int_{Q_{1/p}} |y(\xi)|^2\, d\xi \leq \frac{1}{p^d}\int_{Q_{1/p}} |\mathcal{A}(\xi)y(\xi)|^2\, d\xi \leq \frac{B}{p^d}\int_{Q_{1/p}} |y(\xi)|^2\, d\xi\,. \tag{4.7.2}$$

Now we diagonalize $\mathcal{A}(\xi)^*\mathcal{A}(\xi)$. Since $\mathcal{A}^*\mathcal{A}$ is a measurable matrix-valued function on $\mathbb{T}^d$, its diagonalization can be chosen to be measurable (see Azoff [2]). This means that there exist two measurable matrix-valued functions $\mathcal{U}, \mathcal{D}$ such that $\mathcal{U}(\xi)$ is a unitary matrix, $\mathcal{D}(\xi)$ is of diagonal form and $\mathcal{A}(\xi)^*\mathcal{A}(\xi) = \mathcal{U}(\xi)^*\mathcal{D}(\xi)\mathcal{U}(\xi)$ for all $\xi \in \mathbb{T}^d$.

Hence (4.7.2) is equivalent to

$$A\int_{Q_{1/p}} |\tilde{y}(\xi)|^2\, d\xi \leq \int_{Q_{1/p}} \langle \mathcal{D}(\xi)\tilde{y}(\xi), \tilde{y}(\xi)\rangle\, d\xi \leq B\int_{Q_{1/p}} |\tilde{y}(\xi)|^2\, d\xi \tag{4.7.3}$$

for all vector-valued functions $\tilde{y}(\xi) = \mathcal{U}(\xi)y(\xi)$ with components in $L^2(Q_{1/p})$.

Clearly, inequality (4.7.3) can only hold if $\sigma(\mathcal{D}(\xi)) = \sigma(\mathcal{A}(\xi)^*\mathcal{A}(\xi)) \subseteq [A, B]$ for almost all $\xi \in Q_{1/p}$.

We now apply this lemma to the set $\{e^{2\pi i p l\cdot\xi} Z_{\frac{1}{\beta}}(x + \frac{p}{q\beta}r, \beta\xi) : l \in \mathbb{Z}^d, r \in E_q\}$ and obtain the characterization of Zeevi and Zibulski for rational rectangular lattices [42, 43].

Theorem 4.7.2 *Let $g \in L^2(\mathbb{R}^d)$ and $\alpha\beta = p/q \in \mathbb{Q}$ with $p/q \le 1$. For $x, \xi \in \mathbb{R}^d$ let $\mathcal{Q}(x, \xi)$ be the matrix with entries*

$$\mathcal{Q}(x, \xi)_{r,s} = Z_{\frac{1}{\beta}} g\big(x + \tfrac{p}{\beta q}r, \beta\xi + \tfrac{\beta s}{p}\big) \qquad \forall r \in E_q, s \in E_p .$$

The Gabor family $\mathcal{G}(g, \alpha, \beta)$ is a frame for $L^2(\mathbb{R}^d)$ if and only if the singular values of $\mathcal{Q}(x, \xi)$ are contained in an interval $[A^{1/2}, B^{1/2}] \subseteq (0, \infty)$ for almost all $x, \xi \in \mathbb{R}^d$.

Proof By Theorem 4.6.4, $\mathcal{G}(g, \alpha, \beta)$ is a frame if and only if $\{Z_{\frac{1}{\beta}} g(x + \alpha j, \beta\xi) : j \in \mathbb{Z}^d\} = \{e^{2\pi i p l\cdot\xi} Z_{\frac{1}{\beta}} g(x + \alpha r, \beta\xi) : l \in \mathbb{Z}^d, r \in E_q\}$ is a frame for $L^2(\mathbb{T}^d)$. Now, the claim follows from Lemma 4.7.1 with the functions $h_r(\xi) = Z_{\frac{1}{\beta}} g(x + \alpha r, \beta\xi)$.

The Zak transform has been used frequently to derive theoretical properties of Gabor frames. The Zeevi–Zibulski matrices in particular are very useful for computational issues, and several important counterexamples have been discovered first through numerical tests before being proved rigorously [32, 33]. On the other hand, it seems to be very difficult to apply directly and decide rigorously whether a concrete Gabor system is a frame or not.

4.8 Further Characterizations

So far, we have discussed characterization of Gabor frames that work for arbitrary windows in $L^2(\mathbb{R}^d)$. On a technical level, we have not used more than the Poisson summation formula. Under mild additional conditions that are standard in time–frequency analysis, one can prove further characterizations for Gabor frames. These, however, require additional and more advanced mathematical tools, such as spectral invariance, a noncommutative version of Wiener's lemma or Beurling's method of weak limit. For this reason, we state these characterizations without proofs.

4.8.1 The Wiener Amalgam Space and Irrational Lattices

This condition refines Theorem 4.6.2 for irrational lattices. As the appropriate class of window, we use the Wiener amalgam space $W_0 = W(C, \ell^1)$. It consists of all continuous functions g for which the norm

$$\|g\|_W = \sum_{k\in\mathbb{Z}^d} \sup_{x\in Q_1} |g(x+k)|$$

is finite.

Theorem 4.8.1 ([21]) *Assume that $g \in W_0$ and $\Lambda = \alpha\mathbb{Z}^d \times \beta\mathbb{Z}^d$ with $\alpha\beta \notin \mathbb{Q}$. Then $\mathcal{G}(g, \alpha, \beta)$ is a frame for $L^2(\mathbb{R}^d)$ if and only if there exists some $x_0 \in Q_\alpha$ such that $R(x_0)$ is invertible on $\ell^2(\mathbb{Z}^d)$.*

Thus for irrational lattices, it suffices to check the invertibility of a single Ron–Shen matrix $R(x)$ instead of all matrices. Although this condition looks useful, it has not yet found any applications.

4.8.2 Janssen's Criterion Without Inequalities

Theorem 4.8.2 ([23]) *Assume that $g \in W_0$ and $\Lambda = \alpha\mathbb{Z}^d \times \beta\mathbb{Z}^d$. Then $\mathcal{G}(g, \alpha, \beta)$ is a frame if and only if the pre-Gramian $P(x)$ is one-to-one on $\ell^\infty(\mathbb{Z}^d)$ for all $x \in \mathbb{R}^d$.*

Put differently, to show that $\mathcal{G}(g, \alpha, \beta)$ is a frame, one has to show that

$$\sum_{k\in\mathbb{Z}^d} c_k g(x + \alpha j - \tfrac{k}{\beta}) = 0 \quad \Longrightarrow \quad c \equiv 0\,,$$

with the added subtlety that c is only a bounded sequence, but not necessarily in $\ell^2(\mathbb{Z}^d)$, as is the case in Theorem 4.6.2.

In general, it is easier to verify the injectivity of an operator than to prove its invertibility; therefore, Theorem 4.8.2 is a strong result. It has been applied successfully for the study of totally positive windows of Gaussian type in [23] and carries potential for further applications.

4.8.3 Gabor Frames Without Inequalities

This group of conditions holds for arbitrary lattices and windows in $M^1(\mathbb{R}^d)$. As is well known, the modulation space $M^1(\mathbb{R}^d)$ is a natural condition in many problems in time–frequency analysis, because it is invariant under the Fourier transform and many other transformations. By choosing a suitable norm, $M^1(\mathbb{R}^d)$ becomes a Banach space and its dual space $M^\infty(\mathbb{R}^d)$ consists of all tempered distributions $f \in \mathcal{S}'(\mathbb{R}^d)$ that satisfy

$$\sup_{z\in\mathbb{R}^{2d}} |V_\varphi f(z)| < \infty$$

for some (or equivalently, for all) Schwartz functions φ.

Theorem 4.8.3 ([19]) *Let $g \in M^1(\mathbb{R}^d)$ and $\Lambda \subseteq \mathbb{R}^{2d}$ be a lattice. Then the following are equivalent:*

(i) $\mathcal{G}(g, \Lambda)$ *is a frame for* $L^2(\mathbb{R}^d)$*, i.e.,* $S_{g,\Lambda}$ *is invertible on* $L^2(\mathbb{R}^d)$*.*
(ii) The frame operator $S_{g,\Lambda}$ *is one-to-one on* $M^\infty(\mathbb{R}^d)$*.*
(iii) The analysis operator $C_{g,\Lambda} : f \mapsto \big(\langle f, \pi(\lambda)g\rangle\big)_{\lambda\in\Lambda}$ *is one-to-one from* $M^\infty(\mathbb{R}^d)$ *to* $\ell^\infty(\Lambda)$*.*
(iv) The synthesis operator $D_{g,\Lambda^\circ} : c \mapsto \sum_{\lambda\in\Lambda^\circ} c_\lambda \pi(\lambda)g$ *is one-to-one from* $\ell^\infty(\Lambda^\circ)$ *to* $M^\infty(\mathbb{R}^d)$*.*
(v) The Gramian operator G_{g,Λ° *is one-to-one on* $\ell^\infty(\Lambda^\circ)$*.*

Conceptually, it seems easier to verify that an operator is one-to-one; therefore, one may hope that these conditions will become useful when research on Gabor frames will move from rectangular lattices toward arbitrary ones.

References

1. G. Ascensi, H.G. Feichtinger, N. Kaiblinger, Dilation of the Weyl symbol and Balian-Low theorem. Trans. Amer. Math. Soc. **366**(7), 3865–3880 (2014)
2. E.A. Azoff, Borel measurability in linear algebra. Proc. Amer. Math. Soc. **42**, 346–350 (1974)
3. B. Bekka, Square integrable representations, von Neumann algebras and an application to Gabor analysis. J. Fourier Anal. Appl. **10**(4), 325–349 (2004)
4. J.J. Benedetto, C. Heil, D.F. Walnut, Differentiation and the Balian–Low theorem. J. Fourier Anal. Appl. **1**(4), 355–402 (1995)
5. O. Christensen, An introduction to frames and Riesz bases. Appl. Numer. Harmon. Anal. 2 edn. Birkhäuser/Springer, Cham (2016)
6. O. Christensen, H.O. Kim, R.Y. Kim, Gabor windows supported on $[-1, 1]$ and compactly supported dual windows. Appl. Comput. Harmon. Anal. **28**(1), 89–103 (2010)
7. W. Czaja, A.M. Powell, Recent developments in the Balian-Low theorem, in *Harmonic Analysis and Applications*, Applications Numerical Harmonic Analysis, pp. 79–100. Birkhäuser Boston, Boston, MA (2006)
8. I. Daubechies, *Ten Lectures on Wavelets* (Society for Industrial and Applied Mathematics (SIAM), Philadelphia, PA, 1992)
9. I. Daubechies, A. Grossmann, Y. Meyer, Painless nonorthogonal expansions. J. Math. Phys. **27**(5), 1271–1283 (1986)
10. I. Daubechies, H.J. Landau, Z. Landau, Gabor time-frequency lattices and the Wexler-Raz identity. J. Fourier Anal. Appl. **1**(4), 437–478 (1995)
11. R.J. Duffin, A.C. Schaeffer, A class of nonharmonic Fourier series. Trans. Amer. Math. Soc. **72**, 341–366 (1952)
12. H.G. Feichtinger, K. Gröchenig, Banach spaces related to integrable group representations and their atomic decompositions. I. J. Funct. Anal. **86**(2), 307–340 (1989)
13. H.G. Feichtinger, K. Gröchenig, Gabor Wavelets and the Heisenberg group: gabor expansions and short time Fourier transform from the group theoretical point of view, in *Wavelets: A Tutorial in Theory and Applications*, ed. by C.K. Chui (Academic Press, Boston, MA, 1992), pp. 359–398
14. H.G. Feichtinger, N. Kaiblinger, Varying the time-frequency lattice of Gabor frames. Trans. Amer. Math. Soc. **356**(5), 2001–2023 (electronic) (2004)
15. H. G. Feichtinger, W. Kozek, Quantization of TF lattice-invariant operators on elementary LCA groups, in *Gabor Analysis and Algorithms*, Applications Numerical Harmonic Analysis, pp. 233–266. Birkhäuser Boston, Boston, MA (1998)

16. H.G. Feichtinger, F. Luef, Wiener amalgam spaces for the fundamental identity of Gabor analysis. Collect. Math. (Vol. Extra), 233–253 (2006)
17. D. Gabor, Theory of communication. J. IEE **93**(26), 429–457 (1946)
18. K. Gröchenig, *Foundations of Time-Frequency Analysis*, Applied and Numerical Harmonic Analysis (Birkhäuser Boston Inc, Boston, MA, 2001)
19. K. Gröchenig, Gabor frames without inequalities. Int. Math. Res. Not. IMRN (23), Article ID rnm111, 21 (2007)
20. K. Gröchenig, The mystery of Gabor frames. J. Fourier Anal. Appl. **20**(4), 865–895 (2014)
21. K. Gröchenig, A.J.E.M. Janssen, Letter to the editor: a new criterion for Gabor frames. J. Fourier Anal. Appl. **8**(5), 507–512 (2002)
22. K. Gröchenig, J. Ortega-Cerdà, J.L. Romero, Deformation of Gabor systems. Adv. Math. **277**, 388–425 (2015)
23. K. Gröchenig, J.L. Romero, J. Stöckler, Sampling theorems for shift-invariant spaces, Gabor frames, and totally positive functions. Invent. Math. **211**(3), 1119–1148 (2018)
24. K. Gröchenig, J. Stöckler, Gabor frames and totally positive functions. Duke Math. J. **162**(6), 1003–1031 (2013)
25. C. Heil, History and evolution of the density theorem for Gabor frames. J. Fourier Anal. Appl. **13**(2), 113–166 (2007)
26. M.S. Jakobsen, J. Lemvig, Density and duality theorems for regular Gabor frames. J. Funct. Anal. **270**(1), 229–263 (2016)
27. A.J.E.M. Janssen, Signal analytic proofs of two basic results on lattice expansions. Appl. Comput. Harmon. Anal. **1**(4), 350–354 (1994)
28. A.J.E.M. Janssen, Duality and biorthogonality for Weyl-Heisenberg frames. J. Fourier Anal. Appl. **1**(4), 403–436 (1995)
29. A.J.E.M. Janssen, The duality condition for Weyl-Heisenberg frames, in *Gabor Analysis and Algorithms*, Applications Numerical Harmonic Analysis, pp. 33–84. Birkhäuser Boston, Boston, MA (1998)
30. A.J.E.M. Janssen, On generating tight Gabor frames at critical density. J. Fourier Anal. Appl. **9**(2), 175–214 (2003)
31. A.J.E.M. Janssen, Zak transforms with few zeros and the tie, in *Advances in Gabor analysis*, Applications Numerical Harmonic Analysis, pp. 31–70. Birkhäuser Boston, Boston, MA (2003)
32. J. Lemvig, K. Haahr Nielsen, Counterexamples to the B-spline conjecture for Gabor frames. J. Fourier Anal. Appl. **22**(6), 1440–1451 (2016)
33. Y. Lyubarskii, P.G. Nes, Gabor frames with rational density. Appl. Comput. Harmon. Anal. **34**(3), 488–494 (2013)
34. Y.I. Lyubarskiĭ, Frames in the Bargmann space of entire functions, in *Entire and subharmonic functions*, vol. 11 of Adv. Soviet Math., pp. 167–180. Amer. Math. Soc., Providence, RI (1992)
35. J.V. Neumann, Mathematische Grundlagen der Quantenmechanik. Springer, Berlin, *English translation: "Mathematical Foundations of Quantum Mechanics,"* (Princeton Univ, Press, 1932), p. 1955
36. M.A. Rieffel, Projective modules over higher-dimensional noncommutative tori. Canad. J. Math. **40**(2), 257–338 (1988)
37. A. Ron, Z. Shen, Weyl-Heisenberg frames and Riesz bases in $L_2(\mathbf{R}^d)$. Duke Math. J. **89**(2), 237–282 (1997)
38. K. Seip, Density theorems for sampling and interpolation in the Bargmann-Fock space. I. J. Reine Angew. Math. **429**, 91–106 (1992)
39. E.M. Stein, G. Weiss, *Introduction to Fourier analysis on Euclidean spaces*. Princeton University Press, Princeton, N.J. (1971). Princeton Mathematical Series, No. 32
40. D.F. Walnut, Lattice size estimates for Gabor decompositions. Monatsh. Math. **115**(3), 245–256 (1993)
41. J. Wexler, S. Raz, Discrete Gabor expansions. Signal Process. **21**(3), 207–220 (1990)
42. Y.Y. Zeevi, M. Zibulski, M. Porat, Multi-window Gabor schemes in signal and image representations, in *Gabor Analysis and Algorithms*, Applications Numerical Harmonic Analysis, pp. 381–407. Birkhäuser Boston, Boston, MA (1998)
43. M. Zibulski, Y.Y. Zeevi, Analysis of multiwindow Gabor-type schemes by frame methods. Appl. Comput. Harmon. Anal. **4**(2), 188–221 (1997)

Chapter 5
On the Approximate Unit Distance Problem

Alex Iosevich

Abstract The Erdös unit distance conjecture in the plane says that the number of pairs of points from a point set of size n separated by a fixed (Euclidean) distance is $\leq C_\varepsilon n^{1+\varepsilon}$ for any $\varepsilon > 0$. The best known bound is $Cn^{\frac{4}{3}}$. We show that if the set under consideration is homogeneous, or, more generally, s-adaptable in the sense of [12], and the fixed distance is much smaller than the diameter of the set, then the exponent $\frac{4}{3}$ is significantly improved, even if we consider a small range of distances instead of a fixed value. Corresponding results are also established in higher dimensions. The results are obtained by solving the corresponding continuous problem and using a continuous-to-discrete conversion mechanism. The degree of sharpness of results is tested using the known results on the distribution of lattice points in dilates of convex domains.

5.1 Introduction

One of the hardest longstanding conjectures in extremal combinatorics is the Erdös unit distance conjecture ([2], see also [1]). It says that if P is a planar point set with n points, then the number of pairs of elements of P a fixed Euclidean distance apart is bounded by $C_\varepsilon n^{1+\varepsilon}$ for every $\varepsilon > 0$. The best known bound, obtained by Spencer, Szemeredi and Trotter [16], is $Cn^{\frac{4}{3}}$. An interesting development occurred in 2005 when Pavel Valtr [17] proved that if the Euclidean distance is replaced by a distance induced by the norm defined by a bounded convex set with a smooth boundary and nonvanishing curvature, then the $Cn^{\frac{4}{3}}$ bound is, in general, best possible. A much earlier construction due to Jarnik (see, e.g., [14] and the references contained therein) also yields the same bound, except that the boundary of the Jarnik domain is not smooth.

The purpose of this paper is to show that for a wide variety of point sets, the $Cn^{\frac{4}{3}}$ bound for the number of occurrences of the unit distance can be extended to the

---

A. Iosevich (✉)
Department of Mathematics, University of Rochester, Rochester, NY, USA
e-mail: iosevich@math.rochester.edu; iosevich@gmail.com

A. Aldroubi et al. (eds.), *New Trends in Applied Harmonic Analysis, Volume 2*, Applied and Numerical Harmonic Analysis,
https://doi.org/10.1007/978-3-030-32353-0_5

approximate setting and significantly improved if we count the number of pairs of points separated by a distance that is much smaller than the diameter of the set. Our main result is the following.

Definition 5.1.1 Given a finite set P of size n in $[0, 1]^d, d \geq 2$, we define the discrete s-energy of P by the relation

$$\mathcal{I}_s(P) = n^{-2} \sum_{p \neq p' \in P} |p - p'|^{-s}.$$

Theorem 5.1.2 *Let B be a symmetric bounded convex set in $\mathbb{R}^d$, $d \geq 2$, with a smooth boundary and everywhere nonvanishing Gaussian curvature. Let $P \subset [0, 1]^d$, $d \geq 2$, with $\#P = n$. Suppose that P is δ-separated with $n^{-\frac{2}{d+1}} \leq h \leq \delta \leq n^{-\frac{1}{d}}$. Then*

$$\#\left\{(p, p') \in P \times P : r \leq ||p - p'||_B \leq r + h\right\} \leq Cr^{\frac{d-1}{2}} n^2 h \left(C' + \mathcal{I}_{\frac{d+1}{2}}(P)\right).$$

Remark 5.1.3 Note that the condition $h \leq \delta \leq n^{-\frac{1}{d}}$ above simply comes from the fact that it is impossible make n points in $[0, 1]^d$ separated by more than $cn^{-\frac{1}{d}}$.

Definition 5.1.4 We say that $P \subset [0, 1]^d$ is homogeneous if there exist $C, c > 0$ such that the points of P are $cn^{-\frac{1}{d}}$-separated and every ball of radius $Cn^{-\frac{1}{d}}$ contains at least one point of P. Note that these conditions imply that $\#P \approx n$.

Corollary 5.1.5 *Suppose that $P \subset [0, 1]^d$ is homogeneous. Then*

$$\#\{(p, p') \in P \times P : r \leq |p - p'| \leq r + n^{-\frac{2}{d+1}}\} \leq Cr^{\frac{d-1}{2}} n^{2-\frac{2}{d+1}}.$$

In particular, when $d = 2$,

$$\#\{(p, p') \in P \times P : r \leq |p - p'| \leq r + n^{-\frac{2}{3}}\} \leq Cr^{\frac{1}{2}} n^{\frac{4}{3}}.$$

Corollary 5.1.5 follows from Theorem 5.1.2 and the following calculation that can be found in, for instance, [10, 12]. We shall carry out this calculation at the end of the paper for the sake of completeness.

Lemma 5.1.6 *Let $P \subset [0, 1]^d, d \geq 2$, be a homogeneous set. Then $\mathcal{I}_s(P)$ is bounded with constants independent of n for any $s \in [0, d)$.*

Remark 5.1.7 The discrete $\frac{d+1}{2}$-energy is bounded independently of n for a wide variety of point sets far beyond homogeneous sets. See [12] for a thorough description of these and related issues.

We now describe some consequences of our main result for well-distributed sets, which are those sets that are statistically like lattices but do not necessarily possess any of their arithmetic properties. More precisely, we have the following definition.

Definition 5.1.8 We say that $P \subset \mathbb{R}^d$ of size n is *well-distributed* if there exists $c > 0$ such that $|p - p'| \geq c$ and every unit lattice cube in $\mathbb{R}^d \cap [0, n^{\frac{1}{d}}]^d$ contains exactly one point of P.

While this is essentially a rescaled version of Definition 5.1.4, it provides a natural venue of comparison with lattice point counting problems that we are going to use to test the sharpness of the results of this paper in Sect. 5.1.1. The following result is just a rescaled version of Corollary 5.1.5.

Corollary 5.1.9 *Let B be a symmetric bounded convex set in $\mathbb{R}^d$, $d \geq 2$, with a smooth boundary and everywhere nonvanishing Gaussian curvature. Let P be a well-distributed set of size n. Then for $k \in (1, n^{\frac{1}{d}})$,*

$$\#\left\{(p, p') \in P \times P : k \leq ||p - p'||_B \leq k + n^{-\frac{d-1}{d(d+1)}}\right\} \leq Cn^{2-\frac{2}{d+1}} \cdot \Lambda, \quad (5.1.1)$$

where

$$\Lambda = \left(\frac{k}{n^{\frac{1}{d}}}\right)^{\frac{d-1}{2}}. \quad (5.1.2)$$

In particular, if $d = 2$, the left-hand side of (5.1.1) is bounded by $Cn^{\frac{4}{3}} \cdot \left(\frac{k}{n^{\frac{1}{2}}}\right)^{\frac{1}{2}}$, which is an improvement over the known $Cn^{\frac{4}{3}}$ bound when $k = o(n^{\frac{1}{2}})$.

Remark 5.1.10 It was pointed out to the author by Solymosi that if one considers the exact unit distance problem, Corollary 5.1.9 can be improved. More precisely, a standard cell decomposition and cutting technique can be used to show that

$$\#\left\{(p, p') \in P \times P : ||p - p'||_B = k\right\} \leq Cn^{\frac{4}{3}} \cdot \left(\frac{k}{n^{\frac{1}{2}}}\right)^{\frac{2}{3}},$$

and a corresponding result can be obtained in higher dimensions as well. Zahl pointed out that Solymosi's method with the proof of the two-dimensional case of Corollary 5.1.9 as input allows one to improve the conclusion of Corollary 5.1.9 slightly to the statement

$$\#\left\{(p, p') \in P \times P : k - n^{-\frac{1}{6}} \leq ||p - p'||_B \leq k + n^{-\frac{1}{6}}\right\} \leq Cn^{\frac{4}{3}} \cdot \left(\frac{k}{n^{\frac{1}{2}}}\right)^{\frac{2}{3}}.$$

The resulting connections between the cell decomposition method and analytic techniques employed in this article will be systematically studied in a sequel.

Remark 5.1.11 When $k \approx n^{\frac{1}{d}}$, Theorem 5.1.9 is implicit in the main result in [10], but the key feature here is the dependence on k with the resulting improvement when $k = o(n^{\frac{1}{d}})$. Also, we shall prove below that in the case $k \approx n^{\frac{1}{d}}$, the estimate provided by Theorem 5.1.9 is sharp. See also [13] where the continuous–discrete correspondence is used in reverse in order to obtain sharpness examples for Falconer-type estimates.

Remark 5.1.12 Note that the left-hand side of (5.1.1) is trivially bounded by $Cn \cdot k^{d-1}$. Therefore, the estimate in Theorem 5.1.9 is only interesting when $k >> n^{\frac{1}{d-1} - \frac{4}{(d-1)(d+1)} + \frac{1}{d(d-1)}}$. For example, in dimension two this threshold is $n^{\frac{1}{6}}$.

5.1.1 *Sharpness of Results*

The results associated with the lattice point counting problems provide a useful tool for testing sharpness of Theorem 5.1.9. Let $n \approx q^d$ and $P = \mathbb{Z}^d \cap B(\mathbf{0}, 10q)$, the ball of radius $10q$ centered at the origin. Let $N_d(R)$ denote the number of elements of $\mathbb{Z}^d$ inside the ball of radius R centered at the origin. It is known (see, e.g., [7]) that

$$N_d(R) = \omega_d R^d + D_d(R),$$

where ω_d is the volume of the unit ball, $|D_2(R)| \leq C_\varepsilon R^{\frac{131}{208}+\varepsilon}$ [8], $|D_3(R)| \leq C_\varepsilon R^{\frac{21}{16}+\varepsilon}$ [5], and $|D_d(R)| \leq C_\varepsilon R^{d-2+\varepsilon}$ for $d \geq 4$ [4].

Then

$$\#\left\{(p, p') \in P \times P : q \leq |p - p'| \leq q + q^{-\frac{d-1}{d+1}}\right\} \geq Cq^d \cdot \left(N_d\left(q + q^{-\frac{d-1}{d+1}}\right) - N_d(q)\right).$$

We have

$$N_d\left(q + q^{-\frac{d-1}{d+1}}\right) - N_d(q) = \omega_d\left(\left(q + q^{-\frac{d-1}{d+1}}\right)^d - q^d\right) + D\left(q + q^{-\frac{d-1}{d+1}}\right) - D(q).$$

Using the bounds on $|D(R)|$ described above, we see that

$$N_d\left(q + q^{-\frac{d-1}{d+1}}\right) - N_d(q) \geq cq^{d-1-\frac{d-1}{d+1}},$$

which implies that

$$\#\{(p, p') \in P \times P : q \leq |p - p'| \leq q + q^{-\frac{d-1}{d+1}}\} \geq Cq^{2d-1-\frac{d-1}{d+1}} \approx n^{2-\frac{2}{d+1}},$$

proving that Theorem 5.1.9 is sharp when $k \approx n^{\frac{1}{d}}$.

When $k << q$, we can conclude that

$$N_d\left(k + q^{-\frac{d-1}{d+1}}\right) - N_d(k) \geq ck^{d-1}q^{-\frac{d-1}{d+1}}$$

if the right-hand side is larger than the error term measured in terms of the bounds on $|D_d(R)|$ described above. This happens for a range of k's. In this range,

$$\#\{(p, p') \in P \times P : q \le |p - p'| \le k + q^{-\frac{d-1}{d+1}}\} \ge Ck^{d-1}q^d q^{-\frac{d-1}{d+1}}.$$

The right-hand side is smaller than the bound obtained by Theorem 5.1.9 when $k = o(n^{\frac{1}{d}})$. This may indicate that Theorem 5.1.9 is not sharp in this range, but it is also possible that a more sophisticated sharpness example may be found.

5.2 Proof of Theorem 5.1.2

Define

$$\mu_{n,\delta}(x) = n^{-1}\delta^{-d} \sum_{p \in P} \phi(\delta^{-1}(x - p)),$$

where ϕ is a smooth cutoff function supported in the ball of radius 2 and identically equal to 1 in the ball of radius 1.

This is a natural measure on the δ-neighborhood of P. Our goal is to bound the expression

$$\int\int_{\{(x,y):r \le ||x-y||_B \le r+h\}} d\mu_{n,\delta}(x) d\mu_{n,\delta}(y), \tag{5.2.1}$$

where $||\cdot||_B$ is the norm induced by a bounded symmetric convex set B with a smooth boundary and everywhere nonvanishing curvature, and then relate it to the count for the number of pairs separated by a given distance.

Using a Fourier inversion-type argument (see, e.g., [15] or [18]), the expression (5.2.1) equals a constant multiple of

$$\int |\widehat{\mu}_{n,\delta}(\xi)|^2 \widehat{\chi}_{A_{r,h}}(\xi) d\xi, \tag{5.2.2}$$

where

$$A_{r,h} = \{x \in \mathbb{R}^d : r \le ||x||_B \le r + h\},$$

χ denotes its indicator function, and

$$\widehat{f}(\xi) = \int e^{-2\pi i x \cdot \xi} f(x) dx,$$

defined for functions in $L^2(\mathbb{R}^d)$.

This argument is also carried out explicitly on pp. 59–60 in [15], but the argument is a bit simpler in this case because $\mu_{n,\delta}$ is a actually a smooth function. We have

$$\int\int_{\{(x,y):r \le ||x-y||_B \le r+h\}} d\mu_{n,\delta}(x) d\mu_{n,\delta}(y)$$

$$= \int \int \chi_{A_{r,h}}(x-y)d\mu_{n,\delta}(x)d\mu_{n,\delta}(y)$$

$$= \int \int \int \widehat{\chi}_{A_{r,h}}(\xi)e^{2\pi i(x-y)\cdot\xi}d\mu_{n,\delta}(x)d\mu_{n,\delta}(y)$$

$$= \int \widehat{\chi}_{A_{r,h}}(\xi)|\widehat{\mu}_{n,\delta}(\xi)|^2 d\xi$$

by Fourier inversion and the definition of the Fourier transform, this recovering (5.2.2).

Lemma 5.2.1 ([3]) *With the notation above,*

$$\left|\widehat{\chi}_{A_{r,h}}(\xi)\right| \leq Cr^{\frac{d-1}{2}}|\xi|^{-\frac{d-1}{2}} \min\left\{h, |\xi|^{-1}\right\}, \tag{5.2.3}$$

where C is a universal constant independent of t or q.

Falconer proved this result in [3] in the case when B is the unit ball. The proof of the general case is similar.

Theorem 5.2.2 (Theorem 3.10 in [15]) *Let μ be a compactly supported Borel measure on $\mathbb{R}^d$ and $0 < s < d$. Then*

$$\int \int |x-y|^{-s}d\mu(x)d\mu(y) = c_{d,s}\int |\widehat{\mu}(\xi)|^2|\xi|^{-d+s}d\xi.$$

With Lemma 5.2.1 in tow, we see that

$$\int |\widehat{\mu}_{n,\delta}(\xi)|^2\widehat{\chi}_{A_{r,h}}(\xi)d\xi \leq Cr^{\frac{d-1}{2}}\cdot h\int |\widehat{\mu}_{n,\delta}(\xi)|^2|\xi|^{-\frac{d-1}{2}}d\xi \tag{5.2.4}$$

$$\leq C'r^{\frac{d-1}{2}}h\int |\widehat{\mu}_{n,\delta}(\xi)|^2|\xi|^{-d+\frac{d+1}{2}}d\xi = C''r^{\frac{d-1}{2}}h\int\int |x-y|^{-\frac{d+1}{2}}d\mu_{n,\delta}(x)d\mu_{n,\delta}(y),$$

where the last step follows by Theorem 5.2.2. We have

$$\int\int |x-y|^{-\frac{d+1}{2}}d\mu_{n,\delta}(x)d\mu_{n,\delta}(y)$$

$$= n^{-2}\delta^{-2d}\sum_{p,p'\in P}\int\int \phi(\delta^{-1}(x-p))\phi(\delta^{-1}(y-p'))|x-y|^{-\frac{d+1}{2}}dxdy$$

$$= n^{-2}\delta^{-2d}\sum_{p\in P}\int\int \phi(\delta^{-1}(x-p))\phi(\delta^{-1}(y-p))|x-y|^{-\frac{d+1}{2}}dxdy$$

$$+n^{-2}\delta^{-2d}\sum_{p\neq p'}\int\int \phi(\delta^{-1}(x-p))\phi(\delta^{-1}(y-p'))|x-y|^{-\frac{d+1}{2}}dxdy = I + II.$$

By a direct calculation,

$$I \leq Cn^{-1}h^{-\frac{d+1}{2}} \leq C' \tag{5.2.5}$$

since $\delta \geq n^{-\frac{2}{d+1}}$ by assumption.

Since $p \neq p'$,

$$II \leq Cn^{-2} \sum_{p \neq p'} |p - p'|^{-\frac{d+1}{2}} = C\mathcal{I}_{\frac{d+1}{2}}(P). \tag{5.2.6}$$

We are now ready for the combinatorial conclusion. See [6, 9, 11, 12] where various forms of the continuous-to-discrete conversion mechanisms are developed and applied. Observe that

$$\#\{(p, p') \in P \times P : r \leq ||x - y||_B \leq r + h\}$$

$$\leq Cn^2 \cdot \int \int_{\{(x,y): r \leq ||x-y||_B \leq r+h\}} d\mu_{n,\delta}(x) d\mu_{n,\delta}(y).$$

By (5.2.4), (5.2.5), and (5.2.6), this expression is bounded by

$$Cr^{\frac{d-1}{2}} n^2 h \left(C' + \mathcal{I}_{\frac{d+1}{2}}(P)\right),$$

as claimed.

5.2.1 *Proof of Lemma 5.1.6*

We have

$$\mathcal{I}_s(P) = n^{-2} \sum_{p \neq p' \in P} |p - p'|^{-s} = n^{-2} \sum_{k=0}^{N} \sum_{2^{-k} \leq |p-p'| < 2^{-k+1}} |p - p'|^{-s},$$

where $N \approx \log(n)$. The right-hand side is bounded by

$$Cn^{-2} \sum_{k=0}^{N} 2^{ks} \sum_{2^{-k} \leq |p-p'| < 2^{-k+1}} 1. \tag{5.2.7}$$

By assumption, for any fixed p',

$$\#\{p : 2^{-k} \leq |p - p'| < 2^{-k+1}\} \leq C2^{-kd} n.$$

It follows that the expression in (5.2.7) is bounded by

$$Cn^{-2}\sum_{k=0}^{N}2^{ks}2^{-kd}n^2 = C\sum_{k=0}^{N}2^{k(s-d)}$$

and the series converges with bounds independent of n if $s < d$, as desired. This completes the proof of Lemma 5.1.6.

Acknowledgements The author wishes to thank Adam Sheffer, Josef Solymosi, Josh Zahl, and the anonymous referee for some helpful remarks and suggestions.

References

1. P. Brass, W.O. Moser, J. Pach, *Research Problems in Discrete Geometry* (Springer, 2005)
2. P. Erdös, On sets of distances of n points Amer. Math. Mon. **53**, 248–250 (1946)
3. K.J. Falconer, On the Hausdorff dimensions of distance sets. Mathematika **32**, 206–212 (1986)
4. F. Fricker, *Einfuhrung die Gitterpunktlehre* (Birkhauser, Verlag, 1982)
5. D.R. Heath-Brown, *Lattice Points in the Sphere, Number Theory in Progress*, vol. 2 (Zakopane-Kocielisko, 1997) (de Gruyter, Berlin, 1999), pp. 883–892
6. S. Hofmann, A. Iosevich, Circular averages and Falconer/Erdös distance conjecture in the plane for random metrics Proc. Amer. Mat. Soc. **133**, 133–144 (2005)
7. M.N. Huxley, *Area, Lattice Points, and Exponential Sums, London Mathematical Society Monographs New Series*, vol. 13 (Oxford University, Press, 1996)
8. M.N. Huxley, Exponential sums and lattice points. III Proc. London Math. Soc. **87**(3), 591–609 (2003)
9. A. Iosevich, *Fourier analysis and geometric combinatorics*. Top. Math. Anal. Ser. Anal. Appl. Comput., **3**, in World Scientific, Proceedings of the Padova Lectures in Analysis in 2004 and 2005 (2008)
10. A. Iosevich, H. Jorati, I. Laba, Geometric incidence theorems via Fourier analysis. Trans. Amer. Math. Soc. **361**(12), 6595–6611 (2009)
11. A. Iosevich, I. Łaba, K-distance sets, Falconer conjecture, and discrete analogs
12. A. Iosevich, M. Rudnev, I. Uriarte-Tuero, Theory of dimension for large discrete sets and applications. Math. Model. Nat. Phenom. **9**(5), 148–169 (2014)
13. A. Iosevich, S. Senger, Sharpness of Falconer's $\frac{d+1}{2}$ estimate. Ann. Acad. Sci. Fenn. Math. **41**(2), 713–720 (2016)
14. E. Landau, *Vorlesungen über Zahlentheorie*, vol. I, Part 2, II, III, (German, Chelsea Publing Co., New York, 1969)
15. P. Mattila, *Fourier Analysis and Hausdorff dimension*. Cambridge Studies in Advanced Mathematics, vol. 150 (Cambridge University Press, 2016)
16. J. Spencer, E. Szemerédi, W.T. Trotter. *Unit distances in the Euclidean plane*, in *Graph Theory and Combinatorics*, ed. by B. Bollobás (Academic Press, New York, NY, 1984), pp. 293–303
17. P. Valtr, Strictly convex norms allowing many unit distances and related touching questions, manuscript (2005)
18. T. Wolff, *Lectures on harmonic analysis*, ed. by I. Laba and C. Shubin, vol. 29. University Lecture Series (American Mathematical Society Providence, RI 2003)

Chapter 6
Hausdorff Dimension, Projections, Intersections, and Besicovitch Sets

Pertti Mattila

Abstract This is a survey on recent developments on the Hausdorff dimension of projections and intersections for general subsets of Euclidean spaces, with an emphasis on estimates of the Hausdorff dimension of exceptional sets and on restricted projection families. We shall also discuss relations between projections and Hausdorff dimension of Besicovitch sets.

Subject Classification: 28A75

6.1 Introduction

In this survey, I shall discuss some recent results on integral-geometric properties of Hausdorff dimension and their relations to Kakeya-type problems. More precisely, by integral-geometric properties, I mean properties related to affine subspaces of Euclidean spaces and to rigid motions; orthogonal projections into planes, intersections with planes, and intersections of two sets after a generic rigid motion is applied to one of them. Such questions have been studied for more than 60 years, and there have been a lot of recent activities on them. In particular, I shall discuss estimates on the Hausdorff dimension of exceptional sets of planes and rigid motions, and projections on restricted families of planes. Besicovitch sets are sets of Lebesgue measure zero containing a unit line segment in every direction. They are expected to have full Hausdorff dimension. This problem is related to many topics in modern Fourier analysis, e.g., restriction of the Fourier transform to surfaces, Bochner–Riesz multipliers, local smoothing for PDEs, and L^2-estimates for Dirichlet sums, see Sect. 22.5 of [55]. It is also related to projection theorems, as we shall see at the end of this

The author was supported by the Academy of Finland through the Finnish Center of Excellence in Analysis and Dynamics Research.

P. Mattila (✉)
Department of Mathematics and Statistics, University of Helsinki, Helsinki, Finland
e-mail: pertti.mattila@helsinki.fi

A. Aldroubi et al. (eds.), *New Trends in Applied Harmonic Analysis, Volume 2*, Applied and Numerical Harmonic Analysis,
https://doi.org/10.1007/978-3-030-32353-0_6

survey. In the last section, I shall also discuss (n, k) Besicovitch sets, lines replaced with k-planes, and their relations to projections.

Other recent surveys partially overlapping with this are [21, 45, 54, 78].

Most of the background material can be found, for example, in the books [53, 55].

This survey is partially based on the lectures I gave in the CIMPA2017 conference in Buenos Aires in August 2017. I would like to thank Ursula Molter, Carlos Cabrelli, and the other organizers for that very pleasant and successful event. I am grateful to Tuomas Orponen for many useful comments.

6.2 Hausdorff Dimension, Energy Integrals, and the Fourier Transform

I give here a quick review of the Hausdorff dimension and its relations to energy integrals and the Fourier transform. The details can be found in [53, 55].

The s-dimensional *Hausdorff measure* $\mathcal{H}^s$, $s \geq 0$, is defined for $A \subset \mathbb{R}^n$ by

$$\mathcal{H}^s(A) = \lim_{\delta \to 0} \mathcal{H}^s_\delta(A),$$

where, for $0 < \delta \leq \infty$,

$$\mathcal{H}^s_\delta(A) = \inf\{\sum_{j=1}^{\infty} d(E_j)^s : A \subset \bigcup_{j=1}^{\infty} E_j, d(E_j) < \delta\}.$$

Here $d(E)$ denotes the diameter of the set E.

Then $\mathcal{H}^n$ is a constant multiple of the Lebesgue measure $\mathcal{L}^n$ and the restriction of $\mathcal{H}^{n-1}$ to the unit sphere $S^{n-1} = \{x \in \mathbb{R}^n : |x| = 1\}$ is a constant multiple of the surface measure.

The *Hausdorff dimension* of A is

$$\dim A = \inf\{s : \mathcal{H}^s(A) = 0\} = \sup\{s : \mathcal{H}^s(A) = \infty\}.$$

For $A \subset \mathbb{R}^n$, let $\mathcal{M}(A)$ be the set of Borel measures μ such that $0 < \mu(A) < \infty$ and μ has compact support $\operatorname{spt}\mu \subset A$. We denote by $B(x, r)$ the closed ball with center x and radius r. The following is a useful tool for proving lower bounds for the Hausdorff dimension.

Theorem 6.2.1 (Frostman's lemma) *Let* $0 \leq s \leq n$. *For a Borel set* $A \subset \mathbb{R}^n$, $\mathcal{H}^s(A) > 0$ *if and only there is* $\mu \in \mathcal{M}(A)$ *such that*

$$\mu(B(x, r)) \leq r^s \quad \textit{for all } x \in \mathbb{R}^n, r > 0. \tag{6.2.1}$$

In particular,

$$\dim A = \sup\{s : there is\ \mu \in \mathcal{M}(A)\ such\ that\ (6.2.1)\ holds\}.$$

Such measures μ are often called Frostman measures.
The *s-energy*, $s > 0$, of a Borel measure μ is

$$I_s(\mu) = \iint |x - y|^{-s}\, d\mu x\, d\mu y = \int k_s * \mu\, d\mu,$$

where k_s is the *Riesz kernel*:

$$k_s(x) = |x|^{-s}, \quad x \in \mathbb{R}^n.$$

Integration of Frostman's lemma gives the following theorem.

Theorem 6.2.2 *For a Borel set $A \subset \mathbb{R}^n$,*

$$\dim A = \sup\{s : there\ is\ \mu \in \mathcal{M}(A)\ such\ that\ I_s(\mu) < \infty\}.$$

The *Fourier transform* of $\mu \in \mathcal{M}(\mathbb{R}^n)$ is

$$\widehat{\mu}(\xi) = \int e^{-2\pi i \xi \cdot x}\, d\mu x, \quad \xi \in \mathbb{R}^n.$$

The s-energy of $\mu \in \mathcal{M}(\mathbb{R}^n)$ can be written in terms of the Fourier transform:

$$I_s(\mu) = c(n, s) \int |\widehat{\mu}(x)|^2 |x|^{s-n}\, dx.$$

This comes from Plancherel's theorem and the fact that the Fourier transform, in the distributional sense, of k_s is a constant multiple of k_{n-s}. Thus we have

$$\dim A = \sup\{s < n : \exists \mu \in \mathcal{M}(A) \text{ such that } \int |\widehat{\mu}(x)|^2 |x|^{s-n}\, dx < \infty\}. \quad (6.2.2)$$

Notice that if $I_s(\mu) < \infty$, then $|\widehat{\mu}(x)|^2 < |x|^{-s}$ for most x with large norm. However, this need not hold for all x with large norm.

The upper Minkowski dimension is defined by

$$\dim_M A = \inf\{s \geq 0 : \lim_{\delta \to 0} \delta^{s-n} \mathcal{L}^n(\{x : dist(x, A) < \delta\}) = 0\}.$$

The packing dimension $\dim_P$ can be defined as a modification of this:

$$\dim_P A = \inf\{\sup_i \dim_M A_i : A = \bigcup_{i=1}^{\infty} A_i\}.$$

Then $\dim A \le \dim_P A \le \dim_M A$. We have the following product inequalities:

$$\dim A \times B \ge \dim A + \dim B. \tag{6.2.3}$$

$$\dim_M A \times B \le \dim_M A + \dim_M B. \tag{6.2.4}$$

$$\dim_P A \times B \le \dim_P A + \dim_P B. \tag{6.2.5}$$

There is no Fubini theorem for Hausdorff measures, but we have the following inequality, see [26], 2.10.25. Federer proves this in rather general metric spaces. An easy argument in the Euclidean spaces in the case where $s = m$, and so $\mathcal{H}^m = \mathcal{L}^m$, is given in [53], 7.7.

Proposition 6.2.3 *Let $A \subset \mathbb{R}^{m+n}$ and set $A_x = \{y \in \mathbb{R}^n : (x, y) \in A\}$ for $x \in \mathbb{R}^m$. Then for any nonnegative numbers s and t ($\int^*$ is the upper integral)*

$$\int^* \mathcal{H}^t(A_x)\, d\mathcal{H}^s x \le C(m, n, s, t)\mathcal{H}^{s+t}(A).$$

In particular, if $\dim\{x \in \mathbb{R}^m : \dim A_x \ge t\} \ge s$, *then* $\dim A \ge s + t$.

The latter statement was proved by Marstrand [48].

6.3 Hausdorff Dimension and Exceptional Projections

We shall now discuss the question: how do orthogonal projections affect the Hausdorff dimension? Let $0 < m < n$ be integers and let $G(n, m)$ be the space of all linear m-dimensional subspaces of $\mathbb{R}^n$ and let $\gamma_{n,m}$ be the Borel probability measure on it which is invariant under the orthogonal group $O(n)$ of $\mathbb{R}^n$. For $V \in G(n, m)$ let $P_V : \mathbb{R}^n \to V$ be the orthogonal projection.

The case $m = 1$ and the lines through the origin are simpler and more concrete, and perhaps good to keep in mind. We can parametrize $G(n, 1)$ and the projections onto lines by the unit sphere:

$$P_e(x) = e \cdot x, \quad x \in \mathbb{R}^n, e \in S^{n-1}.$$

Here is the basic projection theorem for the Hausdorff dimension. The first two items of it were proved by Marstrand [47] in 1954 and the third by Falconer and O'Neil [23] in 1999 and by Peres and Schlag [76] in 2000.

Theorem 6.3.1 *Let $A \subset \mathbb{R}^n$ be a Borel set.*

(1) If $\dim A \le m$, *then*

$$\dim P_V(A) = \dim A \quad \textit{for } \gamma_{n,m} \textit{ almost all } V \in G(n, m).$$

(2) If $\dim A > m$*, then*

$$\mathcal{L}^m(P_V(A)) > 0 \quad \textit{for } \gamma_{n,m} \textit{ almost all } V \in G(n,m).$$

(3) If $\dim A > 2m$*, then* $P_V(A)$ *has non-empty interior for* $\gamma_{n,m}$ *almost all* $V \in G(n,m)$.

Proof We only prove this for $m = 1$, the general case can be found in [55]. For $\mu \in \mathcal{M}(A)$, let $\mu_e \in \mathcal{M}(P_e(A))$ be the push-forward of μ under P_e: $\mu_e(B) = \mu(P_e^{-1}(B))$.

To prove (1) let $0 < s < \dim A$ and choose by Theorem 6.2.2 a measure $\mu \in \mathcal{M}(A)$ such that $I_s(\mu) < \infty$. Then

$$\begin{aligned}\int_{S^{n-1}} I_s(\mu_e)\,de &= \int_{S^{n-1}} \iint |P_e(x-y)|^{-s}\,d\mu x\,d\mu y\,de \\ &= \iiint_{S^{n-1}} |P_e(\tfrac{x-y}{|x-y|})|^{-s}\,de|x-y|^{-s}\,d\mu x\,d\mu y = c(s)I_s(\mu) < \infty,\end{aligned}$$

where for $v \in S^{n-1}$, $c(s) = \int_{S^{n-1}} |P_e(v)|^{-s}\,de < \infty$ as $s < 1$. The finiteness of this integral follows from the simple inequality

$$\mathcal{H}^{n-1}(\{e \in S^{n-1} : |P_e(x)| \le \delta\}) \lesssim \delta/|x| \quad \text{for } x \in \mathbb{R}^n \setminus \{0\}, \delta > 0. \tag{6.3.1}$$

Referring again to Theorem 6.2.2, we see that $\dim P_e(A) \ge s$ for almost all $e \in S^{n-1}$. By the arbitrariness of s, $0 < s < \dim A$, we obtain $\dim P_e(A) \ge \dim A$ for almost all $e \in S^{n-1}$. The opposite inequality follows from the fact that the projections are Lipschitz mappings.

To prove (2) choose by (6.2.2) a measure $\mu \in \mathcal{M}(A)$ such that $\int |x|^{1-n}|\widehat{\mu}(x)|^2\,dx < \infty$. Directly from the definition of the Fourier transform, we see that $\widehat{\mu_e}(t) = \widehat{\mu}(te)$ for $t \in \mathbb{R}$, $e \in S^{n-1}$. Integrating in polar coordinates, we obtain

$$\int_{S^{n-1}} \int_{-\infty}^{\infty} |\widehat{\mu_e}(t)|^2\,dt\,de = 2\int_{S^{n-1}} \int_0^{\infty} |\widehat{\mu}(te)|^2\,dt\,de = 2\int |x|^{1-n}|\widehat{\mu}(x)|^2\,dx < \infty.$$

Thus for almost all $e \in S^{n-1}$, $\widehat{\mu_e} \in L^2(\mathbb{R})$ which means that μ_e is absolutely continuous with L^2 density and hence $\mathcal{L}^1(p_e(A)) > 0$.

For the proof of (3), one takes $2 < s < \dim A$ and $\mu \in \mathcal{M}(A)$ such that $I_s(\mu) < \infty$, whence $\int |x|^{s-n}|\widehat{\mu}(x)|^2\,dx < \infty$. Then as above and by the Schwartz inequality

$$\begin{aligned}&\int_{S^{n-1}} \int_{|t|\ge 1} |\widehat{\mu_e}(t)|\,dt\,de = 2\int_{|x|\ge 1} |x|^{1-n}|\widehat{\mu}(x)|\,dx \\ &\le 2\left(\int_{|x|\ge 1} |x|^{2-s-n}\,dx \int_{|x|\ge 1} |x|^{s-n}|\widehat{\mu}(x)|^2\,dx\right)^{1/2} < \infty\end{aligned}$$

since $2 - s - n < -n$. Thus for almost all $e \in S^{n-1}$, $\widehat{\mu_e} \in L^1(\mathbb{R})$ which implies that μ_e is absolutely continuous with continuous density. Hence, $P_e(A)$ has non-empty interior.

Part (2) can rather easily be proven also without the Fourier transform using again inequalities like (6.3.1), see the proof of Theorem 9.7 in [53]. Parts (1) and (2) of Theorem 6.3.1 hold with $\gamma_{n,m}$ replaced with any Borel measure γ on $G(n, m)$ which satisfies

$$\gamma(\{V \in G(n, m) : |P_V(x)| \leq \delta\}) \lesssim (\delta/|x|)^m \quad \text{for } x \in \mathbb{R}^n \setminus \{0\}, \delta > 0.$$

We shall discuss this a bit more later. I do not know any proof for (3) without the Fourier transform.

The conditions $\dim A \leq m$ and $\dim A > m$ in (1) and (2) are of course necessary. The condition $\dim A > 2m$ in (3) is necessary if $m = 1$. I do not know if it is necessary when $m > 1$. In the case $m = 1$, the example in the plane can be obtained with Besicovitch sets, first in the plane, showing that there is no theorem in the plane, and then taking Cartesian products. More precisely, let $B \subset \mathbb{R}^2$ be a Borel set of measure zero which contains a line in every direction. We shall construct such sets in Sect. 6.7. Let $A = \mathbb{R}^2 \setminus \cup_{q \in \mathbb{Q}^2}(B + q)$, where $\mathbb{Q}^2$ is the countable dense set with rational coordinates. Then A has full Lebesgue measure and none of its projections has interior points.

In this section, we shall discuss how much more one can say about the size of the sets of exceptional planes. Kaufman [41] proved in 1968 the first item of the following theorem in the plane (generalized in [50]), Falconer [17] in 1982 the second, and Peres and Schlag [76] in 2000 the third. Recall that the dimension of $G(n, m)$ is $m(n - m)$. To get a better feeling of this, notice that in the case $m = 1$ the three upper bounds are $n - 2 + \dim A$, $n - \dim A$ and $n + 1 - \dim A$.

Theorem 6.3.2 *Let $A \subset \mathbb{R}^n$ be a Borel set.*

(1) If $\dim A \leq m$, then

$$\dim\{V \in G(n, m) : \dim P_V(A) < \dim A\} \leq m(n - m) - m + \dim A.$$

(2) If $\dim A > m$, then

$$\dim\{V \in G(n, m) : \mathcal{L}^m(P_V(A)) = 0\} \leq m(n - m) + m - \dim A.$$

(3) If $\dim A > 2m$, then

$$\dim\{V \in G(n, m) : Int(P_V(A)) = \emptyset\} \leq m(n - m) + 2m - \dim A.$$

The proof of (1) is a rather simple modification of the proof of the corresponding part in Theorem 6.3.1; essentially, one just replaces the measure $\gamma_{n.m}$ with a Frostman measure ν on the exceptional set. The key observation is that instead of (6.3.1) we now have

$$\nu(\{V \in G(n,m) : |P_V(x)| \leq \delta\}) \lesssim (\delta/|x|)^{s-m(n-m-1)} \tag{6.3.2}$$

which easily follows from the Frostman condition $\nu(B(V,r)) \leq r^s$, cf. [55], (5.10) and (5.12). The proofs of (2) and (3) are trickier and require the use of the Fourier transform. They can be found in [55].

Theorem 6.3.2 and much more, for instance, exceptional set estimates for Bernoulli convolutions, is included in the setting of generalized projections developed by Peres and Schlag [76]. Later, these general estimates have been improved in many special cases.

The bounds in (1) and (2) are sharp by the examples which Kaufman and I constructed in 1975 in [42]. I do not know if the bound in (3) is sharp. Another, seemingly very difficult, problem is estimating the dimension of the set in (1) when $\dim A$ is replaced by some $u < \dim A$. We still have by the same proof

$$\dim\{V \in G(n,m) : \dim P_V(A) < u\} \leq m(n-m) - m + u,$$

but this probably is not sharp when $u < \dim A$. In any case, it is far from sharp in the plane when $u = \dim A/2$:

Theorem 6.3.3 *Let $A \subset \mathbb{R}^2$ be a Borel set. Then*

$$\dim\{e \in S^1 : \dim P_e(A) \leq \dim A/2\} = 0. \tag{6.3.3}$$

To get some idea where $\dim A/2$ comes from, notice that the inequality $\dim_M P_e(A) < \dim_M A/2$ is very easy for the upper Minkowski dimension (and also for the packing dimension), and even more is true: there can be at most one direction e for which $\dim_M P_e(A) < \dim_M A/2$. That there cannot be two orthogonal directions follows immediately from the product inequalities (6.2.4) and (6.2.5), and the general case is also easy. However, for the Hausdorff dimension, the exceptional set can always be uncountable, even more: Orponen constructed in [69], Theorem 1.5, a compact set $A \subset \mathbb{R}^2$ such that $\mathcal{H}^1(A) > 0$ and $\dim\{e \in S^1 : \dim P_e(A) = 0\}$ is a dense G_δ subset of S^1. That paper also contains many exceptional set estimates for projections and packing dimension.

Theorem 6.3.3 is due to Bourgain [9, 10]. Bourgain's result is more general and it includes a deep discretized version. The proof uses methods of additive combinatorics. D. M. Oberlin gave a simpler Fourier-analytic proof in [62], but with $\dim P_e(A) \leq \dim A/2$ replaced by $\dim P_e(A) < \dim A/2$. Using combinatorial methods, He [28] proved analogous higher dimensional results.

More generally, it might be true and has been conjectured by Oberlin [62] that Kaufman's estimate

$$\dim\{e \in S^1 : \dim P_e(A) < u\} \leq u \tag{6.3.4}$$

could be extended for $\dim A/2 \leq u \leq \dim A$ to

$$\dim\{e \in S^1 : \dim P_e(A) < u\} \leq 2u - \dim A. \tag{6.3.5}$$

This would be sharp, as the constructions in [42] show. Theorem 6.3.3 is the only case where this is known. However, Orponen improved in the plane theorem in 6.3.2 in [73] for sets A with $\dim A = 1$ but with $\dim P_e(A)$ replaced by the packing dimension of $P_e(A)$: for $0 < t < 1$, there is $\varepsilon(t) > 0$ such that

$$\dim\{e \in S^1 : \dim_P P_e(A) < t\} \leq t - \varepsilon(t). \tag{6.3.6}$$

The following generalization of parts (1) and (2) of Theorem 6.3.1 tells us that a null set of projections can be found first and then the statements hold outside these exceptions for all subsets of positive measure. Statement (2) is due to Marstrand [47]. It means that the push-forward under P_V of the restriction of $\mathcal{H}^s$ to A is absolutely continuous for almost all $V \in G(n,m)$, recall the proof of Theorem 6.3.1(2). Part (1) was proved by Falconer and the author in [22].

Theorem 6.3.4 *Let $A \subset \mathbb{R}^n$ be an $\mathcal{H}^s$-measurable set with $0 < \mathcal{H}^s(A) < \infty$. Then there exists a Borel set $E \subset G(n,m)$ with $\gamma_{n,m}(E) = 0$ such that for all $V \in G(n,m) \setminus E$ and all $\mathcal{H}^s$-measurable sets $B \subset A$ with $\mathcal{H}^s(B) > 0$,*

(1) if $s \leq m$ then $\dim P_V(B) = s$,
(2) if $s > m$ then $\mathcal{L}^m(P_V(B)) > 0$.

The sharper version in the spirit of Theorem 6.3.2 is also valid, see [22].

In the next section, the following theorem will give us information about exceptional plane slices. It was proved by Orponen and the author in [57].

Theorem 6.3.5 *Let A and B be Borel subsets of $\mathbb{R}^n$.*

(1) If $\dim A > m$ and $\dim B > m$, then

$$\gamma_{n,m}\left(\{V \in G(n,m) : \mathcal{L}^m(P_V(A) \cap P_V(B)) > 0\}\right) > 0.$$

(2) If $\dim A > 2m$ and $\dim B > 2m$, then

$$\gamma_{n,m}(\{V \in G(n,m) : Int(P_V(A) \cap P_V(B)) \neq \emptyset\}) > 0.$$

(3) If $\dim A > m$, $\dim B \leq m$, and $\dim A + \dim B > 2m$, then for every $\varepsilon > 0$,

$$\gamma_{n,m}(\{V \in G(n,m) : \dim(P_V(A) \cap P_V(B)) > \dim B - \varepsilon\}) > 0.$$

Proof I only prove (1) when $m = 1$. Choose by (6.2.2) $\mu \in \mathcal{M}(A)$ and $\nu \in \mathcal{M}(B)$ such that $\int |x|^{1-n}|\widehat{\mu}(x)|^2\,dx < \infty$ and $\int |x|^{1-n}|\widehat{\nu}(x)|^2\,dx < \infty$. Let again $\mu_e \in \mathcal{M}(P_e(A))$ and $\nu_e \in \mathcal{M}(P_e(B))$ be the push-forwards of μ and ν under P_e. We know from the proof of Theorem 6.3.1 that for almost all $e \in S^{n-1}$, μ_e and ν_e are absolutely continuous with L^2 densities. Thus as in the proof of Theorem 6.3.1 and by Plancherel's theorem,

$$\iint \mu_e(t)\nu_e(t)\,dt\,de = \iint \widehat{\mu_e}(t)\overline{\widehat{\nu_e}(t)}\,dt\,de = \iint \widehat{\mu}(te)\overline{\widehat{\nu}(te)}\,dt\,de$$
$$= c(n)\int_{\mathbb{R}^n} |x|^{1-n}\widehat{\mu}(x)\overline{\widehat{\nu}(x)}\,dx = c(n,m)\iint |x-y|^{-1}\,d\mu x\,d\nu x > 0.$$

Hence $\int \mu_e(t)\nu_e(t)\,dt > 0$ for positively many e. As $\mu_e\nu_e$ has support in $P_e(A) \cap P_e(B)$, the claim follows.

For other recent projection results, see [3–5, 11, 12].

There are many recent results on projections of various special, for example, self-similar, classes of sets and measures. I shall not discuss them here but [21, 78] give good overviews.

6.4 Restricted Families of Projections

Here we discuss the question: what kind of projection theorems can we get if the whole Grassmannian $G(n,m)$ is replaced by some lower dimensional subset G? A very simple example is the one where $G \subset G(3,1)$ corresponds to a circle in a two-dimensional plane in $\mathbb{R}^3$. For example, we can consider the projections π_θ onto the lines $\{t(\cos\theta, \sin\theta, 0) : t \in \mathbb{R}\}, \theta \in [0,\pi]$. Since $\pi_\theta(A) = \pi_\theta((\pi(A))$ where $\pi(x,y,z) = (x,y)$, and $\dim A \leq \dim \pi(A) + 1$, it is easy to conclude using Marstrand's projection Theorem 6.3.1 that for any Borel set $A \subset \mathbb{R}^3$, for almost all $\theta \in [0,\pi]$,

$$\dim \pi_\theta(A) \geq \dim A - 1 \quad \text{if } \dim A \leq 2,$$
$$\mathcal{L}^1(\pi_\theta(A)) > 0 \quad \text{if } \dim A > 2.$$

This is sharp by trivial examples; consider product sets $A = B \times C$, $B \subset \mathbb{R}^2$, $C \subset \mathbb{R}$. So we only have an essentially trivial result. The situation changes dramatically if we consider the projections p_θ onto the lines $\{t(\cos\theta, \sin\theta, 1) : t \in \mathbb{R}\}$. Then the trivial counterexamples do not work anymore and one can now improve the above estimates. The method used for the proof of Theorem 6.3.1 easily gives that if $A \subset \mathbb{R}^3$ is a Borel set with $\dim A \leq 1/2$, then

$$\dim p_\theta(A) \geq \dim A \quad \text{for almost all } \theta \in [0,\pi].$$

The restriction $1/2$ comes from the fact that instead of (6.3.1) we now have only

$$\mathcal{L}^1(\{\theta : |p_\theta(x)| \leq \delta\}) \lesssim \sqrt{\delta/|x|}. \tag{6.4.1}$$

For $\dim A > 1/2$ this becomes much more difficult. Anyway, we have the following theorem.

Theorem 6.4.1 *Let p_θ and q_θ be the orthogonal projections onto the line $\{t(\cos\theta, \sin\theta, 1) : t \in \mathbb{R}\}, \theta \in [0, \pi]$, and its orthogonal complement. Let $A \subset \mathbb{R}^3$ be a Borel set.*

(1) If $\dim A \leq 1$, then $\dim p_\theta(A) = \dim A$ for almost all $\theta \in [0, \pi]$.
(2) If $\dim A \leq 3/2$, then $\dim q_\theta(A) = \dim A$ for almost all $\theta \in [0, \pi]$.

Käenmäki, Orponen and Venieri proved (1) in [36] and Orponen and Venieri (2) in [75]. They related this problem to circle packing problems and methods of Wolff from [84].

So (1) is the sharp analogue of the corresponding part of Marstrand's projection theorem for these projections. Perhaps, (2) is not sharp in the sense that it might hold with 2 in place of 3/2.

One reason for the possibility of such improvements over the first family of projections considered above, the π_θ, is that the second family, the p_θ, is more curved than the first one. That is, the set of the unit vectors generating the first family is the planar curve $\{(\cos\theta, \sin\theta, 0) : \theta \in [0, \pi]\}$, while for the second it spans the whole space $\mathbb{R}^3$. More precisely, the curve $\gamma(\theta) = (\cos\theta, \sin\theta, 1)/\sqrt{2} \in S^2, \theta \in [0, \pi]$, of the corresponding unit vectors satisfies the curvature condition that for every $\theta \in [0, \pi]$ the vectors $\gamma(\theta), \gamma'(\theta), \gamma''(\theta)$ span the whole space $\mathbb{R}^3$. Partial results were proven earlier by Fässler and Orponen [25, 68] and Oberlin and Oberlin [65] for general C^2 curves on S^2 satisfying this curvature condition. Fässler and Orponen conjectured that the full Marstrand theorem as in Theorem 6.4.1 (with 3/2 replaced by 2) should hold for them.

As we have seen above, if $\rho_e : \mathbb{R}^3 \to \mathbb{R}, e \in S^2$, is a family of linear mappings and σ is a Borel measure on S^2 satisfying

$$\sigma(\{e : |\rho_e(x)| \leq \delta\}) \lesssim \delta/|x|,$$

then the Marstrand statement $\dim \rho_\theta(A) = \min\{\dim A, 1\}$ holds for σ almost all $e \in S^2$. However, such inequality is usually false for less than two-dimensional measures σ. Nevertheless Chen constructed in [12] for all $1 < s < 2$ s-dimensional Ahlfors–David regular random measures for which it holds, and hence also the Marstrand theorem. He had also many other related results in that paper.

Next,we consider projection families in higher dimensions. I state a more general result below but let us start with

$$\pi_t : \mathbb{R}^4 \to \mathbb{R}^2, \pi_t(x, y) = x + ty, x, y \in \mathbb{R}^2, t \in \mathbb{R}.$$

This family is closely connected with Besicovitch sets and the Kakeya conjecture in $\mathbb{R}^3$, as we shall later see. The following theorem is due to Oberlin [63]. It is not explicitly stated there but follows from the proof of Theorem 1.3.

Theorem 6.4.2 *Let $A \subset \mathbb{R}^4$ be a Borel set.*

(1) If $\dim A \leq 3$, then $\dim \pi_t(A) \geq \dim A - 1$ for almost all $t \in \mathbb{R}$.

(2) *If* $\dim A > 3$, *then* $\mathcal{L}^2(\pi_t(A)) > 0$ *for almost all* $t \in \mathbb{R}$.

The bounds here are sharp when $\dim A \geq 2$. To see this let $0 \leq s \leq 1$, $C_s \subset \mathbb{R}$ with $\dim C_s = s$, and $A_s = \{(x, y) \in \mathbb{R}^2 \times \mathbb{R}^2 : x_1 \in C_s, y_1 = 0\}$. Then $\dim A_s = 2 + s$, $\pi_t(A_s) = C_s \times \mathbb{R}$ and $\dim \pi_t(A_s) = 1 + s$. This shows that (1) is sharp. For (2) we can choose C_1 with $\mathcal{L}^1(C_1) = 0$, then $\mathcal{L}^2(\pi_t(A)) = 0$. These bounds are not sharp for all A since we have $\dim \pi_t(A) = \dim A$ for almost all $t \in \mathbb{R}$ if $\dim A \leq 1$. Restricting t to some interval $[c, C]$, $0 < c < C < \infty$, this follows as before from the inequality

$$\mathcal{L}^1(\{t \in [c, C] : |\pi_t(x, y)| \leq \delta\}) \lesssim \delta/|(x, y)|,$$

which is easy to check. If $1 \leq \dim A \leq 2$ we can only say that $\dim \pi_t(A) \geq 1$ for almost all $t \in \mathbb{R}$ since $\pi_t(\mathbb{R} \times \{0\} \times \mathbb{R} \times \{0\}) = \mathbb{R}$.

I give a sketch of the proof of Theorem 6.4.2. Let $\mu \in \mathcal{M}(A)$ with

$$\mu(B(x, r)) \leq r^s \quad \text{for } x \in \mathbb{R}^4, r > 0, \tag{6.4.2}$$

for some $0 < s < 4$. Let $\mu_t \in \mathcal{M}(\pi_t(A))$ be the push-forward of μ under π_t. Then for $\xi \in \mathbb{R}^n$,

$$\widehat{\mu_t}(\xi) = \int e^{-2\pi i \xi \cdot \pi_t(x,y)}\, d\mu(x, y) = \int e^{-2\pi i (\xi, t\xi)\cdot(x,y)}\, d\mu(x, y) = \widehat{\mu}(\xi, t\xi).$$

It is enough to consider t in some fixed bounded interval J. Oberlin proved that for $R > 0$,

$$\int_J \int_{R \leq |\xi| \leq 2R} |\widehat{\mu}(\xi, t\xi)|^2\, d\xi\, dt \lesssim R^{4-s-1}. \tag{6.4.3}$$

This is applied to the dyadic annuli, $R = 2^k, k = 1, 2, \dots$. The sum converges if $s > 3$, and we can choose μ with such s if $\dim A > 3$. This gives $\int_J \int |\widehat{\mu_t}(\xi)|^2\, d\xi\, dt < \infty$ and yields part (2). To prove part (1) let $0 < u < s < \dim A$ and μ as above. Then (6.4.3) yields

$$\int_J \int |\widehat{\mu_t}(\xi)|^2 |\xi|^{u-1-2}\, d\xi\, dt < \infty,$$

so $\dim \pi_t(A) \geq u - 1$ for almost all $t \in J$ and thus $\dim \pi_t(A) \geq \dim A - 1$ for almost all $t \in \mathbb{R}$ by the arbitrariness of J and u.

Let us formulate (6.4.3) as a more general lemma (a special case of Lemma 3.1 in [63]).

Lemma 6.4.3 *Let k and m be positive integers and $N = (k + 1)m$ and let Q be a cube in $\mathbb{R}^k$. Define*

$$T_t\xi = (t_1\xi, \dots, t_k\xi) \in \mathbb{R}^{km} \quad \textit{for } \xi \in \mathbb{R}^m, t \in \mathbb{R}^k.$$

If $\mu \in \mathcal{M}(\mathbb{R}^N)$ with $\mu(B(x, r)) \leq r^s$ for $x \in \mathbb{R}^N$, $r > 0$ for some $0 < s < n$, then

$$\int_Q \int_{R \le |\xi| \le 2R} |\widehat{\mu}(\xi, T_t\xi)|^2 \, d\xi \, dt \lesssim R^{N-s-k}. \tag{6.4.4}$$

We obtain (6.4.3) from this with $k = 1, m = 2$.

In Lemma 3.1 of [63], there is an additional assumption (3.1). This is now trivial: it is applied with λ equal to the Lebesgue measure on Q. See the proof of Theorem 1.3 in [63] for the identification of our Lemma 6.4.3 as a special case of Lemma 3.1 of [63].

To prove Lemma 6.4.3, choose a smooth function g with compact support which equals 1 on the support of μ. Then $\widehat{g\mu} = \widehat{g} * \widehat{\mu}$ and the integral in (6.4.4) equals

$$\int_Q \int_{R \le |\xi| \le 2R} |\widehat{g\mu}(\xi, T_t\xi)|^2 \, d\xi \, dt = \int_Q \int_{R \le |\xi| \le 2R} \left| \int \widehat{g}((\xi, T_t\xi) - y)\widehat{\mu}(y) \, dy \right|^2 \, d\xi \, dt.$$

This can be estimated by standard arguments. When $|y|$ is large as compared to R, $|\widehat{g}((\xi, T_t\xi) - y)|$ is small by the fast decay $\widehat{g}$. For $|y| \lesssim R$ one uses

$$\int_{|y| \le CR} |\widehat{\mu}(y)|^2 \, dy \lesssim R^{s-N},$$

which follows from the assumption $\mu(B(x, r)) \le r^s$, cf. also [55], Sect. 3.8. Of course, I am skipping several technical details here, see [63].

We now formulate a more general version of the above projection theorem. Let k and m be positive integers and $N = (k + 1)m$. Above we had $k = 1, m = 2$. Write

$$x = (x_0^1, \ldots, x_0^m, x_1^1, \ldots, x_1^m, \ldots, x_k^1, \ldots, x_k^m) \in \mathbb{R}^N, t = (t_1, \ldots, t_k) \in \mathbb{R}^k.$$

Consider the linear mappings

$$\begin{aligned}\pi_t : \mathbb{R}^N \to \mathbb{R}^m, \pi_t(x) &= (x_0^1 + \sum_{j=1}^k t_j x_j^1, \ldots, x_0^m + \sum_{j=1}^k t_j x_j^m) \\ &= (x_0^1 + t \cdot x^1, \ldots, x_0^m + t \cdot x^m) = x_0 + t \cdot \tilde{x},\end{aligned}$$

where $x_0 = (x_0^1, \ldots, x_0^m), x^l = (x_1^l, \ldots, x_k^l)$ and $t \cdot \tilde{x} = (t \cdot x^1, \ldots, t \cdot x^m) \in \mathbb{R}^m$. Then for $\mu \in \mathcal{M}(\mathbb{R}^N)$ the push-forward μ_t of μ under π_t has the Fourier transform for $\xi \in \mathbb{R}^m$,

$$\widehat{\mu_t}(\xi) = \int e^{-2\pi i \xi \cdot \pi_t(x)} \, d\mu x = \int e^{-2\pi i (\xi \cdot x_0 + \xi \cdot (t \cdot \tilde{x}))} \, d\mu x = \widehat{\mu}(\xi, T_t\xi),$$

where again $T_t\xi = (t_1\xi, \ldots, t_k\xi) \in \mathbb{R}^{km}$. Lemma 6.4.3 now yields

$$\int_Q \int_{R \le |\xi| \le 2R} |\widehat{\mu_t}(\xi)|^2 \, d\xi \, dt \lesssim R^{N-s-k}, \tag{6.4.5}$$

where $\mu \in \mathcal{M}(\mathbb{R}^N)$ with $\mu(B(x,r)) \leq r^s$ for $x \in \mathbb{R}^N, r > 0$, for some $0 < s < n$. By a similar argument as for Theorem 6.4.2, this leads to the following theorem.

Theorem 6.4.4 *Let $A \subset \mathbb{R}^N$ be a Borel set.*

(1) If $\dim A \leq N - k$, *then* $\dim \pi_t(A) \geq \dim A - k(m-1)$ *for almost all* $t \in \mathbb{R}^k$.
(2) If $\dim A > N - k$, *then* $\mathcal{L}^m(\pi_t(A)) > 0$ *for almost all* $t \in \mathbb{R}^k$.

Part (2) is again sharp. To see this, let A consist of the points $(x_0^1, \dots, x_0^m, x_1^1, \dots, x_1^m, \dots, x_k^1, \dots, x_k^m) \in \mathbb{R}^N$ for which $x_0^1 \in C$, where C has dimension 1 and measure zero, and $x_1^1 = \dots = x_k^1 = 0$. Part (1) is sharp when $m = 1$, but then $k = N - 1$ and the standard Marstrand's projection theorem also applies. It also is sharp, for example, when $m = 2$ for any k with a similar example as in the case $k = 1, m = 2$.

The study of restricted families of projections was started by Järvenpää et al. [35]. This work was continued and generalized by the Järvenpääs and Keleti [34], where they proved sharp inequalities for general smooth nondegenerate families of orthogonal projections onto m-planes in $\mathbb{R}^n$. Now the trivial examples such as $\{t(\cos\theta, \sin\theta, 0) : t \in \mathbb{R}\}, \theta \in [0, \pi]$, are also included, so the bounds are necessarily weaker than in the above special cases. Restricted families appear quite naturally in Heisenberg groups, see [1, 2, 24]. Another motivation for studying them comes from the work of E. Järvenpää, M. Järvenpää, and Ledrappier and their co-workers on measures invariant under geodesic flows on manifolds, see [30, 31].

6.5 Plane Sections and Radial Projections

What can we say about the dimensions if we intersect a subset A of $\mathbb{R}^n$, $\dim A > m$, with $(n-m)$-dimensional planes? Using Proposition 6.2.3 we have for any $V \in G(n, n-m)$,

$$\dim(A \cap (V + x)) \leq \dim A - m \quad \text{for } \mathcal{H}^m \text{ almost all } x \in V^\perp,$$

and for any $x \in \mathbb{R}^n$ (see [50] or [57]),

$$\dim(A \cap (V + x)) \leq \dim A - m \quad \text{for } \gamma_{n,n-m} \text{ almost all } V \in G(n, n-m).$$

The lower bounds are not as obvious, but we have the following result, originally proved by Marstrand in the plane in [47] and then in general dimensions in [50].

Theorem 6.5.1 *Let* $m < s \leq n$ *and let* $A \subset \mathbb{R}^n$ *be* $\mathcal{H}^s$ *measurable with* $0 < \mathcal{H}^s(A) < \infty$. *Then*

(1) For $\mathcal{H}^s$ *almost all* $x \in A$, $\dim(A \cap (V + x)) = s - m$ *for* $\gamma_{n,n-m}$ *almost all* $V \in G(n, n-m)$,

(2) for $\gamma_{n,n-m}$ almost all $V \in G(n, n-m)$,

$$\mathcal{H}^m(\{x \in V^\perp : \dim(A \cap (V + x)) = s - m\}) > 0.$$

These statements are essentially equivalent. Clearly, this generalizes part (2) of Theorem 6.3.1. Now we give exceptional set estimates related to both statements. The first of these is due to Orponen [67].

Theorem 6.5.2 *Let $m < s \leq n$ and let $A \subset \mathbb{R}^n$ be $\mathcal{H}^s$ measurable with $0 < \mathcal{H}^s(A) < \infty$. Then there is a Borel set $E \subset G(n, n-m)$ such that $\dim E \leq m(n-m) + m - s$ and for $V \in G(n, n-m) \setminus E$,*

$$\mathcal{H}^m(\{x \in V^\perp : \dim(A \cap (V + x)) = s - m\}) > 0.$$

The bound $m(n-m) + m - s = \dim G(n, n-m) + m - s$ is the same as in Theorem 6.3.2(2). Since it is sharp there, it also is sharp here.

The second estimate is due to Orponen and the author [57]:

Theorem 6.5.3 *Let $m < s \leq n$ and let $A \subset \mathbb{R}^n$ be $\mathcal{H}^s$ measurable with $0 < \mathcal{H}^s(A) < \infty$. Then there is a Borel set $B \subset \mathbb{R}^n$ such that $\dim B \leq m$ and for $x \in \mathbb{R}^n \setminus B$,*

$$\gamma_{n,n-m}(\{V \in G(n, n-m) : \dim A \cap (V + x) = s - m\}) > 0.$$

This probably is not sharp. I expect that the sharp bound for $\dim B$ in the case $m = n - 1$ would again be $2(n-1) - s$, as for the orthogonal projections and as in Orponen's radial projection Theorem 6.5.4. Moreover, one could hope for an exceptional set estimate including both cases, that is, estimate on the dimension of the exceptional pairs (x, V).

I give a sketch of the proof of Theorem 6.5.3 in the plane. Suppose that it is not true and that there is a set B with $\dim B > 1$ such that through the points of B almost all lines meet A in a set of dimension less than $s - 1$. On the other hand, by Theorem 6.5.1 typical lines through the points of A meet A in a set of dimension $s - 1$. By Fubini-type arguments and using Theorem 6.3.5, we can find such typical lines meeting both A and B leading to a contradiction.

Here, we investigated the dimensions of the intersections of our set with lines through a point. But if we only want to know whether these lines meet the set, we are studying radial projections. For these more can be said. For $x \in \mathbb{R}^n$ define

$$\pi_x : \mathbb{R}^n \setminus \{x\} \to S^{n-1}, \quad \pi_x(y) = \frac{y - x}{|y - x|}.$$

Then by the standard proofs the statements of Marstrand's projection theorem are valid for almost all $x \in \mathbb{R}^n$. Orponen proved in [72, 74] the following sharp estimate for the exceptional set of $x \in \mathbb{R}^n$.

Theorem 6.5.4 *Let $A \subset \mathbb{R}^n$ be a Borel set with $\dim A > n-1$. Then there is a Borel set $B \subset \mathbb{R}^n$ with $\dim B \leq 2(n-1) - \dim A$ such that for every $x \in \mathbb{R}^n \setminus B$, $\mathcal{H}^{n-1}(\pi_x(A)) > 0$. Moreover, if $\mu \in \mathcal{M}(\mathbb{R}^n)$ and $I_s(\mu) < \infty$ for some $n-1 < s < n$, then the push-forward of μ under π_x is absolutely continuous with respect to $\mathcal{H}^{n-1}|S^{n-1}$ for x outside a set of Hausdorff dimension $2(n-1) - s$.*

Orponen proved in [74] also the following rather surprising result. The proof is tricky and technical with a flavor of combinatorial geometry.

Theorem 6.5.5 *Let $A \subset \mathbb{R}^2$ be a Borel set with $\dim A > 0$. Then the set*

$$\{x \in \mathbb{R}^2 : \dim \pi_x(A) < \dim A/2\}$$

has Hausdorff dimension 0 or it is contained in a line.

Obviously, the second alternative is needed, since if A is contained in a line, the above set is the same line.

6.6 General Intersections

The following theorem was proved in [52].

Theorem 6.6.1 *Let s and t be positive numbers with $s + t > n$ and $t > (n+1)/2$. Let A and B be Borel subsets of $\mathbb{R}^n$ with $\mathcal{H}^s(A) > 0$ and $\mathcal{H}^t(B) > 0$. Then for almost all $g \in O(n)$,*

$$\mathcal{L}^n(\{z \in \mathbb{R}^n : \dim A \cap (g(B) + z) \geq s + t - n\}) > 0. \tag{6.6.1}$$

The condition $t > (n+1)/2$ comes from some Fourier transform estimates. Probably, it is not needed.

This was preceded by the papers of Kahane [37] and the author [51] in which it was shown that the above theorem is valid for any $s + t > n$ provided larger transformation groups are used. For example, it suffices to add also typical dilations $x \mapsto rx, r > 0$.

Here we really need the inequality $\dim A \cap (g(B) + z) \geq s + t - n$, the opposite inequality can fail very badly: for any $0 \leq s \leq n$ there exists a Borel set $A \subset \mathbb{R}^n$ such that $\dim A \cap f(A) = s$ for all similarity maps f of $\mathbb{R}^n$. This follows from [19]. The reverse inequality holds if $\dim A \times B = \dim A + \dim B$, see [53], Theorem 13.12. This latter condition is valid if, for example, one of the sets is Ahlfors–David regular, see [53], 8.12. For such reverse inequalities, no rotations g are needed (or, equivalently, they hold for every g).

The following two exceptional set estimates were proven in [56].

Theorem 6.6.2 *Let s and t be positive numbers with $s + t > n + 1$. Let A and B be Borel subsets of $\mathbb{R}^n$ with $\mathcal{H}^s(A) > 0$ and $\mathcal{H}^t(B) > 0$. Then there is a Borel set $E \subset O(n)$ such that*

$$\dim E \leq 2n - s - t + (n-1)(n-2)/2 = n(n-1)/2 - (s + t - (n+1))$$

and for $g \in O(n) \setminus E$,

$$\mathcal{L}^n(\{z \in \mathbb{R}^n : \dim A \cap (g(B) + z) \geq s + t - n\}) > 0. \tag{6.6.2}$$

Notice that $n(n-1)/2$ is the dimension of $O(n)$. The condition $s + t > n + 1$ is not needed in the case where one of the sets has small dimension and in this case we have a better upper bound for $\dim E$, although we then need a slight technical reformulation.

Theorem 6.6.3 *Let A and B be Borel subsets of $\mathbb{R}^n$ with $\dim A = s, \dim B = t$ and suppose that $s \leq (n-1)/2$. If $0 < u < s + t - n$, then there is a Borel set $E \subset O(n)$ with*

$$\dim E \leq n(n-1)/2 - (s + t - n)$$

such that for $g \in O(n) \setminus E$,

$$\mathcal{L}^n(\{z \in \mathbb{R}^n : \dim A \cap (g(B) + z) \geq u\}) > 0. \tag{6.6.3}$$

The formulation in [56] is slightly weaker, but it easily implies the above. What helps here is the following sharp decay estimate for quadratic spherical averages for Fourier transforms of measures with finite energy:

$$\int_{|v|=1} |\widehat{\mu}(rv)|^2 \, dv \leq C(n,s) I_s(\mu) r^{-s}, \quad r > 0, \ 0 < s \leq (n-1)/2.$$

Such an estimate is false for $s > (n-1)/2$. There are sharp estimates in the plane by Wolff [85], and good, but perhaps not sharp, estimates in higher dimensions by Erdoğan [15]. More precisely, for $s \geq n/2$ and $\varepsilon > 0$,

$$\int_{|v|=1} |\widehat{\mu}(rv)|^2 \, dv \leq C(n,s) I_s(\mu) r^{\varepsilon - (n+2s-2)/4}, \quad r > 0. \tag{6.6.4}$$

This is very useful for distance sets, as discussed below, but gives very little for the intersections. The proof uses restriction and Kakeya methods and results. In particular, the case $n \geq 3$ relies on Tao's bilinear restriction theorem. These are discussed in [55].

Let us speculate about the possible sharp estimates in the plane. In Theorem 6.6.2, we have the upper bound $4 - (s + t)$ and in Theorem 6.6.3 we have $3 - (s + t)$. Could the second estimate be valid whenever $s + t > 2$? This would mean that the dimension is 0 when $s + t > 3$. Could the exceptional set even be countable then? I do not think so, but I do not have a counterexample. Anyway, it need not be empty whatever the dimensions are. That is, using only translations we cannot say much for general sets. The following example follows from [51], or see [43] for having $A = B$: there are compact subsets A and B of $\mathbb{R}^n$ such that $\dim A = \dim B = n$ and $A \cap (B + z)$ contains at most one point for every $z \in \mathbb{R}^n$.

A problem related to both projections and intersections is the distance set problem. For $A \subset \mathbb{R}^n$, define the distance set

$$D(A) = \{|x - y| : x, y \in A\} \subset [0, \infty).$$

The following Falconer's conjecture seems plausible.

Conjecture 6.6.4 If $n \geq 2$ and $A \subset \mathbb{R}^n$ is a Borel set with $\dim A > n/2$, then $\mathcal{L}^1(D(A)) > 0$, or even $\mathrm{Int}(D(A)) \neq \emptyset$.

Falconer [20] proved in 1985 that $\dim A > (n+1)/2$ implies $\mathcal{L}^1(D(A)) > 0$, and we also have then $\mathrm{Int}(D(A)) \neq \emptyset$ by Sjölin and myself [58]. Here appears the same bound $(n+1)/2$ as for the intersections, and for the same reason. In both cases for a measure μ with finite s-energy estimates for the measures of the narrow annuli, $\mu(\{y : r < |x - y| < r + \delta\})$, for μ typical centers x are useful. They are rather easily derived with the help of the Fourier transform if $s \geq (n+1)/2$.

The best known result is due to Wolff [85] for $n = 2$ and to Erdoğan [15] for $n \geq 3$.

Theorem 6.6.5 *If $n \geq 2$ and $A \subset \mathbb{R}^n$ is a Borel set with* $\dim A > n/2 + 1/3$, *then* $\mathcal{L}^1(D(A)) > 0$.

The proof is based on the estimate (6.6.4).

The relation to projections appears when we look at the pinned distance sets:

$$D_x(A) = \{|x - y| : y \in A\} \subset [0, \infty), \quad x \in \mathbb{R}^n.$$

Peres and Schlag proved in [76] that these too have positive Lebesgue measure for many x provided $\dim A > (n+1)/2$. We can think of $D_x(A)$ as the image of A under the projection-type mapping $y \mapsto |x - y|$.

Various partial results on distance sets have recently been proved, among others, by Iosevich and Liu [32, 33], Lucá and Rogers [46], Orponen [71] and Shmerkin [80, 81].

6.7 Besicovitch and Furstenberg Sets

We say that a set in $\mathbb{R}^n$, $n \geq 2$, is a *Besicovitch set*, or a Kakeya set, if it has zero Lebesgue measure and it contains a line segment of unit length in every direction. This means that for every $e \in S^{n-1}$ there is $b \in \mathbb{R}^n$ such that $\{te + b : 0 < t < 1\} \subset B$. It is not obvious that Besicovitch sets exist but they do in every $\mathbb{R}^n$, $n \geq 2$.

Theorem 6.7.1 *For any $n \geq 2$ there exists a Borel set $B \subset \mathbb{R}^n$ such that $\mathcal{L}^n(B) = 0$ and B contains a whole line in every direction. Moreover, there exist compact Besicovitch sets in $\mathbb{R}^n$.*

Proof It is enough to prove this in the plane, then $B \times \mathbb{R}^{n-2}$ is fine in $\mathbb{R}^n$. We shall use projections and duality between points and lines. More precisely, parametrize the lines, except those parallel to the y-axis, by $(a, b) \in \mathbb{R}^2$:

$$l(a, b) = \{(x, a + bx) : x \in \mathbb{R}\}.$$

Then if $C \subset \mathbb{R}^2$ is some parameter set and $B = \cup_{(a,b)\in C} l(a, b)$, one checks that

$$B \cap \{(t, y) : y \in \mathbb{R}\} = \{t\} \times \pi_t(C)$$

where

$$\pi_t : \mathbb{R}^2 \to \mathbb{R}^2, \quad \pi_t(a, b) = a + tb,$$

is essentially an orthogonal projection. Suppose that we can find C such that $\pi(C) = [0, 1]$, where $\pi(a, b) = b$, and $\mathcal{L}^1(\pi_t(C)) = 0$ for almost all t. Then $\mathcal{L}^2(B) = 0$ by Fubini's theorem and taking the union of four rotated copies of B gives the desired set. It is not trivial that such sets C exist but they do. For example, a suitably rotated copy of the product of a standard Cantor set with dissection ratio $1/4$ with itself is such, cf., for example, [55], Chap. 10. Restricting x above to a compact subinterval of $\mathbb{R}$ yields a compact Besicovitch set.

The idea to construct Besicovitch sets using duality between lines and points is due to Besicovitch from 1964 in [6], although he gave a geometric construction already in 1919. It was further developed by Falconer [18]. We shall see more of this below.

Conjecture 6.7.2 (Kakeya conjecture) All Besicovitch sets in $\mathbb{R}^n$ have Hausdorff dimension n.

The Kakeya conjecture is open for $n \geq 3$. I shall discuss partial results later, but let us first see how it follows in the plane and how it is related to projection theorems. The following theorem was proved by Davies [13].

Theorem 6.7.3 *For every Besicovitch set $B \subset \mathbb{R}^n$,* $\dim B \geq 2$. *In particular, the Kakeya conjecture is true in the plane.*

The proof of this is, up to some technicalities, reversing the above argument for the proof of Theorem 6.7.1 and using Marstrand's projection Theorem 6.3.1(1), see the proof of Theorem 6.7.4. But let us now look more generally relations between projection theorems and lower bounds for the Hausdorff dimension of Besicovitch sets.

We can parametrize the lines in $\mathbb{R}^n$, except those orthogonal to the x_1-axis, by $(a, b) \in \mathbb{R}^{n-1} \times \mathbb{R}^{n-1}$:

$$l(a, b) = \{(x, a + bx) : x \in \mathbb{R}\}.$$

Then again if $C \subset \mathbb{R}^{2(n-1)}$ is parameter set and $B = \cup_{(a,b)\in C} l(a, b)$ we have for $t \in \mathbb{R}$,

$$B \cap \{(t, y) : y \in \mathbb{R}^{n-1}\} = \{t\} \times \pi_t(C)$$

where

$$\pi_t : \mathbb{R}^{2(n-1)} \to \mathbb{R}^{n-1}, \quad \pi_t(a, b) = a + tb, \quad t \in \mathbb{R}.$$

These are projections of Chap. 4 with $k = 1, m = n - 1$. Suppose now that $\pi(C) = [0, 1]^{n-1}$, where $\pi(a, b) = b$. Then in particular, $\dim C \geq n - 1$. The projection theorem we would need to solve the Kakeya conjecture should tell us that $\dim \pi_t(C) = n - 1$ for almost all $t \in \mathbb{R}$. Then we could conclude by Proposition 6.2.3 that $\dim B = n$. In the plane such projection theorem is true; it is just Marstrand's projection theorem. However, in higher dimensions we do not know of any such projection theorem since we now only have a one-dimensional family of projections. Notice that the space of all orthogonal projections from $\mathbb{R}^{2(n-1)}$ onto $(n - 1)$-planes is $(n - 1)^2$-dimensional. More precisely, we can state the following theorem.

Theorem 6.7.4 *Let $0 < s \leq n - 1$ and $\pi(x, y) = y$ for $x, y \in \mathbb{R}^{n-1}$. Suppose that the following projection theorem holds: For every Borel set $C \subset \mathbb{R}^{2(n-1)}$ with $\mathcal{H}^{n-1}(\pi(C)) > 0$, we have $\dim \pi_t(C) \geq s$ for almost all $t \in \mathbb{R}$. Then for every Besicovitch set $B \subset \mathbb{R}^n$, we have $\dim B \geq s + 1$. In particular, if this projection theorem holds for $s = n - 1$, the Kakeya conjecture is true.*

Proof We may assume that B is a G_δ-set, since any set in $\mathbb{R}^{n-1}$ is contained in a G_δ-set with the same dimension. For $a \in \mathbb{R}^{n-1}$, $b \in [0, 1]^{n-1}$ and $q \in \mathbb{Q}$ denote by $I(a, b, q)$ the line segment $\{(q + t, a + bt) : 0 \leq t \leq 1/2\}$ of length less than 1. Let C_q be the set of (a, b) such that $I(a, b, q) \subset B$. Then each C_q is a G_δ-set, because for any open set G the set of (a, b) such that $I(a, b, q) \subset G$ is open. Since for every $b \in [0, 1]^{n-1}$ some $I(a, b, q) \subset B$, we have $\pi(\cup_{q\in\mathbb{Q}} C_q) = [0, 1]^{n-1}$, so there is $q \in \mathbb{Q}$ for which $\mathcal{H}^{n-1}(\pi(C_q)) > 0$. Then by our assumption, for almost all $t \in \mathbb{R}$, $\dim \pi_t(C_q) \geq s$. We now have for $0 \leq t \leq 1/2$,

$$\{q + t\} \times \pi_t(C_q) = \{(q + t, a + bt) : (a, b) \in C_q\} \subset B \cap \{(x, y) : x = q + t\}.$$

Hence, for a positive measure set of t, vertical t-sections of B have dimension at least s. By Proposition 6.2.3 we obtain that $\dim B \geq s + 1$.

Let us try to apply Oberlin's projection Theorem 6.4.4 together with Theorem 6.7.4. We have to apply it in $\mathbb{R}^{2(n-1)}$ with $k = 1, m = n - 1$. We have $\dim C \geq n - 1$, so we get $\dim \pi_t(C) \geq n-1-(n-2)=1$, thus yielding the lower bound 2 for the Hausdorff dimension of Besicovitch sets. But this also follows by Theorem 6.7.3, and by other methods, see [55]. Unfortunately, no known method seems to give any better projection theorem for the family π_t. From $\mathcal{H}^{n-1}(C)>0$, we could only hope to get $\dim \pi_t(C) \geq (n-1)/2$, at least when n is odd. To see this let $p=(n-1)/2$ and $C=\{(a,b)\in \mathbb{R}^{n-1}\times\mathbb{R}^{n-1} : a_1=\cdots=a_p=b_1=\cdots=b_p=0\}$. Then $\mathcal{H}^{n-1}(C)=\infty$ and $\pi_t(C)=\{x\in\mathbb{R}^{n-1} : x_1=\cdots=x_p=0\}$, so $\dim \pi_t(C)=(n-1)/2$. Even if this estimate were true it would only give the lower bound $(n+1)/2$ for the dimension of Besicovitch sets. This has been known since the 1980s by different methods, see [55], Sect. 23.4. The only hope for better estimates via projections would seem to be that instead of only using the information $\mathcal{H}^{n-1}(C)>0$ we should use that C has positive measure projection on the second factor of $\mathbb{R}^{n-1}\times\mathbb{R}^{n-1}$ Often having one big projection does not help much. However, Fässler and Orponen were able to make use of that in [25], and since we are dealing with a very special family of mappings maybe it could help here too. Moreover, in the known cases the generic dimension of the projections agrees with the largest one.

Yu proved in [88] that the Kakeya conjecture is equivalent to the following: for any Besicovitch set $B\subset\mathbb{R}^n$ and for any $0<m<n$, $\dim P_V(B)$ is constant for $V\in G(n,m)$. The idea is simple but clever: lift your Besicovitch set B from $\mathbb{R}^n$ to $\mathbb{R}^{2n-1}$ in the way it projects back to $\mathbb{R}^n$ as B and it projects to some n-dimensional subspace of $\mathbb{R}^{2n-1}$ as a Besicovitch set where all the defining lines go through the origin. Then this latter projection has positive n-dimensional measure.

So the Kakeya conjecture is true in the plane and open in higher dimensions. The following results give the best known lower bounds for the Hausdorff dimension of Besicovitch sets.

Wolff, based on some earlier work of Bourgain, proved in [83].

Theorem 6.7.5 *The Hausdorff dimension of every Besicovitch set in $\mathbb{R}^n$ is at least $(n+2)/2$.*

Wolff's method is geometric. He proved the following Kakeya maximal function inequality which yields Theorem 6.7.5 rather easily:

$$\|\mathcal{K}_\delta f\|_{L^{\frac{n+2}{2}}(S^{n-1})} \leq C(n,\varepsilon)\delta^{\frac{2-n}{2+n}-\varepsilon}\|f\|_{L^{\frac{n+2}{2}}(\mathbb{R}^n)} \tag{6.7.1}$$

for all $\delta,\varepsilon>0$. Here

$$\mathcal{K}_\delta f(e)=\sup_{a\in\mathbb{R}^n}\frac{1}{\mathcal{L}^n(T_e^\delta(a))}\int_{T_e^\delta(a)}|f|\,d\mathcal{L}^n,$$

where $T_e^\delta(a)$ is the tube with center $a\in\mathbb{R}^n$, direction $e\in S^{n-1}$, width δ and length 1.

Wolff's estimate $\dim B\geq 3$ is still the best known in $\mathbb{R}^4$.

Bourgain introduced in [8] a combinatorial method, further developed by Katz and Tao [38] in [39], which led to the following.

Theorem 6.7.6 *For any Besicovitch set B in $\mathbb{R}^n$,* $\dim B \geq (2-\sqrt{2})(n-4)+3$.

This is the best known lower bound for $n \geq 5$. Quite recently, Katz and Zahl [40] were able to establish an epsilon improvement on Wolff's bound $5/2$ in $\mathbb{R}^3$. Thus in $\mathbb{R}^3$ the best known estimate is the following.

Theorem 6.7.7 *For any Besicovitch set B in $\mathbb{R}^3$, $\dim B \geq 5/2+\varepsilon$ where ε is a small constant.*

The arguments of Katz and Zahl are very involved and complicated combining many earlier ideas. A new feature is the algebraic polynomial method, first used by Dvir [14] to solve the Kakeya conjecture in finite fields. The polynomial methods have recently been used in many connections, and an excellent treatise on these is Guth's book [27]. Orponen applied them to projections in [70].

Let us now look at some relations between unions of lines and line segments. Keleti made the following conjecture in [44].

Conjecture 6.7.8 If A is the union of a family of line segments in $\mathbb{R}^n$ and B is the union of the corresponding lines, then $\dim A = \dim B$.

This is true in the plane, as proved by Keleti.

Theorem 6.7.9 *Conjecture 6.7.8 is true in $\mathbb{R}^2$.*

If Keleti's conjecture is true in $\mathbb{R}^n$ for all $n \geq 3$, it gives a lot of new information on the dimension of Besicovitch sets.

Theorem 6.7.10 (Keleti [44]) *(1) If Conjecture 6.7.8 is true for some n, then, for this n, every Besicovitch set in $\mathbb{R}^n$ has Hausdorff dimension at least $n-1$.*

(2) If Conjecture 6.7.8 is true for all n, then every Besicovitch set in $\mathbb{R}^n$ has upper Minkowski and packing dimension n for all n.

Proof Let F be the projective transformation

$$F(\tilde{x}, x_n) = \frac{1}{x_n}(\tilde{x}, 1), \quad (\tilde{x}, x_n) \in \mathbb{R}^{n-1} \times \mathbb{R}, x_n \neq 0.$$

Then for $e \in S^{n-1}, e_n \neq 0, a \in \mathbb{R}^{n-1}$, F maps the punctured line $l(e,a) = \{te + (a,0) : t \neq 0\}$ onto the punctured line $\{u(a,1) + \frac{1}{e_n}(\tilde{e}, 0) : u \neq 0\}$. If B contains a line segment on $l(e, a_e), e \in S^{n-1}$, then $F(B)$ contains a line segment on $F(l(e, a_e))$, $e \in S^{n-1}$. The line extensions of these latter punctured lines cover $\{x : x_n = 0\}$ so $\dim F(B) \geq n-1$ provided Conjecture 6.7.8 is true. Clearly, F does not change Hausdorff dimension, whence $\dim B \geq n-1$ and (1) holds.

(2) follows by the well-known trick of taking products and by the product inequalities (6.2.4) and (6.2.5). Suppose that Conjecture 6.7.8 is true for all n and there exists

a Besicovitch set B in $\mathbb{R}^n$ with $\dim_P B < n$ for some n. Then $B^k \subset \mathbb{R}^{kn}$ would satisfy by (6.2.5)

$$\dim B^k \leq \dim_P B^k \leq k \dim_P B < kn - 1$$

for large k. This contradicts part (1) since B^k is a Besicovitch set in $\mathbb{R}^{kn}$.

Using Theorem 6.3.4, Falconer and I proved in [22] that in Theorem 6.7.9 line segments can be replaced by sets of positive one-dimensional measure. Later, Héra, Keleti and Máthé [29] proved that sets of dimension one are enough. These methods and results extend to subsets of hyperplanes in $\mathbb{R}^n$, but they do not extend to lower dimensional planes. In particular, they do not apply to Besicovitch sets in higher dimensions.

More generally, we can investigate the following question: suppose E is a Borel family of affine k-planes in $\mathbb{R}^n$. How does the Hausdorff dimension of E (with respect to a natural metric) affect the Lebesgue measure and the Hausdorff dimension of the union $L(E)$ of these planes, or of $B \cap L(E)$ if we know that B intersects every $V \in E$ in a positive measure or in dimension u? Oberlin used in [63] the projection theorems of Sect. 4 to prove that $\dim E > (k+1)(n-k) - k$ implies $\mathcal{L}^n(L(E)) > 0$, and this is sharp. He also proved some lower bounds for the dimension, which are sharp when $k = n - 1$ and $0 < s \leq 1$, and then the lower bound is $n - 1 + s$, but they probably are not always sharp.

Héra et al. studied in [29] questions of the above type and proved many interesting generalizations of the above results. For example, they proved the following.

Theorem 6.7.11 *Let $1 \leq k < n$ be integers and $0 \leq s \leq 1$. If E is a non-empty family of affine k-planes in $\mathbb{R}^n$ with $\dim E = s$ and $B \subset L(E)$ such that $\dim B \cap V = k$ for every $V \in E$, then*

$$\dim B = \dim L(E) = s + k.$$

Again, the right-hand equality can fail if $s > 1$; consider for example more than one-dimensional families of lines in a plane. But the left-hand inequality might hold always. However, it is unknown for $s > 1$.

Furstenberg sets are kind of fractal versions of Besicovitch sets. We consider them only in the plane. For Besicovitch sets, we had a line segment in each direction. We would still have dimension 2 if we would replace line segments with sets of dimension 1. But things get much more difficult if we replace them with lower dimensional sets. We say that $F \subset \mathbb{R}^2$ is a Furstenberg s-set, $0 < s \leq 1$, if for every $e \in S^1$ there is a line L_e in direction e such that $\dim F \cap L_e \geq s$. What can be said about the dimension of F? Wolff [86], Sect. 11.1, showed that

$$\dim F \geq \max\{2s, s + 1/2\} \tag{6.7.2}$$

and that there is such an F with $\dim F = 3s/2 + 1/2$. He conjectured that $\dim F \geq 3s/2 + 1/2$ would hold for all Furstenberg s-sets. When $s = 1/2$, Bourgain [9] improved the lower bound 1 to $\dim F \geq 1 + c$ for some absolute constant $c > 0$.

Oberlin [64] observed a connection to projections, and in particular to dimension estimates for exceptional projections and Conjecture (6.3.5). In this way, he improved Wolff's estimates for some particular Furstenberg sets. Let us see how this goes.

Let $E \subset \mathbb{R}$ be a Borel set with $\dim E = s$ and $C \subset \mathbb{R}^2$ a parameter set for our lines such that $\pi(C) = \mathbb{R}$, $\pi(x, y) = y$, whence $\dim C \geq 1$. Set

$$F = \{(x, a + bx) : x \in E, (a, b) \in C\}.$$

Then F is (essentially; line in y-direction is missing) a Furstenberg s-set. As before for $t \in E$,

$$F \cap \{(t, y) : y \in \mathbb{R}\} = \{t\} \times \pi_t(C)$$

where

$$\pi_t : \mathbb{R}^2 \to \mathbb{R}^2, \quad \pi_t(a, b) = a + tb.$$

Let $0 < u < (s + 1)/2$. If Conjecture (6.3.5) holds, we obtain

$$\dim\{t : \pi_t(C) < u\} \leq 2u - 1 < s = \dim E.$$

Hence, there is $E_1 \subset E$ such that $\dim E_1 = s$ and $\dim \pi_t(C) \geq u$ for $t \in E_1$. It follows by Proposition 6.2.3 that $\dim F \geq s + u$. Letting $u \to (s + 1)/2$, we get $\dim F \geq 3s/2 + 1/2$.

Thus the projection conjecture (6.3.5) implies Wolff's conjecture for these special Furstenberg sets. Even for these no better dimension estimate is known than (6.7.2). Oberlin proved a better estimate, but weaker than the conjectured one, in the case where $C = C_1 \times C_1$ and $C_1 \subset \mathbb{R}$ is the standard symmetric Cantor set of dimension $1/2$. He did this by improving Kaufman's estimate $\dim\{t : \pi_t(C) < u\} \leq u$ in this case.

Orponen has proved (unpublished) that if we have the lower bound $t + (2 - t)s$ for some $t \in [0, 1/2]$ for the Hausdorff dimension of all Furstenberg s-sets $F \subset \mathbb{R}^2$, then

$$\dim\{e \in S^1 : \dim_M P_e(F) \leq u\} \leq \max\left\{\frac{u - t}{1 - t}, 0\right\} \quad \text{for } 0 \leq u \leq 1.$$

Orponen improved in [73] Wolff's bound for the packing dimension.

Theorem 6.7.12 *For $1/2 < s < 1$ there exists a positive constant $\varepsilon(s)$ such that for any Furstenberg s-set $F \subset \mathbb{R}^2$ we have $\dim_P F > 2s + \varepsilon(s)$.*

Recall Orponen's packing dimension estimate for projections (6.3.6). Proofs for these two results are rather similar, and based on combinatorial arguments.

This dimension problem is related to Furstenberg's question on sets invariant under $x \mapsto px(mod 1)$, $x \in \mathbb{R}$, $p \in \mathbb{Z}$. This problem was recently solved, independently and by different methods, by Shmerkin [79] and by Wu [87].

Other recent results on Furstenberg sets are due to Molter and Rela [59–61], and Venieri [82]. Rela has a survey in [77].

One reason for the great interest in Besicovitch sets and Kakeya conjecture is that the restriction conjecture

$$\|\widehat{f}\|_{L^q(\mathbb{R}^n)} \leq C(n,q)\|f\|_{L^\infty(S^{n-1})} \quad \text{for } q > 2n/(n-1),$$

implies the Kakeya conjecture. For more on this, see for example [55, 86].

6.8 (n, k) Besicovitch Sets

We obtain other interesting Besicovitch set problems by replacing lines with higher dimensional planes.

Definition 6.8.1 A set $B \subset \mathbb{R}^n$ is said to be an (n, k) *Besicovitch set* if $\mathcal{L}^n(B) = 0$ and there is a non-empty open set $G \subset G(n, m)$ such that for every $V \in G$ there is $a \in \mathbb{R}^n$ such that $B(a, 1) \cap (V + a) \subset B$.

We say that a set $B \subset \mathbb{R}^n$ is a full (n, k) *Besicovitch set* if $\mathcal{L}^n(B) = 0$ and there is a non-empty open set $G \subset G(n, m)$ such that for every $V \in G$ there is $a \in \mathbb{R}^n$ such that $V + a \subset B$.

We have used the open set G in this definition for later convenience. Our main interest is for what pairs (n, k) such sets exist and for this it is equivalent to use $G = G(n, k)$.

Extending earlier results of Marstrand [49] ($n = 3, k = 2$), Falconer [16] ($k > n/2$), and Bourgain [7] ($2^{k-1} + k \geq n$), Oberlin [66] proved that there exist no (n, k) Besicovitch sets if $(1 + \sqrt{2})^{k-1} + k > n$. For other values of $k \geq 2$, their existence is unknown. Let us now see how this relates to projections.

Mimicking the arguments from the previous section, we only consider affine k-planes in $\mathbb{R}^n$ which are graphs over $\mathbb{R}^k$ identified with the coordinate plane $x_{k+1} = \cdots = x_n = 0$. They can be parametrized as

$$L(l, c) = \{(x, lx + c) : x \in \mathbb{R}^k\}, \quad l \in L(\mathbb{R}^k, \mathbb{R}^{n-k}), c \in \mathbb{R}^{n-k},$$

where $L(\mathbb{R}^k, \mathbb{R}^{n-k})$ is the space of linear maps from $\mathbb{R}^k$ into $\mathbb{R}^{n-k}$, identified with $\mathbb{R}^{k(n-k)}$. Let $\pi : \mathbb{R}^{k(n-k)} \times \mathbb{R}^k \to \mathbb{R}^{k(n-k)}$ with $\pi(l, c) = l$. Suppose we could find a Borel set $C \subset \mathbb{R}^{k(n-k)} \times \mathbb{R}^{n-k}$ for which the interior of $\pi(C)$ is non-empty and $\mathcal{L}^n(B) = 0$ where

$$B = \bigcup_{(l,c) \in C} L(l, c).$$

Then B would be a full (n, k) Besicovitch set. Define

$$\pi_t : L(\mathbb{R}^k, \mathbb{R}^{n-k}) \times \mathbb{R}^{n-k} \to \mathbb{R}^{n-k}, \quad \pi_t(l, c) = lt + c, (l, c) \in L(\mathbb{R}^k, \mathbb{R}^{n-k}) \times \mathbb{R}^{n-k}, t \in \mathbb{R}^k.$$

For $t \in \mathbb{R}^k$ we now have

$$B \cap \{(x, y) \in \mathbb{R}^k \times \mathbb{R}^{n-k} : x = t\} = \{t\} \times \pi_t(C).$$

So by Fubini's theorem $\mathcal{L}^n(B) > 0$ if and only if $\mathcal{L}^{n-k}(\pi_t(C)) > 0$ for t in a set of positive k-dimensional Lebesgue measure.

Hence, the question for which values of n and k the projection properties (P1) and (P2) below are valid is very close to the question of the existence of (n, k) Besicovitch sets:

(P1) If $C \subset \mathbb{R}^{k(n-k)} \times \mathbb{R}^{n-k}$ is a Borel set for which the interior of $\pi(C)$ is nonempty, then $\mathcal{L}^{n-k}(\pi_t(C)) > 0$ for positively many $t \in \mathbb{R}^k$.

(P2) If $C \subset \mathbb{R}^{k(n-k)} \times \mathbb{R}^{n-k}$ is a Borel set with $\mathcal{L}^{k(n-k)}(\pi(C)) > 0$, then $\mathcal{L}^{n-k}(\pi_t(C)) > 0$ for almost all $t \in \mathbb{R}^k$.

(P3) If $C \subset \mathbb{R}^{k(n-k)} \times \mathbb{R}^{n-k}$ is a Borel set with $\mathcal{H}^{k(n-k)}(C) > 0$, then $\mathcal{L}^{n-k}(\pi_t(C)) > 0$ for almost all $t \in \mathbb{R}^k$.

Clearly, (P3) implies (P2) implies (P1). Probably, (P1) and (P2) are equivalent but it may be difficult to show this without really verifying their validity. Notice that (P3) is almost the same as statement (2) in Oberlin's Theorem 6.4.4 in the case $m = n - k$, $N = (k + 1)(n - k)$. We shall come back to that, and we shall see that (P1) does not always imply (P3).

If $k = n - 1$, then the π_t form an $(n - 1)$-dimensional family of linear maps $\mathbb{R}^n \to \mathbb{R}$, which is essentially the same as the full family of orthogonal projections. Thus these statements are true by standard Marstrand's projection theorem and we regain by Proposition 6.8.2 the nonexistence of $(n, n - 1)$ Besicovitch sets. This was proved by Marstrand by a simple geometric method for $n = 3$ and that proof easily generalizes. For other pairs (n, k), the validity of (P1) and (P2) does not seem to have an obvious answer. But we can easily state some connections.

Proposition 6.8.2 *(1) Full (n, k) Besicovitch sets do not exist if and only if (P1) holds.*
(2) (n, k) Besicovitch sets do not exist if (P2) holds.

So if we would know that (P1) and (P2) are equivalent, we would know that the existence of full (n, k) Besicovitch sets and of (n, k) Besicovitch sets is equivalent.

Proof Part (1) was already stated above.

(2) can be proven with an easy modification of the argument that we gave for Theorem 6.7.4. Let $B \subset \mathbb{R}^n$ be a G_δ-set which contains a unit k-ball in every direction. We need to show that $\mathcal{L}^n(B) > 0$. For $q \in \mathbb{Q}^k$, let C_q be the set of (l, c) such that l belongs to the closed unit ball B_L of $L(\mathbb{R}^k, \mathbb{R}^{n-k})$ and $(q + t, lt + c) \in B$ for $t \in B(0, 1/2)\}$. Then $|lt| \leq |t|$ for $t \in \mathbb{R}^k$. Again each C_q is a G_δ-set and $\pi(\cup_{q \in \mathbb{Q}^k} C_q)$

$= B_L$, so there is $q \in \mathbb{Q}^k$ for which $\mathcal{H}^{k(n-k)}(\pi(C_q)) > 0$. Thus by (P2) $\mathcal{L}^{n-k}(\pi_t(C)) > 0$ for almost all $t \in \mathbb{R}^k$. Since for $t \in B(0, 1/2)$,

$$\{q+t\} \times \pi_t(C_q) = \{(q+t, lt+c) : (l, c) \in C_q\} \subset B \cap \{(x, y) : x = q+t\}.$$

we conclude that $\mathcal{L}^n(B) > 0$.

Let us go back to the statement (2) in Oberlin's Theorem 6.4.4 in the case $m = n-k$ and $N = (k+1)(n-k)$. If C is as in (P2), then $\dim C \geq k(n-k)$. If $k(n-k) > (k+1)(n-k) - k$, that is, $k > n/2$, then by Theorem 6.4.4 (P2) holds and we obtain by Proposition 6.8.2 that (n, k) Besicovitch sets do not exist. This was proved by Falconer [16] with a different Fourier-analytic method. As mentioned after Theorem 6.4.4, (P3) fails if $k(n-k) < (k+1)(n-k) - k$. Suppose now that $(1+\sqrt{2})^{k-1} + k \geq n$. Then by the abovementioned results of Bourgain and Oberlin and by Proposition 6.8.2(1), (P1) holds. In particular, we obtain in a rather indirect way a projection theorem from the results of Bourgain and Oberlin. Perhaps, their methods could be used more directly to prove also other interesting projection theorems. We also see now that for pairs (n, k) for which both $k < n/2$ and $(1+\sqrt{2})^{k-1} + k \geq n$, (P3) fails but (P1) holds. It would be interesting to see why this is so just using arguments with projections.

References

1. Z.M. Balogh, E. Durand-Cartagena, K. Fässler, P. Mattila, J.T. Tyson, The effect of projections on dimension in the Heisenberg group. Rev. Mat. Iberoam. **29**, 381–432 (2013)
2. Z.M. Balogh, K. Fässler, P. Mattila, J.T. Tyson, Projection and slicing theorems in Heisenberg groups. Adv. Math. **231**, 569–604 (2012)
3. Z.M. Balogh, A. Iseli, Dimension distortion by projections on Riemannian surfaces of constant curvature. Proc. Amer. Math. Soc. **144**, 2939–2951 (2016)
4. Z.M. Balogh, A. Iseli. Marstrand type projection theorems for normed spaces, to appear in J. Fractal Geom
5. V. Beresnevich, K.J. Falconer, S. Velani, A. Zafeiropoulos, Marstrand's Theorem Revisited: Projecting Sets of Dimension Zero, arXiv:1703.08554
6. A.S. Besicovitch, On fundamental geometric properties of plane-line sets. J. London Math. Soc. **39**, 441–448 (1964)
7. J. Bourgain, Besicovitch type maximal operators and applications to Fourier analysis. Geom. Funct. Anal. **1**, 147–187 (1991)
8. J. Bourgain, On the dimension of Kakeya sets and related maximal inequalities. Geom. Funct. Anal. **9**, 256–282 (1999)
9. J. Bourgain, On the Erdös-Volkmann and Katz-Tao ring conjectures. Geom. Funct. Anal. **13**, 334–365 (2003)
10. J. Bourgain, The discretized sum-product and projection theorems. J. Anal. Math. **112**, 193–236 (2010)
11. C. Chen, Projections in vector spaces over finite fields. Ann. Acad. Sci. Fenn. A Math. **43**, 171–185 (2018)
12. C. Chen, Restricted families of projections and random subspaces, arXiv:1706.03456
13. R.O. Davies, Some remarks on the Kakeya problem. Proc. Cambridge Philos. Soc. **69**, 417–421 (1971)

14. Z. Dvir, On the size of Kakeya sets in finite fields. J. Amer. Math. Soc. **22**, 1093–1097 (2009)
15. M.B. Erdoğan, A bilinear Fourier extension problem and applications to the distance set problem. Int. Math. Res. Not. **23**, 1411–1425 (2005)
16. K.J. Falconer, Sections of sets of zero Lebesgue measure. Mathematika **27**, 90–96 (1980)
17. K.J. Falconer, Hausdorff dimension and the exceptional set of projections. Mathematika **29**, 109–115 (1982)
18. K.J. Falconer, *The Geometry of Fractal Sets* (Cambridge University Press, Cambridge, 1985)
19. K.J. Falconer, Classes of sets with large intersection. Mathematika **32**, 191–205 (1985)
20. K.J. Falconer, On the Hausdorff dimension of distance sets. Mathematika **32**, 206–212 (1985)
21. K.J. Falconer, J. Fraser, X. Jin, *Sixty Years of Fractal Projections*. Fractal geometry and stochastics V, 3–25, Progr. Probab., 70, Birkhäuser/Springer, Cham (2015)
22. K.J. Falconer, P. Mattila, Strong Marstrand theorems and dimensions of sets formed by subsets of hyperplanes. J. Fractal Geom. **3**, 319–329 (2016)
23. K.J. Falconer, T. O'Neil, Convolutions and the geometry of multifractal measures. Math. Nachr. **204**, 61–82 (1999)
24. K. Fässler, R. Hovila, Improved Hausdorff dimension estimate for vertical projections in the Heisenberg group. Ann. Sc. Norm. Super. Pisa Cl. Sci. **15**(5), 459–483 (2016)
25. K. Fässler, T. Orponen, On restricted families of projections in $\mathbb{R}^3$. Proc. London Math. Soc. **109**(3), 353–381 (2014)
26. H. Federer, *Geometric Measure Theory* (Springer, 1969)
27. L. Guth, *Polynomial Methods in Combinatorics* (American Mathematical Society, Provedence, RI, 2016)
28. W. He, Orthogonal projections of discretized sets, arXiv:1710.00759, to appear in J. Fractal Geom
29. K. Héra, T. Keleti, A. Máthé. Hausdorff dimension of union of affine subspaces, arXiv:1701.02299, to appear in J. Fractal Geom
30. R. Hovila, E. Järvenpää, M. Järvenpää, F. Ledrappier, Besicovitch-Federer projection theorem and geodesic flows on Riemann surfaces. Geom. Dedicata **161**, 51–61 (2012)
31. R. Hovila, E. Järvenpää, M. Järvenpää, F. Ledrappier, Singularity of projections of 2-dimensional measures invariant under the geodesic flows on Riemann surfaces. Comm. Math. Phys. **312**, 127–136 (2012)
32. A. Iosevich, B. Liu, Falconer distance problem, additive energy and Cartesian products. Ann. Acad. Sci. Fenn. Math. **41**, 579–585 (2016)
33. A. Iosevich, B. Liu, Pinned distance problem, slicing measures and local smoothing estimates, arXiv:1706.09851
34. E. Järvenpää, M. Järvenpää, T. Keleti, Hausdorff dimension and non-degenerate families of projections. J. Geom. Anal. **24**, 2020–2034 (2014)
35. E. Järvenpää, M. Järvenpää, F. Ledrappier, M. Leikas, One-dimensional families of projections. Nonlinearity **21**(3), 453–463 (2008)
36. A. Käenmäki, T. Orponen, L. Venieri, A Marstrand-type restricted projection theorem in $\mathbb{R}^3$, arXiv:1708.04859
37. J.-P. Kahane, Sur la dimension des intersections, In Aspects of Mathematics and Applications. North-Holland Math. Lib. **34**, 419–430 (1986)
38. N.H. Katz, T. Tao, Bounds on arithmetic projections, and applications to the Kakeya conjecture. Math. Res. Lett. **6**, 625–630 (1999)
39. N.H. Katz, T. Tao, New bounds for Kakeya problems. J. Anal. Math. **87**, 231–263 (2002)
40. N.H. Katz, J. Zahl, An improved bound on the Hausdorff dimension of Besicovitch sets in $\mathbb{R}^3$. J. Amer. Math. Soc., August 2018, published electronically
41. R. Kaufman, On Hausdorff dimension of projections. Mathematika **15**, 153–155 (1968)
42. R. Kaufman, P. Mattila, Hausdorff dimension and exceptional sets of linear transformations. Ann. Acad. Sci. Fenn. A Math. **1**, 387–392 (1975)
43. T. Keleti. A 1-dimensional subset of the reals that intersects each of its translates in at most a single point. Real Anal. Exchange **24**, 843–844 (1998/99)

44. T. Keleti, Are lines much bigger than line segments? Proc. Amer. Math. Soc. **144**, 1535–1541 (2016)
45. T. Keleti, Small union with large set of centers, in Recent developments in Fractals and Related Fields, ed. Julien Barral and Stéphane Seuret, Birkhäuser (2015). arXiv:1701.02762
46. R. Lucá, K. Rogers, Average decay of the Fourier transform of measures with applications, arXiv:1503.00105
47. J.M. Marstrand, Some fundamental geometrical properties of plane sets of fractional dimensions. Proc. London Math. Soc. **4**(3), 257–302 (1954)
48. J.M. Marstrand, The dimension of cartesian product sets. Proc. Cambridge Philos. Soc. **50**(3), 198–202 (1954)
49. J.M. Marstrand, Packing planes in $\mathbb{R}^3$. Mathematika **26**, 180–183 (1979)
50. P. Mattila, Hausdorff dimension, orthogonal projections and intersections with planes. Ann. Acad. Sci. Fenn. A Math. **1**, 227–244 (1975)
51. P. Mattila, Hausdorff dimension and capacities of intersections of sets in n-space. Acta Math. **152**, 77–105 (1984)
52. P. Mattila, On the Hausdorff dimension and capacities of intersections. Mathematika **32**, 213–217 (1985)
53. P. Mattila, *Geometry of Sets and Measures in Euclidean Spaces* (Cambridge University Press, Cambridge, 1995)
54. P. Mattila, *Recent Progress on Dimensions of Projections*. Geometry and analysis of fractals, 283–301, Springer Proc. Math. Stat., 88, Springer, Heidelberg (2014)
55. P. Mattila, *Fourier Analysis and Hausdorff Dimension* (Cambridge University Press, Cambridge, 2015)
56. P. Mattila, Exeptional set estimates for the Hausdorff dimension of intersections. Ann. Acad. Sci. Fenn. A Math. **42**, 611–620 (2017)
57. P. Mattila, T. Orponen, Hausdorff dimension, intersections of projections and exceptional plane sections. Proc. Amer. Math. Soc. **144**, 3419–3430 (2016)
58. P. Mattila, P. Sjölin, Regularity of distance measures and sets. Math. Nachr. **204**, 157–162 (1997)
59. U. Molter, E. Rela, Improving dimension estimates for Furstenberg-type sets. Adv. Math. **223**, 672–688 (2010)
60. U. Molter, E. Rela, Furstenberg sets for a fractal set of directions. Proc. Amer. Math. Soc. **140**, 2753–2765 (2012)
61. U. Molter, E. Rela, Small Furstenberg sets. J. Math. Anal. Appl. **400**, 475–486 (2013)
62. D.M. Oberlin, Restricted Radon transforms and projections of planar sets. Canad. Math. Bull. **55**, 815–820 (2012)
63. D.M. Oberlin, Exceptional sets of projections, unions of k-planes, and associated transforms. Israel J. Math. **202**, 331–342 (2014)
64. D.M. Oberlin, Some toy Furstenberg sets and projections of the four-corner Cantor set. Proc. Amer. Math. Soc. **142**(4), 1209–1215
65. D.M. Oberlin, R. Oberlin, Application of a Fourier restriction theorem to certain families of projections in $\mathbb{R}^3$. J. Geom. Anal. **25**(3), 1476–1491 (2015)
66. R. Oberlin, Two bounds for the X-ray transform. Math. Z. **266**, 623–644 (2010)
67. T. Orponen, Slicing sets and measures, and the dimension of exceptional parameters. J. Geom. Anal. **24**, 47–80 (2014)
68. T. Orponen, Hausdorff dimension estimates for some restricted families of projections in $\mathbb{R}^3$. Adv. Math. **275**, 147–183 (2015)
69. T. Orponen, On the packing dimension and category of exceptional sets of orthogonal projections. Ann. Mat. Pura Appl. **194**(4), 843–880 (2015)
70. T. Orponen, Projections of planar sets in well-separated directions. Adv. Math. **297**, 1–25 (2016)
71. T. Orponen, On the distance sets of AD-regular sets. Adv. Math. **307**, 1029–1045 (2017)
72. T. Orponen, A sharp exceptional set estimate for visibility. Bull. London. Math Soc. **50**, 1–6 (2018)

73. T. Orponen, An improved bound on the packing dimension of Furstenberg sets in the plane, arXiv:1611.09762, to appear in J. Eur. Math. Soc
74. T. Orponen, On the dimension and smoothness of radial projections, arXiv:1710.11053, to appear in Anal. PDE
75. T. Orponen, L. Venieri, Improved bounds for restricted families of projections to planes in $\mathbb{R}^3$, appeared online in Int (Math. Res, Notices, 2018)
76. Y. Peres, W. Schlag, Smoothness of projections, Bernoulli convolutions, and the dimension of exceptions. Duke Math. J. **102**, 193–251 (2000)
77. E. Rela, Refined size estimates for Furstenberg sets via Hausdorff measures: a survey of some recent results, arXiv:1305.3752
78. P. Shmerkin, Projections of self-similar and related fractals: a survey of recent developments, arXiv:1501.00875, Fractal geometry and stochastics V, 53–74, Progr. Probab., 70, Birkhäuser/Springer, Cham (2015)
79. P. Shmerkin, On Furstenberg's intersection conjecture, self-similar measures, and the L^q norms of convolutions, arXiv:1609.07802
80. P. Shmerkin, On distance sets, box-counting and Ahlfors-regular sets. Discrete Anal. Paper No. 9, 22 pp (2017)
81. P. Shmerkin, On the Hausdorff dimension of pinned distance sets, arXiv:1706.00131
82. L. Venieri, Dimension estimates for Kakeya sets defined in an axiomatic setting, arXiv:1703.03635, Ann. Acad. Sci Fenn. Disserationes 161 (2017)
83. T.W. Wolff, An improved bound for Kakeya type maximal functions. Rev. Mat. Iberoam. **11**, 651–674 (1995)
84. T.W. Wolff, A Kakeya-type problem for circles. Amer. J. Math. **119**, 985–1026 (1997)
85. T.W. Wolff, Decay of circular means of Fourier transforms of measures. Int. Math. Res. Not. **10**, 547–567 (1999)
86. T.W. Wolff, *Lectures on Harmonic Analysis*, vol. 29 (American Mathematical Society, 2003)
87. M. Wu, A proof of Furstenberg's conjecture on the intersections of $\times p$ and $\times q$ invariant sets, arXiv:1609.08053
88. H. Yu, Kakeya books and projections of Kakeya sets, arXiv:1704.04488

Chapter 7
Dyadic Harmonic Analysis and Weighted Inequalities: The Sparse Revolution

María Cristina Pereyra

Abstract We will introduce the basics of dyadic harmonic analysis and how it can be used to obtain weighted estimates for classical Calderón–Zygmund singular integral operators and their commutators. Harmonic analysts have used dyadic models for many years as a first step toward the understanding of more complex continuous operators. In 2000, Stefanie Petermichl discovered a representation formula for the venerable Hilbert transform as an average (over grids) of dyadic shift operators, allowing her to reduce arguments to finding estimates for these simpler dyadic models. For the next decade, the technique used to get sharp weighted inequalities was the Bellman function method introduced by Nazarov, Treil, and Volberg, paired with sharp extrapolation by Dragičević et al. Other methods where introduced by Hytönen, Lerner, Cruz-Uribe, Martell, Pérez, Lacey, Reguera, Sawyer, and Uriarte-Tuero, involving stopping time and median oscillation arguments, precursors of the very successful domination by positive sparse operators methodology. The culmination of this work was Tuomas Hytönen's 2012 proof of the A_2 conjecture based on a representation formula for any Calderón–Zygmund operator as an average of appropriate dyadic operators. Since then domination by sparse dyadic operators has taken central stage and has found applications well beyond Hytönen's A_p theorem. We will survey this remarkable progression and more in these lecture notes.

7.1 Introduction

These notes are based on lectures delivered by the author on August 7–9, 2017 at the CIMPA 2017 *Research School—IX Escuela Santaló: Harmonic Analysis, Geometric Measure Theory and Applications*, held in Buenos Aires, Argentina. The course was titled "Dyadic Harmonic Analysis and Weighted Inequalities".

M. C. Pereyra (✉)
Department of Mathematics and Statistics, 1 University of New Mexico, 311 Terrace St. NE, MSC01 1115, Albuquerque, NM 87131-0001, USA
e-mail: crisp@math.unm.edu

A. Aldroubi et al. (eds.), *New Trends in Applied Harmonic Analysis, Volume 2*, Applied and Numerical Harmonic Analysis,
https://doi.org/10.1007/978-3-030-32353-0_7

The main question of interest in these notes is to decide for a given operator or class of operators and a pair of weights (u, v), if there is a positive constant, $C_p(u, v, T)$, such that

$$\|Tf\|_{L^p(v)} \leq C_p(u, v, T)\, \|f\|_{L^p(u)} \quad \text{for all functions} \quad f \in L^p(u).$$

The main goals in these lectures are twofold. First, given an operator T (or family of operators), identify and classify pairs of weights (u, v) for which the operator(s) T is(are) bounded on weighted Lebesgue spaces, more specifically from $L^p(u)$ to $L^p(v)$—*qualitative bounds*. Second, understand the nature of the constant $C_p(u, v, T)$—*quantitative bounds*.

We concentrate on one-weight L^p inequalities for $1 < p < \infty$, that is, the case when $u = v = w$, for the prototypical operators, dyadic models, and their commutators, although we will state some of the known two-weight results. The operators we will focus on are the Hardy–Littlewood maximal function; Calderón–Zygmund operators T, such as the Hilbert transform H; and their dyadic analogues, specifically the dyadic maximal function, the martingale transform, the dyadic square function, the Haar shift multipliers, the dyadic paraproducts, and the sparse dyadic operators.

The question now reduces to the following: Given weight w and $1 < p < \infty$, is there a constant $C_p(w, T) > 0$ such that for all functions $f \in L^p(w)$

$$\|Tf\|_{L^p(w)} \leq C_p(w, T)\, \|f\|_{L^p(w)}\ ?$$

We have known since the 70s that the maximal function is bounded on $L^p(w)$ if an only if the weight w is in the Muckenhoupt A_p class [141]; similar result holds for the Hilbert transform [88]. General Calderón–Zygmund operators and dyadic analogues are bounded on $L^p(w)$ [39] when the weight $w \in A_p$ and the same holds for their commutators with functions in the space of bounded mean oscillation (BMO) [8, 25]. The quantitative versions of these results were obtained several decades later, in 1993 for the maximal function [27], in 2007 for the Hilbert transform [162], and in 2012 for Calderón–Zygmund singular integral operators [90] and for their commutators [37]. We will say more about A_p weights and the quantitative versions of these classical results in the following pages.

We will show or at least describe, for the model operators T, the validity of a weighted L^2 inequality that is linear on $[w]_{A_2}$, the A_2 characteristic of the weight, namely, there is a constant $C > 0$ such that for all weights $w \in A_2$ and for all functions $f \in L^2(w)$

$$\|Tf\|_{L^2(w)} \leq C[w]_{A_2} \|f\|_{L^2(w)}.$$

That this holds for all Calderón–Zygmund singular integrals operators was the A_2 conjecture. We will also describe several approaches for the corresponding quadratic estimate for the commutator $[b, T] = bT - Tb$ where b is a function in BMO, namely,

$$\|[b, T]f\|_{L^2(w)} \leq C[w]_{A_2}^2 \|b\|_{\mathrm{BMO}} \|f\|_{L^2(w)}.$$

Dyadic models have been used in harmonic analysis and other areas of mathematics for a long time; Terry Tao has an interesting post on his blog[1] regarding the ubiquitous "dyadic model." For a presentation suitable for beginners, see the lecture notes by the author [154], which describe the status quo of dyadic harmonic analysis and weighted inequalities as of 2000. This millennium has seen new dyadic techniques evolve, become mainstream, and help settle old problems; these lecture notes try to illustrate some of this progress. In particular, averaging and sparse domination techniques with and by dyadic operators have allowed researchers to transfer results from the dyadic world to the continuous world. No longer the dyadic models are just toy models in harmonic analysis, they can truly inform the continuous models. Here are some examples where this dyadic paradigm has been useful.

The dyadic maximal function controls the maximal function (the converse is immediate) by means of the one-third trick. Estimates for the dyadic maximal function are easier to obtain and transfer to the maximal function painlessly.

The Walsh model is the dyadic counterpart to Fourier analysis. The first real progress toward proving boundedness of the bilinear Hilbert transform [125], result that earned Christoph Thiele and Michael Lacey the 1996 Salem Prize,[2] was made by Thiele in his 1995 Ph.D. thesis proving the Walsh model version of such result [172].

Stefanie Petermichl showed in 2000 that one can write the Hilbert transform as an "average of dyadic shift operators" over random dyadic grids [161]. She achieved this using the well-known symmetry properties that characterize the Hilbert transform. Namely, the Hilbert transform commutes with translations and dilations, and anticommutes with reflections. A linear and bounded operator on $L^2(\mathbb{R})$ with those properties must be a constant multiple of the Hilbert transform. Similarly, the Riesz transforms [163] can be written as averages of suitable dyadic operators. Petermichl proved the A_2 conjecture for these dyadic operators using Bellman function techniques [162, 163]. These results added a very precise new dyadic perspective to such classic and well-studied operators in harmonic analysis and earned Petermichl the 2006 Salem Prize; first time this prize was awarded to a female mathematician.

The Martingale transform was considered the dyadic toy model "par excellence" for Calderón–Zygmund singular integral operators. For many years, one would test the martingale transform first and, if successful, then worry about the continuous versions. In 2000, Janine Wittwer proved the A_2 conjecture for the martingale transform using Bellman functions [186]. The Beurling transform can be written as an average of martingale transforms in the complex plane, and this allowed Stefanie Petermichl and Sasha Volberg [165] to prove in 2002 linear weighted inequalities on $L^p(w)$ for $p \geq 2$, and as a consequence deduce an important end point result in

[1] https://terrytao.wordpress.com/2007/07/27/dyadic-models/.

[2] The Salem Prize, founded by the widow of Raphael Salem, is awarded every year to a young mathematician judged to have done outstanding work in Salem's field of interest, primarily the theory of Fourier series. The prize is considered highly prestigious and many Fields Medalists previously received Salem prize (Wikipedia).

the theory of quasiconformal mappings that had been conjectured by Kari Astala, Tadeusz Iwaniec, and Eero Saksman [9].

Surprisingly, all Calderón–Zygmund singular integral operators can be written as averages of Haar shift dyadic operators of arbitrary complexity and dyadic paraproducts as proven by Tuomas Hytönen [90]. In 2008, Oleksandra Beznosova proved the A_2 conjecture for the dyadic paraproduct [21] and, together with Hytönen's dyadic representation theorem, this lead to Hytönen's proof of the full A_2 conjecture [90].

Leading toward Hytönen's result, there were a number of breakthroughs that have recently coalesced under the umbrella of "domination by finitely many sparse positive dyadic operators." Andrei Lerner's early results [130] played a central role in this development. It is usually straightforward to verify that these sparse operators have desired (quantitative) estimates; it is harder to prove appropriate domination results for each particular operator and function it acts on. This methodology has seen an explosion of applications well beyond the original A_2 conjecture where it originated. Identifying the sparse collections associated to a given operator and function is the most difficult part of the argument and it involves using weak-type inequalities, stopping time techniques, and adjacent dyadic grids.

We will explore some of these examples in the lecture notes with emphasis on quantitative weighted estimates. We will illustrate in a few case studies different techniques that have evolved as a result of these investigations such as Bellman functions, quantitative extrapolation and transference theorems, and reduction to studying dyadic operators either by averaging or by sparse domination.

The structure of the lecture notes remains faithful to the lectures delivered by the author in Buenos Aires except for some minor reorganization. Some themes are touched at the beginning, to wet the appetite of the audience, and are expanded on later sections. Most objects are defined as they make their first appearance in the story. Naturally, more details are provided than in the actual lectures, and some details were in the original slides but had to be skipped or fast-forwarded; those topics are included in these lecture notes. The sections are peppered with historical remarks and references, but inevitably some will be missing or could be inaccurate despite the time and effort spent by the author on them. Thus, the author apologizes in advance for any inaccuracy or omission, and gratefully would like to hear about any corrections for future reference.

In Sect. 7.2, we introduce the basic model operators: the Hilbert transform and the maximal function, and we discuss their classical L^p and weighted L^p boundedness properties. We show that A_p is a necessary condition for the boundedness of the maximal function on weighted Lebesgue spaces L^p. We describe why are we interested in weighted estimates, and more recently on quantitative weighted estimates. In particular, we describe the linear weighted L^2 estimates saga leading toward the resolution of the A_2 conjecture and how to derive quantitative weighted L^p estimates using sharp extrapolation. We finalize the section with a brief summary of the two-weight results known for the Hilbert transform and the maximal function.

In Sect. 7.3, we introduce the elements of dyadic harmonic analysis and the basic dyadic maximal function. More precisely, we discuss dyadic grids (regular, random, adjacent) and Haar functions (on the line, on $\mathbb{R}^d$, on spaces of homogeneous type). As a first example, illustrating the power of the dyadic techniques, we present Lerner's proof of Buckley's quantitative L^p estimates for the maximal function, which reduces, using the one-third trick, to estimates for the dyadic maximal function. We also describe, given dyadic cubes on spaces of homogeneous type, how to construct corresponding Haar bases and briefly describe the Auscher–Hytönen "wavelets" in this setting.

In Sect. 7.4, we discuss the basic dyadic operators: the martingale transform, the dyadic square function, the Haar shifts multipliers (Petermichl's and those of arbitrary complexity), and the dyadic paraproducts. These are the ingredients needed to state Petermichl's and Hytönen's representation theorems for the Hilbert transform and Calderón–Zygmund operators, respectively. For each of these dyadic model operators, we describe the known L^p and weighted L^p theory and we state both Petermichl's and Hytönen's representation theorems.

In Sect. 7.5, we sketch Beznosova's proof of the A_2 conjecture for the dyadic paraproduct; this is a Bellman function argument. As a first approach, we get a 3/2 estimate, and with a refinement the linear estimate for the dyadic paraproduct is obtained. Along the way, we introduce weighted Carleson sequences, a weighted Carleson embedding lemma, some Bellman function lemmas: the Little lemma and the α-Lemma, and weighted Haar functions needed in the argument; we also sketch the proofs of these auxiliary results.

In Sect. 7.6, we discuss weighted inequalities in a case study: the commutator of the Hilbert transform H with a function b in BMO. We summarize chronologically the weighted norm inequalities known for the commutator. We sketch the dyadic proof of the quantitative weighted L^2 estimate for the commutator $[b, H]$ due to Daewon Chung, yielding the optimal quadratic dependence on the A_2 characteristic of the weight. We discuss a very useful transference theorem of Daewon Chung, Carlos Pérez, and the author, and present its proof based on the celebrated Coifman–Rochberg–Weiss argument. The transference theorem allows to deduce quantitative weighted L^p estimates for the commutator of a linear operator with a BMO function, from given weighted L^p estimates for the operator.

In Sect. 7.7, we introduce the sparse domination by positive dyadic operators paradigm that has emerged and proven to be very powerful with applications in many areas not only weighted inequalities. We discuss a characterization of sparse families of cubes via Carleson families of dyadic cubes due to Andrei Lerner and Fedja Nazarov. We present the beautiful proof of the A_2 conjecture for sparse operators due to David Cruz-Uribe, Chema Martell, and Carlos Pérez. We illustrate with one toy model example, the martingale transform, how to achieve the pointwise domination by sparse operators following an argument by Michael Lacey. Finally, we briefly discuss a sparse domination theorem for commutators valid for (rough) Calderón–Zygmund singular integral operators due to Andrei Lerner, Sheldon Ombrosi, and Israel Rivera-Ríos that yields a new quantitative two-weight estimates of Bloom type, and recovers all known weighted results for the commutators.

Finally, in Sect. 7.8, we present a summary and briefly discuss some very recent progress.

Throughout the lecture notes, a constant $C > 0$ might change from line to line. The notation $A := B$ or $B =: A$ means that A is defined to be B. The notation $A \lesssim B$ means that there is a constant $C > 0$ such that $A \le CB$. The notation $A \sim B$ means that $A \lesssim B$ and $B \lesssim A$. The notation $A \lesssim_{r,s} B$ means that the constant $C > 0$ in the implied inequality depends only on the parameters r, s.

7.2 Weighted Norm Inequalities

In this section, we introduce some basic notation and the model operators: the Hilbert transform and the maximal function and we discuss their classical L^p and weighted L^p boundedness properties. We show that A_p is a necessary condition for the boundedness of the maximal function on weighted L^p. We describe why are we interested in weighted estimates, and more recently on quantitative weighted estimates. In particular, we describe the linear weighted L^2 estimates saga leading toward the resolution of the A_2 conjecture and how to derive quantitative weighted L^p estimates using sharp extrapolation. We finalize the section with a brief summary of the two-weight results known for the Hilbert transform and the maximal function.

7.2.1 Some Basic Notation and Prototypical Operators

We introduce some basic notation used throughout the lecture notes. We remind the reader the basic spaces (weighted L^p and bounded mean oscillation, BMO), and the prototypical continuous operators to be studied, namely, the maximal function, the Hilbert transform, and its commutator with functions in BMO. We briefly recall some of the settings where these operators appear.

The *weights* u and v are locally integrable functions on $\mathbb{R}^d$, namely, $u, v \in L^1_{loc}(\mathbb{R}^d)$, that are almost everywhere positive functions.

Given a weight u, a measurable function f is in $L^p(u)$ if and only if

$$\|f\|_{L^p(u)} := \left(\int_{\mathbb{R}^d} |f(x)|^p \, u(x) \, dx \right)^{1/p} < \infty.$$

When $u \equiv 1$, we denote $L^p(\mathbb{R}^d) = L^p(u)$ and $\|f\|_{L^p} := \|f\|_{L^p(\mathbb{R}^d)}$.

Given $f, g \in L^1(\mathbb{R}^d)$ their *convolution* is given by

$$f * g(x) = \int_{\mathbb{R}^d} f(x - y) \, g(y) \, dy. \tag{7.2.1}$$

A locally integrable function b is in the space of *bounded mean oscillation*, namely, $b \in \mathrm{BMO}$, if and only if

$$\|b\|_{\text{BMO}} := \sup_{Q} \frac{1}{|Q|} \int_{Q} |b(x) - \langle b \rangle_{Q}| \, dx < \infty, \text{ where } \langle b \rangle_{Q} = \frac{1}{|Q|} \int_{Q} b(t) \, dt, \tag{7.2.2}$$

where $Q \subset \mathbb{R}^d$ are cubes with sides parallel to the axes, $|Q|$ denotes the volume of the cube Q, and more generally, $|E|$ denotes the Lebesgue measure of a measurable set E in $\mathbb{R}^d$. Note that $L^\infty(\mathbb{R}^d)$, the space of essentially bounded functions on $\mathbb{R}^d$, is a proper subset of BMO (e.g., $\log|x|$ is a function in BMO but not in $L^\infty(\mathbb{R})$).

We will consider linear or sublinear operators $T : L^p(u) \to L^p(v)$. Among the linear operators, the Calderón–Zygmund singular integral operators and their dyadic analogues will be most important for us.

The prototypical *Calderón–Zygmund singular integral operator* is the *Hilbert transform* on $\mathbb{R}$, given by convolution with the distributional *Hilbert kernel* $k_H(x) :=$ p.v.$\big(1/(\pi x)\big)$

$$Hf(x) := k_H * f(x) = \text{p.v.}\frac{1}{\pi} \int \frac{f(y)}{x-y} \, dy := \lim_{\varepsilon \to 0} \frac{1}{\pi} \int_{|x-y|>\varepsilon} \frac{f(y)}{x-y} \, dy. \tag{7.2.3}$$

The Hilbert transform and its periodic analogue naturally appear in complex analysis and in the study of convergence on L^p of partial Fourier sums/integrals. The Hilbert transform siblings, the Riesz transforms on $\mathbb{R}^d$ and the Beurling transform on $\mathbb{C}$, are intimately connected to partial differential equations and to quasiconformal theory, respectively. Its cousin, the Cauchy integral on curves and higher dimensional analogues, is connected to rectifiability and geometric measure theory.

A prototypical sublinear operator is the *Hardy–Littlewood maximal function*

$$Mf(x) := \sup_{Q : x \in Q} \frac{1}{|Q|} \int_{Q} |f(y)| \, dy, \tag{7.2.4}$$

where the supremum is taken over all cubes $Q \subset \mathbb{R}^d$ containing x and with sides parallel to the axes. The maximal function naturally controls many singular integral operators and approximations of the identity; its weak-boundedness properties on $L^1(\mathbb{R}^d)$ imply the Lebesgue differentiation theorem. Another sublinear operator that we will encounter in these lectures is the dyadic square function, see Sect. 7.4.2.

Given T a linear or sublinear operator, its *commutator* with a function b is given by

$$[b, T](f) := b\,T(f) - T(bf).$$

The commutators are important in the study of factorization for Hardy spaces and to characterize the space of bounded mean oscillation (BMO). They also play a central role in the theory of partial differential equations (PDEs).

We refer the reader to [80, 81, 171] for encyclopedic presentations of classical harmonic analysis, [68] for a more succinct yet deep presentation, and [156] for an elementary presentation emphasizing the dyadic point of view.

7.2.2 Hilbert Transform

We now recall familiar facts about the Hilbert transform, including its L^p and one-weight (quantitative) L^p boundedness properties.

The Hilbert transform is defined by (7.2.3) on the underlying space and on frequency space the following representation as a *Fourier multiplier* with *Fourier symbol* m_H, holds:

$$\widehat{Hf}(\xi) = m_H(\xi)\,\widehat{f}(\xi), \quad \text{where } m_H(\xi) := -i\,\mathrm{sgn}(\xi). \tag{7.2.5}$$

To connect the two representations for the Hilbert transform, on the underlying space and on the frequency space, remember that multiplication on the Fourier side corresponds to convolution on the underlying space. Therefore, k_H, the Hilbert kernel, is given by the inverse Fourier transform of the Fourier symbol m_H,

$$Hf(x) = k_H * f(x), \quad \text{where } k_H(x) := (m_H)^\vee(x) = \mathrm{p.v.}\frac{1}{\pi x},$$

which is precisely the content of (7.2.3). Here, the *Fourier transform* and *inverse Fourier transform* of a Schwartz function f on $\mathbb{R}$ are defined by

$$\widehat{f}(\xi) := \int_{\mathbb{R}} f(x)\,e^{-2\pi i\xi x}\,dx, \qquad (f)^\vee(x) := \int_{\mathbb{R}} f(\xi)\,e^{2\pi i\xi x}\,d\xi.$$

The Fourier transform is a bijection and an L^2 isometry on the Schwartz class that can be extended to be an isometry on $L^2(\mathbb{R})$, that is, $\|\widehat{f}\|_{L^2(\mathbb{R})} = \|f\|_{L^2(\mathbb{R})}$ (*Plancherel's identity*), and it can also be extended to be a bijection on the space of tempered distributions. The convolution $f * g$ is a well-defined function on $L^r(\mathbb{R})$ when $f \in L^p(\mathbb{R})$ and $g \in L^p(\mathbb{R})$, provided $\frac{1}{p} + \frac{1}{q} = \frac{1}{r} + 1$ and $p, q, r \in [1, \infty]$. Moreover, on the same range, *Young's inequality* holds:

$$\|g * f\|_{L^r} \leq \|g\|_{L^q}\|f\|_{L^p}. \tag{7.2.6}$$

In these lecture notes, we will explore, in Sect. 7.4.3, a third representation for the Hilbert transform in terms of dyadic shift operators discovered by Stefanie Petermichl [161] in 2000.

7.2.2.1 L^p Boundedness Properties of H

Fourier theory ensures boundedness on $L^2(\mathbb{R})$ for the Hilbert transform H. In fact, applying Plancherel's identity twice and using the fact that $|m_H(\xi)| = 1$ a.e., one immediately verifies that H is an isometry on $L^2(\mathbb{R})$, namely,

$$\|Hf\|_{L^2} = \|\widehat{Hf}\|_{L^2} = \|\widehat{f}\|_{L^2} = \|f\|_{L^2}.$$

Young's inequality (7.2.6) for $p \geq 1, q = 1$ (hence, $r = p$), implies that if $g \in L^1(\mathbb{R})$ and $f \in L^p(\mathbb{R})$ then $g * f \in L^p(\mathbb{R})$; moreover,

$$\|g * f\|_{L^p} \leq \|g\|_{L^1}\|f\|_{L^p}.$$

This would imply boundedness on $L^p(\mathbb{R})$ for the Hilbert transform IF the Hilbert kernel, k_H, were integrable, but is not. Despite this fact, the following boundedness properties for the Hilbert transform hold (shared by all Calderón–Zygmund singular integral operators).

The Hilbert transform is not bounded on $L^1(\mathbb{R})$; it is of *weak-type* (1,1) (Kolmogorov 1927), that is, there is a constant $C > 0$ such that for all $\lambda > 0$ and for all $f \in L^1(\mathbb{R})$

$$|\{x \in \mathbb{R} : |Hf(x)| > \lambda\}| \leq \frac{C}{\lambda}\|f\|_{L^1}.$$

The Hilbert transform is bounded on $L^p(\mathbb{R})$ for all $1 < p < \infty$ (M. Riesz 1927), namely, there is a constant $C_p > 0$ such that for all $f \in L^p(\mathbb{R})$

$$\|Hf\|_{L^p} \leq C_p\|f\|_{L^p} \quad \text{(best constant was found by Pichorides in 1972).}$$

Note that for $1 < p < 2$ the L^p boundedness can be obtained by Marcinkiewicz interpolation theorem, from the weak-type (1,1) and the L^2 boundedness. Then, for $2 < p < \infty$, the boundedness on $L^p(\mathbb{R})$ can be obtained by a duality argument, suffices to observe that the adjoint of H is $-H$, that is, the Hilbert transform is almost self-adjoint. However, the Marcinkiewicz interpolation did not exist in 1927. Riesz proved instead that boundedness on $L^p(\mathbb{R})$ implied boundedness on $L^{2p}(\mathbb{R})$, and hence boundedness on $L^2(\mathbb{R})$ implied boundedness on $L^4(\mathbb{R})$, then on $L^8(\mathbb{R})$ and by induction on $L^{2^n}(\mathbb{R})$. Strong interpolation, which already existed, then gave boundedness on $L^p(\mathbb{R})$ for $2^n \leq p \leq 2^{n+1}$ and for all $n \geq 1$, that Is, for all $2 \leq p < \infty$. Finally, a duality argument took care of $1 < p < 2$. In Sect. 7.4.3, we will deduce the L^p boundedness of the Hilbert transform from the L^p boundedness of dyadic shift operators, see Sect. 7.2.6.

Interpolation is an extremely powerful tool in analysis that allows to deduce intermediate norm inequalities given two end point (weak)norm inequalities. We will not discuss interpolation further in these notes; instead, we will focus on extrapolation that allows us to deduce weighted L^p norm inequalities for all $1 < p < \infty$ given weighted L^r norm inequalities for one index $r > 1$.

Finally, it is important to note that the Hilbert transform is not bounded on $L^\infty(\mathbb{R})$; however, it is bounded on the larger space BMO of functions of bounded mean oscillation (C. Fefferman 1971).

To illustrate the lack of boundedness on $L^\infty(\mathbb{R})$ and on $L^1(\mathbb{R})$, it is helpful to calculate the Hilbert transform for some simple functions, showing in fact that the Hilbert transform does not map either $L^1(\mathbb{R})$ or $L^\infty(\mathbb{R})$ into themselves. This immediately eliminates the possibility for the Hilbert transform being bounded on either space.

Example 1 (*Hilbert transform of an indicator function*)

$$H\mathbb{1}_{[a,b]}(x) = (1/\pi)\,\log\big(|x-a|/|x-b|\big),$$

where the indicator $\mathbb{1}_{[a,b]}(x) := 1$ when $x \in [a,b]$ and zero otherwise, a bounded and integrable function, that is, $\mathbb{1}_{[a,b]} \in L^1(\mathbb{R}) \cap L^\infty(\mathbb{R})$. However, $\log|x|$ is neither in $L^\infty(\mathbb{R})$ nor in $L^1(\mathbb{R})$, but it is a function of bounded mean oscillation. The functions f in $L^1(\mathbb{R})$ whose Hilbert transforms Hf are also in $L^1(\mathbb{R})$ constitute the Hardy space $H^1(\mathbb{R})$; such functions need to have some cancellation ($\int_{\mathbb{R}} f(x)\,dx = 0$), clearly not shared by the indicator function $\mathbb{1}_{[a,b]}$.

7.2.2.2 One-Weight Inequalities for *H*

The one-weight theory à la Muckenhoupt for the Hilbert transform is well understood, the qualitative theory has been known since 1973 [88], and the quantitative estimates were settled by Stefanie Petermichl in 2007 [162]. The two-weight problem, on the other hand, was studied for a long time but the necessary and sufficient conditions à la Muckenhoupt for pairs of weights (u, v) that ensure boundedness of the Hilbert transform from $L^p(u)$ into $L^p(v)$ were only settled in 2014 by Michael Lacey, Chun-Yen Shen, Eric Sawyer, and Ignacio Uriarte-Tuero [113, 123].

Theorem 1 (Hunt, Muckenhoupt, Wheeden 1973) *The Hilbert transform is bounded on $L^p(w)$ for $1 < p < \infty$ if and only if the weight $w \in A_p$. In either case, there is a constant $C_p(w) > 0$ depending on p and on the weight w such that*

$$\|Hf\|_{L^p(w)} \leq C_p(w)\|f\|_{L^p(w)} \quad \textit{for all } f \in L^p(w).$$

At this point, we remind the reader that a weight w is in the *Muckenhoupt A_p class* if and only if $[w]_{A_p} < \infty$, where the *A_p characteristic* of the weight w is defined to be

$$[w]_{A_p} := \sup_Q \left(\frac{1}{|Q|}\int_Q w(x)\,dx\right)\left(\frac{1}{|Q|}\int_Q w^{\frac{-1}{p-1}}(x)\,dx\right)^{p-1} \quad \text{for } 1 < p < \infty,$$

the supremum is taken over all cubes Q in $\mathbb{R}^d$ with sides parallel to the axes. We will denote integral averages with respect to Lebesgue measure on cubes or on measurable sets E by $\langle f\rangle_E := \frac{1}{|E|}\int_E f(x)\,dx$. Also given w, a weight, $w(E)$ will denote the w-mass of the measurable set E, that is, $w(E) = \int_E w(x)\,dx$. With this notation

$$[w]_{A_2} := \sup_Q \langle w\rangle_Q \langle w^{-1}\rangle_Q.$$

Note that $w \in A_2$ if and only if $w^{-1} \in A_2$.

Example 2 Power weights offer examples of A_p weights on $\mathbb{R}^d$, $w(x) = |x|^\alpha$ is in A_p if and only if $-d \le \alpha \le d(p-1)$ for $1 < p < \infty$.

In Theorem 1, the optimal dependence of the constant $C_p(w)$ on the A_p characteristic $[w]_{A_p}$ of the weight w was found more than 30 years later.

Theorem 2 (Petermichl 2007) *Given* $1 < p < \infty$, *for all* $w \in A_p$ *and for all* $f \in L^p(w)$, *we have that*

$$\|Hf\|_{L^p(w)} \lesssim_p [w]_{A_p}^{\max\{1, \frac{1}{p-1}\}} \|f\|_{L^p(w)}.$$

Note that the estimate is *linear* on $[w]_{A_p}$ for $p \ge 2$, and of power $\frac{1}{p-1}$ for $1 < p < 2$.

Proof (Cartoon of the proof) The following is a very brief sketch of Petermichl's argument. First, write H as an average over dyadic grids of dyadic shift operators [161]. Second, find linear estimates, uniform (on the dyadic grids), for the dyadic shift operators on $L^2(w)$ [162]. Deduce from the first two steps linear estimates on $L^2(w)$ for the Hilbert transform, namely, estimates valid for all $w \in A_2$ and for all $f \in L^2(w)$ of the form

$$\|Hf\|_{L^2(w)} \lesssim [w]_{A_2} \|f\|_{L^2(w)}.$$

Third, use a sharp extrapolation theorem [67] to get estimates for $p \neq 2$ from the linear $L^2(w)$ estimate.

Same estimates hold for ALL Calderón–Zygmund singular integral operators, solving the famous A_2 conjecture, which was proven by Tuomas Hytönen in 2012, see [90]. We will say more about Petermichl's and Hytönen's landmark results as well as about sharp extrapolation later in Sect. 7.2.6 and in Sect. 7.4.

7.2.3 Maximal Function

We summarize the L^p and one-weight (quantitative) L^p boundedness properties for the maximal function. We also show that the A_p condition on the weight w is a necessary condition for boundedness of the maximal function on $L^p(w)$.

7.2.3.1 L^p Boundedness Properties of M

From its definition (7.2.4), it is clear that the maximal function is bounded on $L^\infty(\mathbb{R}^d)$ with norm one. The maximal function is not bounded on $L^1(\mathbb{R}^d)$; however, it is of weak-type (1,1) (Hardy, Littlewood 1930). The next example shows that the maximal function does not map $L^1(\mathbb{R})$ onto itself.

Example 3 The characteristic function $\mathbb{1}_{[0,1]}$ is integrable; However, its image, under the maximal function, $M\mathbb{1}_{[0,1]}$, is not. The diligent reader can verify that $M\mathbb{1}_{[0,1]}(x) = 1/(1-x)$ if $x < 0$, $M\mathbb{1}_{[0,1]}(x) = 1$ if $0 \le x \le 1$, and $M\mathbb{1}_{[0,1]}(x) = 1/x$ if $x > 1$.

Marcinkiewicz interpolation gives boundedness of the maximal function on $L^p(\mathbb{R}^d)$ for $1 < p < \infty$ from the strong L^∞ and the weak-type (1, 1) boundedness results. We will present an alternate argument in Sect. 7.3.2.2 that will cover the weighted L^p estimates as well without reference to neither interpolation nor extrapolation.

7.2.3.2 One-Weight L^p Inequalities for M

The maximal function is of weak $L^p(w)$ type if and only if $w \in A_p$; moreover, the following quantitative result was proven in 1972 by Benjamin Muckenhoupt [141], for $p \ge 1$ and for all $w \in A_p$,

$$\|M\|_{L^p(w)\to L^{p,\infty}(w)} \lesssim_p [w]_{A_p}^{1/p}, \tag{7.2.7}$$

where the quantity on the left-hand side, $\|M\|_{L^p(w)\to L^{p,\infty}(w)}$, denotes the smallest constant $C > 0$ such that for all $\lambda > 0$ and for all $f \in L^p(w)$

$$w\big(\{x \in \mathbb{R}^d : Mf(x) > \lambda\}\big) \le \left(\frac{C}{\lambda}\|f\|_{L^p(w)}\right)^p.$$

We say a weight w is in the *Muckenhoupt A_1 class* if and only if there is a constant $C > 0$ such that

$$Mw(x) \le Cw(x) \quad \text{for a.e. } x \in \mathbb{R}^d.$$

The infimum over all possible such constants C is denoted $[w]_{A_1}$. The A_1 class of weights is contained in all A_p classes of weights for $p > 1$.

The maximal function is bounded on $L^p(w)$; moreover, the following quantitative result was proven in 1993 by Stephen Buckley [27] valid for $p > 1$ and for all $w \in A_p$ and $f \in L^p(w)$,

$$\|Mf\|_{L^p(w)} \lesssim_p [w]_{A_p}^{1/(p-1)}\|f\|_{L^p(w)}. \tag{7.2.8}$$

Buckley deduced these estimates from quantitative self-improvement integrability results known for A_p weights, the weak $L^{p\pm\varepsilon}(w)$ boundedness of the maximal function, and Marcinkiewicz interpolation. More precisely, $w \in A_p$ implies $w \in A_{p-\varepsilon}$ with $\varepsilon \sim [w]_{A_p}^{1-p'}$ and $[w]_{A_{p-\varepsilon}} \le 2[w]_{A_p}$, on the other hand, Hölder's inequality implies $A_p \subset A_{p+\varepsilon}$ and $[w]_{A_{p+\varepsilon}} \le [w]_{A_p}$. Interpolating between weak $L^{p-\varepsilon}(w)$ and weak $L^{p+\varepsilon}(w)$ estimates and keeping track of the constants, one gets Buckley's quantitative estimate (7.2.8).

In particular, when $p = 2$ the maximal function obeys a linear estimate on $L^2(w)$ with respect to the A_2 characteristic of the weight, namely, for all $w \in A_2$ and $f \in$

$L^2(w)$

$$\|Mf\|_{L^2(w)} \lesssim [w]_{A_2}\|f\|_{L^2(w)}.$$

A beautiful proof of Buckley's quantitative estimate for the maximal function was presented in 2008 by Andrei Lerner [126], mixed A_p-A_∞ estimates in 2011 by Tuomas Hytönen and Carlos Pérez [98], and extensions to spaces of homogeneous type in 2012 by Tuomas Hytönen and Anna Kairema [94]. We will present Lerner's proof of Buckley's inequality (7.2.8) in Sect. 7.3.2.2.

7.2.3.3 A_p is a Necessary Condition for $L^p(w)$ Boundedness of M

We would like to demystify the appearance of the A_p weights in the theory by showing that $w \in A_p$ is a necessary condition for the maximal function to be bounded on $L^p(w)$ when $p > 1$.

We will show that IF the maximal function is bounded on $L^p(w)$ THEN the weight w must be in the Muckenhoupt A_p class.

Proof By hypothesis, there is a constant $C > 0$ such that for all $f \in L^p(w)$,

$$\|Mf\|_{L^p(w)} \leq C\|f\|_{L^p(w)}.$$

For all $\lambda > 0$, let E_λ^{Mf} be the λ-level set for the maximal function Mf, that is,

$$E_\lambda^{Mf} := \{x \in \mathbb{R}^d : Mf(x) \geq \lambda\},$$

then, by Chebychev's inequality,[3] and using the hypothesis we conclude that

$$w(E_\lambda^{Mf}) = \int_{E_\lambda^{Mf}} w(x)\,dx \leq \frac{1}{\lambda^p}\int_{\mathbb{R}^d} |Mf(x)|^p w(x)\,dx \leq \frac{C^p}{\lambda^p}\|f\|_{L^p(w)}^p.$$

Fix a cube $Q \subset \mathbb{R}^d$, for any integrable function $f \geq 0$, supported on the cube Q, let $\lambda := \frac{1}{|Q|}\int_Q f(y)\,dy$. Then $Mf(x) \geq \lambda$ for all $x \in Q$, hence, $Q \subset E_\lambda^{Mf}$; moreover,

$$\Big(\frac{1}{|Q|}\int_Q f(x)\,dx\Big)^p w(Q) \leq \lambda^p\, w(E_\lambda^{Mf}) \leq C^p \int_Q f^p(x)\,w(x)\,dx. \tag{7.2.9}$$

Consider the specific function $f = w^{\frac{-1}{p-1}}\mathbb{1}_Q$ supported on Q and chosen so that both integrands coincide, namely, $f = f^p w$. Substitute this specific function f into (7.2.9) to obtain the following inequality only pertaining the weight w and the cube Q,

[3] Namely, for $g \in L^1(\mu)$ it holds that $\mu\{x \in \mathbb{R} : |g(x)| > \lambda\}| \leq \frac{1}{\lambda}\|g\|_{L^1(\mu)}$ for all $\lambda > 0$, in other words if $g \in L^1(\mu)$ then $g \in L^{1,\infty}(\mu)$, where $g \in L^{p,\infty}(\mu)$ means $\|g\|_{L^{p,\infty}(\mu)} := \sup_{\lambda>0}\lambda\mu^{1/p}\{x \in \mathbb{R}^d : |g(x)| > \lambda\} < \infty$.

$$\frac{1}{|Q|^p}\Big(\int_Q w^{\frac{-1}{p-1}}(x)\,dx\Big)^{p-1} w(Q) \le C^p.$$

Distribute $|Q|$ and take the supremum over all cubes Q to conclude that $[w]_{A_p} \le C^p$, and hence $w \in A_p$. There is one technicality; the chosen function may not be integrable; choose instead $f_\varepsilon = \mathbb{1}_Q(w+\varepsilon)^{\frac{-1}{p-1}}$, run the argument for each $\varepsilon > 0$ then let ε go to zero.

We just showed that if the maximal function M is bounded on $L^p(w)$ then it is of weak $L^p(w)$ type. Moreover, $[w]_{A_p}^{1/p} \le \|M\|_{L^p(w)\to L^{p,\infty}(w)}$; therefore, Muckenhoupt's weak $L^p(w)$ bound (7.2.7) is optimal.

7.2.4 Why Are We Interested in These Estimates?

We record a few instances where L^p and weighted L^p estimates are of importance in analysis.

- Fourier Analysis: Boundedness of the periodic Hilbert transform on $L^p(\mathbb{T})$ implies convergence on $L^p(\mathbb{T})$ of the partial Fourier sums.
- Complex Analysis: Hf is the boundary value of the harmonic conjugate of the Poisson extension to the upper half-plane of a function $f \in L^p(\mathbb{R})$.
- Factorization: Theory of (holomorphic) Hardy spaces H^p. Elements of H^p can be defined as those distributions whose image under properly defined maximal functions (or other suitable singular operators or square functions) are in L^p.
- Approximation Theory: Boundedness properties of the martingale transform (a dyadic analogue of the Hilbert transform) show that Haar functions and other wavelet families are unconditional bases of several functional spaces.
- PDEs: Boundedness of the *Riesz transforms* (analogues of the Hilbert transform on $\mathbb{R}^d$) and their commutators have deep connections to partial differential equations.
- Quasiconformal Theory: Boundedness of the *Beurling transform* (singular integral operator on $\mathbb{C}$) on $L^p(w)$ for $p > 2$ and with linear estimates on $[w]_{A_p}$ implies borderline regularity result.
- Operator Theory: Weighted inequalities appear naturally in the theory of *Hankel and Toeplitz operators*, perturbation theory, etc.

We expand on the weighted estimate needed in quasiconformal theory which propelled the interest in quantitative weighted estimates. This was worked by Kari Astala, Tadeusz Iwaniec, and Eero Saksman in 2001; we refer to their paper [9] for appropriate definitions. They showed that for $1 < K < \infty$ every weakly K-quasi-regular mapping, contained in a Sobolev space $W^{1,q}_{\text{loc}}(\Omega)$ with $2K/(K+1) < q \le 2$, is quasi-regular on Ω, that is to say, it belongs to $W^{1,2}_{\text{loc}}(\Omega)$. For each $q < 2K/(K+1)$, there are weakly K-quasi-regular mappings $f \in W^{1,q}_{\text{loc}}(\mathbb{C})$ which are not quasi-regular. The only value of q that remained unresolved was the end point; they conjectured that all weakly K-quasi-regular mappings $f \in W^{1,q}_{\text{loc}}$ with $q = 2K/(K+1)$

are in fact quasi-regular. They reduced the conjecture to showing [9, Proposition 22] that the Beurling transform T satisfies linear bounds in $L^p(w)$ for "$p > 1$", namely,

$$\|Tg\|_{L^p(w)} \lesssim_p [w]_{A_p}\|g\|_{L^p(w)}, \quad \text{for all } w \in A_p \text{ and } g \in L^p(w).$$

Fortunately, the values of interest for q are $1 < q < 2$ and $p = q' > 2$. Linear bounds for the Beurling transform and $p \geq 2$ were proven in 2002 by Stefanie Petermichl and Sasha Volberg [165]. As a consequence, the regularity at the borderline case $q = 2K/(K+1)$ was established. For $1 < p < 2$, the correct estimate for the Beurling transform is of the form

$$\|Tg\|_{L^p(w)} \lesssim_p [w]_{A_p}^{1/(p-1)}\|g\|_{L^p(w)}, \quad \text{for all } w \in A_p \text{ and } g \in L^p(w),$$

as shown in [67].

7.2.5 *First Linear Estimates*

Interest in quantitative weighted estimates exploded in this millennium. A chronology of the early linear estimates on $L^2(w)$ for the weight w in the Muckenhoupt A_2 class, namely, $\|Tf\|_{L^2(w)} \leq C[w]_{A_2}\|f\|_{L^2(w)}$, is as follows:

- *Maximal function* (Buckley '93 [27]).
- *Martingale transform* (Wittwer '00 [186]).
- *(Dyadic) square function* (Hukovic, Treil, Volberg '00 [87] ; Wittwer '02 [187]).
- *Beurling transform* (Petermichl, Volberg '02 [165]).
- *Hilbert transform* (Petermichl '07 [162]).
- *Riesz transforms* (Petermichl '08 [163]).
- *Dyadic paraproduct* (Beznosova '08 [21]).

Except for the maximal function, all these linear estimates were obtained using Bellman functions and (bilinear) Carleson estimates for certain dyadic operators (Petermichl dyadic shift operators, martingale transform, dyadic paraproducts, dyadic square function), and then either the operator under study was one of them or had enough symmetries that it could be represented as a suitable average of dyadic operators (Beurling, Hilbert, and Riesz transforms). The Bellman function method was introduced in the 90s to the harmonic analysis by Fedja Nazarov, Sergei Treil, and Sasha Volberg [145, 147], although they credit Donald Burkholder in his celebrated work finding the exact L^p norm for the martingale transform [30]. With their students and collaborators, they have been able to use the Bellman function method to obtain a number of astonishing results not only in this area, see Volberg's INRIA lecture notes [178] and references. In Volberg's own words[4] "the Bellman function method makes apparent the hidden multiscale properties of Harmonic Analysis problems."

---

[4]http://www-sop.inria.fr/apics/ahpi/summerschool11/bellman_lectures_volberg-1.pdf.

A flurry of work ensued and other techniques were brought into play including stopping time techniques (corona decompositions) and median oscillation techniques. These techniques became the precursors of what is now known as the method of domination by dyadic sparse operators, with important contributions from David Cruz-Uribe, Chema Martell, Carlos Pérez, Andrei Lerner, Tuomas Hytönen, Michael Lacey, Mari Carmen Reguera, Stefanie Petermichl, Fedja Nazarov, Sergei Treil, Sasha Volberg, and others. We will say more about sparse domination in Sect. 7.7.

The culmination of this work was the celebrated resolution of the A_2 conjecture by Tuomas Hytönen [90] in 2012 where he showed that first every Calderón–Zygmund operator could be written as an average of dyadic shift operators of arbitrary complexity, dyadic paraproducts, and their adjoints; second the weighted L^2 norm of the dyadic shifts depended linearly on the A_2 characteristic of the weight and polynomially on the complexity; and third these ingredients implied that the Calderón–Zygmund operator obeyed linear bounds on $L^2(w)$. How about weighted L^p estimates for $1 < p < \infty$?

7.2.6 *Extrapolation and Hytönen's A_p Theorem*

There is a, by now, classical technique to obtain weighted L^p estimates from weighted L^2 estimates or more generally from weighted L^r estimates, called extrapolation. In this section, we recall the classical Rubio de Francia extrapolation theorem, a quantitative version, due to Oliver Dragičević et al, called "sharp extrapolation," and deduce from the later Hytönen's A_p theorem.

7.2.6.1 Rubio de Francia Extrapolation Theorem

José Luis Rubio de Francia introduced in the 80s his celebrated extrapolation result, a theorem that allowed to transfer estimates from weighted L^r (provided it held for all A_r weights) to weighted L^p for all $1 < p < \infty$ and all A_p weights.

Theorem 3 (Rubio de Francia 1981) *Given T a sublinear operator and $r \in \mathbb{R}$ with $1 < r < \infty$. If for all $w \in A_r$, there is a constant $C_{T,r,d,w} > 0$ such that*

$$\|Tf\|_{L^r(w)} \leq C_{T,r,d,w} \|f\|_{L^r(w)} \text{ for all } f \in L^r(w).$$

Then for each $1 < p < \infty$ and for all $w \in A_p$, there is a constant $C_{T,p,r,d,w} > 0$ such that

$$\|Tf\|_{L^p(w)} \leq C_{T,p,r,d,w} \|f\|_{L^p(w)} \text{ for all } f \in L^p(w).$$

If we choose $r = 2$, paraphrasing Antonio Córdoba[5] we will conclude that

[5]See page 8 in José García-Cuerva's eulogy for José Luis Rubio de Francia (1949–1988) [75].

There is no L^p just weighted L^2.

(Since $w \equiv 1 \in A_p$ for all p.)

There are books dedicated to the subject that cover this and many useful variants of this theorem; the classical reference is the out-of-print 1985 book by García-Cuerva and Rubio de Francia [76]. A modern presentation, including quantitative versions of this theorem, is the 2011 book by David Cruz-Uribe, Chema Martell, and Carlos Pérez [52].

7.2.6.2 Sharp Extrapolation

In the 80s and 90s, the interest was on qualitative weighted estimates. Once the interest on quantitative weighted estimates was sparked, it was natural to consider quantitative extrapolation theorems, what we call "sharp extrapolation theorems." This is precisely what Stefanie Petermichl and Sasha Volberg did [165] to obtain linear estimates for the Beurling transform and $p \geq 2$; they missed the range $1 < p < 2$ because it was of no interest, and their calculation was very specific to the martingale transforms that properly averaged yielded the Beurling transform. It was soon realized that a general principle was at work [67]. We state a simplified version of what a quantitative extrapolation theorem says, useful for the purposes of this survey.

Theorem 4 (Dragičević et al. 2005) *Let T be a sublinear operator, $r \in \mathbb{R}$ with $1 < r < \infty$. If for all $w \in A_r$ there are constants $\alpha, C_{T,r,d} > 0$ such that*

$$\|Tf\|_{L^r(w)} \leq C_{T,r,d}[w]_{A_r}^{\alpha} \|f\|_{L^r(w)} \text{ for all } f \in L^r(w).$$

Then for each $1 < p < \infty$ and for all $w \in A_p$, there is a constant $C_{T,p,r,d} > 0$ such that

$$\|Tf\|_{L^p(w)} \leq C_{T,p,r,d}[w]_{A_p}^{\alpha \max\{1, \frac{r-1}{p-1}\}} \|f\|_{L^p(w)} \text{ for all } f \in L^p(w).$$

The proof follows by now standard arguments involving the celebrated Rubio de Francia algorithm, and inserting whenever possible Buckley's quantitative bounds (7.2.8) for the maximal function [27].

An alternative, streamlined proof of the sharp extrapolation theorem, was presented by Javier Duoandikoetxea in [69], extending the result to more general settings including off-diagonal and partial range extrapolation. It was observed [52] that one can replace the pair (Tf, f) by a pair of functions (g, f) in the extrapolation theorem, in particular, one could consider the pair (f, Tf) instead, as long as one has the corresponding initial weighted inequalities required to jump-start the theorem.

Sharp extrapolation is sharp in the sense that no better power for $[w]_{A_p}$ can appear in the conclusion that will work for all operators. For some operators, it is known that the extrapolated $L^p(w)$ bounds from the known optimal $L^r(w)$ estimates are themselves optimal for all $1 < p < \infty$. However, it is not necessarily optimal for a particular given operator. Here are some examples illustrating this phenomenon.

Example 4 Start with Buckley's sharp estimate on $L^r(w)$, $\alpha = \frac{1}{r-1}$; for the maximal function, extrapolation will give sharp bounds only for $1 < p \leq r$.

Example 5 Sharp extrapolation from $r = 2$, $\alpha = 1$, is sharp for the Hilbert, Beurling, and Riesz transforms for all $1 < p < \infty$ (for $p > 2$ [162, 163, 165]; $1 < p < 2$ [67]).

Example 6 Extrapolation from linear bound on $L^2(w)$ is sharp for the dyadic square function only when $1 < p \leq 2$ ("sharp" [67], "only" [127]). However, extrapolation from square root bound on $L^3(w)$ is sharp for all $p > 1$ [53].

7.2.6.3 Hytönen's A_p Theorem

Sharp extrapolation was used by Tuomas Hytönen to prove the celebrated A_p theorem, the quantitative weighted L^p estimates for Calderón–Zygmund operators [90].

Theorem 5 (Hytönen 2012) *Let* $1 < p < \infty$ *and let* T *be any Calderón–Zygmund singular integral operator on* $\mathbb{R}^d$, *then for all* $w \in A_p$ *and* $f \in L^p(w)$

$$\|Tf\|_{L^p(w)} \lesssim_{T,d,p} [w]_{A_p}^{\max\{1, \frac{1}{p-1}\}} \|f\|_{L^p(w)}.$$

Proof (*Cartoon of the proof*) Enough to prove the $p = 2$ case, thanks to sharp extrapolation. To prove the linear weighted L^2 estimate, two important steps were required.

First, prove a representation theorem in terms of Haar shift operators of arbitrary complexity, dyadic paraproducts, and their adjoints on random dyadic grids introduced in [148]. This representation hinges on certain reductions obtained in [160].

Second, prove linear estimates on $L^2(w)$ with respect to the A_2 characteristic for paraproducts [21] and Haar shift operators [122] but with polynomial dependence on the complexity (independent of the dyadic grid) [90].

We will say more about random dyadic grids, Haar shift operators, and paraproducts, the ingredients in Hytönen's theorem, in Sects. 7.3 and 7.4. It is now well understood that the $L^2(w)$ bounds for the Haar shift operators not only depend linearly on the A_2 characteristic of w but also depend linearly on the complexity [174].

Sharp extrapolation has also been used to obtain quantitative estimates in other settings. For example, Sandra Pott and Mari Carmen Reguera used sharp extrapolation when studying the Bergman projection on weighted Bergman spaces in terms of the Békollé constant [167]. They proved the base estimate on $L^2(w)$ for certain sparse dyadic operators and then showed that the Bergman projection could be dominated with these sparse dyadic operators.

7.2.7 Two-Weight Problem for the Hilbert Transform and the Maximal Function

We briefly state a necessarily incomplete chronological list of two-weight results for the Hilbert transform, the maximal function, and allied dyadic operators.

7.2.7.1 Two-Weight Problem for H and Its Dyadic Model the Martingale Transform

In the '80s, Mischa Cotlar and Cora Sadosky found necessary and sufficient conditions à la Helson–Szegö solving the two-weight problem for the Hilbert transform. The methods used involved complex analysis and had applications to operator theory [48, 49]. Afterward, various sets of sufficient conditions à la Muckenhoupt were found to be valid also in the matrix-valued context; one of the earliest such sets appeared in 1997 in joint work with Nets Katz [107], see also the 2005 unpublished manuscript [150]. Necessary and sufficient conditions for (uniform and individual) martingale transform and well-localized dyadic operators were found in 1999 and 2008, respectively, by Fedja Nazarov, Sergei Treil, Sasha Volberg [146, 149]; using Bellman function techniques, we will say more about this in Sect. 7.4.1. Long-time sought necessary and sufficient conditions for two-weight boundedness of the Hilbert transform were found in 2014 by Michael Lacey, Eric Sawyer, Chun-Yen Shen, and Ignacio Uriarte-Tuero [113, 123] for pairs of weights that do not share a common point mass. Corresponding quantitative estimates were obtained using very delicate stopping time arguments. See also [114]. Improvements have since been obtained, relaxing the conditions on the weights, by the same authors and Tuomas Hytönen [92].

7.2.7.2 Two-Weight Estimates for the Maximal Function

In 1982, Eric Sawyer showed in [168] that the maximal function M is bounded from $L^2(u)$ into $L^2(v)$ if and only if the following testing conditions[6] hold for the weights u and v: there is a constant $C_{u,v} > 0$ such that for all cubes Q

$$\int_Q \big(M(\mathbb{1}_Q u^{-1})(x)\big)^2 v(x)\,dx \leq C_{u,v} u^{-1}(Q) \text{ and } \int_Q \big(M(\mathbb{1}_Q v)(x)\big)^2 u^{-1}(x)\,dx \leq C_{u,v} v(Q).$$

Sawyer also identified necessary and sufficient conditions for two-weight inequalities for certain positive operators, the fractional and Poisson integrals [169]; these results were of qualitative type. In 2009, Kabe Moen presented the first quantitative result [139]; he proved that the two-weight operator norm of M is comparable to the constants $C_{u,v}$ in Sawyer's result. Note that Sawyer's testing conditions imply the

[6] Nowadays called "Sawyer's testing conditions".

following *joint* $\mathcal{A}_2$ *condition*:

$$[u, v]_{\mathcal{A}_2} := \sup_Q \langle u^{-1}\rangle_Q \langle v\rangle_Q < \infty, \quad \text{where} \quad \langle v\rangle_Q := v(Q)/|Q|$$

In 2015, Carlos Pérez and Ezequiel Rela [159] considered a particular case when $(u, v) \in \mathcal{A}_2$ and $u^{-1} \in A_\infty$ and showed the following so-called mixed-type estimate

$$\|M\|_{L^2(u)\to L^2(v)} \lesssim [u, v]_{\mathcal{A}_2}^{\frac{1}{2}} [u^{-1}]_{A_\infty}^{\frac{1}{2}}.$$

In the one-weight setting, when $u = v = w$, one gets the following improved mixed-type estimate:

$$\|M\|_{L^2(w)\to L^2(w)} \lesssim [w]_{A_2}^{\frac{1}{2}} [w^{-1}]_{A_\infty}^{\frac{1}{2}} \le [w]_{A_2}.$$

The A_∞ *class* of weights is defined to be the union of all the A_p classes of weights for $p > 1$; the *classical* A_∞ *characteristic* is given by

$$[w]_{A_\infty^{\mathrm{cl}}} := \sup_Q \langle w\rangle_Q \exp(-\langle \log w\rangle_Q).$$

A weight w is in A_∞ if and only if $[w]_{A_\infty^{\mathrm{cl}}} < \infty$. An equivalent characterization is obtained using instead the Fujii–Wilson characteristic, defined by

$$[w]_{A_\infty} := \sup_Q \frac{1}{w(Q)} \int_Q M(w\chi_Q)(x)\, dx.$$

The Fujii–Wilson A_∞ characteristic is smaller than the classical one [23]. For mixed-type estimates of similar nature for Calderón–Zygmund singular integral operators, see [98].

For sharp weighted inequalities for fractional integral operators, see [121].

7.3 Dyadic Harmonic Analysis

In this section, we introduce the elements of dyadic harmonic analysis and the basic dyadic maximal function. More precisely, we discuss dyadic grids (regular, random, adjacent) and Haar functions on the line, on $\mathbb{R}^d$, and on spaces of homogeneous type. As a first example, illustrating the power of the dyadic techniques, we present Lerner's proof of Buckley's quantitative L^p estimates for the maximal function, which reduces, using the one-third trick, to estimates for the dyadic maximal function. We also describe, given dyadic cubes on spaces of homogeneous type, how to construct corresponding Haar bases, and briefly describe the Auscher–Hytönen "wavelets" in this setting.

7.3.1 *Dyadic Intervals, Dyadic Maximal Functions*

In this section, we recall the dyadic intervals and the weighted dyadic maximal function on the line, as well as basic L^p estimates for the dyadic maximal function.

7.3.1.1 Dyadic Intervals

The *standard dyadic grid* $\mathcal{D}$ on $\mathbb{R}$ is the collection of intervals of the form $[k2^{-j}, (k+1)2^{-j})$, for all integers $k, j \in \mathbb{Z}$. The dyadic intervals are organized by generations: $\mathcal{D} = \cup_{j\in\mathbb{Z}}\mathcal{D}_j$, where $I \in \mathcal{D}_j$ if and only if $|I| = 2^{-j}$. Note that the Larger the j is, the smaller the intervals are. For each interval $J \in \mathcal{D}$ denote by $\mathcal{D}(J)$ the collection of dyadic intervals I contained in J.

The standard dyadic intervals satisfy the following properties,

- (Partition Property) Each generation $\mathcal{D}_j$ is a partition of $\mathbb{R}$.
- (Nested property) If $I, J \in \mathcal{D}$ then $I \cap J = \emptyset, \quad I \subseteq J$, or $J \subset I$.
- (One parent property) If $I \in \mathcal{D}_j$ then there is a unique interval $\widetilde{I} \in \mathcal{D}_{j-1}$, called the *parent* of I, such that $I \subset \widetilde{I}$. The parent is twice as long as the child, that is, $|\widetilde{I}| = 2|I|$.
- (Two children property) Given $I \in \mathcal{D}_j$, there are two disjoint intervals $I_r, I_l \in \mathcal{D}_{j+1}$ (the right and left children), such that $I = I_l \cup I_r$.
- (Tower of dyadic intervals) Each point $x \in \mathbb{R}$ belongs to exactly one dyadic interval $I_j(x) \in \mathcal{D}_j$. The family $\{I_j(x)\}_{j\in\mathbb{Z}}$ forms a "tower" or "cone" over x. The union of the intervals in a "tower", $\cup_{j\in\mathbb{Z}}I_j(x)$, is a "quadrant".
- (Two-quadrant property) The origin, 0, separates the positive and the negative dyadic interval, creating two "quadrants".

More generally, a *dyadic grid* on $\mathbb{R}$ is a collection of intervals organized in generations with the partition, nested, and two children properties. In this subsection, we reserve the name $\mathcal{D}$ for the standard dyadic grid; however, later on we will use $\mathcal{D}$ to denote a general dyadic grid.

The partition and nested properties are common to all dyadic grids; the one parent property is a consequence of these properties. The two children property is responsible for the name "dyadic", the equal-length property is a consequence of choosing to subdivide into halves, and is in general not so important; one could subdivide into two children of different lengths; if the ratio is uniformly bounded, we have a homogeneous or doubling dyadic grid. One can manufacture dyadic grids on the line where each interval has two equal-length children but there is no distinguished point and only one quadrant. This is because given an interval in the grid, its descendants are completely determined; however, we have two choices for the parent, and hence four choices for the grandparent, etc. In [133], dyadic grids are defined to have one quadrant; such grids have the additional useful property that given any compact set there will be a dyadic interval containing it.

There are many variants, for example, we could subdivide each interval into a uniformly bounded number of children or into arbitrarily finitely many children.

In fact, there are regular dyadic structures on $\mathbb{R}^d$ where the role of the intervals is played by cubes with sides parallel to the axes. In this case, each cube in the dyadic grid is subdivided into 2^d congruent children, see Sect. 7.3.4.2. We will also see that there are dyadic structures in spaces of homogeneous type, where each "cube" may have no more than a fixed number of children, but sometimes it will only have one child (itself) for several generations, see Sect. 7.3.5.2. In all cases, the dyadic grids provide a hierarchical structure that allows for simplified arguments in this setting, the so-called "induction on scale arguments."

7.3.1.2 Dyadic Maximal Function

Given a dyadic grid $\mathcal{D}$ on $\mathbb{R}^d$ and a weight u, the (weighted) dyadic maximal function $M_u^{\mathcal{D}}$ is defined as the maximal function M except that instead of taking the supremum over all cubes in $\mathbb{R}^d$ with sides parallel to the axes we restrict to the dyadic cubes. This is often how one transitions from continuous to dyadic models.

More precisely, the *weighted dyadic maximal function* with respect to a weight u and a dyadic grid $\mathcal{D}$ on $\mathbb{R}^d$ is defined by

$$M_u^{\mathcal{D}} f(x) := \sup_{Q\in\mathcal{D}, Q\ni x} \frac{1}{u(Q)} \int_Q |f(y)|\, u(y)\, dy.$$

Here $u(Q) := \int_Q u(x)\, dx$. When $u = 1$ a.e. then $M_1^{\mathcal{D}} =: M^{\mathcal{D}}$.

The dyadic maximal function inherits boundedness properties from the regular maximal function. This is clear once one notices that the dyadic maximal function is trivially pointwise dominated by the maximal function. However, these properties are much easier to verify for the dyadic maximal function. We now list three basic boundedness properties of the dyadic maximal function, with a word or two as to how one can verify each one of them.

First, the dyadic maximal function, $M_u^{\mathcal{D}}$, is of weak $L^1(u)$ type, with constant one (independent of dimension). This is an immediate corollary of the Calderón–Zygmund lemma (a stopping time); no covering lemmas are required unlike the usual arguments for M.

Second, clearly $M_u^{\mathcal{D}}$ is bounded on $L^\infty(u)$ with constant one. Interpolation between the weak $L^1(u)$ and the $L^\infty(u)$ estimates shows that $M_u^{\mathcal{D}}$ is bounded on $L^p(u)$ for all $p > 1$. Moreover, the following estimate holds with a constant independent of the weight v and the dimension d,

$$\|M_u^{\mathcal{D}} f\|_{L^p(u)} \lesssim p' \|f\|_{L^p(u)} \quad \text{where } \frac{1}{p} + \frac{1}{p'} = 1 \text{ and } p > 1. \tag{7.3.1}$$

Third, the dyadic maximal function is pointwise comparable to the maximal function. We explain in Sect. 7.3.2.1 why this domination holds in the one-dimensional case ($d = 1$).

7.3.2 *One-Third Trick and Lerner's Proof of Buckley's Result*

We present the one-third trick on $\mathbb{R}$ and how it can be used to dominate the maximal function by a sum of dyadic maximal functions. The one-third trick appeared in print in 1991 in Kate Okikiolu characterization of subsets of rectifiable curves in $\mathbb{R}^d$ [152, Lemma 1(b)], see [50, Footnote p.32] for fascinating historical remarks on the one-third trick. This was probably well known among the John Garnett's school of thought see, for example, [78], and also by the Polish school specifically by Tadeusz Figiel [74]. We illustrate how this principle can be used to recover Buckley's quantitative weighted L^p estimate for the maximal function.

7.3.2.1 One-Third Trick

The families of intervals $\mathcal{D}^i := \cup_{j\in\mathbb{Z}}\mathcal{D}^i_j$, for $i = 0, 1, 2$, where

$$\mathcal{D}^i_j := \{2^{-j}\big([0,1) + m + (-1)^j \tfrac{i}{3}\big) : m \in \mathbb{Z}\},$$

are dyadic grids satisfying partition, nested, and two equal children properties. We make four observations. First, when $i = 0$ we recover the standard dyadic grid, $\mathcal{D}^0 = \mathcal{D}$. Second, the grids $\mathcal{D}^1$ and $\mathcal{D}^2$ are nested but there is only one quadrant (the line $\mathbb{R}$). Third, the grids, $\mathcal{D}^i$, for $i = 0, 1, 2$ are as "far away" as possible from each other, to be made more precise in Example 8. Fourth, given any finite interval $I \subset \mathbb{R}$, for *at least two values* of $i = 0, 1, 2$, there are $J^i \in \mathcal{D}^i$ such that $I \subset J^i$, $3|I| \le |J^i| \le 6|I|$. In particular, this implies that given $i \neq k$, $i, k = 0, 1, 2$, there is at least one interval $J \in \mathcal{D}^i \cup \mathcal{D}^k$ such that $I \subset J$ and $3|I| \le |J| \le 6|I|$, and furthermore

$$\frac{1}{|I|}\int_I |f(y)|\,dy \le \frac{6}{|J|}\int_J |f(y)|\,dy.$$

This last observation allows us to dominate the maximal function M by its dyadic counterpart. In fact, the following estimate holds,

$$Mf(x) \le 6\big(M^{\mathcal{D}}f(x) + M^{\mathcal{D}^1}f(x)\big). \tag{7.3.2}$$

More precisely, for $i \neq k$

$$\begin{aligned} Mf(x) &= \sup_{I\ni x}\frac{1}{|I|}\int_I |f(y)|\,dy \le 6 \sup_{J\in\mathcal{D}^i\cup\mathcal{D}^k : J\ni x}\frac{1}{|J|}\int_J |f(y)|\,dy \\ &\le 6\max\{M^{\mathcal{D}^i}f(x), M^{\mathcal{D}^k}f(x)\} \le 6\Big[M^{\mathcal{D}^i}f(x) + M^{\mathcal{D}^k}f(x)\Big]. \end{aligned}$$

In particular, setting $i = 0$ and $k = 1$, we obtain (7.3.2).

There is an analogue of the one-third trick in higher dimensions. In $\mathbb{R}^d$, one can get by with 3^d grids as is very well explained in [133, Sect. 3], with 2^d grids [98], or, with $d+1$ grids and this is optimal, by cleverly choosing the grids, for $\mathbb{R}$ and for the d-torus see [138], for $\mathbb{R}^d$ and $d>1$ see [43].

7.3.2.2 Buckley's A_p Estimate for the Maximal Function

We illustrate the use of dyadic techniques paired with domination to recover Stephen Buckley's quantitative weighted L^p estimate for the maximal function [27]. Namely, for all $w \in A_p$ and $f \in L^p(w)$

$$\|Mf\|_{L^p(w)} \lesssim [w]_{A_p}^{\frac{1}{p-1}} \|f\|_{L^p(w)}.$$

The beautiful argument we present is due to Andrei Lerner [126].

Proof (Lerner's Proof) By the one-third trick suffices to check that for $1<p<\infty$ there is a constant $C_p>0$ such that for all $w \in A_p$ and for all $f \in L^p(w)$ then

$$\|M^{\mathcal{D}} f\|_{L^p(w)} \le C_p [w]_{A_p}^{\frac{1}{p-1}} \|f\|_{L^p(w)},$$

independently of the dyadic grid $\mathcal{D}$ chosen on $\mathbb{R}^d$.

For any dyadic cube $Q \in \mathcal{D}$, let $A_p(Q) = w(Q)\big(\sigma(Q)\big)^{p-1}/|Q|^p$, where we denote by $\sigma := w^{\frac{-1}{p-1}}$ the dual weight of w, then

$$\begin{aligned}\frac{1}{|Q|}\int_Q |f(x)|\,dx &= A_p(Q)^{\frac{1}{p-1}}\left[\frac{|Q|}{w(Q)}\Big(\frac{1}{\sigma(Q)}\int_Q |f(x)|\,\sigma^{-1}(x)\,\sigma(x)\,dx\Big)^{p-1}\right]^{\frac{1}{p-1}} \\ &\le [w]_{A_p}^{\frac{1}{p-1}}\left[\frac{1}{w(Q)}\int_Q \big(M_\sigma^{\mathcal{D}}(f\sigma^{-1})(x)\,w^{-1}(x)\,w(x)\,dx\big)^{p-1}\right]^{\frac{1}{p-1}}.\end{aligned}$$

Taking the supremum over $Q \in \mathcal{D}$, we obtain

$$M^{\mathcal{D}} f(x) \le [w]_{A_p}^{\frac{1}{p-1}}\Big[M_w^{\mathcal{D}}(M_\sigma^{\mathcal{D}}(f\sigma^{-1})^{p-1}w^{-1})(x)\Big]^{\frac{1}{p-1}}.$$

Computing the $L^p(w)$ norm on both sides, recalling that $(p-1)p' = p$ where $\frac{1}{p}+\frac{1}{p'}=1$ and carefully peeling off the maximal functions, we get

$$\begin{aligned}
\|M^{\mathcal{D}} f\|_{L^p(w)} &\leq [w]_{A_p}^{\frac{1}{p-1}} \|M_w^{\mathcal{D}}(M_\sigma^{\mathcal{D}}(f\sigma^{-1})^{p-1}w^{-1})\|_{L^{p'}(w)}^{\frac{1}{p-1}} \\
&\leq [w]_{A_p}^{\frac{1}{p-1}} \|M_w^{\mathcal{D}}\|_{L^{p'}(w)}^{\frac{1}{p-1}} \|M_\sigma^{\mathcal{D}}(f\sigma^{-1})\|_{L^p(\sigma)} \\
&\leq [w]_{A_p}^{\frac{1}{p-1}} \|M_w^{\mathcal{D}}\|_{L^{p'}(w)}^{\frac{1}{p-1}} \|M_\sigma^{\mathcal{D}}\|_{L^p(\sigma)} \|f\sigma^{-1}\|_{L^p(\sigma)} \\
&\leq p^{\frac{1}{p-1}} p' [w]_{A_p}^{\frac{1}{p-1}} \|f\|_{L^p(w)},
\end{aligned}$$

where we used in the last line the uniform bounds (7.3.1) of $M_w^{\mathcal{D}}$ on $L^{p'}(w)$ and $M_\sigma^{\mathcal{D}}$ on $L^p(\sigma)$.

Notice that in this argument neither extrapolation nor interpolation is used. For extensions to two-weight inequalities and to the fractional maximal function, see [139].

7.3.3 Random Dyadic Grids on $\mathbb{R}$

For the purpose of this section, a dyadic grid on $\mathbb{R}$ is a collection of intervals that are organized in generations; each generation provides a partition of $\mathbb{R}$ and the family has the nested, one parent, and two equal-length children per interval properties. Shifted and scaled regular dyadic grid are dyadic grids. These are not the only ones; there are other dyadic grids, such as the ones defined for the one-third trick: $\mathcal{D}^1$ and $\mathcal{D}^2$. The following parametrization will capture ALL dyadic grids in $\mathbb{R}$ [89].

Lemma 1 (Hytönen 2008) *For each scaling parameter r with $1 \leq r < 2$, and shift parameter $\beta \in \{0,1\}^{\mathbb{Z}}$, meaning $\beta = \{\beta_i\}_{i\in\mathbb{Z}}$ with $\beta_i = 0$ or 1, then $\mathcal{D}^{r,\beta} := \cup_{j\in\mathbb{Z}}\mathcal{D}_j^{r,\beta}$ is a dyadic grid, where*

$$\mathcal{D}_j^{r,\beta} := r\mathcal{D}_j^{\beta}, \quad \text{and} \quad \mathcal{D}_j^{\beta} := x_j + \mathcal{D}_j, \quad \text{with} \quad x_j = \sum_{i>j} \beta_i 2^{-i}.$$

We shift by a different parameter x_j at each level j, in a way that is consistent and preserves the nested property of the grid. Moreover, the shift parameter $\beta_j = 0, 1$ for $j \in \mathbb{Z}$ encodes the information whether a base interval at level j will be the right or the left half of its parent.

Example 7 Shifted and scaled regular grids correspond to the shift parameter $\beta_i = 0$ for all $i < N$ (or $\beta_i = 1$ for all $i < N$) for some integer N. These are the grids with two quadrants. Comparatively speaking, this set of dyadic grids is negligible, since it corresponds to a set of measure zero in parameter space described below.

Example 8 The 1/3-shifted dyadic grids introduced in the previous section correspond to Hytönen's dyadic grids for $r = 1$. More precisely,

$$\mathcal{D}^i = \mathcal{D}^{1,\beta^i} \quad \text{for } i \in \{0, 1, 2\},$$

where for all $j \in \mathbb{Z}$, $\beta_j^0 \equiv 0$ (or $\equiv 1$), $\beta_j^1 = \mathbb{1}_{2\mathbb{Z}}(j)$, and $\beta_j^2 = \mathbb{1}_{2\mathbb{Z}+1}(j)$.

We call these grids *random dyadic grids* because we view the parameters β_j and r as independent identically distributed random variables. There is a very natural probability space, say $(\Omega, \mathbb{P})$ associated to the parameters, $\Omega = [1, 2) \times \{0, 1\}^{\mathbb{Z}}$. Averaging in this context means calculating the expectation in this probability space, that is,

$$E_\Omega f = \int_\Omega f(\omega)\, d\mathbb{P}(\omega) = \int_1^2 \int_{\{0,1\}^{\mathbb{Z}}} f(r, \beta)\, d\mu(\beta)\, \frac{dr}{r},$$

where μ stands for the canonical probability measure on $\{0, 1\}^{\mathbb{Z}}$ which makes the coordinate functions β_j independent with $\mu(\beta_j = 0) = \mu(\beta_j = 1) = 1/2$.

Random dyadic grids have been used, for example, in the study of $T(b)$ theorems on metric spaces with non-doubling measures [97, 148] and of BMO from dyadic BMO on the bidisc and product spaces of spaces of homogeneous type [32, 166], inspired by celebrated work of John Garnett and Peter Jones from the 80s [78]. They have also been used in Hytönen's representation theorem [90] and in the resolution of the two-weight problem for the Hilbert transform [113, 123].

7.3.4 Haar Bases

Associated to dyadic intervals (or dyadic cubes), there is a very important collection of step functions, the Haar functions. In this section, we recall the Haar bases on $\mathbb{R}$ and on $\mathbb{R}^d$, and some of their well-known properties.

7.3.4.1 Haar Basis on $\mathbb{R}$

The *Haar function* associated to an interval $I \subset \mathbb{R}$ is defined to be

$$h_I(x) := |I|^{-1/2}\big(\mathbb{1}_{I_r}(x) - \mathbb{1}_{I_l}(x)\big),$$

where I_r and I_l are the right and left halves, respectively, of I, and the characteristic function $\mathbb{1}_I(x) = 1$ if $x \in I$, zero otherwise. Haar functions have mean zero, that is, $\int_{\mathbb{R}} h_I = 0$, and they are normalized on $L^2(\mathbb{R})$.

The Haar functions indexed on any dyadic grid $\mathcal{D}$, $\{h_I\}_{I\in\mathcal{D}}$, form a *complete orthonormal system* of $L^2(\mathbb{R})$ (Haar 1910). In particular for all $f \in L^2(\mathbb{R})$, with $\langle f, g\rangle := \int_{\mathbb{R}} f(x)\, \overline{g(x)}\, dx$,

$$f = \sum_{I\in\mathcal{D}} \langle f, h_I\rangle\, h_I.$$

You can find a complete proof of this statement in [156, Chap. 9].

The Haar basis is an unconditional basis of $L^p(\mathbb{R})$ and of $L^p(w)$ if $w \in A_p$ for $1 < p < \infty$ [175]. This is deduced from the boundedness properties of the martingale transform; we will say more about this dyadic operator in Sect. 7.4.1.

The Haar basis constitutes the first example of a wavelet basis[7] and its corresponding Haar multiresolution analysis provides the canonical example of a multiresolution analysis [156, Chaps. 9–11].

7.3.4.2 Dyadic Cubes and Haar Basis on $\mathbb{R}^d$

In d-dimensional Euclidean space, the regular dyadic cubes are Cartesian products of regular dyadic intervals of the same generation. More precisely, a cube $Q \in \mathcal{D}_j(\mathbb{R}^d)$ if and only if $Q = I_1 \times \cdots \times I_d$, where $I_n \in \mathcal{D}_j(\mathbb{R})$ for $n = 1, 2, \ldots, d$. Each generation $\mathcal{D}_j(\mathbb{R}^d)$ is a partition of $\mathbb{R}^d$ and they form a nested grid; each cube has one parent and 2^d congruent children, and there are 2^d quadrants. If we had used dyadic intervals with just one quadrant, then the corresponding dyadic cubes in $\mathbb{R}^d$ will also have only one quadrant. We denote $\mathcal{D}(\mathbb{R}^d)$ the collection of all dyadic cubes in all generations, that is, $\mathcal{D}(\mathbb{R}^d) = \cup_{j\in\mathbb{Z}}\mathcal{D}_j(\mathbb{R}^d)$. For $Q \in \mathcal{D}(\mathbb{R}^d)$, we denote $\mathcal{D}(Q)$ the set of dyadic cubes contained in Q.

For each dyadic cube Q in $\mathbb{R}^d$, we can associate 2^d step functions, constant on each children of Q by taking appropriate tensor products. More precisely, for $Q \in \mathcal{D}(\mathbb{R}^d)$ and $\varepsilon = (\varepsilon_1, \ldots, \varepsilon_d)$, with $\varepsilon_n = 0$ or 1, let

$$h_Q^{\varepsilon}(x_1, \ldots, x_d) := h_{I_1}^{\varepsilon_1}(x_1) \times \cdots \times h_{I_d}^{\varepsilon_d}(x_d),$$

where for each dyadic interval I we denote $h_I^0 := h_I$ and $h_I^1 = |I|^{-1/2}\mathbb{1}_I$. Note that $h_Q^{\mathbf{1}} = |Q|^{-1/2}\mathbb{1}_Q$, where $\mathbf{1} = (1, 1, \ldots, 1)$. The remaining $(2^d - 1)$ functions are the Haar functions associated to the cube Q. The tensor product Haar functions h_Q^{ε}, for $\varepsilon \neq \mathbf{1}$, are supported on the corresponding dyadic cube Q, they have mean zero, L^2 norm one, and they are constant on Q's children. The collection $\{h_Q^{\varepsilon} : \varepsilon \neq \mathbf{1},\ Q \in \mathcal{D}(\mathbb{R}^d)\}$ is an orthonormal basis of $L^2(\mathbb{R}^d)$, and an unconditional basis of $L^p(\mathbb{R}^d), 1 < p < \infty$ (the Haar basis). Figures 7.1 and 7.2 illustrate the Haar functions associated to a square in $\mathbb{R}^2$ and to a cube in $\mathbb{R}^3$, respectively.

The tensor product construction just described seems very rigid, and it is very dependent on the geometry of the cubes and on the group structure of the Euclidean space $\mathbb{R}^d$. Can we do dyadic analysis on other settings? The answer is a resounding YES!!!! One such setting is on spaces of homogeneous type introduced by Coifman and Weiss in the early 70s. In Sect. 7.3.5, we will describe how to construct Haar basis on spaces of homogeneous type given suitable collections of "dyadic cubes"

---

[7] An *orthonormal wavelet basis* of $L^2(\mathbb{R})$ is an orthonormal basis where all its elements are translations and dilations of a fixed function ψ, called the wavelet. More precisely, a function $\psi \in L^2(\mathbb{R})$ is a *wavelet* if and only if the functions $\psi_{j,k}(x) = 2^{j/2}\psi(2^j x - k)$ for $j, k \in \mathbb{Z}$ form an orthonormal basis of $L^2(\mathbb{R})$.

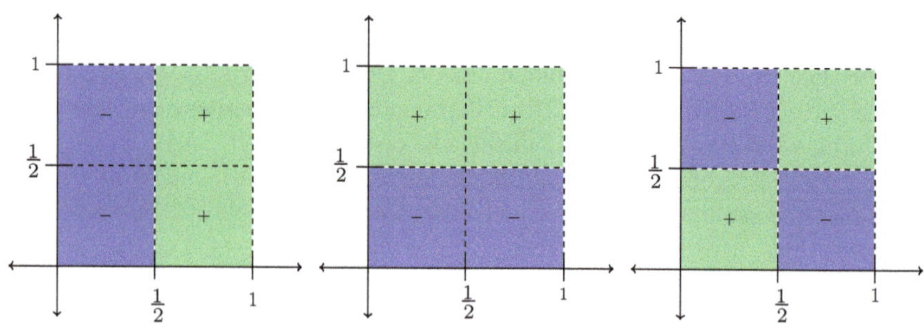

Fig. 7.1 The three Haar functions associated to the unit square in $\mathbb{R}^2$. Figure kindly provided by David Weirich [182]

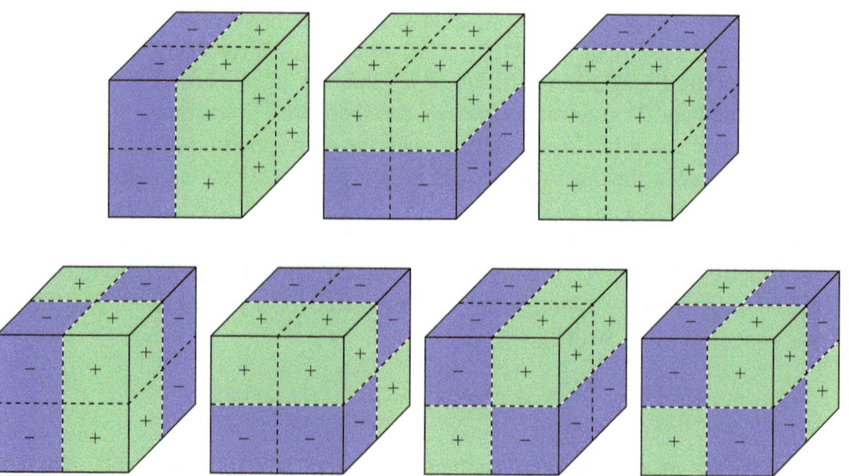

Fig. 7.2 The seven Haar functions associated to a cube in $\mathbb{R}^3$. Figure kindly provided by David Weirich [182]

and argüe why they constitute an orthonormal basis. This argument can be used to show that the Haar functions introduced in this section constitute an orthonormal basis of $L^2(\mathbb{R}^d)$.

7.3.5 Dyadic Analysis on Spaces of Homogeneous Type

In this section, we will define spaces of homogeneous type. We will present a generalization of the dyadic cubes adapted to this setting. Given dyadic cubes, we will show how to construct corresponding Haar functions, and briefly discuss the Auscher–Hytönen wavelets on spaces of homogeneous type.

Before we start, we would like to quote Yves Meyer.

One is amazed by the dramatic changes that occurred in analysis during the twentieth century. In the 1930s complex methods and Fourier series played a seminal role. After many improvements, mostly achieved by the Calderón–Zygmund school, the action takes place today on spaces of homogeneous type. No group structure is available, the Fourier transform is missing, but a version of harmonic analysis is still present. Indeed the geometry is conducting the analysis.

Yves Meyer[8] in his preface to [66].

7.3.5.1 Spaces of Homogeneous Type (SHT)

Let us first define what is a space of homogeneous type in the sense of Coifman and Weiss [42].

Definition 1 (*Coifman, Weiss 1971*) For a set X, a triple (X,ρ,μ) is a space of homogeneous type (SHT) in Coifman–Weiss's sense if

(i) $\rho : X\times X \longrightarrow [0,\infty)$ is a quasi-metric on X, more precisely the following hold:

  (a) (*positive definite*) $\rho(x,y)=0$ if and only if $x=y$;
  (b) (*symmetry*) $\rho(x,y)=\rho(y,x)\geq 0$ for all $x,y\in X$;
  (c) (*quasi-triangle inequality*) there exists constant $A_0\geq 1$ such that

$$\rho(x,y)\leq A_0\big(\rho(x,z)+\rho(z,y)\big) \quad \text{for all} \quad x,y,z\in X.$$

(ii) μ is a nonzero Borel regular[9] measure with respect to the topology induced by the quasi-metric.[10]
(iii) Quasi-metric balls are μ-measurable. A quasi-metric ball is the set $B(x,r):=\{y\in X:\rho(x,y)<r\}$, where $x\in X$ and $r>0$.
(iv) μ is a doubling measure, namely, there exists a constant $D_\mu\geq 1$ (the doubling constant of the measure μ) such that for each quasi-metric ball $B(x,r)$

$$0<\mu(B(x,2r))\leq D_\mu\,\mu(B(x,r))<\infty \quad \text{for all} \quad x\in X, r>0.$$

Notice that Condition (4) implies that there are constants $\omega>0$ (known as an *upper dimension* of μ) and $C\geq 1$ such that for all $x\in X$, $\lambda\geq 1$ and $r>0$

$$\mu(B(x,\lambda r))\leq C\lambda^{\omega}\mu(B(x,r)).$$

---

[8] Recipient of the 2017 Abel Prize.

[9] A measurable set E of finite measure is *Borel regular* if there is a Borel set B such that $E\subset B$ and $\mu(E)=\mu(B)$.

[10] The topology induced by a quasi-metric is the largest topology $\mathcal{T}$ such that for each $x\in X$ the quasi-metric balls centered at x form a fundamental system of neighborhoods of x. Equivalently, a set Ω is *open*, $\Omega\in\mathcal{T}$, if for each $x\in\Omega$ there exists $r>0$ such that the quasi-metric ball $B(x,r)\subset\Omega$. A set in X is *closed* if it is the complement of an open set.

In fact, we can choose $C = D_\mu \geq 1$ and $\omega = \log_2 D_\mu$.

The quasi-metric balls may NOT be open in the topology induced by the quasi-metric, as Example 9 shows. Therefore, the assumption that the quasi-metric balls are μ-measurable is not redundant. The following example illustrates this phenomenon [94].

Example 9 Consider the set $X = \{-1\} \cup [0, \infty)$, the map $\rho : X \times X \to [0, \infty)$ given by $\rho(-1, 0) = \rho(0, -1) = 1/2$ and $\rho(x, y) = |x - y|$; otherwise, and the measure $\mu(E) = \delta_{-1}(E) + m\big(E \cap [0, \infty)\big)$, where m is the Lebesgue measure and δ_{-1} the point mass at $x = -1$, that is, $\delta_{-1}(E) = 0$ if $-1 \notin E$ and $\delta_{-1}(E) = 1$ if $-1 \in E$. Then ρ is not a metric since $\rho(1, -1) = 2 > 3/2 = 1 + 1/2 = \rho(1, 0) + \rho(0, -1)$; however, ρ is a quasi-metric and the measure μ is doubling. It is a good exercise to compute both the quasi-triangle constant of ρ and the doubling constant of μ. Finally, the ball $B(-1, 1) = \{-1, 0\}$ is not open because it does not contain any ball centered at 0 with positive radius r, since $[0, r) \subset B(0, r)$ and the interval $[0, r)$ is not contained in $B(-1, 1)$.

A couple of further remarks are in order.

First, a given quasi-metric ρ may NOT be Hölder regular. Recall that ρ is a *Hölder regular quasi-metric* if there are constants $0 < \theta < 1$ and $C_0 > 0$ such that

$$|\rho(x, y) - \rho(x', y)| \leq C_0 \rho(x, x')^\theta \big[\rho(x, y) + \rho(x, y')\big]^{1-\theta} \quad \forall x, x', y \in X.$$

Metrics are Hölder regular for any $0 < \theta \leq 1$, $C_0 = 1$. The quasi-metric in Example 9 is not continuous let alone Hölder regular. Quasi-metric balls for Hölder regular quasi-metrics are always open.

Second, Roberto Macías and Carlos Segovia showed in 1979 [137] that given a space of homogeneous type (X, ρ, μ) there is an equivalent Hölder regular quasi-metric ρ' on X and some $\theta \in (0, 1)$, and for which the measure μ is 1-*Ahlfors regular*, more precisely,

$$\mu\big(B_{\rho'}(x, r)\big) \sim r^1.$$

Here are some examples of spaces of homogeneous type.

- $\mathbb{R}^n$, with the Euclidean metric and the Lebesgue measure.
- $\mathbb{R}^n$ with the Euclidean metric and an absolutely continuous measure with respect to the Lebesgue measure $d\mu = w\,dx$ where w is a doubling weight (for example, w could be an A_∞ weight).
- Quasi-metric spaces with *d-Ahlfors regular measure*: $\mu(B(x, r)) \sim r^d$ (e.g., Lipschitz surfaces, fractal sets, n-thick subsets of $\mathbb{R}^n$). More concretely, consider, for example, X the four-corner Cantor set with the Euclidean metric and the one-dimensional Hausdorff measure, or consider X the graph of a Lipschitz function $F : \mathbb{R}^n \to \mathbb{R}$ with the induced Euclidean metric and measure the volume of the set's "shadow", $\mu(E) = m\Big(\{x \in \mathbb{R}^n : \big(x, F(x)\big) \subset E\}\Big)$ where m is the Lebesgue measure on $\mathbb{R}^n$.
- C^∞ manifolds with doubling volume measure for geodesic balls.

- Nilpotent Lie groups G with the left-invariant Riemannian metric and the induced measure (e.g., Heisenberg group where X is the boundary of the unit ball in $\mathbb{C}^n$, $\rho(z, w) = 1 - \overline{z} \cdot w$ and with surface measure).

The 2015 book by Ryan Alvarado and Marius Mitrea [7] discusses in more detail many of these examples and relies heavily on the Macías-Segovia philosophy, meaning they consider equivalent classes of quasi-metrics knowing that among them they can choose a representative that is Hölder regular and for which the measure is Ahlfors regular.

7.3.5.2 **Dyadic Cubes in** SHT

Systems of "dyadic cubes" were built by Hugo Aimar and Roberto Macías, Eric Sawyer and Richard Wheeden, and Guy David in the 80s [6, 60, 170], and by Michael Christ in the 90s [33] on spaces of homogeneous type, and by Tuomas Hytönen and Anna Kairema in 2012 on geometrically doubling quasi-metric spaces [94] without reference to a measure.

A *geometrically doubling quasi-metric space* (X, d) is one such that every quasi-metric ball of radius r can be covered with at most N quasi-metric balls of radius $r/2$ for some natural number N.

Example 10 Spaces of homogeneous type in the Coifman–Weiss's sense are geometrically doubling [42].

Systems $\mathcal{D}$ of dyadic cubes in spaces of homogeneous type or, more generally, on geometrically doubling spaces, are organized in disjoint generations $\mathcal{D}_k$, $k \in \mathbb{Z}$, such that $\mathcal{D} = \cup_{k\in\mathbb{Z}}\mathcal{D}_k$ and the following qualitative properties hold.

(a) Each generation $\mathcal{D}_k$ is a partition of X, so the cubes in a generation are pairwise disjoint and form a covering of X.
(b) The generations are nested, that is, there is no partial overlap across generations.
(c) As a consequence, each cube has unique ancestors in earlier generations.
(d) Dyadic cubes have at most M children for some positive natural number M (this is a consequence of the geometric doubling property).
(e) There exists a constant $\delta \in (0, 1)$ such that for every dyadic cube in $\mathcal{D}_k$ there are inner and outer balls of radius roughly δ^k (the "sidelength" of the cube).
(f) The outer ball corresponding to a dyadic cube's child is inside its parent's outer ball.

Note that since $\delta \in (0, 1)$ the larger k is the smaller in diameter the cubes are. If $Q \in \mathcal{D}_k$ then its parent will be the unique cube $\widetilde{Q} \in \mathcal{D}_{k-1}$ such that $Q \subset \widetilde{Q}$.

Furthermore, cubes can be constructed to have a "small boundary property" [33, 94] which is very useful in applications.

A quantitative and more precise statement of the defining properties for a dyadic system of cubes on geometric doubling metric spaces is encapsulated in the following construction that appeared in [94, Theorem 2.2].

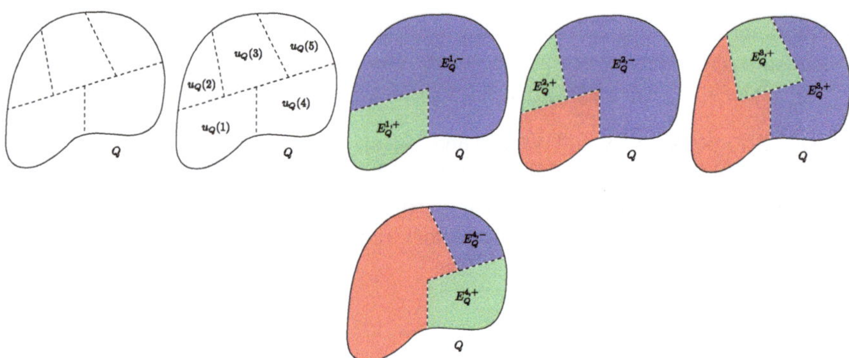

Fig. 7.3 The four Haar functions for a cube with five children in SHT. Figures kindly provided by David Weirich [182]

Theorem 6 (Hytönen, Kairema 2012) *Given (X, d) a geometrically doubling quasi-metric space. Suppose the constants $C_0 \geq c_0 > 1$ and $\delta \in (0, 1)$ satisfy $12A_0^3 C_0 \delta \leq c_0$. Given a set of points $\{z_\alpha^k : \alpha \in \mathcal{A}_k\}$, where $\mathcal{A}_k$ is a countable set of indexes, with the properties that*

$$d(z_\alpha^k, z_\beta^k) \geq c_0 \delta_k \ (\alpha \neq \beta), \quad \min_{\alpha \in \mathcal{A}_k} d(x, z_\alpha^k) < C_0 \delta_k \text{ for all } x \in X.$$

For each $k \in \mathbb{Z}$ and $\alpha \in \mathcal{A}_k$, there exist sets $Q_\alpha^{k,\circ} \subseteq Q_\alpha^k \subseteq \overline{Q}_\alpha^k$—called open, half-open, and closed dyadic cubes—such that

- *(i)* *$Q_\alpha^{k,\circ}$ and $\overline{Q}_\alpha^k$ are the interior and closure of Q_α^k, respectively;*
- *(ii)* (nested) *if $\ell \geq k$, then either $Q_\beta^\ell \subseteq Q_\alpha^k$ or $Q_\alpha^k \cap Q_\beta^\ell = \emptyset$;*
- *(iii)* (partition) $X = \bigcup_{\alpha \in \mathcal{A}_k} Q_\alpha^k$ *for all $k \in \mathbb{Z}$;*
- *(iv)* (inner/outer balls) $B(z_\alpha^k, c_1\delta^k) \subseteq Q_\alpha^k \subseteq B(z_\alpha^k, C_1\delta^k)$ *where $c_1 := (3A_0^2)^{-1} c_0$ and $C_1 := 2A_0 C_0$;*
- *(v)* *if $\ell \geq k$ and $Q_\beta^\ell \subseteq Q_\alpha^k$, then $B(z_\beta^\ell, C_1\delta^\ell) \subseteq B(z_\alpha^k, C_1\delta^k)$.*

The open and closed cubes $Q_\alpha^{k,\circ}$ and $\overline{Q}_\alpha^k$ depend only on the points z_β^ℓ for $\ell \geq k$. The half-open cubes Q_α^k depend on z_β^ℓ for $\ell \geq \min(k, k_0)$, where $k_0 \in \mathbb{Z}$ is a pre-assigned number entering the construction.

The geometrically doubling condition implies that sets of points $\{x_\alpha^k : k \in \mathbb{Z}, \alpha \in \mathcal{A}_k\}$ with the required separation properties exist and that the set $\mathcal{A}_k$ is a countable set of indices for each $k \in \mathbb{Z}$. The cubes in this construction are built as countable unions of quasi-metric balls; hence, once a space of homogeneous type is given, the cubes will be measurable sets.

7.3.5.3 Haar Basis on SHT

Given a space of homogeneous type (X, ρ, μ) with a dyadic structure $\mathcal{D}$ given by Theorem 6, we can construct a system of Haar functions that will be an orthonormal basis of $L^2(X, \mu)$.

Given a cube $Q \in \mathcal{D}$, denote by $\text{ch}(Q)$ the collection of dyadic children of Q, and by $N(Q)$ its cardinality, that is, Q has $N(Q)$ children. Let S_Q be the subspace of $L^2(X, \mu)$ spanned by those square integrable functions that are supported on Q and are constant on the children of Q. The subspace S_Q has dimension $N(Q)$ as the characteristic functions of the children cubes normalized with respect to the L^2 norm, namely, $\{\mathbb{1}_{Q'}/\sqrt{\mu(Q')} : Q' \in \text{ch}(Q)\}$, form an orthonormal basis for S_Q. The subspace S_Q^0 of S_Q consisting of those functions that have mean zero, that is, $\int_Q f(x)\,dx = 0$, will have one fewer dimension, namely, $\dim(S_Q^0) = N(Q) - 1$.

Given an enumeration of the children of Q, that is, a bijection $u_Q : \{1, 2, \ldots, N(Q)\} \to ch(Q)$, we will define recursively subsets of Q that are unions of children of Q. More precisely at each stage, we will remove one child according to the given enumeration, let $E_Q^1 := Q$, given $E_Q^k \subset Q$, let $E_Q^{k+1} = E_Q^k \setminus u_Q(k)$ for $k = 1, 2, \ldots, N(Q) - 1$. We can split each of these sets into two disjoint pieces, $E_Q^i := E_Q^{i,+} \cup E_Q^{i,-}$ where $E_Q^{i,+} = u_Q(i)$, the child removed (light grey in Fig. 7.3) and $E_Q^{i,-} = E_Q^{i+1}$ (dark grey in Fig. 7.3). With this notation, the Haar functions associated to the cube Q and the enumeration u_Q as illustrated in Fig. 7.3, are supported on Q and are constant on the shaded regions: positive on the light grey regions, negative on the dark grey regions, and zero on the grey regions, and thus they are given by

$$h_Q^i(x) = a\mathbb{1}_{E_Q^{i,+}}(x) - b\mathbb{1}_{E_Q^{i,-}}(x), \quad 1 \le i \le N(Q) - 1,$$

where the positive constants a and b, dependent on the base cube Q and the label i, are chosen to enforce L^2 normalization and mean zero. More precisely, the unknowns a, b must satisfy the system of two equations:

$$\int_Q |h_Q^i(x)|^2\,d\mu = a^2\mu(E_Q^{i,+}) + b^2\mu(E_Q^{i,-}) = 1$$

$$\int_Q h_Q^i(x)\,d\mu = a\,\mu(E_Q^{i,+}) - b\,\mu(E_Q^{i,-}) = 0.$$

Solving the system of equations, we get the positive solutions

$$a = \sqrt{\mu(E_Q^{i,-})/\big(\mu(E_Q^i)\,\mu(E_Q^{i,+})\big)}, \quad b = \sqrt{\mu(E_Q^{i,+})/\big(\mu(E_Q^i)\,\mu(E_Q^{i,-})\big)}.$$

Note that the doubling condition on the measure μ ensures $\mu(Q) > 0$ for all $Q \in \mathcal{D}$, and hence also $\mu(E_Q^i) > 0$ for all labels i.

The Haar basis consists of all functions h_Q^i where $Q \in \mathcal{D}$ and $i = 1, 2, \ldots, N(Q) - 1$. Note that a cube may not subdivide for a while, meaning that it could have just one child, itself, for several generations or forever. In the former case, we wait until we subdivide to define the subspace S_Q^0; in the latter case, we let S_Q^0 be the trivial subspace.

By construction for each $Q \in \mathcal{D}$, the collection $\{h_Q^i : i = 1, \dots, N(Q) - 1\}$ is normalized on $L^2(X, \mu)$; each Haar function has mean zero, and by the nested property of the dyadic cubes it is easy to verify this is an orthonormal family. No matter what enumeration for $\mathrm{ch}(Q)$ we use we will get each time an orthonormal basis of S_Q^0. The orthogonal projection onto S_Q^0 of a square integrable function f is independent of the orthonormal basis chosen on S_Q^0. Given $x \in Q$, choose an enumeration so that $x \in u_Q(1) =: R \in \mathrm{ch}(Q)$ then

$$\mathrm{Proj}_{S_Q} f(x) = \langle f, h_Q^1 \rangle_\mu h_Q^1(x) = \langle f \rangle_R^\mu - \langle f \rangle_Q^\mu,$$

where $\langle f, g \rangle_\mu$ denotes the inner product in $L^2(X, \mu)$ and $\langle f \rangle_Q^\mu$ denotes the μ-average of f. The first equality holds by support considerations, since $h^i(x) = 0$ for all $i > 1$ by the choice of the enumeration; the second equality is now a simple calculation by substitution.

Using a telescoping sum argument, one can verify that completeness of the Haar basis on $L^2(\mu)$ hinges on the following limits holding in the $L^2(\mu)$ sense:

$$\lim_{j \to \infty} E_j^\mu f = f,$$

$$\lim_{j \to \infty} E_j^\mu f = 0,$$

where $E_j f := \langle f \rangle_Q^\mu$, with $x \in Q \in \mathcal{D}_j$, or $E_j f = \sum_{Q \in \mathcal{D}_j} \langle f \rangle_Q^\mu \mathbb{1}_Q$. That the limits do hold can be justified by martingale theory [91, 142]; in fact, they do hold in $L^p(X, \mu)$ for $1 < p < \infty$. The pointwise convergence a.e. of the averages to f as j goes to infinity is a consequence of the Lebesgue differentiation theorem which holds because the measure is assumed to be Borel regular, see [7, Sect. 3.3].

Haar-type bases for $L^2(X, \mu)$ have been constructed in general metric spaces, and the construction, along the lines described here, is well known to experts. Haar-type wavelets associated to nested partitions in abstract measure spaces were constructed in 1997 by Girardi and Sweldens [79]. For the case of spaces of homogeneous type, there is a lot of work related to Haar bases done in Argentina this millennium, specifically by Hugo Aimar and collaborators Osvaldo Gorosito, Ana Bernardis, Bibiana Iaffei, and Luis Nowak [1–5], all descendants of Eleonor Harboure. Haar functions have been used in geometrically doubling metric spaces [144]. For the case of a geometrically doubling quasi-metric space (X, ρ), with a positive Borel regular measure μ, see [105].

7.3.5.4 Random Dyadic Grids, Adjacent Dyadic Grids, and Wavelets on SHT

The counterparts of the random dyadic grids and the one-third trick have been identified in the general setting of geometrically doubling quasi-metric spaces by Tuomas Hytönen and his students and collaborators. Using them, Pascal Auscher and Tuomas Hytönen constructed in 2013 a remarkable orthonormal basis of $L^2(X, \mu)$ [10, 11].

A notion of random dyadic grids can be introduced on geometrically doubling quasi-metric spaces (X, d) by randomizing the order relations in the construction of the Hytönen–Kairema cubes [94, 97]. In 2014, Tuomas Hytönen and Olli Tapiola modified the randomization to improve upon Auscher–Hytönen wavelets in metric spaces [101]. A different randomization can be found in [144].

One can find finitely many adjacent families of Hytönen–Kairema dyadic cubes, $\mathcal{D}^t$ for $t = 1, \dots, T$, with the same parameters, that play the role of the 1/3-shifted dyadic grids in $\mathbb{R}$. The main property the adjacent families of dyadic cubes have is that given any ball $B(x,r) \in X$, with $r \sim \delta^k$, then there is $t \in \{1, 2, \dots, T\}$ and a cube in the t-grid and in the kth generation, $Q \in \mathcal{D}^t_k$, such that $B(x,r) \subset Q \subset B(x, Cr)$, where $C > 0$ is a geometric constant only dependent on the quasi-metric and geometric doubling parameters of X [94]. Furthermore, given a σ-finite measure μ on X, the adjacent dyadic systems can be chosen so that all cubes have small boundaries: $\mu(\partial Q) = 0$ for all $Q \in \cup_{t=1}^T \mathcal{D}^t$ [94].

Given nested maximal sets $\mathcal{X}^k$ of δ^k-separated points in X for $k \in \mathbb{Z}$, let $\mathcal{Y}^k := \mathcal{X}^{k+1} \setminus \mathcal{X}^k$ and relabel points in $\mathcal{Y}^k$ by y^k_α. To each point y^k_α, Auscher and Hytönen associate a wavelet function ψ^k_α (a linear spline) of regularity $0 < \eta < 1$ that is morally supported near y^k_α at scale δ^k, with mean zero and some smoothness. More precisely, these functions are not compactly supported but have exponential decay away from the base cube Q^k_α, and they have Hölder regularity exponent $\eta > 0$, where η depends only on δ and on some finite quantities needed for extra labeling of the random dyadic grids used in the construction of the wavelets. The number of indexes α so that $y^k_\alpha \in \mathcal{Y}_k$ for each Q^k_α is exactly $N(Q^k_\alpha) - 1$, where recall that $N(Q^k_\alpha)$ denotes the number of children of Q^k_α. This is the right number of wavelets per cube Q^k_α if our intuition is to be guided by the constructions of the Haar functions. The precise nature of these wavelets is detailed in [10, Theorem 7.1].

Furthermore, the functions $\{\psi^k_\alpha\}_{k\in\mathbb{Z}, \alpha\in\mathcal{Y}^k}$ form an unconditional basis on $L^p(X)$ for all $1 < p < \infty$ and the following wavelet expansion is valid in $L^p(X)$,

$$f(x) = \sum_{k\in\mathbb{Z}} \sum_{y^k_\alpha \in \mathcal{Y}^k} \langle f, \psi^k_\alpha \rangle\, \psi^k_\alpha(x).$$

Hytönen and Tapiola were able to build such wavelets for all $0 < \eta < 1$ in the context of metric spaces [101]. It is still an open problem to construct smooth wavelets that are compactly supported. These wavelets have been used to study Hardy and BMO spaces on product spaces of homogeneous type, as well as their dyadic counterparts [105].

7.4 Dyadic Operators, Weighted Inequalities, and Hytönen's Representation Theorem

In this section, we introduce the model dyadic operators: the martingale transform, the dyadic square function, the dyadic paraproduct, Petermichl's Haar shift operator, and Haar shift operators of arbitrary complexity, all ingredients in Hytönen's proof of the A_2 conjecture [90]. We will state the known quantitative one- and two-weight inequalities for these dyadic operators. We end the section with Hytönen's representation theorem in terms of Haar shift operators of arbitrary complexity, dyadic paraproducts, and adjoints of dyadic paraproducts over random dyadic grids, valid for all Calderón–Zygmund operators and key to the resolution of the A_2 conjecture.

7.4.1 Martingale Transform

Let $\mathcal{D}$ denote a dyadic grid on $\mathbb{R}$, the Martingale transform is the linear operator formally defined as

$$T_\sigma f(x) := \sum_{I\in\mathcal{D}} \sigma_I \langle f, h_I\rangle h_I(x), \qquad \text{where} \quad \sigma_I = \pm 1.$$

This is a constant Haar multiplier in analogy to Fourier multipliers, where here the Haar coefficients are modified multiplying them by uniformly bounded constants, the *Haar symbol* $\{\sigma_I : I \in \mathcal{D}\}$ (in this case, arbitrary changes of sign). The martingale transform is bounded on $L^2(\mathbb{R})$; in fact, it is an isometry on $L^2(\mathbb{R})$ by Plancherel's identity, that is, $\|T_\sigma f\|_{L^2} = \|f\|_{L^2}$.

The martingale transform is a good toy model for Calderón–Zygmund singular integral operators such as the Hilbert transform. Suffices to recall that on Fourier side the Hilbert transform is a Fourier multiplier with Fourier symbol $m_H(\xi) = -i\,\mathrm{sgn}(\xi)$. Compare the Fourier transform of the Hilbert transform and the "Haar transform" of the martingale transform, namely,

$$\widehat{Hf}(\xi) = -i\,\mathrm{sgn}(\xi)\,\widehat{f}(\xi) \quad \text{and} \quad \langle T_\sigma f, h_I\rangle = \sigma_I \langle f, h_I\rangle.$$

Unconditionality of the Haar basis on $L^p(\mathbb{R})$ follows from uniform (on the choice of signs σ) boundedness of the martingale transform T_σ on $L^p(\mathbb{R})$. More precisely for all $f \in L^p(\mathbb{R})$

$$\sup_\sigma \|T_\sigma f\|_{L^p} \lesssim_p \|f\|_{L^p}.$$

This was proven by Donald Burkholder in 1984; he also found the optimal constant C_p in work that can be described as the precursor of the (exact) Bellman function method [30].

Unconditionality of the Haar basis on $L^p(w)$ when $w \in A_p$ follows from the uniform boundedness of T_σ on $L^p(w)$, this was proven in 1996 by Sergei Treil and Sasha Volberg [175].

7.4.1.1 Quantitative Weighted Inequalities for the Martingale Transform

Quantitative one- and two-weight inequalities are known for the martingale transform. In fact, the A_2 conjecture (linear bound) was proven by Janine Wittwer in 2000 and necessary and sufficient conditions for two-weight uniform (on the symbol σ) L^2 boundedness were identified by Fedja Nazarov, Sergei Treil, and Sasha Volberg in 1999. We present now the precise statements.

Sharp linear bounds on $L^2(w)$ when w is an A_2 weight are known [186]. More precisely, for all σ, there is $C > 0$ such that for all $w \in A_2$ and all $f \in L^2(w)$

$$\|T_\sigma f\|_{L^2(w)} \leq C[w]_{A_2}\|f\|_{L^2(w)}.$$

Sharp extrapolation gives optimal bounds on $L^p(w)$ when w is an A_p weight [67]. More precisely, for all σ there is a constant $C_p > 0$ such that for all $w \in A_p$ and $f \in L^p(w)$

$$\|T_\sigma f\|_{L^p(w)} \leq C_p[w]_{A_p}^{\max\{1,\frac{1}{p-1}\}}\|f\|_{L^p(w)}.$$

Necessary and sufficient conditions on pairs of weights (u, v) are known ensuring two-weight boundedness [146].

Theorem 7 (Nazarov, Treil, Volberg 1999) *The martingale transforms T_σ are uniformly (on σ) bounded from $L^2(u)$ to $L^2(v)$ if and only if the following conditions hold simultaneously:*

(i) (u, v) *is in* joint dyadic $\mathcal{A}_2$. *Namely,* $[u, v]_{\mathcal{A}_2} := \sup_{I\in\mathcal{D}}\langle u^{-1}\rangle_I\langle v\rangle_I < \infty$.
(ii) $\{|I|\,|\Delta_I(u^{-1})|^2\langle v\rangle_I\}_{I\in\mathcal{D}}$ *is a* u^{-1}-Carleson sequence.
(iii) $\{|I|\,|\Delta_I v|^2\langle u^{-1}\rangle_I\}_{I\in\mathcal{D}}$ *is a* v-Carleson sequence *(dual condition).*
(iv) The positive dyadic operator T_0 is bounded from $L^2(u)$ into $L^2(v)$. Where

$$T_0 f(x) := \sum_{I\in\mathcal{D}}\frac{\alpha_I}{|I|}\langle f\rangle_I\,\mathbb{1}_I(x),$$

with $\alpha_I := \big(|\Delta_I v|/\langle v\rangle_I\big)\big(|\Delta_I(u^{-1})|/\langle u^{-1}\rangle_I\big)|I|$, *and* $\Delta_I v := \langle v\rangle_{I_+} - \langle v\rangle_{I_-}$.

A sequence $\{\lambda_I\}_{I\in\mathcal{D}}$ is v-*Carleson* if and only if there is constant $B > 0$ such that $\sum_{I\in\mathcal{D}(J)}\lambda_I \leq Bv(J)$ for all $J \in \mathcal{D}$. The smallest constant B is called the *intensity* of the sequence. When $u = v = w \in A_2$ then (i)–(iii) hold, and by Example 15 the sequence $\{\alpha_I\}_{I\in\mathcal{D}}$ is a 1-Carleson sequence implying (iv).

In 2008, Nazarov, Treil, and Volberg found necessary and sufficient conditions for two-weight boundedness of individual martingale transforms and other well-localized operators [149], see also [180].

7.4.2 *Dyadic Square Function*

The dyadic square function is the sublinear operator formally defined as

$$(S^{\mathcal{D}} f)(x) := \Big(\sum_{I \in \mathcal{D}} \frac{|\langle f, h_I \rangle|^2}{|I|} \mathbb{1}_I(x) \Big)^{1/2}.$$

The dyadic square function is an isometry on $L^2(\mathbb{R})$, as a calculation quickly reveals, namely, $\|S^{\mathcal{D}} f\|_{L^2} = \|f\|_{L^2}$. It is also bounded on $L^p(\mathbb{R})$ for $1 < p < \infty$; furthermore,

$$\|S^{\mathcal{D}} f\|_{L^p} \sim \|f\|_{L^p}.$$

This result plays the role of *Plancherel on* L^p (Littlewood–Paley theory). It readily implies boundedness of T_σ on $L^p(\mathbb{R})$ since $S^{\mathcal{D}}(T_\sigma f) = S^{\mathcal{D}} f$, as follows:

$$\|T_\sigma f\|_{L^p} \sim \|S^{\mathcal{D}}(T_\sigma f)\|_{L^p} = \|S^{\mathcal{D}} f\|_{L^p} \sim \|f\|_{L^p}.$$

A somewhat convoluted argument can be done to prove the L^p boundedness of the dyadic square function. First prove $L^2(w)$ estimates for A_2 weights w, second extrapolate to get $L^p(w)$ estimates for A_p weights w, and third set $w \equiv 1 \in A_p$. Stephen Buckley has a very nice and elementary argument showing boundedness of the dyadic square function on $L^2(w)$ when w is an A_2 weight [28] or see [154, Sect. 2.5.1]. One can track the dependence on the weight and get a 3/2 power on the A_2 characteristic of the weight [22, Sect. 5], far from the optimal linear dependence discussed in Sect. 7.4.2.1.

A seminal paper on weighted inequalities for the dyadic square function is [184], see also the book [185], both authored by Mike Wilson.

7.4.2.1 One-Weight Estimates for $S^{\mathcal{D}}$

Quantitative one-weight inequalities are known for the dyadic square function. The A_2 conjecture (linear bound) was proven by Sanja Hukovic, Sergei Treil, and Sasha Volberg in 2000 [87] and the reverse estimate was proven by Stefanie Petermichl and Sandra Pott in 2002 [164].

We present now the precise statements. For all weights $w \in A_2$ and functions $f \in L^2(w)$

$$[w]_{A_2}^{-\frac{1}{2}} \|f\|_{L^2(w)} \lesssim \|S^{\mathcal{D}} f\|_{L^2(w)} \lesssim [w]_{A_2} \|f\|_{L^2(w)}$$

The direct and reverse estimates on $L^2(w)$ for the dyadic square function play the role of Plancherel on $L^2(w)$. We can use these inequalities to obtain $L^2(w)$ bounds for the martingale transform T_σ of the form $[w]_{A_2}^{3/2}$. However, the optimal bound is linear [186], as we already mentioned in Sect. 7.4.1.1.

Boundedness on $L^2(w)$ for all weights $w \in A_2$ implies by extrapolation boundedness on $L^p(\mathbb{R})$ (and on $L^p(w)$ for all $w \in A_p$). However, sharp extrapolation will only yield the optimal power for $1 < p \leq 2$, if one starts with the optimal linear bound on $L^2(w)$. Not only $S^{\mathcal{D}}$ is bounded on $L^p(w)$ if $w \in A_p$, moreover, for $1 < p < \infty$ and for all $w \in A_p$ and $f \in L^p(w)$

$$\|S^{\mathcal{D}} f\|_{L^p(w)} \lesssim_p [w]_{A_p}^{\max\{\frac{1}{2},\frac{1}{p-1}\}} \|f\|_{L^p(w)}.$$

The power $\max\{1/2, 1/(p-1)\}$ is optimal. It corresponds to *sharp extrapolation* starting at $r = 3$ with square root power [53]. More precisely, for all $w \in A_3$ and $f \in L^3(w)$,

$$\|S^{\mathcal{D}} f\|_{L^3(w)} \lesssim [w]_{A_3}^{\frac{1}{2}} \|f\|_{L^3(w)}.$$

This estimate is valid more generally for *Wilson's intrinsic square function* [127, 185].

Sharp extrapolation from the reverse estimate on $L^2(w)$ also yields the following reverse estimate on $L^p(w)$ for all $w \in A_p$ and $f \in L^p(w)$,

$$\|f\|_{L^p(w)} \lesssim_p [w]_{A_p}^{\frac{1}{2}\max\{1,\frac{1}{p-1}\}} \|S^{\mathcal{D}} f\|_{L^p(w)}.$$

This estimate can be improved, using deep estimates of Chang, Wilson, and Wolff [31] for all $p > 1$ to the following $1/2$ power of the smaller Fujii–Wilson A_∞ characteristic,

$$\|f\|_{L^p(w)} \lesssim_p [w]_{A_\infty}^{\frac{1}{2}} \|S^{\mathcal{D}} f\|_{L^p(w)}.$$

This estimate is better in the range $1 < p < 2$ where the power is $1/2$ instead of $1/2(p-1)$.

For future reference, we can compute precisely the weighted L^2 norm of $S^{\mathcal{D}} f$ as follows:

$$\|S^{\mathcal{D}} f\|_{L^2(w)}^2 = \sum_{I \in \mathcal{D}} |\langle f, h_I\rangle|^2 \langle w\rangle_I.$$

7.4.2.2 Two-Weight Estimates for $S^{\mathcal{D}}$

Two-weight inequalities are understood for the dyadic square function. The necessary and sufficient conditions for two-weight L^2 boundedness are known [146]. Qualitative (mixed) estimates have been found by different authors, and these estimates reduce to the linear estimate in the one-weight case. We present now the precise statements.

Theorem 8 (Nazarov, Treil, Volberg 1999) *The dyadic square function $S^{\mathcal{D}}$ is bounded from $L^2(u)$ into $L^2(v)$ if and only if the following conditions hold simultaneously:*

(i) $(u, v) \in \mathcal{A}_2$ *(joint dyadic $\mathcal{A}_2$).*
(ii) $\{|I|\,|\Delta_I u^{-1}|^2 \langle v\rangle_I\}_{I\in\mathcal{D}}$ *is a u^{-1}-Carleson sequence with intensity $C_{u,v}$.*

Notice (ii) is a localized "testing condition" on test functions $u^{-1}\mathbb{1}_J$. Also, note that the necessary and sufficient conditions (i)-(iii) in Theorem 7 for the martingale transform can be now replaced by

(i) $S^{\mathcal{D}}$ is bounded from $L^2(u)$ into $L^2(v)$.
(ii) $S^{\mathcal{D}}$ is bounded from $L^2(v^{-1})$ into $L^2(u^{-1})$.

This is because $(u, v) \in \mathcal{A}_2$ if and only if $(v^{-1}, u^{-1}) \in \mathcal{A}_2$.

A quantitative version of the boundedness estimate in terms of the constants appearing in the necessary and sufficient conditions is the following:

$$\|S^{\mathcal{D}}\|_{L^2(u)\to L^2(v)} \lesssim ([u, v]_{\mathcal{A}_2^d} + C_{u,v})^{1/2}.$$

There are similar two-weight L^p estimates for continuous square function [117, 118], see also [22, Theorem 6.2].

If the weights $(u, v) \in \mathcal{A}_2$ and $u^{-1} \in A_\infty$, then they satisfy the necessary and sufficient conditions in Theorem 8 and the following estimate holds [22]:

$$\|S^{\mathcal{D}}\|_{L^2(u)\to L^2(v)} \lesssim ([u, v]_{\mathcal{A}_2} + [u, v]_{\mathcal{A}_2}[u^{-1}]_{A_\infty})^{1/2}.$$

Setting $u = v = w \in A_2$, this improves the known linear bound to a mixed-type bound

$$\|S^{\mathcal{D}}\|_{L^2(w)} \lesssim ([w]_{A_2}[w^{-1}]_{A_\infty})^{1/2} \lesssim [w]_{A_2}.$$

For mixed $L^p - L^\infty$ or mixed $L^p - L^r$ estimates for the square function and general Calderón-Zygmund operators see [95, 96, 128, 132].

Same one-weight estimate have been shown to hold for the dyadic square function and for matrix-valued weights [99]. Quantitative weighted estimates from $L^p(u)$ into $L^q(v)$ in terms of *quadratic testing condition* are known [181].

7.4.3 Petermichl's Dyadic Shift Operator

Given parameters $(r, \beta) \in \Omega = [1, 2) \times \{0, 1\}^{\mathbb{Z}}$, the *Petermichl's dyadic shift operator* $Ш^{r,\beta}$ (pronounced "Sha") associated to the random dyadic grid $\mathcal{D}^{r,\beta}$ is defined for functions $f \in L^2(\mathbb{R})$ by

$$\text{Ш}^{r,\beta} f(x) := \sum_{I \in \mathcal{D}^{r,\beta}} \langle f, h_I \rangle \, H_I(x) = \sum_{I \in \mathcal{D}^{r,\beta}} 2^{-1/2} \sigma(I) \, \langle f, h_{\widetilde{I}} \rangle \, h_I(x),$$

where $H_I = 2^{-1/2}(h_{I_r} - h_{I_l})$ and $\sigma(I)$ is ± 1 depending whether I is the right or left child of I's parent $\widetilde{I}$. More precisely, $\sigma(I) = 1$ if $I = (\widetilde{I})_r$ and $\sigma(I) = -1$ if $I = (\widetilde{I})_l$.

When $r = 1$ and $\beta_j = 0$ for all $j \in \mathbb{Z}$, the corresponding grid is the regular dyadic grid and we denote the associated daydic shift operator simply Ш.

Petermichl's dyadic shift operators are isometries on $L^2(\mathbb{R})$, that is, for all $r, \beta \in \Omega$, $\|\text{Ш}^{r,\beta} f\|_{L^2} = \|f\|_{L^2}$, and they are bounded operators on $L^p(\mathbb{R})$, as can be readily seen using Plancherel's identity and dyadic square function estimates.

Each operator $\text{Ш}^{r,\beta}$ is a good dyadic model for the Hilbert transform H. The images under $\text{Ш}^{r,\beta}$ of the Haar functions are the normalized differences of the Haar functions on its children, namely, $\text{Ш}^{r,\beta} h_J(x) = H_J(x)$. The functions h_J and H_J can be viewed as localized sines and cosines, in the sense that if we were to extend them periodically, with period the length of the support, we will see two square waves shifted by half the length of the period. More evidence comes from the way the family $\{\text{Ш}^{r,\beta}\}_{(r,\beta)\in\Omega}$ interacts with translations, dilations, and reflections. Each dyadic shift operator does not have symmetries that characterize the Hilbert transform,[11] but an average over all random dyadic grids $\mathcal{D}^{r,\beta}$ does. It is a good exercise to figure out how each individual shift interacts with these rigid motions; they almost commute except that the dyadic grid changes. For example, regarding reflections, it can be seen that if we denote by $R(x) = -x$ then $R\text{Ш}^{r,\beta} = -\text{Ш}^{r,-\beta} R$, where $-\beta = \{1 - \beta_i\}_{i\in\mathbb{Z}}$. The corresponding rules for translations and dilations are slightly more complicated, but what matters is that there is a one-to-one correspondence between the dyadic grids so that when averaging over all dyadic grids the average will have the desired properties, and hence it will be a constant multiple of the Hilbert transform. This is precisely what Stefanie Petermichl proved in 2000, a groundbreaking and unexpected new result for the Hilbert transform [161]. More precisely, she showed that

$$H = -\frac{8}{\pi} \mathbb{E}_\Omega \text{Ш}^{r,\beta} = -\frac{8}{\pi} \int_\Omega \text{Ш}^{r,\beta} d\mathbb{P}(r, \beta).$$

The result follows after verifying that the averages have the invariance properties that characterize the Hilbert transform [89, 161]. Because the shift operators $\text{Ш}^{r,\beta}$ are uniformly bounded on $L^p(\mathbb{R})$ for $1 < p < \infty$, this representation will immediately imply that the Hilbert transform H is bounded on $L^p(\mathbb{R})$ in the same range, a result first proved by Marcel Riesz in 1928. Similarly, once uniform (on the dyadic grids $\mathcal{D}^{r,\beta}$) weighted inequalities are verified for $\text{Ш}^{r,\beta}$, the inequalities will be inherited by the Hilbert transform. Petermichl proved the linear bounds on $L^2(w)$ for the shift operators using Bellman function methods, and hence she proved the A_2 conjecture for the Hilbert transform [162].

[11] Recall that any bounded linear operator on $L^2(\mathbb{R})$ that commutes with dilations and translations and anticommutes with reflexions must be a constant multiple of the Hilbert transform.

These results added a very precise new dyadic perspective to such a classic and well-studied operator as the Hilbert transform. Similar representations hold for the *Beurling* and the *Riesz* transforms [163, 165], these operators have many invariance properties as the Hilbert transform does. For a while, it was believed that such invariances were responsible for these representation formulas. It came as a surprise when Tuomas Hytönen proved in 2012 that there is a representation formula valid for ALL Calderón–Zygmund singular integral operators [90]. To state Hytönen's result, we need to introduce Haar shift operators of arbitrary complexity and paraproducts.

7.4.4 Haar Shift Operators of Arbitrary Complexity

The *Haar shift operators of complexity* (m, n) associated to a dyadic grid $\mathcal{D}$ were introduced by Michael Lacey, Stefanie Petermichl, and Mari Carmen Reguera [122]; they are defined on $L^2(\mathbb{R})$ as follows:

$$\text{Ш}_{m,n} f(x) := \sum_{L\in\mathcal{D}} \sum_{I\in\mathcal{D}_m(L), J\in\mathcal{D}_n(L)} c_{I,J}^L \langle f, h_I\rangle h_J(x),$$

where the coefficients $|c_{I,J}^L| \leq \frac{\sqrt{|I||J|}}{|L|}$, and $\mathcal{D}_m(L)$ denotes the dyadic subintervals of L with length $2^{-m}|L|$.

The cancellation property of the Haar functions and the normalization of the coefficients ensures that $\|\text{Ш}_{m,n} f\|_{L^2} \leq \|f\|_{L^2}$, and square function estimates ensure boundedness on $L^p(\mathbb{R})$ for all $1 < p < \infty$. The martingale transform, T_σ, is a Haar shift operator of complexity $(0, 0)$. Petermichl's $\text{Ш}^{r,\beta}$ operators are Haar shift operators of complexity $(0, 1)$. The dyadic paraproduct, π_b, to be introduced in Sect. 7.4.5, is not one of these and nor is its adjoint π_b^*.

The following estimates are known for dyadic shift operators of arbitrary complexity. First, Michael Lacey, Stefanie Petermichl, and Mari Carmen Reguera proved the A_2 conjecture for the Haar shift operators of arbitrary complexity with constant depending exponentially on the complexity [122]. Unlike their predecessors, they did not use Bellman functions; instead, they used stopping time techniques and a two-weight theorem for "well-localized operators" of [149]. Second, David Cruz-Uribe, Chema Martell, and Carlos Pérez [53] used a local median oscillation technique introduced by Andrei Lerner [126, 127]. The local median oscillation method was quite flexible; they obtained new results such as the sharp bounds for the square function for $p > 2$, for the dyadic paraproduct, also for vector-valued maximal operators, as well as two-weight results; however, for the dyadic shift operators the weighted estimates still depended exponentially on the complexity. Third, Tuomas Hytönen [90] obtained the linear estimates with polynomial dependence on the complexity, needed to prove the A_2 conjecture for Calderón–Zygmund singular integral operators.

7.4.5 Dyadic Paraproduct

Quoting from an article on "What is....a Paraproduct?" for the broader public by Arpad Bényi, Diego Maldonado, and Virginia Naibo [17]:

> The term paraproduct is nowadays used rather loosely in the literature to indicate a bilinear operator that, although noncommutative, is somehow better behaved than the usual product of functions. Paraproducts emerged in J.-M. Bony's theory of paradifferential operators [26], which stands as a milestone on the road beyond pseudodifferential operators pioneered by R. R. Coifman and Y. Meyer in [40]. Incidentally, the Greek word $\pi\alpha\rho\alpha$ (para) translates as *beyond* in English and *au délà de* in French, just as in the title of [40]. The defining properties of a paraproduct should therefore go beyond the desirable properties of the product.

The *dyadic paraproduct* associated to a dyadic grid $\mathcal{D}$ and to $b \in \mathrm{BMO}^{\mathcal{D}}$ is an operator acting on square integrable functions f as follows:

$$\pi_b f(x) := \sum_{I\in\mathcal{D}} \langle f\rangle_I \, \langle b, h_I\rangle \, h_I(x),$$

where $\langle f\rangle_I = \frac{1}{|I|}\int_I f(x)\,dx = \langle f, \mathbb{1}_I/|I|\rangle$. A function b is in the *space of dyadic bounded mean oscillation*, $\mathrm{BMO}^{\mathcal{D}}$ if and only if

$$\|b\|_{\mathrm{BMO}^{\mathcal{D}}} := \sup_{J\in\mathcal{D}} \left(\frac{1}{|J|}\int_J |b(x) - \langle b\rangle_J|^2 dx\right)^{1/2} < \infty.$$

Notice that we are using an L^2 mean oscillation instead of the L^1 mean oscillation used in (7.2.2) in the definition of BMO, and of course, we are restricting to dyadic intervals. As it turns out, one could use an L^p mean oscillation for any $1 \le p$ and obtain equivalent norms in BMO, thanks to the celebrated John–Nirenberg lemma [104].

Formally, expanding f and b in the Haar basis, multiplying and separating the terms into upper triangular, diagonal, and lower triangular parts, one gets that

$$bf = \pi_b f + \pi_b^* f + \pi_f b,$$

in doing so is important to note that $\sum_{I\in\mathcal{D}:I\supset J}\langle f, h_I\rangle\, h_I = \langle f\rangle_J$. It is well known that multiplication by a function b is a bounded operator on $L^p(\mathbb{R})$ if and only if the function is essentially bounded, that is, $b \in L^\infty(\mathbb{R})$. However, the paraproduct is a bounded operator on $L^p(\mathbb{R})$ if and only if $b \in \mathrm{BMO}^{\mathcal{D}}$, which is a space strictly larger than $L^\infty(\mathbb{R})$. The L^2 estimate can be obtained using, for example, the Carleson embedding lemma, see Sect. 7.5.1.

Using a weighted Carleson embedding lemma, one can check that the paraproduct is bounded on $L^2(w)$ for all $w \in A_2$ [154]. Furthermore, Beznosova proved the A_2 conjecture for paraproducts [21], namely,

$$\|\pi_b f\|_{L^2(w)} \le C[w]_{A_2}\|b\|_{\mathrm{BMO}^{\mathcal{D}}}\|f\|_{L^2(w)}.$$

By extrapolation, one concludes that the paraproduct is bounded on $L^p(w)$ for all $w \in A_p$ and $1 < p < \infty$, in particular, it is bounded on $L^p(\mathbb{R})$. In Sect. 7.5, we will present Beznosova's Bellman function argument proving the A_2 conjecture for the dyadic paraproduct. This argument was generalized to $\mathbb{R}^d$ in [36] and to spaces of homogeneous type in [182]. It was pointed out to us recently [183] that the paraproduct is a well-localized operator (for trivial reasons) in the sense of [149], and therefore it falls under their theory.

To finish this brief introduction to the paraproduct, we would like to mention its intimate connection to the $T(1)$ and $T(b)$ theorems of Guy David, Jean-Lin Journé, and Stephen Semmes [61, 62]. These theorems give (necessary and sufficient) conditions to verify boundedness on $L^2(\mathbb{R})$ for singular integral operators T with a Calderón–Zygmund kernel when Fourier analysis, almost-orthogonality (Cotlar's lemma), or other more standard techniques fail. In the $T(1)$ theorem, the conditions amount to checking some weak-boundedness property which is a necessary condition, and checking that the function 1 is "mapped" under the operator and its adjoint, $T(1)$ and $T^*(1)$, into BMO. Once this is verified, the operator can be decomposed into a "simpler" operator S with the property that $S(1) = S^*(1) = 0$, a paraproduct, $\pi_{T(1)}$, and the adjoint of a paraproduct, $\pi^*_{T^*(1)}$. The paraproduct terms are bounded on $L^2(\mathbb{R})$, the operator S can be verified to be bounded on $L^2(\mathbb{R})$, and as a consequence so will be the operator T.

We have defined all these model operators in the one-dimensional Case; there are corresponding Haar shift operators and dyadic paraproducts defined on $\mathbb{R}^d$ as well as $T(1)$ and $T(b)$ theorems.

7.4.6 Hytönen's Representation Theorem

Let us remind the reader that a bounded operator on $L^2(\mathbb{R}^d)$ is a Calderón–Zygmund singular integral operator with smoothness parameter $\alpha > 0$ if it has an integral representation

$$Tf(x) = \int_{\mathbb{R}^d} K(x, y) f(y)\, dy, \qquad x \notin \operatorname{supp} f,$$

for a kernel $K(x, y)$ defined for all $(x, y) \in \mathbb{R}^d \times \mathbb{R}^d$ such that $x \neq y$, and verifying the standard size and smoothness estimates, respectively, $|K(x, y)| \leq C/|x - y|^d$ and

$$|K(x + h, y) - K(x, y)| + |K(x, y + h) - K(x, y)| \leq C|h|^\alpha/|x - y|^{d+\alpha},$$

for all $|x - y| > 2|h| > 0$ and some fixed $\alpha \in [0, 1]$.

It is worth remembering that such Calderón–Zygmund singular integral operators are bounded on $L^p(\mathbb{R}^d)$ for all $1 < p < \infty$, they are of weak-type $(1, 1)$, and they map BMO into itself.

We now have all the ingredients to state the celebrated Hytönen's representation theorem [90] at least in the one-dimensional case.

Theorem 9 (Hytönen's 2012) *Let T be a Calderón–Zygmund singular integral operator with smoothness parameter $\alpha > 0$, then*

$$T(f) = \mathbb{E}_\Omega \left(\sum_{(m,n)\in\mathbb{N}^2} e^{-(m+n)\alpha/2} Ш^{r,\beta}_{m,n}(f) + \pi^{r,\beta}_{T(1)}(f) + (\pi^{r,\beta}_{T^*(1)})^*(f) \right).$$

Where for each pair of random parameters $(r, \beta) \in \Omega$, the operator $Ш^{r,\beta}_{m,n}$ is a Haar shift operators of complexity (m, n), the operator $\pi^{r,\beta}_{T(1)}$ is a dyadic paraproduct, and the operator $(\pi^{r,\beta}_{T^*(1)})^*$ is the adjoint of a dyadic paraproduct, all defined on the random dyadic grid $\mathcal{D}^{r,\beta}$. The paraproducts and their adjoints in the decomposition depend on the operator T via $T(1)$ and $T^*(1)$. The Haar shift operators in the decomposition also depend on T although it is not obvious in the notation we used. Indeed, the coefficients $c^L_{I,J}$, in the definition of the Haar shift multiplier of complexity (m, n) (see Sect. 7.4.4), will depend on the given operator T for each $L \in \mathcal{D}^{r,\beta}$, $I \in \mathcal{D}^{r,\beta}_m(L)$, and $J \in \mathcal{D}^{r,\beta}_n(L)$. Notice that the exponential nature of the coefficients in the expansion explains why the Haar shift multipliers of arbitrary complexity will need to be bounded with a bound depending at most polynomially on the complexity.

To the author, this is a remarkable result providing a dyadic decomposition theorem for a large class of operators. Once you have such decomposition and $L^2(w)$ estimates for each of the components (Haar shift operators, paraproducts, and their adjoints) that are linear on $[w]_{A_2}$ and that are uniform on the dyadic grids, then the A_2 conjecture is resolved in the positive, as Tuomas Hytönen did in his celebrated paper [90].

7.5 A_2 Theorem for the Dyadic Paraproduct: A Bellman Function Proof

As a model example, we will present in this section Beznosova's argument proving the A_2 conjecture for the dyadic paraproduct [21]. The goal is to show that for all weights $w \in A_2$, functions $b \in \mathrm{BMO}^{\mathcal{D}}$, and functions $f \in L^2(w)$ the following estimate holds:

$$\|\pi_b f\|_{L^2(w)} \lesssim [w]_{A_2} \|b\|_{BMO^{\mathcal{D}}} \|f\|_{L^2(w)}.$$

We remind the reader that the *dyadic paraproduct* associated to $b \in \mathrm{BMO}^{\mathcal{D}}$ is defined by $\pi_b f(x) := \sum_{I\in\mathcal{D}} \langle f\rangle_I \, b_I \, h_I(x)$, where $b_I = \langle b, h_I\rangle$ and $\langle f\rangle_I = (1/|I|) \int_I f(y)\,dy$.

To achieve a preliminary estimate, where instead of the linear bound on $[w]_{A_2}$ we get a 3/2 bound, namely, $[w]^{3/2}_{A_2}$, we need to introduce a few ingredients: (weighted)

Carleson sequences, Beznosova's Little lemma, and the weighted Carleson lemma. Afterward, in Sect. 7.5.4, we will refine the argument to get the desired linear bound. To achieve the linear bound, we will need a few additional ingredients, including the α-Lemma, introduced by Oleksandra Beznosova in her original proof [20]. Both the Little lemma and the α-Lemma are proved using Bellman functions, and we sketch their proofs, as well as the proof of the weighted Carleson lemma.

7.5.1 *Weighted Carleson Sequences, Weighted Carleson Lemma, and Little Lemma*

In this section, we introduce weighted an unweighted Carleson sequences and the weighted Carleson embedding lemma. We also present Bezonosova's Little lemma that enables us to ensure that given a weight w and a Carleson sequence $\{\lambda_I\}_{I\in\mathcal{D}}$, we can create a w-weighted Carleson sequence by multiplying each term of the given sequence by the reciprocal of $w^{-1}(I)$.

7.5.1.1 Weighted Carleson Sequences and Lemma

Given a weight w, a positive sequence $\{\lambda_I\}_{I\in\mathcal{D}}$ is w-*Carleson* if there is a constant $A>0$ such that

$$\sum_{I\in\mathcal{D}(J)} \lambda_I \leq A w(J) \quad \text{for all } J\in\mathcal{D},$$

where $w(J)=\int_J w(x)\,dx$. The smallest constant $A>0$ is called the *intensity* of the sequence. When $w=1$ a.e., we say that the sequence is Carleson (not 1-Carleson).

Example 11 If $b\in \mathrm{BMO}^{\mathcal{D}}$, then the sequence $\{b_I^2\}_{I\in\mathcal{D}}$ is Carleson with intensity $\|b\|^2_{\mathrm{BMO}^{\mathcal{D}}}$. Indeed, for any $J\in\mathcal{D}$, the collection of Haar functions corresponding to dyadic intervals $I\subset J$, $\{h_I\}_{I\in\mathcal{D}(J)}$, forms an orthonormal basis on $L^2_0(J)=\{f\in L^2(J):\int_J f(x)\,dx=0\}$. The function $(b-\langle b\rangle_J)\big|_J$ belongs to $L^2_0(J)$, therefore by Plancherel's inequality,

$$\sum_{I\in\mathcal{D}(J)} b_I^2 = \sum_{I\in\mathcal{D}(J)} |\langle b,h_I\rangle|^2 = \int_J |b(x)-\langle b\rangle_J|^2\,dx \leq \|b\|^2_{\mathrm{BMO}^{\mathcal{D}}}|J|.$$

The following weighted Carleson lemma that appeared in [146] will be extremely useful in our estimates; you can find a proof in [140] that we reproduce in Sect. 7.5.5.3.

Lemma 2 (Weighted Carleson Lemma) *Given a weight v, then $\{\lambda_I\}_{I\in\mathcal{D}}$ is a v-Carleson sequence with intensity A if and only if for all nonnegative $F\in L^1(v)$, we have*

$$\sum_{I\in\mathcal{D}} \lambda_I \inf_{x\in I} F(x) \le A \int_{\mathbb{R}} F(x)\, v(x)\, dx.$$

The following particular instance of the weighted Carleson lemma will be useful.

Example 12 Let $\{\lambda_I\}_{I\in\mathcal{D}}$ be a v-Carleson sequence with intensity A, let $f \in L^2(v)$ and set $F(x) = (M_v^{\mathcal{D}} f(x))^2$ where $M_v^{\mathcal{D}}$ is the weighted dyadic maximal function, namely,

$$M_v^{\mathcal{D}} f(x) := \sup_{I\in\mathcal{D}:x\in I} \langle |f|\rangle_I^v \quad \text{where } \langle |f|\rangle_I^v := \langle |f|v\rangle_I / \langle v\rangle_I .$$

By definition of the dyadic maximal function, $\langle |f|\rangle_I^v \le \inf_{x\in I} M_v^{\mathcal{D}} f(x)$. Then by the weighted Carleson lemma (Lemma 2) and the boundedness of $M_v^{\mathcal{D}}$ on $L^2(v)$ with operator bound independent of the weight, we conclude that

$$\sum_{I\in\mathcal{D}} \lambda_I \,(\langle |f|\rangle_I^v)^2 \le A\|M_v^{\mathcal{D}} f\|_{L^2(v)}^2 \lesssim A\|f\|_{L^2(v)}^2.$$

Specializing even further we get another useful result that establishes the boundedness of the dyadic paraproduct on $L^2(\mathbb{R})$ when $b \in \mathrm{BMO}^{\mathcal{D}}$.

Example 13 In particular, if $v \equiv 1$ and $b \in \mathrm{BMO}^{\mathcal{D}}$, then $\lambda_I := b_I^2$ for $I \in \mathcal{D}$ defines a Carleson sequence with intensity $\|b\|_{\mathrm{BMO}^{\mathcal{D}}}^2$; hence,

$$\|\pi_b f\|_{L^2}^2 = \sum_{I\in\mathcal{D}} |\langle \pi_b f, h_I\rangle|^2 \le \sum_{I\in\mathcal{D}} b_I^2 \,\langle |f|\rangle_I^2 \lesssim \|b\|_{\mathrm{BMO}^{\mathcal{D}}}^2 \|f\|_{L^2}^2.$$

7.5.1.2 Beznosova's Little Lemma

We will need to create w-Carleson sequences from given Carleson sequences. The following lemma will come in handy [21].

Lemma 3 (Little Lemma) *Let w be a weight, such that w^{-1} is also a weight. Let $\{\lambda_I\}_{I\in\mathcal{D}}$ be a Carleson sequence with intensity A, the sequence $\{\lambda_I/\langle w^{-1}\rangle_I\}_{I\in\mathcal{D}}$ is w-Carleson with intensity $4A$. In other words, for all $J \in \mathcal{D}$*

$$\sum_{I\in\mathcal{D}(J)} \frac{\lambda_I}{\langle w^{-1}\rangle_I} \le 4A\; w(J). \tag{7.5.1}$$

The proof uses a Bellman function argument that we will present in Sect. 7.5.5. Note that the weight w in the Little lemma is not required to be in the Muckenhoupt A_2 class. It does require that the reciprocal w^{-1} is a weight, of course if $w \in A_2$ then w^{-1} is a weight in A_2.

Example 14 Let $b \in \mathrm{BMO}^{\mathcal{D}}$ and $w \in A_2$.The sequence $\{b_I^2/\langle w\rangle_I\}_{I\in\mathcal{D}}$ is a w^{-1}-Carleson, with intensity $4\|b\|^2_{\mathrm{BMO}^{\mathcal{D}}}$. By Example 11, the sequence $\{b_I^2\}_{I\in\mathcal{D}}$ is a Carleson sequence with intensity $\|b\|^2_{\mathrm{BMO}^{\mathcal{D}}}$, and then applying Lemma 3 with the roles of w and w^{-1} interchanged, we get the stated result.

7.5.2 The 3/2 Bound for the Paraproduct on Weighted L^2

We now show that the paraproduct π_b is bounded on $L^2(w)$ when $w \in A_2$ and $b \in \mathrm{BMO}^{\mathcal{D}}$, with bound $[w]_{A_2}^{3/2}\|b\|_{\mathrm{BMO}^{\mathcal{D}}}$, not yet the optimal linear bound.

Proof By duality suffices to show that for all $f \in L^2(w)$ and $g \in L^2(w^{-1})$,

$$|\langle \pi_b f, g\rangle| \lesssim [w]_{A_2}^{3/2}\|b\|_{\mathrm{BMO}^{\mathcal{D}}}\|f\|_{L^2(w)}\|g\|_{L^2(w^{-1})}.$$

By definition of the dyadic paraproduct and the triangle inequality,

$$|\langle \pi_b f, g\rangle| \leq \sum_{I\in\mathcal{D}} \langle |f|\rangle_I \, |b_I| \, |\langle g, h_I\rangle|.$$

First, using the Cauchy–Schwarz inequality, we can estimate as follows:

$$|\langle \pi_b f, g\rangle| \leq \left(\sum_{I\in\mathcal{D}} \frac{\langle |f|\rangle_I^2 \, b_I^2}{\langle w^{-1}\rangle_I}\right)^{1/2} \left(\sum_{I\in\mathcal{D}} |\langle g, h_I\rangle|^2 \langle w^{-1}\rangle_I\right)^{1/2}.$$

Second, using the fact that $\|S^{\mathcal{D}}g\|^2_{L^2(w^{-1})} = \sum_{I\in\mathcal{D}} |\langle g, h_I\rangle|^2\langle w^{-1}\rangle_I$ and the linear bound on $L^2(v)$ for the square function for $v = w^{-1} \in A_2$ with $[w^{-1}]_{A_2} = [w]_{A_2}$, we further estimate by

$$\begin{aligned}
|\langle \pi_b f, g\rangle| &\leq \left(\sum_{I\in\mathcal{D}} \left(\frac{\langle |f| w w^{-1}\rangle_I}{\langle w^{-1}\rangle_I}\right)^2 \frac{b_I^2}{\langle w\rangle_I} \langle w\rangle_I \, \langle w^{-1}\rangle_I\right)^{1/2} \|S^{\mathcal{D}} g\|_{L^2(w^{-1})} \\
&\lesssim [w]_{A_2}^{1/2} \left(\sum_{I\in\mathcal{D}} \left(\langle |f| w\rangle_I^{w^{-1}}\right)^2 \frac{b_I^2}{\langle w\rangle_I}\right)^{1/2} [w]_{A_2}\|g\|_{L^2(w^{-1})}.
\end{aligned}$$

Third, using the weighted Carleson lemma (Lemma 2) for the w^{-1}-Carleson sequence $\{b_I^2/\langle w\rangle_I\}_{I\in\mathcal{D}}$ with intensity $4\|b\|^2_{\mathrm{BMO}^{\mathcal{D}}}$ (see Example 14), together with the fact that $\|f\|_{L^2(w)} = \|fw\|_{L^2(w^{-1})}$, we get that

$$\begin{aligned}
|\langle \pi_b f, g\rangle| &\lesssim [w]_{A_2}^{3/2} \, 2\|b\|_{\mathrm{BMO}^{\mathcal{D}}}\|M^{\mathcal{D}}_{w^{-1}}(fw)\|_{L^2(w^{-1})}\|g\|_{L^2(w^{-1})} \\
&\lesssim [w]_{A_2}^{3/2}\|b\|_{\mathrm{BMO}^{\mathcal{D}}}\|f\|_{L^2(w)}\|g\|_{L^2(w^{-1})},
\end{aligned}$$

where in the last line we used the boundedness of the dyadic weighted maximal function $M_v^{\mathcal{D}}$ on $L^2(v)$ with an operator norm independent of the weight v. This implies that

$$\|\pi_b f\|_{L^2(w)} \lesssim [w]_{A_2}^{3/2} \|b\|_{\mathrm{BMO}^{\mathcal{D}}} \|f\|_{L^2(w)}.$$

This is precisely what we set out to prove.

7.5.3 Algebra of Carleson Sequences, α-Lemma, and Weighted Haar Bases

To get a linear bound instead of the 3/2 power bound, we just obtained, we will need a couple more ingredients, some algebra with Carleson sequences, the α-Lemma, and weighted Haar bases.

7.5.3.1 Algebra of Carleson Sequences

Given weighted Carleson sequences, we can create new weighted Carleson sequences by linear operations or by taking geometric means.

Lemma 4 (Algebra of Carleson sequences) *Given a weight v, let $\{\lambda_I\}_{I\in\mathcal{D}}$ and $\{\gamma_I\}_{I\in\mathcal{D}}$ be two v-Carleson sequences with intensities A and B, respectively, then for any $c, d > 0$,*

(i) The sequence $\{c\lambda_I + d\gamma_I\}_{I\in\mathcal{D}}$ is a v-Carleson sequence with intensity at most $cA + dB$.
(ii) The sequence $\{\sqrt{\lambda_I\gamma_I}\}_{I\in\mathcal{D}}$ is a v-Carleson sequence with intensity at most $\sqrt{AB}$.

The proof is a simple exercise which we leave to the interested reader. We do need some specific Carleson sequences, and we record them in the next example.

Example 15 Let $u, v \in A_\infty$ and $\Delta_I v := \langle v\rangle_{I_+} - \langle v\rangle_{I_-}$. Then,

(i) The sequence $\left\{ |\Delta_I v|/\langle v\rangle_I|^2\, |I|\right\}_{I\in\mathcal{D}}$ is a Carleson sequence, with intensity $C\log[w]_{A_\infty}$.
(ii) Let $\alpha_I = (|\Delta_I v|/\langle v\rangle_I)(|\Delta_I u|/\langle u\rangle_I)|I|$. The sequence $\{\alpha_I\}_{I\in\mathcal{D}}$ is a Carleson sequence.
(iii) When $v \in A_2$, $u = v^{-1}$ (also in A_2), the sequence $\{\alpha_I\}_{I\in\mathcal{D}}$ defined in item (ii) has intensity at most $\log[v]_{A_2}$.

Example 15(i) was discovered by Robert Fefferman, Carlos Kenig,[12] and Jill Pipher,[13] in 1991, see [72]. The sharp constant $C = 8$ was obtained by Vasily Vasyunin using the Bellman function method [176]. In fact, this example provides a characterization of A_∞ by summation conditions; for many more such characterizations for other weight classes, see [23, 28]. Examples 15(ii)–(iii) follow from Example 15(i) and from Lemma 4(ii).

7.5.3.2 The α-Lemma

The key to dropping from power 3/2 to linear power in the weighted L^2 estimate for the paraproduct is the following lemma, discovered by Beznosova, like the Little lemma, in the course of writing her Ph.D. Dissertation [20], see also [21, 140]. Both lemmas were proved using Bellman functions, and we will sketch the arguments in Sect. 7.5.5.

Lemma 5 (α-Lemma) *If $w \in A_2$ and $\alpha > 0$, then the sequence*

$$\mu_I := \langle w\rangle_I^\alpha \langle w^{-1}\rangle_I^\alpha \, |I| \left(\frac{|\Delta_I w|^2}{\langle w\rangle_I^2} + \frac{|\Delta_I w^{-1}|^2}{\langle w^{-1}\rangle_I^2} \right) \quad I \in \mathcal{D}$$

is a Carleson sequence with intensity at most $C_\alpha [w]_{A_2}^\alpha$, and $C_\alpha = \max\{72/(\alpha - 2\alpha^2), 576\}$.

Notice that the algebra of Carleson sequences encoded in Lemma 4 together with the R. Fefferman–Kenig–Pipher Example 15(iii) give, for μ_I, an intensity of $[w]_{A_2}^\alpha \log[w]_{A_2}$, which is larger by a logarithmic factor than the one claimed in the α-Lemma. This lesser estimate will improve the 3/2 estimate to a linear times logarithmic estimate [20], the stronger α-Lemma will yield the desired linear estimate.

Example 16 Let $w \in A_2$ and $b \in \mathrm{BMO}^{\mathcal{D}}$. By the α-Lemma and the algebra of Carleson sequences, we conclude that

(i) $\{\nu_I := |\Delta_I w|^2 \langle w^{-1}\rangle_I^2 |I|\}_{I \in \mathcal{D}}$ is Carleson with intensity $C_{1/4}[w]_{A_2}^2$, $C_{1/4} = 576$.
(ii) $\{b_I \sqrt{\nu_I}\}_{I \in \mathcal{D}}$ is Carleson with intensity $24[w]_{A_2} \|b\|_{\mathrm{BMO}^{\mathcal{D}}}$.

7.5.3.3 Weighted Haar Basis

The last ingredient before we present the proof of the A_2 conjecture for the dyadic paraproduct is the weighted Haar basis.

[12] Carlos Kenig, an Argentinian mathematician, was elected President of the International Mathematical Union in July 2018 in the International Congress of Mathematicians (ICM) held in Brazil and for the first time in the Southern hemisphere.

[13] Jill Pipher is the president-elect of the American Mathematical Society (AMS), and will begin a 2-year term in 2019.

Given a doubling weight w and an interval I, the *weighted Haar function* h_I^w is given by

$$h_I^w(x) := \sqrt{w(I_-)}/\sqrt{w(I)w(I_+)}\,\mathbb{1}_{I_+}(x) - \sqrt{w(I_+)}/\sqrt{w(I)w(I_-)}\,\mathbb{1}_{I_-}(x).$$

The collection $\{h_I^w\}_{I\in\mathcal{D}}$, of weighted Haar functions indexed on $\mathcal{D}$—a system of dyadic intervals—is an orthonormal system of $L^2(w)$. In fact, the weighted Haar functions are the Haar functions corresponding to the space of homogeneous type $X = \mathbb{R}$ with the Euclidean metric, the doubling measure $d\mu = w\,dx$, and the dyadic structure $\mathcal{D}$, defined in Sect. 7.3.5.3.

There is a very simple formula relating the weighted Haar function and the regular Haar function. More precisely, given $I \in \mathcal{D}$ there exist numbers α_I^w, β_I^w such that

$$h_I(x) = \alpha_I^w h_I^w(x) + \beta_I^w \mathbb{1}_I(x)/\sqrt{|I|}.$$

The coefficients can be calculated precisely, and they have the following upper bounds:

(i) $|\alpha_I^w| \le \sqrt{\langle w\rangle_I}$, (ii) $|\beta_I^w| \le |\Delta_I w|/\langle w\rangle_I$ where $\Delta_I w := \langle w\rangle_{I_+} - \langle w\rangle_{I_-}$.

7.5.4 A_2 Conjecture for the Dyadic Paraproduct

We present a proof of Beznosova's theorem, namely, for all $b \in \mathrm{BMO}^{\mathcal{D}}$, $w \in A_2$, and $f \in L^2(w)$

$$\|\pi_b f\|_{L^2(w)} \lesssim \|b\|_{\mathrm{BMO}^{\mathcal{D}}}[w]_{A_2}\|f\|_{L^2(w)}.$$

The proof uses the same ingredients introduced by Oleksandra Beznosova [21], and a beautiful argument by Fedja Nazarov, Sasha Reznikov, and Sasha Volberg that yields polynomial in the complexity bounds for Haar shift operators on geometric doubling metric spaces [144]. An extension of their result to *paraproducts with arbitrary complexity* can be found in joint work with Jean Moraes [140].

Proof Suffices by duality to prove that

$$|\langle \pi_b f, g\rangle| \le C\|b\|_{\mathrm{BMO}^{\mathcal{D}}}[w]_{A_2}\|f\|_{L^2(w)}\|g\|_{L^2(w^{-1})}.$$

We introduce weighted Haar functions to obtain two terms to be estimated separately,

$$|\langle \pi_b f, g\rangle| \le \sum_{I\in\mathcal{D}} |b_I|\,\langle |f|ww^{-1}\rangle_I\,|\langle gw^{-1}w, h_I\rangle| \le \Sigma_1 + \Sigma_2.$$

Explicitly, the sums Σ_1 and Σ_2 are obtained replacing $h_I = \alpha_I^w h_I^w + \beta_I^w \mathbb{1}_I/\sqrt{|I|}$, and using the estimates on the coefficients α_I, β_I, to get

$$\Sigma_1 := \sum_{I\in\mathcal{D}} |b_I|\, \langle |f|ww^{-1}\rangle_I\, |\langle gw^{-1}w, h_I^w\rangle| \sqrt{\langle w\rangle_I},$$

$$\Sigma_2 := \sum_{I\in\mathcal{D}} |b_I|\, \langle |f|ww^{-1}\rangle_I\, \langle |g|w^{-1}w\rangle_I \frac{|\Delta_I w|}{\langle w\rangle_I}\sqrt{|I|}.$$

7.5.4.1 First Sum Σ_1

Denote the $L^2(w)$ pairing $\langle h,k\rangle_{L^2(w)} := \langle hw,k\rangle$. To estimate the first sum, we observe that the weighted average (with respect to the weight w^{-1}) of the function $|f|w$ over a dyadic interval is bounded by the corresponding dyadic weighted maximal function evaluated at any point on the interval, hence by the infimum over the interval, more precisely, $\langle |f|ww^{-1}\rangle_I / \langle w^{-1}\rangle_I \le \inf_{x\in I} M^{\mathcal{D}}_{w^{-1}}(fw)(x)$. Then using the definition of A_2 and the Cauchy–Schwarz inequality, we get

$$\begin{aligned}
\Sigma_1 &= \sum_{I\in\mathcal{D}} \frac{|b_I|}{\sqrt{\langle w\rangle_I}} \frac{\langle |f|ww^{-1}\rangle_I}{\langle w^{-1}\rangle_I} |\langle gw^{-1}, h_I^w\rangle_{L^2(w)}|\, \langle w\rangle_I \langle w^{-1}\rangle_I \\
&\le [w]_{A_2} \sum_{I\in\mathcal{D}} \frac{|b_I|}{\sqrt{\langle w\rangle_I}} \inf_{x\in I} M^{\mathcal{D}}_{w^{-1}}(fw)(x)\, |\langle gw^{-1}, h_I^w\rangle_{L^2(w)}| \\
&\le [w]_{A_2} \Bigg(\sum_{I\in\mathcal{D}} \frac{|b_I|^2}{\langle w\rangle_I} \inf_{x\in I} |M^{\mathcal{D}}_{w^{-1}}(fw)(x)|^2\Bigg)^{\frac12} \Bigg(\sum_{I\in\mathcal{D}} |\langle gw^{-1}, h_I^w\rangle_{L^2(w)}|^2\Bigg)^{\frac12}.
\end{aligned}$$

Using the weighted Carleson lemma (Lemma 2) with $F(x) = |M^{\mathcal{D}}_{w^{-1}}(fw)(x)|^2$, with weight $v = w^{-1}\in A_2$ recalling $[w]_{A_2} = [w^{-1}]_{A_2}$, and with w^{-1}-Carleson sequence $\{b_I^2/\langle w\rangle_I\}_{I\in\mathcal{D}}$ with intensity $4\|b\|^2_{\mathrm{BMO}^{\mathcal{D}}}$ by the Little lemma (Lemma 7.5.1), we get that

$$\begin{aligned}
\Sigma_1 &\le 2[w]_{A_2}\|b\|_{\mathrm{BMO}^{\mathcal{D}}} \left(\int_{\mathbb{R}} |M^{\mathcal{D}}_{w^{-1}}(fw)(x)|^2\, w^{-1}(x)\,dx\right)^{\frac12} \|gw^{-1}\|_{L^2(w)} \\
&\le 4[w]_{A_2}\|b\|_{\mathrm{BMO}^{\mathcal{D}}}\|f\|_{L^2(w)}\|g\|_{L^2(w^{-1})}.
\end{aligned}$$

where we used in the last inequality the estimate (7.3.1) for the weighted dyadic maximal function, and noting that $h\in L^2(w)$ if and only if $hw\in L^2(w^{-1})$, moreover, $\|hw\|_{L^2(w^{-1})} = \|h\|_{L^2(w)}$ (we used this twice, for $h=f$ and for $h=gw^{-1}$).

7.5.4.2 Second Sum Σ_2

Using similar arguments to those used for Σ_1, we get

$$\begin{aligned}
\Sigma_2 &\le \sum_{I\in\mathcal{D}} |b_I| \frac{\langle |f|ww^{-1}\rangle)}{\langle w^{-1}\rangle_I} \frac{\langle |g|w^{-1}w\rangle_I}{\langle w\rangle_I} \sqrt{|\Delta_I w|^2\langle w^{-1}\rangle_I^2 \|I|} \\
&\le \sum_{I\in\mathcal{D}} |b_I| \sqrt{\nu_I}\, \inf_{x\in I} M^{\mathcal{D}}_{w^{-1}}(fw)(x)\, M^{\mathcal{D}}_w(gw^{-1})(x),
\end{aligned}$$

where $|b_I|^2$ and $\nu_I := |\Delta_I w|^2 \langle w^{-1}\rangle_I^2 |I|$ are Carleson sequences with intensities $\|b\|^2_{\mathrm{BMO}^{\mathcal{D}}}$ and $[w]^2_{A_2}$, respectively, by Example 11 and Example 16(i). Then by the algebra of Carleson Sequences, the sequence $|b_I|\sqrt{\nu_I}$ is a Carleson sequence with intensity $\|b\|_{\mathrm{BMO}^{\mathcal{D}}}[w]_{A_2}$. Using the weighted Carleson lemma (Lemma 2) with $F(x) = M^{\mathcal{D}}_{w^{-1}}(fw)(x)\, M^{\mathcal{D}}_{w}(gw^{-1})(x)$ and with $v = 1$, we conclude that

$$\Sigma_2 \le [w]_{A_2} \|b\|_{\mathrm{BMO}^{\mathcal{D}}} \int_{\mathbb{R}} M^{\mathcal{D}}_{w^{-1}}(fw)(x)\, M^{\mathcal{D}}_{w}(gw^{-1})(x)\, dx.$$

To finish, we use the Cauchy–Schwarz inequality, the fact that $w^{\frac{1}{2}}(x)\, w^{-\frac{1}{2}}(x) = 1$, and estimate (7.3.1) for the weighted dyadic maximal functions, to get that

$$\begin{aligned}
\Sigma_2 &\le [w]_{A_2} \|b\|_{\mathrm{BMO}^{\mathcal{D}}} \left[\int_{\mathbb{R}} (M^{\mathcal{D}}_{w^{-1}}(fw)(x))^2 w^{-1}(x)\, dx\right]^{\frac{1}{2}} \left[\int_{\mathbb{R}} (M^{\mathcal{D}}_{w}(gw^{-1})(x))^2 w(x)\, dx\right]^{\frac{1}{2}} \\
&= [w]_{A_2} \|b\|_{\mathrm{BMO}^{\mathcal{D}}} \|M^{\mathcal{D}}_{w^{-1}}(fw)\|_{L^2(w^{-1})} \|M^{\mathcal{D}}_{w}(gw^{-1})\|_{L^2(w)} \\
&\le 4[w]_{A_2} \|b\|_{\mathrm{BMO}^{\mathcal{D}}} \|f\|_{L^2(w)} \|g\|_{L^2(w^{-1})}.
\end{aligned}$$

All together, this implies that $\|\pi_b f\|_{L^2(w)} \le 8[w]_{A_2}\|b\|_{\mathrm{BMO}^{\mathcal{D}}}\|f\|_{L^2(w)}$, proving the A_2 conjecture for the dyadic paraproduct.

7.5.5 *Auxiliary Lemmas*

We now present the Bellman function proofs (or at least the main ideas) for the Little lemma (Lemma 3) and the α-Lemma (Lemma 5), to illustrate the method in a very simple setting. For completeness, we also present the proof of the weighted Carleson lemma (Lemma 2). The lemmas in this section hold on $\mathbb{R}^d$ and also on geometrically doubling metric spaces [36, 144].

7.5.5.1 Proof of Beznosova's Little Lemma

We wish to prove Lemma 3. The proof uses a Bellman function argument, which we now describe. As usual, the argument proceeds in two steps. First, Lemma 6 encodes what now is called an *induction on scales argument*. If we can find a Bellman function with certain properties, then we will solve our problem by induction on scales. This type of arguments shows that if we can find a function with certain size, domain, and dyadic convexity properties tailored to the inequality of interest, we will be able to induct on scales and obtain the desired inequality. Second, Lemma 7 will show that such Bellman function exists.

Lemma 6 (Beznosova 2008) *Suppose there exists a real-valued function of three variables $B(x) = B(u, v, l)$, whose domain $\mathfrak{D}$ contains points $x = (u, v, l)$*

$$\mathfrak{D} := \{(u, v, l) \in \mathbb{R}^3 : u, v > 0, \quad uv \geq 1 \quad \textit{and} \quad 0 \leq l \leq 1\},$$

whose range is given by $0 \leq B(x) \leq u$, and such that the following convexity property holds:

$$B(x) - (B(x_+) + B(x_-))/2 \ \geq \ \alpha/4v, \ \ \textit{for all } x, x_\pm \in \mathfrak{D} \textit{ with } x - \frac{x_+ + x_-}{2} = (0, 0, \alpha). \tag{7.5.2}$$

Then Lemma 3 will be proven; more precisely, (7.5.1) *holds.*

Proof Without loss of generality, we may assume that the intensity A of the Carleson sequence $\{\lambda_I\}_{I \in \mathcal{D}}$ in Lemma 3 is one, $A = 1$.

Fix a dyadic interval J. Let $u_J := \langle w \rangle_J, v_J := \langle w^{-1} \rangle_J$ and $\ell_J := \frac{1}{|J|} \sum_{I \in \mathcal{D}(J)} \lambda_I$, then $x_J := (u_J, v_J, \ell_J) \in \mathfrak{D}$. Recall that $\mathcal{D}(J)$ denotes the intervals $I \in \mathcal{D}$ such that $I \subset J$.

Let $x_\pm := x_{J^\pm} \in \mathfrak{D}$, then

$$x_J - \frac{x_{J^+} + x_{J^-}}{2} \ = \ (0, 0, \alpha_J), \quad \text{where } \ \alpha_J := \frac{\lambda_J}{|J|}.$$

Hence, by the size and convexity property (7.5.2), and $|J^+| = |J^-| = |J|/2$,

$$|J| \, \langle w \rangle_J \ \geq \ |J| \ B(x_J) \geq |J^+| B(x_{J^+}) + |J^-| B(x_{J^-}) + \lambda_J / 4 \langle w^{-1} \rangle_J.$$

Repeat the argument this time for $|J^+| B(x_{J^+})$ and $|J^-| B(x_{J^-})$, use that $B \geq 0$ on $\mathfrak{D}$, and keep repeating to get, after dividing by $|J|$, that

$$\langle w \rangle_J \geq \frac{1}{4|J|} \sum_{I \in D(J)} \frac{\lambda_I}{\langle w^{-1} \rangle_I}$$

which implies (7.5.1) after multiplying through by $4|J|$. The lemma is proved.

The previous induction on scales argument is conditioned on the existence of a function with certain properties, a Bellman function. We now establish the existence of such function. Both lemmas appeared in [21].

Lemma 7 (Beznosova 2008) *The function $B(u, v, l) := u - \frac{1}{v(1+l)}$ is* (i) *defined on the domain $\mathfrak{D}$ introduced in Lemma 6,* (ii) $0 \leq B(x) \leq u$ *for all $x = (u, v, l) \in \mathfrak{D}$, and* (iii) *obeys the following differential estimates on $\mathfrak{D}$:*

$$(\partial B / \partial l)(u, v, l) \ \geq \ 1/(4v) \ \ \textit{and} \quad - (du, dv, dl) \, d^2 B(u, v, l) \, (du, dv, dl)^t \ \geq \ 0,$$

where $d^2B(u,v,l)$ denotes the Hessian matrix of the function B evaluated at (u,v,l). Moreover, these imply the dyadic convexity condition $B(x)-(B(x_+)+B(x_-))/2 \ge \alpha/(4v)$.

Proof Differential conditions can be checked by a direct calculation that we leave as an exercise for the reader. By the mean value theorem and some calculus,

$$B(x)-\frac{B(x_+)+B(x_-)}{2}=\frac{\partial B}{\partial l}(u,v,l')\alpha-\frac{1}{2}\int_{-1}^{1}(1-|t|)b''(t)dt \ge \frac{\alpha}{4v},$$

where $b(t):=B(x(t))$ and $x(t):=\frac{1+t}{2}x_+ + \frac{1-t}{2}x_-$ for $-1\le t\le 1$.

Note that $x(t)\in\mathfrak{D}$ whenever x_+ and x_- are in the domain, since $\mathfrak{D}$ is a convex domain and $x(t)$ is a point on the line segment between x_+ and x_-, and l' is a point between l and $\frac{l_+ + l_-}{2}$. This proves the lemma.

These two lemmas prove Beznosova's Little lemma (Lemma 3).

7.5.5.2 α-Lemma

We present a very brief sketch of the argument leading to the proof of the α-Lemma (Lemma 5), see [21] for $0<\alpha<1/2$, and [140] for $\alpha\ge 1/2$. Recall that we wish to show that if $w\in A_2$ and $0<\alpha$, then the sequence

$$\mu_I := \langle w\rangle_I^{\alpha}\langle w^{-1}\rangle_I^{\alpha}|I|\left(\frac{|\Delta_I w|^2}{\langle w\rangle_I^2}+\frac{|\Delta_I w^{-1}|^2}{\langle w^{-1}\rangle_I^2}\right) \quad \text{for } I\in\mathcal{D}$$

is a Carleson sequence with intensity at most $C_\alpha[w]_{A_2}^{\alpha}$, and $C_\alpha=\max\{72/(\alpha-2\alpha^2), 576\}$.

Proof (*Sketch of the Proof*) Use the Bellman function method. Figure out the domain, range, and dyadic convexity conditions needed to run an induction on scale argument that will yield the inequality. Verify that the Bellman function $B(u,v)=(uv)^{\alpha}$ satisfies those conditions (or at least a differential version that can then be seen implies the dyadic convexity) for $0<\alpha<1/2$. For $\alpha\ge 1/2$, just observe that one can factor out $\langle w\rangle_I^{\alpha-1/4}\langle w^{-1}\rangle_I^{\alpha-1/4}\le [w]_{A_2}^{\alpha-1/4}$ and then use the already proven lemma when $\alpha=1/4<1/2$.

7.5.5.3 Weighted Carleson Lemma

Finally, we present a proof of the weighted Carleson lemma (Lemma 2), which states that if v is a weight, $\{\alpha_L\}_{L\in\mathcal{D}}$ a v-Carleson sequence with intensity A, and F a positive measurable function on $\mathbb{R}$, then

$$\sum_{L \in \mathcal{D}} \alpha_L \inf_{x \in L} F(x) \le A \int_{\mathbb{R}} F(x)\, v(x)\, dx.$$

The weighted Carleson lemma we present here is a variation in the spirit of other weighted Carleson embedding theorems that appeared before in the literature [146]. The converse is immediately true by choosing $F(x) = \mathbb{1}_J(x)$.

Proof Assume that $F \in L^1(v)$; otherwise, the first statement is automatically true. Setting $\gamma_L = \inf_{x \in L} F(x)$, we can write

$$\sum_{L \in \mathcal{D}} \alpha_L \gamma_L = \sum_{L \in \mathcal{D}} \alpha_L \int_0^\infty \chi(L,t)\, dt = \int_0^\infty \Big(\sum_{L \in \mathcal{D}} \chi(L,t)\, \alpha_L \Big) dt, \qquad (7.5.3)$$

where $\chi(L,t) = 1$ for $t < \gamma_L$ and zero otherwise, and where we used the monotone convergence theorem in the last equality. Define the level set $E_t = \{x \in \mathbb{R} : F(x) > t\}$. Since $F \in L^1(v)$ then E_t is a v-measurable set for every t and we have, by Chebychev's inequality, that the v-measure of E_t is finite for all $t > 0$. Moreover, there is a collection of maximal disjoint dyadic intervals $\mathcal{P}_t$ that will cover E_t except for at most a set of v-measure zero. Finally, observe that $L \subset E_t$ if and only if $\chi(L,t) = 1$. All together we can rewrite the integrand in the right-hand side of (7.5.3) as

$$\sum_{L \in \mathcal{D}} \chi(L,t)\, \alpha_L = \sum_{L \subset E_t} \alpha_L \le \sum_{L \in \mathcal{P}_t} \sum_{I \in \mathcal{D}(L)} \alpha_I \le A \sum_{L \in \mathcal{P}_t} v(L) = A\, v(E_t),$$

where we used in the second inequality the fact that $\{\alpha_J\}_{I \in \mathcal{D}}$ is a v-Carleson sequence with intensity A. Thus, we can estimate

$$\sum_{L \in \mathcal{D}} \alpha_L \inf_{x \in L} F(x) = \sum_{L \in \mathcal{D}} \alpha_L \gamma_L \le A \int_0^\infty v(E_t)\, dt = A \int_{\mathbb{R}} F(x)\, v(x)\, dx,$$

where the last equality follows from the layer cake representation.

7.6 Case Study: Commutator of Hilbert Transform and Function in BMO

In this section, we summarize chronologically the weighted norm inequalities known for the commutator $[b, T]$ where T is a linear operator and b a function in BMO. In particular, we will consider $T = H$ the Hilbert transform. We sketch a dyadic proof of the first quantitative weighted estimate for the commutator $[b, H]$ due to Daewon Chung [35], yielding the optimal quadratic dependence on the A_2 characteristic of the weight. We discuss a very useful transference theorem of Chung, Pérez and the author [37], and present its proof based on the celebrated Coifman–Rochberg–Weiss argument. The transference theorem allows to deduce quantitative weighted L^p estimates for the commutator of a linear operator with a BMO function, from given quantitative weighted L^p estimates for the operator.

7.6.1 L^p Theory for $[b, H]$

Recall that the commutator of a function $b \in \mathrm{BMO}$ and H the Hilbert Transform is defined to be

$$[b, H](f) := b\,(Hf) - H(bf).$$

The commutator $[b, H]$ is bounded on $L^p(\mathbb{R})$ for $1 < p < \infty$ if and only if $b \in \mathrm{BMO}$ [41]. Moreover, the following estimate is known to hold for all $b \in \mathrm{BMO}$ and $f \in L^p(\mathbb{R})$

$$\|[b, H](f)\|_{L^p} \lesssim_p \|b\|_{\mathrm{BMO}} \|f\|_{L^p}.$$

In fact, the operator norm $\|[b, H]\|_{L^2 \to L^2} \sim \|b\|_{\mathrm{BMO}}$. Observe that bH and Hb are NOT necessarily bounded on $L^p(\mathbb{R})$ when $b \in \mathrm{BMO}$. The commutator introduces some key cancellation. This is very much connected to the celebrated H^1-BMO duality theorem by Fefferman and Stein [73], where the Hardy space H^1 can be defined as those functions f in $L^1(\mathbb{R})$ such that their maximal function Mf is also in $L^1(\mathbb{R})$.

The commutator $[b, H]$ is more singular than H, as evidenced by the fact that, unlike the Hilbert transform, the commutator is NOT of weak-type $(1, 1)$ [157]. In particular, the commutator is not a Calderón–Zygmund operator, if it were it would be of weak-type $(1, 1)$, and is not.

7.6.2 Weighted Inequalities

The first two-weight results for the commutator that we present are of a qualitative nature. The first one is a two-weight result due to Steven Bloom for the commutator of the Hilbert transform with a function in weighted BMO when both weights are in A_p [25].

Theorem 10 (Bloom 1985) *If $u, v \in A_p$, then $[b, H] : L^p(u) \to L^p(v)$ is bounded if and only if $b \in \mathrm{BMO}_\mu$ where $\mu = u^{-1/p} v^{1/p}$. Where $b \in \mathrm{BMO}_\mu$ if and only if*

$$\|b\|_{\mathrm{BMO}_\mu} := \sup_{I \in \mathbb{R}} \frac{1}{\mu(I)} \int_I |b(x) - \langle b \rangle_I|\, dx < \infty. \tag{7.6.1}$$

What is important in this setting is that the hypothesis $u, v \in A_p$ implies that $\mu \in A_2$ as can be seen by a direct calculation using Hölder's inequality. The weighted BMO space defined by (7.6.1) was first introduced by Eric Sawyer and Richard Wheeden [170], and it has been called in the literature, somewhat misleadingly, Bloom BMO. For a "modern" dyadic proof of Bloom's result, see [84, 85].

The second result is a one-weight result for very general linear operators T obtained by Josefina Álvarez, Richard Bagby, Doug Kurtz, and Carlos Pérez [8]; they also prove two-weight estimates.

Theorem 11 (Álvarez, Bagby, Kurtz, Pérez 1993) *Let T be a linear operator on the set of real-valued Lebesgue measurable functions defined on $\mathbb{R}^d$, with a domain of definition which contains every compactly supported function in a fixed L^p space. If $w \in A_p$ and $b \in$ BMO, then there is a constant $C_p(w) > 0$ such that for all $f \in L^p(w)$ the following inequality holds:*

$$\|[b, T](f)\|_{L^p(w)} \leq C_p(w)\|b\|_{\mathrm{BMO}}\|f\|_{L^p(w)}.$$

The proof uses a classical argument by Raphy Coifman,[14] Richard Rochberg, and Guido Weiss [41]. In Sect. 7.6.4.2, we will present a quantitative version of this argument [37]. For a proof of Bloom's result using this type of argument, see [93].

The next result is a quantitative weighted inequality obtained by Daewon Chung in his Ph.D. Dissertation [34, 35].

Theorem 12 (Chung 2010) *For all $b \in$ BMO, $w \in A_2$ and $f \in L^2(w)$, the following holds:*

$$\|[b, H](f)\|_{L^2(w)} \lesssim \|b\|_{\mathrm{BMO}}[w]^2_{A_2}\|f\|_{L^2(w)}.$$

The quadratic power on the A_2 characteristic and the linear bound on the BMO norm are both optimal powers. The quadratic dependence on the A_2 characteristic is another indication that this operator is more singular than the Calderón–Zygmund singular integral operators for whom the dependence is linear [90], as we have emphasized throughout these lectures.

7.6.3 Dyadic Proof of Chung's Theorem

We now sketch Chung's dyadic proof of the quadratic estimate for the commutator [35].

Proof (*Sketch of proof*) Chung's "dyadic" proof is based on using Petermichl's dyadic shift operators $Ш^{r,\beta}$ instead of H [161] and proving uniform (on the dyadic grids $\mathcal{D}^{r,\beta}$) quadratic estimates for the corresponding commutators $[Ш^{r,\beta}, b]$. To ease

[14] As I am writing these notes, it has been announced that Coifman won the 2018 Schock Prize in Mathematics for his "fundamental contributions to pure and applied harmonic analysis."

notation, we drop the superscripts r, β and simply write Ш for $Ш^{r,\beta}$, and the estimates will be independent of the parameters r and β.

To achieve this, we first recall the decomposition of a product bf in terms of paraproducts and their adjoints,

$$bf = \pi_b f + \pi_b^* f + \pi_f b,$$

notice that the first two terms are bounded on $L^p(w)$ when $b \in \mathrm{BMO}$ and $w \in A_p$; the enemy is the third term. Decomposing the commutator accordingly, we get

$$[b, Ш](f) = [\pi_b, Ш](f) + [\pi_b^*, Ш](f) + \big(\pi_{Ш f}(b) - Ш(\pi_f b)\big). \tag{7.6.2}$$

Known linear bounds on $L^2(w)$ for the dyadic paraproduct π_b, its adjoint π_b^*, and for Petermichl's dyadic shift operator Ш, see [21, 162], immediately given by iteration, quadratic bounds for the first two terms on the right-hand side of (7.6.2). Surprisingly, the third term is better; it obeys a linear bound, and so do halves of the first two commutators, as shown in [35] using *Bellman function* techniques, namely,

$$\|\pi_{Ш f}(b) - X(\pi_f b)\|_{L^2(w)} + \|Ш\pi_b(f)\|_{L^2(w)} + \|\pi_b^* Ш(f)\|_{L^2(w)} \leq C\|b\|_{\mathrm{BMO}}[w]_{A_2}\|f\|_{L^2(w)}.$$

All together providing uniform (on the random dyadic grids $\mathcal{D}^{r,\beta}$) quadratic bounds for the commutators $[b, Ш^{r,\beta}]$, and hence, averaging over the random grids, we get the desired quadratic estimate for $[b, H]$.

The quadratic estimate and the corresponding extrapolated estimates, namely, for all $b \in \mathrm{BMO}$, $w \in A_p$, and $f \in L^p(w)$

$$\|[b, H](f)\|_{L^p(w)} \lesssim_p [w]_{A_p}^{2\max\{1, \frac{1}{p-1}\}} \|b\|_{\mathrm{BMO}}\|f\|_{L^p(w)}, \tag{7.6.3}$$

are optimal for all $1 < p < \infty$, as can be seen considering appropriate power functions and power weights [37].

The "bad guys" are the nonlocal terms $\pi_b Ш$, $Ш\pi_b^*$. A posteriori one realizes the pieces that obey linear bounds are generalized Haar Shift operators, and hence their linear bounds can be deduced from general results for those operators.

As a byproduct of Chung's dyadic proof, we get that the extrapolated bounds for the dyadic paraproduct are optimal [155], namely, for all $b \in \mathrm{BMO}$, $w \in A_p$, and $f \in L^p(w)$

$$\|\pi_b f\|_{L^p(w)} \lesssim_p [w]_{A_p}^{\max\{1, \frac{1}{p-1}\}} \|b\|_{\mathrm{BMO}}\|f\|_{L^p(w)}.$$

Proof By contradiction, if not for some p then $[b, H]$ will have a better bound in $L^p(w)$ than the known optimal bound given by (7.6.3) for that p.

7.6.4 A Quantitative Transference Theorem

The following theorem provides a mechanism for transferring known quantitative weighted estimates for linear operators to their commutators with BMO functions [37, 155].

Theorem 13 (Chung, Pereyra, Pérez 2012) *Given linear operator T and $1 < r < \infty$, such that for all $w \in A_r$ and $f \in L^r(w)$, the following estimate holds*

$$\|Tf\|_{L^r(w)} \lesssim_{T,d} [w]^{\alpha}_{A_r} \|f\|_{L^r(w)},$$

then the commutator of T with $b \in \mathrm{BMO}$ is such that for all $w \in A_r$ and $f \in L^r(w)$

$$\|[b,T](f)\|_{L^r(w)} \lesssim_{r,T,d} [w]^{\alpha+\max\{1,\frac{1}{r-1}\}}_{A_r} \|b\|_{\mathrm{BMO}} \|f\|_{L^r(w)}.$$

The proof follows the classical Coifman–Rochberg–Weiss argument using (i) the Cauchy integral formula; (ii) the following quantitative Coifman–Fefferman result: $w \in A_r$ implies $w \in RH_q$ with $q = 1 + c_d/[w]_{A_r}$ and $[w]_{RH_q} \leq 2$; and (iii) a quantitative version of the estimate: $b \in \mathrm{BMO}$ implies $e^{\alpha b} \in A_r$ for α small enough with control on $[e^{\alpha b}]_{A_r}$. We will present the whole argument in the case $r = 2$ in Sect. 7.6.4.2. Here, the *Reverse Hölder-q weight class* (RH_q) for $1 < q < \infty$ is defined to be all those weights w such that

$$[w]_{RH_q} := \sup_Q \langle w^q \rangle_Q^{1/q} \langle w \rangle_Q^{-1} < \infty,$$

where the supremum is taken over all cubes in $\mathbb{R}^d$ with sides parallel to the axes.

A variation on the argument yields corresponding estimates for the *higher order commutators* $T^k_b := [b, T^{k-1}_b]$ for $k \geq 1$ and $T^0_b := T$. More precisely, given the initial estimate $\|T^0_b f\|_{L^r(w)} \lesssim [w]^{\alpha}_{A_r}$, valid for all $w \in A_r$, then the following estimate holds for all $k \geq 1$, $b \in \mathrm{BMO}$, $w \in A_r$, and $f \in L^r(w)$

$$\|T^k_b f\|_{L^r(w)} \lesssim_{r,T,d} [w]^{\alpha+k\max\{1,\frac{1}{r-1}\}}_{A_r} \|b\|^k_{\mathrm{BMO}} \|f\|_{L^r(w)}.$$

Transference theorems for commutators are useless unless there are operators known to obey an initial $L^r(w)$ bound valid for all $w \in A_r$. We have already mentioned that the class of Calderón–Zygmund singular integral operators obey linear bounds on $L^2(w)$, thanks to Hytönen's A_2 theorem [90]. We conclude that for all Calderón–Zygmund singular integral operators T their commutators obey a quadratic bound on $L^2(w)$, more precisely,

$$\|[b,T]f\|_{L^2(w)} \lesssim_{T,d} [w]^2_{A_2} \|b\|_{\mathrm{BMO}} \|f\|_{L^2(w)}.$$

With a slight modification of the argument, one can see [37] that the correct estimate for the iterated commutators of Calderón–Zygmund singular integral operators and function $b \in$ BMO is

$$\|[T_b^k f\|_{L^2(w)} \lesssim_{T,d} [w]_{A_2}^{1+k} \|b\|_{\mathrm{BMO}}^k \|f\|_{L^2(w)}.$$

There are operators (for example, the Hilbert, Riesz, and Beurling transforms) for whom these estimates are optimal in terms of the powers for both the A_2 characteristic and the BMO norm. This can be seen testing power functions and weights [37].

7.6.4.1 Some Generalizations

There are extensions to commutators with fractional integral operators, two-weight problem, and more [50, 51]. There are mixed A_2-A_∞ estimates, where recall that $A_\infty = \cup_{p>1} A_p$ and $[w]_{A_\infty} \leq [w]_{A_2}$ [98, 153], more precisely estimates of the form,

$$\|[b, T]\|_{L^2(w)} \lesssim [w]_{A_2}^{\frac{1}{2}} \big([w]_{A_\infty} + [w^{-1}]_{A_\infty}\big)^{\frac{3}{2}} \|b\|_{\mathrm{BMO}}.$$

There are generalizations to commutators of matrix-valued operators and BMO [103] as well as to the two-weight setting (both weights in A_p, à la Bloom) [84, 85], and also for biparameter Journé operators [86]. See also the comprehensive paper [18] where a systematic use of the Coifman–Rochberg–Weiss trick recovers all known results and some new ones such as boundedness of the commutator of the bilinear Hilbert transform and a function in BMO. Pointwise control by sparse operators adapted to the commutator, improving weak-type, Orlicz bounds, and quantitative two-weight Bloom bounds was recently obtained [134, 135]. We will say more about this generalization in Sect. 7.7.

7.6.4.2 Proof of the Transference Theorem

We now present the proof of the quantitative transference theorem when $r = 2$, following the lines of the Coifman–Rochberg–Weiss argument [41] with a few quantitative ingredients. For $r \neq 2$, see [155].

Proof (*Proof in* [37]) "Conjugate" the operator as follows: for any $z \in \mathbb{C}$, define

$$T_z(f) = e^{zb}\, T\,(e^{-zb} f).$$

A computation together with the Cauchy integral theorem give (for "nice" functions)

$$[b, T](f) = \frac{d}{dz} T_z(f)|_{z=0} = \frac{1}{2\pi i} \int_{|z|=\varepsilon} \frac{T_z(f)}{z^2}\, dz, \quad \varepsilon > 0.$$

Now, by Minkowski's integral inequality

$$\|[b,T](f)\|_{L^2(w)} \leq \frac{1}{2\pi\,\varepsilon^2}\int_{|z|=\varepsilon} \|T_z(f)\|_{L^2(w)}|dz|, \quad \varepsilon>0.$$

The key point is to find an appropriate radius $\varepsilon > 0$. To that effect, we look at the inner norm and try to find bounds depending on z. More precisely,

$$\|T_z(f)\|_{L^2(w)} = \|T(e^{-zb}f)\|_{L^2(w\,e^{2\mathrm{Re}\,z\,b})}.$$

We use the main hypothesis, namely, that T is bounded on $L^2(v)$ if $v\in A_2$ with $\|T\|_{L^2(v)} \leq C[v]_{A_2}$, for $v = w\,e^{2\mathrm{Re}\,z\,b}$. We must check that if $w\in A_2$ then $v\in A_2$ for $|z|$ small enough. Indeed,

$$[v]_{A_2} = \sup_Q \left(\frac{1}{|Q|}\int_Q w(x)\,e^{2\mathrm{Re}\,z\,b(x)}\,dx\right)\left(\frac{1}{|Q|}\int_Q w^{-1}(x)\,e^{-2\mathrm{Re}\,z\,b(x)}\,dx\right).$$

It is well known that if $w\in A_2$ then $w\in RH_q$ for some $q>1$ [39]. There is a quantitative version of this result [158], namely, if $q = 1+\frac{1}{2^{d+5}[w]_{A_2}}$ then

$$\left(\frac{1}{|Q|}\int_Q w^q(x)\,dx\right)^{\frac{1}{q}} \leq \frac{2}{|Q|}\int_Q w(x)\,dx,$$

similarly for $w^{-1}\in A_2$ and for the same q, since $[w]_{A_2} = [w^{-1}]_{A_2}$, we have that

$$\left(\frac{1}{|Q|}\int_Q w^{-q}(x)\,dx\right)^{\frac{1}{q}} \leq \frac{2}{|Q|}\int_Q w^{-1}(x)\,dx\,.$$

In what follows $q = 1+1/(2^{d+5}[w]_{A_2})$. Using these estimates and Holder's inequality, we have for an arbitrary cube Q

$$\begin{aligned}
&\left(\frac{1}{|Q|}\int_Q w(x)e^{2\mathrm{Re}\,z\,b(x)}\,dx\right)\left(\frac{1}{|Q|}\int_Q w(x)^{-1}e^{-2\mathrm{Re}\,z\,b(x)}\,dx\right)\\
&\leq \left(\frac{1}{|Q|}\int_Q w^q(x)\,dx\right)^{\frac{1}{q}}\left(\frac{1}{|Q|}\int_Q e^{2\mathrm{Re}zq'b(x)}\,dx\right)^{\frac{1}{q'}}\left(\frac{1}{|Q|}\int_Q w^{-q}(x)\,dx\right)^{\frac{1}{q}}\left(\frac{1}{|Q|}\int_Q e^{-2\mathrm{Re}zq'b(x)}\,dx\right)^{\frac{1}{q'}}\\
&\leq 4\left(\frac{1}{|Q|}\int_Q w(x)\,dx\right)\left(\frac{1}{|Q|}\int_Q w^{-1}(x)\,dx\right)\left(\frac{1}{|Q|}\int_Q e^{2\mathrm{Re}\,z\,q'\,b(x)}\,dx\right)^{\frac{1}{q'}}\left(\frac{1}{|Q|}\int_Q e^{-2\mathrm{Re}\,z\,q'\,b(x)}\,dx\right)^{\frac{1}{q'}}\\
&\leq 4\,[w]_{A_2}\,[e^{2\mathrm{Re}\,z\,q'\,b}]_{A_2}^{\frac{1}{q'}}.
\end{aligned}$$

Taking the supremum over all cubes, we conclude that

$$[v]_{A_2} = [w\,e^{2\mathrm{Re}\,z\,b}]_{A_2} \leq 4\,[w]_{A_2}\,[e^{2\mathrm{Re}\,z\,q'\,b}]_{A_2}^{\frac{1}{q'}}.$$

Now, since $b\in \mathrm{BMO}$ there are $0<\alpha_d<1$ and $\beta_d>1$ such that if $|2\mathrm{Re}\,z\,q'| \leq \alpha_d/\|b\|_{\mathrm{BMO}}$ then $[e^{2\mathrm{Re}\,z\,q'\,b}]_{A_2} \leq \beta_d$, see [37, Lemma 2.2]. Hence, for these z,

$$[v]_{A_2} \leq 4\,[w]_{A_2}\,\beta_d^{\frac{1}{q'}} \leq 4\,[w]_{A_2}\,\beta_d.$$

We have shown that if $|z| \leq \alpha_d/(2q'\|b\|_{\mathrm{BMO}})$ then $[v]_{A_2} \leq 4[w]_{A_2}\,\beta_d$ and

$$\|T_z(f)\|_{L^2(w)} = \|T(e^{-zb}f)\|_{L^2(v)} \lesssim [v]_{A_2}\|f\|_{L^2(w)} \leq 4[w]_{A_2}\,\beta_d\,\|f\|_{L^2(w)}.$$

Here, the first inequality holds since $\|e^{-zb}f\|_{L^2(v)} = \|e^{-zb}f\|_{L^2(we^{2\mathrm{Re}\,z\,b})} = \|f\|_{L^2(w)}$.

Thus choose the radius $\varepsilon := \alpha_d/(2q'\|b\|_{\mathrm{BMO}})$, and get

$$\begin{aligned}\|[b,T](f)\|_{L^2(w)} &\leq \frac{1}{2\pi\,\varepsilon^2}\int_{|z|=\varepsilon}\|T_z(f)\|_{L^2(w)}|dz|\\ &\leq \frac{1}{2\pi\,\varepsilon^2}\int_{|z|=\varepsilon}4[w]_{A_2}\,\beta_d\,\|f\|_{L^2(w)}|dz| = \frac{1}{\varepsilon}\,4[w]_{A_2}\,\beta_d\,\|f\|_{L^2(w)}.\end{aligned}$$

Note that $\varepsilon^{-1} \approx [w]_{A_2}\|b\|_{\mathrm{BMO}}$, because $q' = 1 + 2^{d+5}[w]_{A_2} \approx 2^d[w]_{A_2}$,

$$\|[b,T](f)\|_{L^2(w)} \leq C_d\,[w]_{A_2}^2\,\|b\|_{\mathrm{BMO}},$$

which is exactly what we wanted to prove.

7.7 Sparse Operators and Sparse Families of Dyadic Cubes

In this section, we discuss the sparse domination by finitely many positive dyadic operators' paradigm that has recently emerged as a byproduct of the study of weighted inequalities. This sparse domination paradigm has proven to be very powerful with applications in areas other than weighted norm inequalities. In this section, we introduce the sparse operators and the sparse families of cubes. We discuss a characterization of sparse families of cubes via Carleson families of dyadic cubes due to Andrei Lerner and Fedja Nazarov; however, this was well known 20 years earlier by Igor Verbitsky [177, Corollary 2, p. 23], see also [82]. We present the beautiful proof of the A_2 conjecture for sparse operators due to David Cruz-Uribe, Chema Martell, and Carlos Pérez. We record the sparse domination results for the operators discussed in these notes. We present how to dominate pointwise the martingale transform by a sparse operator following Michael Lacey's argument, illustrating the technique in a toy model. Finally, we briefly discuss a sparse domination theorem for commutators valid for (rough) Calderón–Zygmund singular integral operators due to Andrei Lerner, Sheldy Ombrosi, and Israel Rivera-Ríos that yields new quantitative two-weight estimates of Bloom type and recovers all known weighted results for the commutators.

7.7.1 Sparse Operators

David Cruz-Uribe, Chema Martell, and Carlos Pérez showed in [53] the A_2 conjecture in a few lines for *sparse operators* $\mathcal{A}_{\mathcal{S}}$ defined as follows

$$\mathcal{A}_{\mathcal{S}} f(x) = \sum_{Q\in\mathcal{S}} \langle f\rangle_Q \, \mathbb{1}_Q(x).$$

Here, $\mathcal{S}$ is a sparse collection of dyadic cubes. A collection of dyadic cubes $\mathcal{S}$ in $\mathbb{R}^d$ is η*-sparse*, $0 < \eta < 1$ if there are pairwise disjoint measurable sets E_Q for each $Q \in \mathcal{S}$ such that

$$E_Q \subset Q \;\text{ with }\; |E_Q| \ge \eta|Q| \quad \text{for all} \;\; Q \in \mathcal{S}.$$

A primary example for us are the Calderón–Zygmund singular integral operators; they and the "rough" Calderón–Zygmund operators have been shown to be pointwise dominated by a finite number of sparse operators [47, 115, 129, 133]. A quantitative form of these estimates can be found in [100, 131]. More recently, see sparse domination principles for rough Calderón–Zygmund singular integral operators [44, 55, 65, 100].

7.7.2 Sparse Versus Carleson Families of Dyadic Cubes

We have seen in Sect. 7.4 how Carleson sequences and Carleson embedding lemmas come handy when proving weighted inequalities. There is an intimate connection between Carleson families of cubes and sparse families of cubes. A family of dyadic cubes $\mathcal{S}$ in $\mathbb{R}^d$ is called Λ*-Carleson* for $\Lambda > 1$ if

$$\sum_{P\in\mathcal{S},\,P\subset Q} |P| \le \Lambda|Q| \quad \forall Q \in \mathcal{D}.$$

Notice that a family of cubes being Λ-Carleson is equivalent to the sequence $\{|P|\mathbb{1}_{\mathcal{S}}(P)\}_{P\in\mathcal{D}}$ being Carleson with intensity Λ. Furthermore, the notion is equivalent to the family of cubes being $1/\Lambda$-sparse. These types of conditions are also called *Carleson packing conditions*.

Lemma 8 (Verbitsky 1996, Lerner, Nazarov 2014) *Let* $\Lambda > 1$. *The family of dyadic cubes* $\mathcal{S}$ *in* $\mathbb{R}^d$ *is* Λ*-Carleson if and only if* $\mathcal{S}$ *is* $1/\Lambda$*-sparse.*

Proof We sketch the beautiful argument in [133].
($\Leftarrow$) The family of cubes $\mathcal{S}$ being $1/\Lambda$-sparse means that for all cubes $P \in \mathcal{S}$ there are pairwise disjoint subsets $E_P \subset P$ that have a considerable portion of the total mass of the cube, more precisely $\Lambda|E_P| \ge |P|$. Hence,

$$\sum_{P\in\mathcal{S},P\subset Q} |P| \le \Lambda \sum_{P\in\mathcal{S},P\subset Q} |E_P| \le \Lambda|Q|.$$

Here, the last inequality holds because the sets $E_P \subset Q$ and are pairwise disjoint. Therefore, the family of cubes $\mathcal{S}$ is Λ-Carleson.

($\Rightarrow$) Assume now that $\mathcal{S}$ is a Λ-Carleson family. We say that a *family $\mathcal{S}$ has a bottom layer $\mathcal{D}_K$* if for all $Q\in\mathcal{S}$ we have $Q\in\mathcal{D}_k$ for some $k\le K$. Assume $\mathcal{S}$ HAS A BOTTOM LAYER $\mathcal{D}_K$. Then consider all cubes in the bottom layer, $Q\in\mathcal{S}\cap\mathcal{D}_K$, and choose any sets $E_Q\subset Q$ with $|E_Q|=\frac{1}{\Lambda}|Q|$. This choice is always possible, because of the nature of the Lebesgue measure, and the sets will automatically be pairwise disjoint because the cubes in a fixed generation $\mathcal{D}_K$ are pairwise disjoint. Then, go up layer by layer, meaning we have already selected sets $E_R\subset R$ for all $R\in\mathcal{S}\cap\mathcal{D}_j$ and $k<j\le K$ with the property that $|E_R|=\frac{1}{\Lambda}|R|$, then for each $Q\in\mathcal{D}_k$, $k<K$, choose any $E_Q\subset Q\setminus\cup_{R\in\mathcal{S},R\subsetneq Q}E_R$ with $|E_Q|=\frac{1}{\Lambda}|Q|$. Such choice is always possible because for every $Q\in\mathcal{S}$ we have

$$\left|\cup_{R\in\mathcal{S},R\subsetneq Q} E_R\right| \le \frac{1}{\Lambda}\sum_{R\in\mathcal{S},R\subsetneq Q}|R| \le \frac{\Lambda-1}{\Lambda}|Q| = \left(1-\frac{1}{\Lambda}\right)|Q|,$$

where we used in the inequality the hypothesis that $\mathcal{S}$ is a Λ-Carleson family. Therefore,

$$|Q\setminus\cup_{R\in\mathcal{S},R\subsetneq Q}E_R| \ge \frac{1}{\Lambda}|Q|,$$

Hence, there is enough mass left in Q, after removing the sets E_R corresponding to R in $\mathcal{S}$ and proper subcubes of Q, to select a subset E_Q of Q with the aformentioned property. Moreover, by construction the sets E_Q are pairwise disjoint, and we are done.

BUT, WHAT IF THERE IS NO BOTTOM LAYER? The idea is to run the construction for each $K\ge 0$ and pass to the limit! One has to be a bit careful! As Lerner and Nazarov put it: "*All we have to do is replace 'free choice' with 'canonical choice'.*" The diligent reader can find the details of the argument, including a very illuminating picture, in [133, Lemma 6.3 and Fig. 8].

7.7.3 A$_2$ Theorem for Sparse Operators

We now present David Cruz-Uribe, Chema Martell, and Carlos Pérez's beautiful proof of the A_2 conjecture for sparse operators [53].

Theorem 14 (Cruz-Uribe, Martell, Pérez 2012) *Let $\mathcal{S}$ be an η-sparse family of cubes, then For all $w\in A_2$ and $f\in L^2(w)$ the following inequality holds*

$$:\|\mathcal{A}_{\mathcal{S}}f\|_{L^2(w)} \lesssim [w]_{A_2}\|f\|_{L^2(w)}. \tag{7.7.1}$$

Proof For $w \in A_2$, $\mathcal{S}$ and η-sparse family with $\eta \in (0,1)$, showing (7.7.1) is equivalent by duality to showing that for all $f \in L^2(w)$, $g \in L^2(w^{-1})$

$$|\langle \mathcal{A}_{\mathcal{S}} f, g\rangle| \lesssim [w]_{A_2} \|f\|_{L^2(w)} \|g\|_{L^2(w^{-1})}.$$

By the Cauchy–Schwarz inequality $|E_Q| = \int_{E_Q} w^{\frac{1}{2}} w^{-\frac{1}{2}} \leq (w(E_Q))^{\frac{1}{2}} (w^{-1}(E_Q))^{\frac{1}{2}}$. Using the definition of the sparse operator, some algebra and the definition of an η-sparse family of cubes, namely, $|Q| \leq (1/\eta)|E_Q|$ we get that

$$\begin{aligned}
|\langle \mathcal{A}_{\mathcal{S}} f, g\rangle| &\leq \sum_{Q\in\mathcal{S}} \langle |f|\rangle_Q \,\langle |g|\rangle_Q\, |Q| \\
&\leq \frac{1}{\eta} \sum_{Q\in\mathcal{S}} \frac{\langle |f| w w^{-1}\rangle_Q}{\langle w^{-1}\rangle_Q} \,\frac{\langle |g| w^{-1} w\rangle_Q}{\langle w\rangle_Q}\, \langle w\rangle_Q \,\langle w^{-1}\rangle_Q\, |E_Q| \\
&\leq \frac{[w]_{A_2}}{\eta} \sum_{Q\in\mathcal{S}} \frac{\langle |f| w w^{-1}\rangle_Q}{\langle w^{-1}\rangle_Q} (w^{-1}(E_Q))^{\frac{1}{2}} \,\frac{\langle |g| w^{-1} w\rangle_Q}{\langle w\rangle_Q} (w(E_Q))^{\frac{1}{2}}.
\end{aligned}$$

Using once more the Cauchy–Schwarz inequality and the fact that for all $x \in E_Q \subset Q$ it holds that $\langle |h|v\rangle_Q/\langle v\rangle_Q \leq M_v^{\mathcal{D}} h(x)$, therefore $|\langle |h|v\rangle_Q/\langle v\rangle_Q|^2 v(E_Q) \leq \int_{E_Q} |M_v^{\mathcal{D}} h(x)|^2\, v(x)\, dx$, we conclude that

$$\begin{aligned}
|\langle \mathcal{A}_{\mathcal{S}} f, g\rangle| &\leq \frac{[w]_{A_2}}{\eta} \Big[\sum_{Q\in\mathcal{S}} \frac{\langle |f| w w^{-1}\rangle_Q^2}{\langle w^{-1}\rangle_Q^2} w^{-1}(E_Q)\Big]^{\frac{1}{2}} \Big[\sum_{Q\in\mathcal{S}} \frac{\langle |g| w^{-1} w\rangle_Q^2}{\langle w\rangle_Q^2} w(E_Q)\Big]^{\frac{1}{2}} \\
&\leq \frac{[w]_{A_2}}{\eta} \Big[\sum_{Q\in\mathcal{S}} \int_{E_Q} |M_{w^{-1}}^{\mathcal{D}}(fw)(x)|^2 w^{-1}(x)\, dx\Big]^{\frac{1}{2}} \Big[\sum_{Q\in\mathcal{S}} \int_{E_Q} |M_w^{\mathcal{D}}(gw^{-1})(x)|^2 w(x)\, dx\Big]^{\frac{1}{2}} \\
&\leq \frac{[w]_{A_2}}{\eta} \|M_{w^{-1}}^{\mathcal{D}}(fw)\|_{L^2(w^{-1})} \|M_w^{\mathcal{D}}(gw^{-1})\|_{L^2(w)} \\
&\lesssim [w]_{A_2} \|fw\|_{L^2(w^{-1})} \|gw^{-1}\|_{L^2(w)} = [w]_{A_2} \|f\|_{L^2(w)} \|g\|_{L^2(w^{-1})}.
\end{aligned}$$

where, in the line before the last, we used the fact that the sets E_Q for $Q \in \mathcal{S}$ are pairwise disjoint and, in the last line, we used estimate (7.3.1) for the weighted dyadic maximal functions.

Similar argument yields linear bounds on $L^p(w)$ for $p > 2$ and by duality (sparse operators are self-adjoint) we get bounds like $[w]_{A_p}^{1/(p-1)} = [w^{-1/(p-1)}]_{A_{p'}}$ when $1 < p < 2$, see [139]. In other words, we can get directly the same $L^p(w)$ bounds that sharp extrapolation will give if we were to extrapolate from the linear $L^2(w)$ bounds, namely, for all $w \in A_p$ and $f \in L^p(w)$

$$\|\mathcal{A}_{\mathcal{S}} f\|_{L^p(w)} \lesssim [w]_{A_p}^{\max\{1, \frac{1}{p-1}\}} \|f\|_{L^p(w)}. \tag{7.7.2}$$

7.7.4 Domination by Sparse Operators

Many operators can be dominated by finitely many sparse operators, pointwise, in norm, or by forms. The collections $\mathcal{S}$, $\mathcal{S}_i$ are sparse families tailored to the operator and the particular function f the operator is acting on. Identifying these sparse families is where most of the work lies, usually done using some sort of weak-$(1, 1)$ inequality that is available a priori, or a specific stopping time designed for the problem at hand. We will illustrate this process for the martingale transform in Sect. 7.7.5. Here is the status, in terms of sparse domination, of the operators we have been discussing in these lecture notes. In particular, quantitative weighted estimates for corresponding sparse operators, such as (7.7.2), immediately transfer to the dominated operators, providing new and streamlined proofs of the quantitative weighted inequalities we have been focusing on previous sections.

The martingale transforms and the dyadic paraproduct are locally pointwise dominated by sparse operators [115]. More precisely, given a cube Q_0 and $f \in L^1(\mathbb{R})$ there are sparse families $\mathcal{S}$, $\mathcal{S}'$ such that

$$|\mathbb{1}_{Q_0} T_\sigma(f \mathbb{1}_{Q_0})| \lesssim \mathcal{A}_{\mathcal{S}}|f|, \quad |\mathbb{1}_{Q_0} \pi_b(f \mathbb{1}_{Q_0})| \lesssim \mathcal{A}_{\mathcal{S}'}|f|.$$

We will say more about the martingale transform in Sect. 7.7.5.

Calderón–Zygmund operators are pointwise dominated by finitely many sparse operators [47, 131, 133]. More precisely, given T and f there are finitely many sparse families $\mathcal{S}_i$, for $i = 1, \ldots, N_d$, such that

$$|Tf| \le \sum_{i=1}^{N_d} \mathcal{A}_{\mathcal{S}_i} f.$$

The dyadic square function is pointwise dominated by finitely many sparse-like operators [118]. More precisely, given f there are finitely many sparse families $\mathcal{S}_i$, for $i = 1, \ldots, N_d$, such that

$$|S^{\mathcal{D}} f|^2 \le \sum_{i=1}^{N_d} \sum_{Q \in \mathcal{S}_i} \langle |f| \rangle_Q^2 \mathbb{1}_Q.$$

Notice that the sparse-like operators have been adapted to the square function.

Commutator $[b, T]$ for T an ω-Calderón–Zygmund operator with ω satisfying a Dini condition, $b \in L^1_{\text{loc}}(\mathbb{R})$, can be pointwise dominated by finitely many sparse-like operators and their adjoints [134, 135]. We will say more about this in Sect. 7.7.6.

The finitely many sparse families come from the analogue of the one-third trick for the dyadic grids; usually, $N_d = 3^d$ will suffice.

7.7.5 Domination of Martingale Transform D'après Lacey

We would like to illustrate how to achieve domination by sparse operators for a toy model operator, the martingale transform T_σ on $L^2(\mathbb{R})$, following an argument of Michael Lacey [115, Sect. 3].

Given interval $I_0 \in \mathcal{D}$ and function $f \in L^1(\mathbb{R})$ supported on I_0, we need to find a 1/2-sparse family $\mathcal{S} \subset \mathcal{D}$, such that for all choices of signs σ, there is a constant $C > 0$ such that

$$|\mathbb{1}_{I_0} T_\sigma f| \leq C \mathcal{A}_{\mathcal{S}} |f|.$$

Proof Without loss of generality, we can assume that $f \in L^1(\mathbb{R})$ is not only supported on I_0 but also $\int_{I_0} |f(x)|dx > 0$. We will need the following well-known weak-type estimates.

First, the sharp truncation $T_\sigma^\sharp$ is of weak-type (1, 1) [29], with a constant independent of the choice of signs σ, thus

$$\sup_{\lambda>0} \lambda \big|\{x \in \mathbb{R} : T_\sigma^\sharp f(x) > \lambda\}\big| \leq C \|f\|_{L^1(\mathbb{R})},$$

where $T_\sigma^\sharp f = \sup_{I' \in \mathcal{D}} \big| \sum_{I \in \mathcal{D}, I \supset I'} \sigma_I \langle f, h_I \rangle h_I \big|$.

Second, the maximal function M is also of weak-type (1, 1), therefore

$$\sup_{\lambda>0} \lambda \big|\{x \in \mathbb{R} : Mf(x) > \lambda\}\big| \leq C \|f\|_{L^1(\mathbb{R})},$$

As a consequence, there exists a constant $C_0 > 0$ such that the subset of I_0 defined by

$$F_{I_0} := \{x \in I_0 : \max\{Mf(x), T_\sigma^\sharp f(x)\} > \frac{1}{2} C_0 \langle |f| \rangle_{I_0}\}$$

has no more than half the mass of I_0, that is, $|F_{I_0}| \leq \frac{1}{2}|I_0|$. In fact, suppose no such constant would exist, then for all $C_0 > 0$ it would hold that

$$|F_{I_0}| = \Big|\{x \in I_0 : \max\{Mf(x), T_\sigma^\sharp f(x)\} > \frac{1}{2} C_0 \langle |f| \rangle_{I_0}\}\Big| > \frac{1}{2}|I_0|,$$

therefore for each $C_0 > 0$, it must be that either $|\{x \in I_0 : Mf(x) > \frac{1}{2} C_0 \langle |f| \rangle_{I_0}\}| > \frac{1}{4}|I_0|$ or $|\{x \in I_0 : T_\sigma^\sharp f(x) > \frac{1}{2} C_0 \langle |f| \rangle_{I_0}\}| > \frac{1}{4}|I_0|$. But either of these sets has measure bounded above by $2C\|f\|_{L^1(\mathbb{R})}/(C_0 \langle |f| \rangle_{I_0})$, choosing C_0 large enough, so that $2C\|f\|_{L^1(\mathbb{R})}/(C_0 \int_{I_0} |f(y)|dy) < 1/4$, a contradiction will be reached. It sems as if the constant C_0 depends on the interval I_0; however, once we recall that the function f is supported on I_0, then all is required is that $2C/C_0 < 1/4$.

Let $\mathcal{E}_{I_0}$ be the collection of maximal dyadic intervals $I \in \mathcal{D}$ contained in the set F_{I_0}, then we claim that

$$|T_\sigma f(x)|\,\mathbb{1}_{I_0}(x) \le C_0\langle |f|\rangle_{I_0} + \sum_{I\in\mathcal{E}_{I_0}} |T_\sigma^I f(x)| \tag{7.7.3}$$

where $T_\sigma^I f := \sigma_{\widetilde{I}}\langle f\rangle_I \mathbb{1}_I + \sum_{J:J\subset I}\sigma_J\langle f, h_J\rangle h_J$, and $\widetilde{I}$ is the parent of I.

Repeat for each $I\in\mathcal{E}_{I_0}$ and the function $T_\sigma^I f$ which is supported on I, then repeat for each $I'\in\mathcal{E}_I$, etc. Let $\mathcal{S}_0=\{I_0\}$, and $\mathcal{S}_j := \cup_{I\in\mathcal{S}_{j-1}}\mathcal{E}_I$. Finally, let $\mathcal{S} := \cup_{j=0}^{\infty}\mathcal{S}_j$. For each $I\in\mathcal{S}$, let $E_I = I\setminus F_I$, by construction the sets $E_I\subset I$ are pairwise disjoint and $|E_I|\ge \frac{1}{2}|I|$, and therefore $\mathcal{S}$ is a $\frac{1}{2}$-sparse family. Moreover,

$$|\mathbb{1}_{I_0} T_\sigma f| \le C_0 \mathcal{A}_{\mathcal{S}}|f|,$$

which is what we set out to prove. We are done modulo verifying the claimed inequality (7.7.3), which we now prove. Note that $|T_\sigma f(x)|\le 2T_\sigma^\sharp f(x)$. Thus, if $x\in I_0\setminus F_{I_0}$ then $|T_\sigma f(x)|\le C_0\langle |f|\rangle_{I_0}$, and (7.7.3) is satisfied.

If $x\in F_{I_0}$, then there is unique $I\in\mathcal{S}_1=\mathcal{E}_{I_0}$ with $x\in I$, and recalling that $\langle f, h_{\widetilde{I}}\rangle h_{\widetilde{I}}(x) = \langle f\rangle_I - \langle f\rangle_{\widetilde{I}}$, we conclude that

$$\begin{aligned} T_\sigma f(x) &= \sum_{J\supsetneq \widetilde{I}} \sigma_J\langle f, h_J\rangle h_J(x) + \sum_{J\subset\widetilde{I}} \sigma_J\langle f, h_J\rangle h_J(x) \\ &= \sum_{J\supsetneq \widetilde{I}} \sigma_J\langle f, h_J\rangle h_J(x) - \sigma_{\widetilde{I}}\langle f\rangle_{\widetilde{I}} + T_\sigma^I f(x). \end{aligned}$$

Therefore, we find that when $x\in F_{I_0}$ and for all $y\in\widetilde{I}$ the following inequality holds:

$$|T_\sigma f(x)| \le T_\sigma^\sharp f(y) + Mf(y) + \sum_{I\in\mathcal{E}_{I_0}} T_\sigma^I f(x). \tag{7.7.4}$$

In particular, because I is a maximal dyadic interval in F_{I_0}, there must be $y_0\in\widetilde{I}\setminus I$ such that $y_0\notin F_{I_0}$ and therefore $T_\sigma^\sharp f(y_0) + Mf(y_0)\le \frac{1}{2}C_0\langle |f|\rangle_{I_0}$. Substituting $y=y_0$ in (7.7.4), and using this estimate proves the claimed inequality (7.7.3), and therefore the pointwise localized domination by sparse operators for the martingale transform is proven.

7.7.6 Case Study: Sparse Operators Versus Commutators

Carlos Pérez and Israel Rivera-Ríos proposed the following $L\log L$-sparse operator as a candidate for sparse domination of the commutator.

$$B_{\mathcal{S}} f(x) = \sum_{Q\in\mathcal{S}} \|f\|_{L\log L, Q}\mathbb{1}_Q(x).$$

The reason for this choice is that $M^2 \sim M_{L\log L}$ is the correct maximal function for the commutator. However, they showed that these operators cannot bound pointwise the commutator $[b, T]$ in [159].

Andrei Lerner, Sheldy Ombrosi, and Israel Rivera-Ríos proposed the following sparse-like operator and its adjoint adapted to the commutator with locally integrable function b,

$$\mathcal{T}_{\mathcal{S},b} f(x) := \sum_{Q \in \mathcal{S}} |b(x) - \langle b \rangle_Q| \, \langle |f| \rangle_Q \, \mathbb{1}_Q(x),$$

$$\mathcal{T}^*_{\mathcal{S},b} f(x) := \sum_{Q \in \mathcal{S}} \langle |b - \langle b \rangle_Q| \, |f| \rangle_Q \, \mathbb{1}_Q(x).$$

They showed, in [134], that finitely many of these operators will provide pointwise domination for the commutator, $[b, T]$, where T is a rough Calderón–Zygmund operator and b a locally integrable function.

Theorem 15 (Lerner, Ombrosi, Rivera-Ríos 2017) *Let T be an ω-Calderón–Zygmund singular integral operator with ω satisfying a Dini condition, $b \in L^1_{\rm loc}(\mathbb{R}^d)$. For every compactly supported $f \in L^\infty(\mathbb{R}^d)$, there are 3^n dyadic lattices $\mathcal{D}^{(k)}$ and $\frac{1}{2\cdot 9^n}$-sparse families $\mathcal{S}_k \subset \mathcal{D}^{(k)}$ such that for a.e. $x \in \mathbb{R}^d$*

$$|[b, T](f)(x)| \lesssim_{d,T} \sum_{k=1}^{3^n} \big(\mathcal{T}_{\mathcal{S}_k,b}|f|(x) + \mathcal{T}^*_{\mathcal{S}_k,b}|f|(x)\big).$$

Quadratic bounds on $L^2(w)$ for the commutator $[b, T]$ will follow from quadratic bounds for these adapted sparse operators [134]. The following quadratic bounds on $L^2(w)$ for $\mathcal{T}_{\mathcal{S},b}$, $\mathcal{T}^*_{\mathcal{S},b}$ hold:

$$\|\mathcal{T}_{\mathcal{S},b} f\|_{L^2(w)} + \|\mathcal{T}^*_{\mathcal{S},b} f\|_{L^2(w)} \lesssim [w]^2_{A_2} \|b\|_{\rm BMO} \|f\|_{L^2(w)}.$$

These quadratic bounds, the corresponding extrapolated bounds on $L^p(w)$

$$\|\mathcal{T}_{\mathcal{S},b} f\|_{L^p(w)} + \|\mathcal{T}^*_{\mathcal{S},b} f\|_{L^p(w)} \lesssim_p [w]_{A_p}^{2\max\{1,\frac{1}{p-1}\}} \|b\|_{\rm BMO} \|f\|_{L^p(w)},$$

and much more follow from a key lemma that we now state.

Lemma 9 (Lerner, Ombrosi, Rivera-Ríos 2017) *Given $\mathcal{S}$ an η-sparse family in $\mathcal{D}$, $b \in L^1_{\rm loc}(\mathbb{R}^d)$ then there is a larger collection $\widetilde{\mathcal{S}} \in \mathcal{D}$ which is an $\frac{\eta}{2(1+\eta)}$-sparse family, $\mathcal{S} \subset \widetilde{\mathcal{S}}$, such that for all $Q \in \widetilde{\mathcal{S}}$, the following estimate holds:*

$$|b(x) - \langle b \rangle_Q| \le 2^{d+2} \sum_{R \in \widetilde{\mathcal{S}}, R \subset Q} \Omega(b; R) \mathbb{1}_R(x), \quad a.e.\ x \in Q,$$

where $\Omega(b; R) := \frac{1}{|R|}\int_R |b(x) - \langle b\rangle_R|\,dx$, *the mean oscillation of b on the dyadic cube R.*

From this lemma, we immediately deduce quantitative Bloom bounds for the sparse-like adjoint operator associated to the commutator [134]. A similar result holds for $\mathcal{T}_{\mathcal{S},b}$.

Corollary 1 (Quantitative Bloom) *Let* $u, v \in A_p$, $\mu = u^{1/p}v^{-1/p}$ *and* $b \in \mathrm{BMO}_\mu$, *then there is a constant* $c_{d,p} > 0$ *such that for all* $f \in L^p(u)$ *the following inequality holds:*

$$\|\mathcal{T}^*_{\mathcal{S},b}|f|\|_{L^p(v)} \le c_{d,p}\|b\|_{\mathrm{BMO}_\mu}\big([v]_{A_p}[u]_{A_p}\big)^{\max\{1,\frac{1}{p-1}\}}\|f\|_{L^p(u)}.$$

Similarly, for $\mathcal{T}_{\mathcal{S},b}$.

Proof First notice that since $\|b\|_{\mathrm{BMO}_\mu} = \sup_Q |Q|\,\Omega(b;Q)/\mu(Q)$,

$$\mathcal{T}^*_{\widetilde{\mathcal{S}},b}|f|(x) \le c_d\|b\|_{\mathrm{BMO}_\mu}\mathcal{A}_{\widetilde{\mathcal{S}}}\big(\mathcal{A}_{\widetilde{\mathcal{S}}}(|f|)\mu\big)(x),$$

where $\widetilde{\mathcal{S}}$ is the larger sparse family given by Lemma 9.

Taking $L^p(v)$ norm on both sides, and unfolding we conclude that

$$\begin{aligned}\|\mathcal{T}^*_{\widetilde{\mathcal{S}},b}|f|\|_{L^p(v)} &\le c_{d,p}\|b\|_{\mathrm{BMO}_\mu}\|\mathcal{A}_{\widetilde{\mathcal{S}}}\|_{L^p(v)}\|\mathcal{A}_{\widetilde{\mathcal{S}}}\|_{L^p(u)}\|f\|_{L^p(u)}\\ &\le c_{d,p}\|b\|_{\mathrm{BMO}_\mu}\big([v]_{A_p}[u]_{A_p}\big)^{\max\{1,\frac{1}{p-1}\}}\|f\|_{L^p(u)},\end{aligned}$$

where in the last line we used the one-weight estimates on both $L^p(u)$ and $L^p(v)$ for the sparse operator $\mathcal{A}_{\widetilde{\mathcal{S}}}$ given that u and v are A_p weights by assumption. Observing that $\mathcal{T}^*_{\mathcal{S},b}|f|(x) \le \mathcal{T}^*_{\widetilde{\mathcal{S}},b}|f|(x)$, we get the desired estimate.

Setting $u = v = w \in A_p$, then $\mu \equiv 1$, $b \in \mathrm{BMO}$, and we recover the expected one-weight quantitative L^p estimates for the sparse-like operators dominating the commutator, and hence for the commutator itself, without using extrapolation,

$$\|\mathcal{T}_{\mathcal{S},b}|f|\|_{L^p(w)} + \|\mathcal{T}^*_{\mathcal{S},b}|f|\|_{L^p(w)} \le c_{n,p}\|b\|_{\mathrm{BMO}}[w]_{A_p}^{2\max\{1,\frac{1}{p-1}\}}\|f\|_{L^p(w)}.$$

7.8 Summary and Recent Progress

In these lecture notes, we have studied weighted norm inequalities through the dyadic harmonic analysis lens. We focused on classical operators such as the Hilbert transform and the maximal function, and dyadic operators such as the dyadic maximal function, the martingale transform, the dyadic square function, Haar shift multipliers, the dyadic paraproduct, and the latest "kid in the block" the dyadic sparse operator. To carry on our program, we discussed dyadic tools such as dyadic cubes (regular, random, adjacent) and Haar functions on $\mathbb{R}$, $\mathbb{R}^d$, and more generally on spaces of homogenenous type.

In this millennium, the interest shifted from qualitative weighted norm inequalities to quantitative weighted norm inequalities. New techniques were developed to obtain quantitative estimates, including Bellman function and median oscillation techniques, quantitative extrapolation and transference theorems, corona decompositions and stopping times, representation of operators as averages of dyadic operators, and, most recently, domination by dyadic sparse operators. One important landmark in this quest was the proof of the A_2 conjecture. Some of these techniques are amenable to generalizations to other settings that support dyadic structures such as spaces of homogeneous type.

We tried to illustrate the power of the dyadic methods studying in detail the maximal function and the commutator of the Hilbert transform with a function in BMO via their dyadic counterparts, in both cases obtaining the optimal estimates on weighted Lebesgue spaces. We presented a self-contained Bellman function proof of the A_2 conjecture for the dyadic paraproduct, in order to illustrate these techniques. We showed how to pointwise dominate the martingale transform by sparse operators, and we presented the beautiful and simple proof of the A_2 conjecture for sparse operators. We illustrated the power of pointwise domination techniques by sparse-like operators through a case study: the commutator of Calderón–Zygmund singular integral operators and locally integrable functions, recovering all the quantitative weighted norm inequalities discussed in the notes, and some new ones.

The methods developed in this millennium, initially to study quantitative weighted inequalities for operators defined on $\mathbb{R}^d$, have proven to be quite flexible and far-reaching. There are extensions to metric spaces with geometrically doubling condition, spaces of homogeneous type, and beyond doubling even in a noncommutative setting of operator-valued dyadic harmonic analysis [45, 70, 92, 105, 136, 144, 173]. There are off-diagonal sharp two-weight estimates for sparse operators [71]. There are generalizations to matrix-valued operators [103], so far the best weighted L^2 estimates in this setting are 3/2 powers for the matrix-valued paraproducts, shift operators, and Calderón–Zygmund operators satisfying a Dini condition [143], and linear for the square function [99]. The validity of the A_2 conjecture in the matrix setting is unknown. Two-weight estimates have been obtained for well-localized operators with matrix weights [24], and a weighted Carleson embedding theorem with matrix weights is known and proved using a "Bellman function with a parameter" [59] . Researchers are busy working toward increasing our knowledge on this setting, see, for example, [58] where a bilinear Carleson embedding theorem with matrix weight and scalar measure is proved using Bellman function techniques.

More importantly, out of these investigations a domination paradigm by sparse positive dyadic operators has emerged and proven to be very powerful with applications in many areas not only weighted inequalities. The following is a partial and ever-growing list of such applications to (maximal) rough singular integrals [44, 55, 65, 100]; singular nonintegral operators [19]; multilinear maximal and singular integral operators [15, 56, 133, 188]; nonhomogeneous spaces and operator-valued singular integral operators [46, 179]; uncentered variational operators [63]; variational Carleson operators [64]; Walsh–Fourier multipliers [54]; Bochner–Riesz multipliers [14, 108, 120]; oscillatory and random singular operators [110, 112, 124];

spherical maximal function [116]; Radon transform [151]; Hilbert transform along curves [38]; pseudodifferential operators [13]; the lattice Hardy–Littlewood maximal operator [83]; fractional operator with $L^{\alpha,r'}$-Hörmander conditions [102]; and Rubio de Francia's Littlewood–Paley square function [77]. Sparse $T(1)$ theorems [119] and applications in the discrete setting [57, 109, 111] have been found as well as logarithmic bounds for maximal sparse operators [106].

We are starting to understand why in certain settings this philosophy does not work. For example, very recently it was shown that dominating the dyadic strong maximal function by (1,1)-type sparse forms based on rectangles with sides parallel to the axes is impossible [12]; this is in the realm of multiparameter analysis where many questions still need to be answered. Perhaps, a new type of sparse domination in this setting will have to be dreamed.

Not only the methodology is tried on each author's favorite operator, far-reaching extensions and broader understanding are being gained. For example, the convex body domination paradigm [143] shows that if a scalar operator can be dominated by a sparse operator, then its vector version can be dominated by a convex body-valued sparse operator, a transference theorem. Similarly, multiple vector-valued extensions of operators and more can be explained through the very general helicoidal method [16], yet another far-reaching transference methodology.

This is a very active area of research and we hope these lecture notes have helped to impress on the reader its vitality.

Acknowledgements I would like to thank Ursula Molter, Carlos Cabrelli, and all the organizers of the CIMPA 2017 *Research School—IX Escuela Santaló: Harmonic Analysis, Geometric Measure Theory and Applications*, held in Buenos Aires, Argentina from July 31 to August 11, 2017, for the invitation to give the course on which these lecture notes are based. It meant a lot to me to teach in the "Pabellón 1 de la Facultad de Ciencias Exactas", having grown up hearing stories about the mythical Universidad de Buenos Aires (UBA) from my parents, Concepción Ballester and Victor Pereyra, and dear friends (such as Julián and Amanda Araoz, Manolo Bemporad, Mischa and Yanny Cotlar, Rebeca Guber, Mauricio and Gloria Milchberg, Cora Ratto and Manuel Sadosky, Cora Sadosky and Daniel Goldstein, and Cristina Zoltan, some sadly no longer with us.) who, like us, were welcomed in Venezuela in the late 60s and 70s, and to whom I would like to dedicate these lecture notes. Unfortunately, the flow is now being reversed as many Venezuelans of all walks of life are fleeing their country and many, among them mathematicians and scientists, are finding a home in other South American countries, in particular in Argentina. I would also like to thank the enthusiastic students and other attendants, as always; there is no course without an audience; you are always an inspiration for us. I thank the kind referee, who made many comments that greatly improved the presentation, and my former Ph.D. student David Weirich, who kindly provided the figures. Last but not least, I would like to thank my husband, who looked after our boys while I was traveling, and my family in Buenos Aires who lodged and fed me.

References

1. H. Aimar, Construction of Haar type bases on quasi-metric spaces with finite Assouad dimension. Anal. Acad. Nac. Cs. Ex., F. y Nat., Buenos Aires **54** (2002)

2. H. Aimar, A. Bernardis, B. Iaffei, Multiresolution approximations and unconditional bases on weighted Lebesgue spaces on spaces of homogeneous type. J. Approx. Theory **148**(1), 12–34 (2007)
3. H. Aimar, A. Bernardis, L. Nowak, Dyadic Fefferman-Stein inequalities and the equivalence of Haar bases on weighted Lebesgue spaces. Proc. R. Soc. Edinburgh Sect. A **141**(1), 1–21 (2011)
4. H. Aimar, A. Bernardis, L. Nowak, Equivalence of Haar bases associated with different dyadic systems. J. Geom. Anal. **21**(2), 288–304 (2011)
5. H. Aimar, O. Gorosito, Unconditional Haar bases for Lebesgue spaces on spaces of homogeneous type, in *Proceedings of the SPIEE, Wavelet Applications in Signal and Image Processing VIII*, vol. 4119 (2000), pp. 556–563
6. H. Aimar, R.A. Macías, Weighted norm inequalities for the Hardy-Littlewood maximal operator on spaces of homogeneous type. Proc. Am. Math. Soc. **91**(2), 213–216 (1984)
7. R. Alvarado, M. Mitrea, *Hardy Spaces on Ahlfors-Regular Quasi Metric Spaces. A Sharp Theory*. Springer Lecture Notes in Mathematics, vol. 2142 (2015)
8. J. Álvarez, R.J. Bagby, D.S. Kurtz, C. Pérez, Weighted estimates for commutators of linear operators. Studia Math. **104**(2), 195–209 (1993)
9. K. Astala, T. Iwaniec, E. Saksman, Beltrami operators in the plane. Duke Math. J. **107**(1), 27–56 (2001)
10. P. Auscher, T. Hytönen, Orthonormal bases of regular wavelets in spaces of homogeneous type. Appl. Comput. Harmon. Anal. **34**(2), 266–296 (2013)
11. P. Auscher, T. Hytönen, Addendum to Orthonormal bases of regular wavelets in spaces of homogeneous type. Appl. Comput. Harmon. Anal. **39**(3), 568–569 (2015)
12. A. Barron, J. Conde-Alonso, Y. Ou, G. Rey, Sparse domination and the strong maximal function. arXiv:1811.01243
13. D. Beltran, L. Cladek, Sparse bounds for pseudodifferential operators. J. Anal. Math. arXiv:1711.02339
14. C. Benea, F. Bernicot, T. Luque, Sparse bilinear forms for Bochner Riesz multipliers and applications. arXiv:1605.06401
15. C. Benea, C. Muscalu, Multiple vector-valued inequalities via the helicoidal method. Anal. PDE **9**(8), 1931–1988 (2016)
16. C. Benea, C. Muscalu, Sparse domination via the helicoidal method. arXiv:1707.05484
17. A. Bényi, D. Maldonado, V. Naibo, What is... a Paraproduct?. Not. Am. Math. Soc. **57**(7), 858–860 (2010)
18. A. Bényi, J.M. Martell, K. Moen, E. Stachura, R. Torres, Boundedness results for commutators with BMO functions via weighted estimates: a comprehensive approach. arXiv:1710.08515
19. F. Bernicot, D. Frey, S. Petermichl, Sharp weighted norm estimates beyond Calderón-Zygmund theory. Anal. PDE **9**, 1079–1113 (2016)
20. O. Beznosova, Bellman functions, paraproducts, Haar multipliers, and weighted inequalities. Ph.D. Dissertation, University of New Mexico, 2008
21. O. Beznosova, Linear bound for the dyadic paraproduct on weighted Lebesgue space $L^2(w)$. J. Func. Anal. **255**, 994–1007 (2008)
22. O. Beznosova, D. Chung, J. Moraes, M. C. Pereyra, On two weight estimates for dyadic operators, in *Harmonic Analysis, Partial Differential Equations, Complex Analysis, Banach Spaces, and Operator Theory*, vol. 2, Association for Women in Mathematics Series, vol. 5. (Springer, Cham, 2017), pp. 135–169
23. O. Beznosova, A. Reznikov, Equivalent definitions of dyadic Muckenhoupt and Reverse Holder classes in terms of Carleson sequences, weak classes, and comparability of dyadic $L \log L$ and A_∞ constants. Rev. Mat. Iberoam. **30**(4), 1190–1191 (2014)
24. K. Bickel, A. Culiuc, S. Treil, B. Wick, Two weight estimates for well localized operators with matrix weights. Trans. Am. Math. Soc. **371**, 6213–6240 (2019)
25. S. Bloom, A commutator theorem and weighted BMO. Trans. Am. Math. Soc. **292**(1), 103–122 (1985)

26. J.-M. Bony, Calcul symbolique et propagation des singularités pour les équations aux dérivées partielles non linéaires. Ann. Sci. École Norm. Sup. **14**, 209–246 (1981)
27. S.M. Buckley, Estimates for operator norms on weighted spaces and reverse Jensen inequalities. Trans. Am. Math. Soc. **340**(1), 253–272 (1993)
28. S.M. Buckley, Summation condition on weights. Mich. Math. J. **40**, 153–170 (1993)
29. D.L. Burkholder, Martingale transforms. Ann. Math. Stat. **37**(6), 1494–1504 (1966)
30. D.L. Burkholder, Boundary value problems and sharp inequalities for martingale transforms. Ann. Probab. **12**, 647–702 (1984)
31. S.-Y.A. Chang, J.M. Wilson, T.H. Wolff, Some weighted norm inequalities concerning the Schrödinger operators. Comment. Math. Helv. **60**(2), 217–246 (1985)
32. P. Chen, J. Li, L.A. Ward, BMO from dyadic BMO via expectations on product spaces of homogeneous type. J. Func. Anal. **265**(10), 2420–2451 (2013)
33. M. Christ, A $T(b)$ theorem with remarks on analytic capacity and the Cauchy integral. Colloq. Math. **60/61**(2), 601–628 (1990)
34. D. Chung, Commutators and dyadic paraproducts on weighted Lebesgue spaces. Ph.D. Dissertation, University of New Mexico, 2010
35. D. Chung, Sharp estimates for the commutators of the Hilbert, Riesz and Beurling transforms on weighted Lebesgue spaces. Indiana U. Math. J. **60**(5), 1543–1588 (2011)
36. D. Chung, Weighted inequalities for multivariable dyadic paraproducts. Publ. Mat. **55**(2), 475–499 (2011)
37. D. Chung, M.C. Pereyra, C. Pérez, Sharp bounds for general commutators on weighted Lebesgue spaces. Trans. Am. Math. Soc. **364**, 1163–1177 (2012)
38. L. Cladek, Y. Ou, Sparse domination of Hilbert transforms along curves. Math. Res. Lett. **25**(2), 415–436 (2018)
39. R. Coifman, C. Fefferman, Weighted norm inequalities for maximal functions and singular integrals. Studia Math. **51**, 241–250 (1974)
40. R.R. Coifman, Y. Meyer, Au délà des opérateurs pseudo-différentiels. Astérisque **57** (1979)
41. R.R. Coifman, R. Rochberg, G. Weiss, Factorization theorems for Hardy spaces in several variables. Ann. Math. **103**, 611–635 (1976)
42. R.R. Coifman, G. Weiss, *Analyse harmonique non-commutative sur certains espaces homogènes. Etude de certaines intégrales singulières.* Lecture Notes in Mathematics, vol. 242 (Springer, Berlin, 1971)
43. J.M. Conde, A note on dyadic coverings and nondoubling Calderón-Zygmund theory. J. Math. Anal. Appl. **397**(2), 785–790 (2013)
44. J.M. Conde-Alonso, A. Culiuc, F. Di Plinio, Y. Ou, A sparse domination principle for rough singular integrals. Anal. PDE **10**(5), 1255–1284 (2017)
45. J.M. Conde-Alonso, L.D. López-Sánchez, Operator-valued dyadic harmonic analysis beyond doubling measures. Proc. Am. Math. Soc. **144**(9), 3869–3885 (2016)
46. J. M. Conde-Alonso, J. Parcet, *Nondoubling Calderón-Zygmund theory -a dyadic approach-*. J. Fourier Anal. Appl. arXiv:1604.03711
47. J.M. Conde-Alonso, G. Rey, A pointwise estimate for positive dyadic shifts and some applications. Math. Ann. **365**(3–4), 1111–1135 (2016)
48. M. Cotlar, C. Sadosky, On the Helson-Szegö theorem and a related class of modified Toeplitz kernels, in *Harmonic Analysis in Euclidean spaces*, ed. by G.Weiss, S. Wainger, Proceedings of Symposia in Pure Mathematics, vol. 35 (Providence, R.I. 1979), 383–407. American Mathematical Society
49. M. Cotlar, C. Sadosky, On some L^p versions of the Helson-Szegö theorem, in *Conference on Harmonic Analysis in honor of Antoni Zygmund*, vol. I, II (Chicago, Ill., 1981), 306–317. Wadsworth Mathematics Series, Wadsworth, Belmont, CA (1983)
50. D. Cruz-Uribe, Two weight norm inequalities for fractional integral operators and commutators. Advanced Courses of Mathematical Analysis V **I**, 25–85 (2017)
51. D. Cruz-Uribe, K. Moen, Sharp norm inequalities for commutators of classical operators. Publ. Mat. **56**, 147–190 (2012)

52. D. Cruz-Uribe, J.M. Martell, C. Peréz, *Weights, Extrapolation and the Theory of Rubio the Francia* (Birkhäuser, 2011)
53. D. Cruz-Uribe, J.M. Martell, C. Pérez, Sharp weighted estimates for classical operators. Adv. Math. **229**, 408–441 (2012)
54. A. Culiuc, F. Di Plinio, M.T. Lacey, Y. Ou, Endpoint sparse bound for Walsh-Fourier multipliers of Marcinkiewicz type. Rev. Mat. Iberoam. arXiv:1805.06060
55. A. Culiuc, F. Di Plinio, Y. Ou, Uniform sparse domination of singular integrals via dyadic shifts. Math. Res. Lett. **25**(1), 21–42 (2018)
56. A. Culiuc, F. Di Plinio, Y. Ou, Domination of multilinear singular integrals by positive sparse forms. J. Lond. Math. Soc. **98**(2), 369–392 (2018)
57. A. Culiuc, R. Kesler, M.T. Lacey, Sparse Bounds for the discrete cubic Hilbert transform. Anal. PDE. arXiv:1612:08881
58. A. Culiuc, S. Petermichl, S. Pott, A matrix weighted bilinear Carleson lemma and maximal function. arXiv:1811.05838
59. A. Culiuc, S. Treil, The Carleson Embedding Theorem with matrix weights. Int. Math. Res. Not. https://doi.org/10.1093/imrn/rnx222, arXiv:1508.01716
60. G. David, Morceaux de graphes lipschitziens et intégrales singulières sur une surface. Rev. Mat. Iberoam. **4**(1), 73–114 (1988)
61. G. David, S. Semmes, A boundedness criterion for generalized Calderón-Zygmund operators. Ann. Math. **20**, 371–397 (1984)
62. G. David, J.-L. Journé, S. Semmes, Opérateurs de Calderón-Zygmund, fonctions paraaccrétives et interpolation. Rev. Mat. Iberoam. **1**(4), 1–56 (1985)
63. F.C. de França Silva, P. Zorin-Kranich, Sparse domination of sharp variational truncations. arXiv:1604.05506
64. F. Di Plinio, Y. Do, G. Uraltsev, Positive sparse domination of variational Carleson operators. Ann. Sc. Norm. Super. Pisa Cl. Sci. (5) **18**(4), 1443–1458 (2018)
65. F. Di Plinio, T. Hytönen, K. Li, Sparse bounds for maximal rough singular integrals via the Fourier transform. Ann. Inst. Fourier. (Grenoble). arXiv:1706.07111
66. D.G. Deng, Y. Han, Harmonic analysis on spaces of homogeneous type (Springer, 2009)
67. O. Dragičević, L. Grafakos, M.C. Pereyra, S. Petermichl, Extrapolation and sharp norm estimates for classical operators in weighted Lebesgue spaces. Publ. Mat. **49**, 73–91 (2005)
68. J. Duoandicoetxea, *Fourier Analysis. Graduate Studies in Mathematics*, vol. 29, American Mathematical Society (Providence, RI, 2001)
69. J. Duoandicoetxea, Extrapolation of weights revisited: new proofs and sharp bounds. J. Func. Anal. **260**(6, 15), 1886–1901 (2011)
70. X.T. Duong, R. Gong, M.-J.S. Kuffner, J. Li, B. Wick, D. Yang, Two weight commutators on spaces of homogeneous type and applications. arXiv:1809.07942
71. S. Fackler, T. Hytönen, Off-diagonal sharp two-weight estimates for sparse operators. N. Y. J. Math. **24**, 21–42 (2018)
72. R. Fefferman, C. Kenig, J. Pipher, The theory of weights and the Dirichlet problem for elliptic equations. Ann. Math. (2) **134**(1), 65–124 (1991)
73. C. Fefferman, E. Stein, H^p spaces of several variables. Acta Math. **129**(3–4), 137–193 (1972)
74. T. Fiegel, *Singular Integral Operators: A Martingale Approach, in: Geometry of Banach Spaces*. London Mathematical Society. Lecture Notes Series, vol. 158 (Cambridge University Press, 1990), pp. 95–110
75. J. García-Cuerva, José Luis Rubio de Francia (1949–1988). Collect. Math. **38**, 3–15 (1987)
76. J. García-Cuerva, J.L. Rubio de Francia, Weighted norm inequalities and related topics. North-Holland Mathematics Studies, vol. 116 (Amsterdam, 1981)
77. R. Garg, L. Roncal, S. Shrivastava, Quantitative weighted estimates for Rubio de Francia's Littlewood–Paley square function. arxiv:1809.02937
78. J. Garnett, P.W. Jones, BMO from dyadic BMO. Pac. J. Math. **99**(2), 351–371 (1982)
79. M. Girardi, W. Sweldens, A new class of unbalanced Haar wavelets that form an unconditional basis for L^p on general measure spaces. J. Fourier Anal. Appl. **3**(4), 45–474 (1997)

80. L. Grafakos, *Classical Fourier Analysis*. Graduate Texts in Mathematics, vol. 249, 3rd edn. (Springer, New York, 2014)
81. L. Grafakos, *Modern Fourier Analysis*. Graduate Texts in Mathematics, vol. 250, 3rd edn. (Springer, New York, 2014)
82. T.S. Hänninen, Equivalence of sparse and Carleson coefficients for general sets, arXiv:1709.10457
83. T.S. Hänninen, E. Lorist, Sparse domination for the lattice Hardy-Littlewood maximal operator. arXiv:1712.02592
84. I. Holmes, M.T. Lacey, B. Wick, Commutators in the two-weight setting. Math. Ann. **367**, 5–80 (2017)
85. I. Holmes, M.T. Lacey, B. Wick, Bloom's inequality: commutators in a two-weight setting. Arch. Math. (Basel) **106**(1), 53–63 (2016)
86. I. Holmes, S. Petermichl, B. Wick, Weighted little bmo and two-weight inequalities for Journé commutators. Anal. PDE **11**(7), 1693–1740 (2018)
87. S. Hukovic, S. Treil, A. Volberg, The Bellman function and sharp weighted inequalities for square functions, in *Complex Analysis, Operators and Related Topics*. Operator Theory: Advances and Applications, vol. 113, (Birkaüser Basel, 2000), pp. 97–113
88. R. Hunt, B. Muckenhoupt, R. Wheeden, Weighted norm inequalities for the conjugate function and the Hilbert transform. Trans. Am. Math. Soc. **176**, 227–252 (1973)
89. T. Hytönen, On Petermichl's dyadic shift and the Hilbert transform. C. R. Math. **346**, 1133–1136 (2008)
90. T. Hytönen, The sharp weighted bound for general Calderón-Zygmund operators. Ann. Math. (2) **175**(3), 1473–1506 (2012)
91. T. Hytönen, Martingales and harmonic analysis (2013). http://www.ctr.maths.lu.se/media/MATP29/2016vt2016/maha-eng_-1.pdf
92. T. Hytönen, The two-weight inequality for the Hilbert transform with general measures. Proc. Lond. Math. Soc. **117**(3), 483–526 (2018)
93. T. Hytönen, The Holmes-Wick theorem on two-weight bounds for higher order commutators revisited. Archiv der Mathematik **107**(4), 389–395 (2016)
94. T. Hytönen, A. Kairema, Systems of dyadic cubes in a doubling metric space. Colloq. Math. **126**(1), 1–33 (2012)
95. T. Hytönen, M.T. Lacey, The A_p-A_∞ inequality for general Calderón-Zygmund operators. Indiana Univ. Math. J. **61**, 2041–2052 (2012)
96. T. Hytönen, K. Li, Weak and strong A_p-A_∞ estimates for square functions and related operators. Proc. Am. Math. Soc. **146**(6), 2497–2507 (2018)
97. T. Hytönen, H. Martikainen, Non-homogeneous Tb theorem and random dyadic cubes on metric measure spaces. J. Geom. Anal. **22**(4), 1071–1107 (2012)
98. T. Hytönen, C. Pérez, Sharp weighted bounds involving A_∞. Anal. PDE **6**(4), 777–818 (2013)
99. T. Hytönen, S. Petermichl, A. Volberg, The sharp square function estimate with matrix weight. arXiv:1702.04569
100. T. Hytönen, L. Roncal, O. Tapiola, Quantitative weighted estimates for rough homogeneous singular integrals. Israel J. Math. **218**(1), 133–164 (2017)
101. T. Hytönen, O. Tapiola, Almost Lipschitz-continuous wavelets in metric spaces via a new randomization of dyadic cubes. J. Approx. Theory **185**, 12–30 (2014)
102. G.H. Ibañez-Firnkorn, M.S. Riveros, R.E. Vidal, Sharp bounds for fractional operator with $L^{\alpha,r'}$-Hörmander conditions. arXiv:1804.09631
103. J. Isralowitz, H.K. Kwon, S. Pott, Matrix weighted norm inequalities for commutators and paraproducts with matrix symbols. J. Lond. Math. Soc. (2) **96**(1), 243–270 (2017)
104. F. John, L. Nirenberg, On functions of bounded mean oscillation. Commun. Pure Appl. Math. **14**, 415–426 (1961)
105. A. Kairema, J. Li, M.C. Pereyra, L.A. Ward, Haar bases on quasi-metric measure spaces, and dyadic structure theorems for function spaces on product spaces of homogeneous type. J. Func. Anal. **271**(7), 1793–1843 (2016)

106. G.A. Karagulyan, M.T. Lacey, On logarithmic bounds of maximal sparse operators. arXiv:1802.00954
107. N.H. Katz, M.C. Pereyra, On the two weight problem for the Hilbert transform. Rev. Mat. Iberoam. **13**(01), 211–242 (1997)
108. R. Kesler, M.T. Lacey, Sparse endpoint estimates for Bohner-Riesz multipliers on the plane. Collec. Math. **69**(3), 427–435 (2018)
109. R. Kesler, D. Mena, Uniform sparse bounds for discrete quadratic phase Hilbert transform. Anal. Math. Phys. (2017). https://doi.org/10.1007/s13324-017-0195-3
110. B. Krause, M.T. Lacey, Sparse bounds for maximal monomial oscillatory Hilbert transforms. Studia Math. **242**(3), 217–229 (2018)
111. B. Krause, M.T. Lacey, Sparse bounds for random discrete Carleson theorems, in *50 Years with Hardy Spaces*. Operator Theory: Advances and Applications, vol. 261 (Birkhäuser/Springer, Cham, 2018), pp. 317–332
112. B. Krause, M.T. Lacey, Sparse bounds for maximally truncated oscillatory singular integrals. Ann. Sci. Scuola Norm. Sup. 20 (2018). https://doi.org/10.2422/2036-2145.201706_023
113. M.T. Lacey, Two weight Inequality for the Hilbert transform: a real variable characterization. II. Duke Math. J. **163**(15), 2821–2840 (2014)
114. M.T. Lacey, The two weight inequality for the Hilbert transform: a primer, in *Harmonic Analysis, Partial Differential Equations, Banach Spaces, and Operator Theory*. Association for Women in Mathematics Series 5, vol. 2 (Springer, Cham, 2017), pp. 11–84
115. M.T. Lacey, An elementary proof of the A_2 bound. Israel J. Math. **217**, 181–195 (2017)
116. M.T. Lacey, Sparse bounds for spherical maximal functions. J. d'Analyse Math. arXiv:1702.08594
117. M.T. Lacey, K. Li, Two weight norm inequalities for the g function. Math. Res. Lett. **21**(03), 521–536 (2014)
118. M.T. Lacey, K. Li, On A_p-A_∞ estimates for square functions. Math. Z. **284**, 1211–1222 (2016)
119. M.T. Lacey, D. Mena, The sparse $T1$ theorem. Houst. J. Math. **43**(1), 111–127 (2017)
120. M.T. Lacey, D. Mena, M.C. Reguera, Sparse bounds for Bochner-Riesz multipliers. J. Fourier Anal. Appl. (2017). https://doi.org/10.1007/s00041-017-9590-2
121. M.T. Lacey, K. Moen, C. Pérez, R.H. Torres, Sharp weighted bounds for fractional integral operators. J. Funct. Anal. **259**, 107–1097 (2010)
122. M.T. Lacey, S. Petermichl, M.C. Reguera, Sharp A_2 inequality for Haar shift operators. Math. Ann. **348**, 127–141 (2010)
123. M.T. Lacey, E. Sawyer, C.-Y. Shen, I. Uriarte-Tuero, The two weight inequality for the Hilbert transform, coronas and energy conditions. Duke Math. J. **163**(15), 2795–2820 (2014)
124. M.T. Lacey, S. Spencer, Sparse bounds for oscillatory and random singular integrals. N. Y. J. Math. **23**, 119–131 (2017)
125. M.T. Lacey, C. Thiele, L^p bounds for the bilinear Hilbert transform. Ann. Math. **146**, 693–724 (1997)
126. A.K. Lerner, An elementary approach to several results on the Hardy-Littlewood maximal operator. Proc. Am. Math. Soc. **136**(8), 2829–2833 (2008)
127. A.K. Lerner, Sharp weighted norm inequalities for Littlewood-Paley operators and singular integrals. Adv. Math. **226**, 3912–3926 (2011)
128. A.K. Lerner, Mixed A_p-A_r inequalities for classical singular integrals and Littlewood-Paley operators. J. Geom. Anal. **23**, 1343–1354 (2013)
129. A.K. Lerner, On an estimate of Calderón-Zygmund operators by dyadic positive operators. J. Anal. Math. **121**, 141–161 (2013)
130. A.K. Lerner, A simple proof of the A_2 conjecture. Int. Math. Res. Not. **14**, 3159–3170 (2013)
131. A.K. Lerner, On pointwise estimates involving sparse operators. N. Y. J. Math. **22**, 341–349 (2016)
132. A.K. Lerner, K. Moen, Mixed A_p-A_∞ estimates with one supremum. Studia Math. **219**(3), 247–267 (2013)
133. A.K. Lerner, F. Nazarov, Intuitive dyadic calculus: the basics. Expo. Math. arXiv:1508.05639

134. A.K. Lerner, S. Ombrosi, I. Rivera-Ríos, On pointwise and weighted estimates for commutators of Calderón-Zygmund operators. Adv. Math. **319**, 153–181 (2017)
135. A.K. Lerner, S. Ombrosi, I. Rivera-Ríos, Commutators of singular integrals revisited. Bull. Lond. Math. Soc. (2018). https://doi.org/10.1112/blms.12216
136. L.D. López-Sánchez, J.M. Martell, J. Parcet, Dyadic harmonic analysis beyond doubling measures. Adv. Math. **267**, 44–93 (2014)
137. R.A. Macías, C. Segovia, Lipschitz functions on spaces of homogeneous type. Adv. Math. **33**, 257–270 (1979)
138. T. Mei, BMO is the intersection of two translates of dyadic BMO. C. R. Math. Acad. Sci. Paris **336**(12), 1003–1006 (2003)
139. K. Moen, Sharp one-weight and two-weight bounds for maximal operators. Studia Math. **194**(2), 163–180 (2009)
140. J.C. Moraes, M.C. Pereyra, Weighted estimates for dyadic Paraproducts and t-Haar multiplies with complexity (m, n). Publ. Mat. **57**, 265–294 (2013)
141. B. Muckenhoupt, Weighted norm inequalities for the Hardy-Littlewood maximal function. Trans. Am. Math. Soc. **165**, 207–226 (1972)
142. P. F. X. Müller, *Isomorphisms Between H^1 Spaces*. Mathematics Institute of the Polish Academy of Sciences. Mathematical Monographs (New Series), vol. 66 (Birkhäuser Verlag, Basel, 2005)
143. F. Nazarov, S. Petermichl, S. Treil, A. Volberg, Convex body domination and weighted estimates with matrix weights. Adv. Math. **318**, 279–306 (2017)
144. F. Nazarov, A. Reznikov, A. Volberg, The proof of A_2 conjecture in a geometrically doubling metric space. Indiana Univ. Math. J. **62**(5), 1503–1533 (2013)
145. F. Nazarov, S. Treil, The hunt for a Bellman function: applications to estimates for singular integral operators and to other classical problems of harmonic analysis. St. Petersburg Math. J. **8**, 721–824 (1997)
146. F. Nazarov, S. Treil, A. Volberg, The Bellman functions and the two-weight inequalities for Haar multipliers. J. Am. Math. Soc. **12**, 909–928 (1999)
147. F. Nazarov, S. Treil, A. Volberg, *Bellman Function in Stochastic Optimal Control and Harmonic Analysis (How our Bellman Function got its name)*. Operator Theory: Advances and Applications, vol. 129 (2001), pp. 393–424
148. F. Nazarov, S. Treil, A. Volberg, The Tb-theorem on non-homogeneous spaces. Acta Math. **190**, 151–239 (2003)
149. F. Nazarov, S. Treil, A. Volberg, Two weight inequalities for individual Haar multipliers and other well localized operators. Math. Res. Lett. **15**(3), 583–597 (2008)
150. F. Nazarov, S. Treil, A. Volberg, Two weight estimate for the Hilbert transform and corona decomposition for non-doubling measures. Preprint 2005 posted in 2010. arXiv:1003.1596
151. R. Oberlin, Sparse bounds for a prototypical singular Radon transform. Canadian Math. Bull. **62**(2), 405–415 (2019)
152. K. Okikiolu, Characterization of subsets of rectifiable curves in $\mathbb{R}^n$. J. Lond. Math. Soc. (2) **46**(2), 336–348 (1992)
153. C. Ortiz-Caraballo, C. Pérez, E. Rela, *Improving Bounds for Singular Operators via Sharp Reverse Hölder Inequality for A_∞*. Operator Theory: Advances and Applications, vol. 229 (2013), pp. 303–321
154. M.C. Pereyra, Lecture notes on dyadic harmonic analysis. Contemp. Math. **289**, 1–60 (2001)
155. M.C. Pereyra, Weighted inequalities and dyadic harmonic analysis, in *Excursions in Harmonic Analysis*. Applied and Numerical Harmonic Analysis, vol. 2 (Birkhauser/Springer, New York, 2013), pp. 281–306
156. M.C. Pereyra, L.A. Ward, *Harmonic Analysis: From Fourier to Wavelets*. Student Mathematical Library Series, American Mathematical Society, vol. 63 (2012)
157. C. Pérez, Endpoint estimates for commutators of singular integral operators. J. Func. Anal. (1) **128**, 163–185 (1995)
158. C. Pérez, A course on singular integrals and weights, in *Harmonic and Geometric Analysis*. Advanced courses in Mathematics C.R.M. Barcelona (Birkauser, Basel, 2015)

159. C. Pérez, E. Rela, A new quantitative two weight theorem for the Hardy-Littlewood maximal operator. Proc. Am. Math. Soc. **143**, 641–655 (2015)
160. C. Pérez, S. Treil, A. Volberg, Sharp weighted estimates for dyadic shifts and the A_2 conjecture. J. Reine Angew. Math. (Crelle's J.) **687**, 43–86 (2014)
161. S. Petermichl, Dyadic shift and a logarithmic estimate for Hankel operators with matrix symbol. C. R. Acad. Sci. Paris Sér. I Math. **330**(6), 455–460 (2000)
162. S. Petermichl, The sharp bound for the Hilbert transform on weighted Lebesgue spaces in terms of the classical A_p characteristic. Am. J. Math. **129**, 1355–1375 (2007)
163. S. Petermichl, The sharp weighted bound for the Riesz transforms. Proc. Am. Math. Soc. **136**(04), 1237–1249 (2007)
164. S. Petermichl, S. Pott, An estimate for weighted Hilbert transform via square functions. Trans. Am. Math. Soc. **354**, 281–305 (2002)
165. S. Petermichl, A. Volberg, Heating of the Ahlfors-Beurling operator: weakly quasiregular maps on the plane are quasiregular. Duke Math. J. **112**(2), 281–305 (2002)
166. J. Pipher, L.A. Ward, BMO from dyadic BMO on the bidisc. J. Lond. Math. Soc. **77**(2), 524–544 (2008)
167. S. Pott, M.C. Reguera, Sharp Bekolle estimates for the Bergman projection. J. Func. Anal. **265**(12), 3233–3244 (2013)
168. E. Sawyer, A characterization of a two weight norm inequality for maximal functions. Studia Math. **75**(1), 1–11 (1982)
169. E. Sawyer, A characterization of two weight norm inequalities for fractional and Poisson integrals. Trans. Am. Math. Soc. **308**(2), 533–545 (1988)
170. E. Sawyer, R.L. Wheeden, Weighted inequalities for fractional integrals on Euclidean and homogeneous spaces. Am. J. Math. **114**(4), 813–874 (1992)
171. E. Stein, *Harmonic Analysis: Real-variable Methods, Orthogonality, and Oscillatory Integrals*, 1st edn. (Princeton University Press, 1993)
172. C. Thiele, Time-frequency analysis in the discrete phase plane. Ph.D. Thesis Yale, 1995
173. C. Thiele, S. Treil, A. Volberg, Weighted martingale multipliers in the non-homogeneous setting and outer measure spaces. Adv. Math. **285**, 1155–1188 (2015)
174. S. Treil, Sharp A_2 estimates of Haar shifts via Bellman function, in *Recent Trends in Analysis*. Theta Series in Advanced Mathematics (Theta, Bucharest, 2013), pp. 187–208
175. S. Treil, A. Volberg, Wavelets and the angle between past and future. J. Func. Anal. **143**(2), 269–308 (1997)
176. V. Vasyunin, in *Cincinnati Lectures on Bellman Functions* ed. by L. Slavin. arXiv:1508.07668
177. I. Verbitsky, Imbedding and multiplier theorems for discrete Littlewood-Paley spaces. Pac. J. Math. **176**(2), 529–556 (1996)
178. A. Volberg, *Bellman Function Technique in Harmonic Analysis*. Lectures of INRIA Summer School in Antibes (2011), pp. 1–58. arXiv:1106.3899
179. A. Volberg, P. Zorin-Kranish, Sparse domination on non-homogeneous spaces with an application to A_p weights. Rev. Mat. Iberoam. **34**(3), 1401–1414 (2018)
180. E. Vuorinen, $L^p(\mu) \to L^q(\nu)$ characterization for well localized operators. J. Fourier Anal. Appl. **22**(5), 1059–1075 (2016)
181. E. Vuorinen, Two weight L^p-inequalities for dyadic shifts and the dyadic square function. Studia Math. **237**(1), 25–56 (2017)
182. D. Weirich, Weighted inequalities for dyadic operators over spaces of homogeneous type. Ph.D. Dissertation, University of New Mexico, 2018
183. B. Wick, *Personal Communication* (April 2016)
184. M. Wilson, Weighted inequalities for the dyadic square function without dyadic A_∞. Duke Math. J. **55**, 19–49 (1987)
185. M. Wilson, *Weighted Littlewood-Paley Theory and Exponential-Square Integrability*. Lecture Notes in Mathematics, vol. 1924 (Springer, Berlin, 2008)
186. J. Wittwer, A sharp estimate on the norm of the martingale transform. Math. Res. Lett. **7**, 1–12 (2000)

187. J. Wittwer, A sharp estimate on the norm of the continuous square function. Proc. Am. Math. Soc. **130**(8), 2335–2342 (2002)
188. P. Zorin-Kranish, $A_p - A_\infty$ estimates for multilinear maximal and sparse operators. J. Anal. Math

Chapter 8
Sharp Quantitative Weighted BMO Estimates and a New Proof of the Harboure–Macías–Segovia's Extrapolation Theorem

Alberto Criado, Carlos Pérez and Israel P. Rivera-Ríos

Abstract In this paper, we are concerned with quantitative weighted BMO-type estimates. We provide a new quantitative proof for a result due to Harboure, Macías and Segovia (Amer J Math 110 (1988), 383–397, [15]) that also allows to slightly weaken the hypothesis. We also obtain some sharp weighted L_c^∞ – BMO-type estimates for Calderón–Zygmund operators.

8.1 Introduction and Main Results

The celebrated extrapolation theorem of Rubio de Francia [27] (see also [7, 9, 10, 12]) establishes that if T is a bounded operator, not necessarily linear, on $L^{p_0}(w)$ for some $p_0 \in (1, \infty)$ and for every $w \in A_{p_0}$, then for any $p \in (1, \infty)$, T is also a bounded operator on $L^p(w)$ for every $w \in A_p$. Although in many applications the exponent $p_0 = 2$ is a natural initial extrapolation hypothesis, other natural assumptions are of interest. For instance, it is well known that the same conclusion holds if we assume that T is sublinear and it is bounded on the main natural endpoint. Indeed, if T is

A. Criado · C. Pérez (✉) · I. P. Rivera-Ríos
Department of Mathematics, University of the Basque Country, Leioa, Spain
e-mail: cperez@bcamath.org

A. Criado
e-mail: alberto.criado@gmail.com

I. P. Rivera-Ríos
e-mail: israel.rivera@uns.edu.ar

C. Pérez
IKERBASQUE (Basque Foundation for Science) and BCAM –Basque Center for Applied Mathematics, Bilbao, Spain

I. P. Rivera-Ríos
BCAM –Basque Center for Applied Mathematics, Bilbao, Spain

Department of Mathematics, Universidad Nacional del Sur, Bahía Blanca, Argentina

INMABB, CONICET, Bahía Blanca, Argentina

A. Aldroubi et al. (eds.), *New Trends in Applied Harmonic Analysis, Volume 2*, Applied and Numerical Harmonic Analysis,
https://doi.org/10.1007/978-3-030-32353-0_8

a weak type (1, 1) operator with respect to any A_1 weight, namely, $T : L^1(w) \to L^{1,\infty}(w)$, is bounded for any $w \in A_1$, then if $p \in (1, \infty)$, T is of strong type (p, p) with respect to any A_p weight, namely, $T : L^p(w) \to L^p(w)$ is bounded for any $w \in A_p$. Harboure et al. [15] obtained another highly interesting extrapolation theorem from the other endpoint which corresponds to the classical situation of T being bounded from $L^\infty_c(\mathbb{R}^n)$ to $\mathrm{BMO}(\mathbb{R}^n)$. Being precise, they obtained the following theorem.

Theorem A (Harboure–Macías–Segovia) *Let T be a sublinear operator defined on $C_0^\infty(\mathbb{R}^n)$ which satisfies*

$$\fint_Q |Tf - (Tf)_Q| \leq C \left\| \frac{f}{w} \right\|_{L^\infty} \operatorname*{ess\,inf}_Q w \tag{8.1.1}$$

for any cube $Q \subset \mathbb{R}^n$ and any weight w such that $w \in A_1$ and where the constant C depends on T and the A_1 constant of w. Then, if $p \in (1, \infty)$,

$$T : L^p(w) \to L^p(w)$$

is a bounded operator for any $w \in A_p$.

We are using here the standard notation $f_Q = \fint_Q f$ for the average of the function f over the cube Q.

Interestingly enough, estimates like (8.1.1) were considered earlier by Muckenhoupt and Wheeden in [26].

Theorem B *If $w \in A_1$, then*

$$\fint_Q |Hf - (Hf)_Q| \leq C_w \left\| \frac{f}{w} \right\|_{L^\infty} \operatorname*{ess\,inf}_Q w \tag{8.1.2}$$

for any interval $Q \subset \mathbb{R}$ and any weight w, where H stands for the Hilbert transform, namely,

$$Hf(x) = \lim_{\varepsilon \to 0} \int_{|x-y|>\varepsilon} \frac{f(y)}{x - y} dy.$$

Conversely, if (8.1.2) *holds then $w \in A_1$.*

Most probably, this result was the source of inspiration for Theorem A.

Another interesting fact is that (8.1.1) is related to the "asymmetric" or mixed weighted BMO that we will denote as BMO_w, introduced in [26]. Indeed, if (8.1.1) holds then

$$\|Tf\|_{\mathrm{BMO}_w} \leq C_w \left\| \frac{f}{w} \right\|_{L^\infty} \tag{8.1.3}$$

where

$$\|f\|_{\mathrm{BMO}_w} = \sup_Q \frac{1}{w(Q)} \int_Q |f - f_Q|.$$

and hence (8.1.3) holds in the case of the Hilbert transform.

Later on, Bloom [2] considered a variant of this class BMO_w to characterize a very particular two-weighted estimate for the commutator of Coifman–Rochberg–Weiss [6],

$$[b, H]f = bHf - H(bf)$$

in terms of the weighted BMO_w. See [13, 16, 24, 25] for more results in that direction.

8.2 Main Results

8.2.1 An Extrapolation Result Revisited

Now we turn our attention to our contribution. Our first result is a quantitative extension of the aforementioned extrapolation theorem of Harboure, Macías, and Segovia, Theorem A.

Theorem 8.2.1 *Let T a sublinear operator and $\delta \in (0, 1]$ such that, for every $u \in A_1$,*

$$\inf_{c\in\mathbb{R}} \left(\frac{1}{|Q|} \int_Q |Tf - c|^\delta \right)^{\frac{1}{\delta}} \leq c_{n,\delta}\varphi([u]_{A_1}) \inf_{z\in Q} u(z) \left\| \frac{f}{u} \right\|_{L^\infty}. \tag{8.2.1}$$

Then if $w \in A_p$,

$$\|T\|_{L^p(w)} \leq c_n \varphi(\|M\|_{L^p(w)}) \, \|M\|_{L^{p'}(\sigma)},$$

where $\sigma = w^{-\frac{1}{p-1}}$.

Since $\|M\|_{L^p(w)} \leq c_n p' \left([w]_{A_p}[\sigma]_{A_\infty}\right)^{\frac{1}{p}}$ (see [18]) and $[\sigma]_{A_{p'}} = [w]_{A_p}^{\frac{1}{p-1}}$ the preceding estimate yields that

$$\|T\|_{L^p(w)} \leq c_n p \, \varphi\left(c_n p' \left([w]_{A_p}[\sigma]_{A_\infty}\right)^{\frac{1}{p}}\right) \left([w]_{A_p}^{\frac{1}{p-1}} [w]_{A_\infty}\right)^{\frac{1}{p'}}$$

8.2.2 A Sharp Quantitative Weighted L^∞ – BMO Estimate

Our next result generalizes the sufficiency in Theorem B to general Calderón–Zygmund operators providing as well the sharp dependence on the constant of the weight. The precise statement is the following.

Theorem 8.2.2 *Let T be a Calderón–Zygmund operator, w be a weight and $f \in L^p$ for some $p \in [1, \infty)$ a function such that $|f| \lesssim w$ a.e. Then for all $r > 1$ and any cube Q with sides parallel to the axes, one has*

$$\fint_Q |Tf - (Tf)_Q| \leq C_T\, r' \left\| \frac{f}{w} \right\|_{L^\infty} \operatorname*{ess\,inf}_Q M_r w, \tag{8.2.2}$$

where C_T is a constant only depending on the kernel K and the right-hand side is finite provided $w \in L^r_{loc}$. If additionally $w \in A_\infty$,

$$\fint_Q |Tf - (Tf)_Q| \leq C_T\, [w]_{A_\infty} \left\| \frac{f}{w} \right\|_{L^\infty} \operatorname*{ess\,inf}_Q Mw. \tag{8.2.3}$$

and hence, if $w \in A_1$,

$$\fint_Q |Tf - (Tf)_Q| \leq C_T\, [w]_{A_\infty}[w]_{A_1} \left\| \frac{f}{w} \right\|_{L^\infty} \operatorname*{ess\,inf}_Q w, \tag{8.2.4}$$

Inequalities (8.2.3) and (8.2.4) are sharp in the sense that neither $[w]_{A_\infty}$ nor $[w]_{A_1}$ can be replaced by $\phi([w]_{A_\infty})$ or $\psi([w]_{A_1})$ with $\phi(t), \psi(t) = o(t)$ as $t \to \infty$.

Remark 1 We observe that we can restate (8.2.2), (8.2.3), and (8.2.4) as norm inequalities as follows:

$$\left\| \frac{M^\sharp(Tf)}{M_r w} \right\|_{L^\infty} \leq C_T\, r' \left\| \frac{f}{w} \right\|_{L^\infty}, \tag{8.2.5}$$

$$\left\| \frac{M^\sharp(Tf)}{Mw} \right\|_{L^\infty} \leq C_T\, [w]_{A_\infty} \left\| \frac{f}{w} \right\|_{L^\infty}, \tag{8.2.6}$$

$$\left\| \frac{M^\sharp(Tf)}{w} \right\|_{L^\infty} \leq C_T\, [w]_{A_1}[w]_{A_\infty} \left\| \frac{f}{w} \right\|_{L^\infty}. \tag{8.2.7}$$

8.2.3 An Improved Version

In this section, we will provide a natural counterpart of Theorem 8.2.2 in terms of slightly smaller oscillations. This idea was considered in [1] (motivated by [20]) having as a goal to provide a simpler proof of the classical Coifman–Fefferman theorem [5] relating any Calderón–Zygmund operator and the Hardy–Littlewood maximal function.

For $0 < \varepsilon < 1$, we define the following modification of the sharp maximal operator:

$$M_\varepsilon^\sharp f(x) := \left(M^\sharp(|f|^\varepsilon)(x)\right)^{1/\varepsilon} = \sup_{x\in Q}\left(\fint_Q \Big||f(y)|^\varepsilon - (|f|^\varepsilon)_Q\Big| dy\right)^{1/\varepsilon}.$$

Since

$$\fint_Q |f-(f)_Q| \simeq \inf_c \fint_Q |f-c|$$

In particular we have that

$$\left(\fint_Q \big||f|^\varepsilon - (|f|^\varepsilon)_Q\big|\right)^{\frac{1}{\varepsilon}} \simeq \inf_c \left(\fint_Q \big||f|^\varepsilon - c\big|\right)^{\frac{1}{\varepsilon}}$$

From that equivalence it follows that

$$\begin{aligned} M_\varepsilon^\sharp f(x) &\lesssim \sup_{x\in Q}\left(\fint_Q \Big||f(y)|^\varepsilon - |f_Q|^\varepsilon\Big| dy\right)^{1/\varepsilon} \\ &\lesssim \sup_{x\in Q}\left(\fint_Q \Big|f(y)-f_Q\Big|^\varepsilon dy\right)^{1/\varepsilon} \le 2^{\frac{1}{\varepsilon}} M^\sharp f(x) \end{aligned}$$

by the numeric inequality $||a|^\varepsilon - |b|^\varepsilon| \le |a-b|^\varepsilon$ and Jensen inequality. We also note that

$$\begin{aligned} \left(\fint_Q \Big||f(y)|^\varepsilon - (|f|^\varepsilon)_Q\Big| dy\right)^{1/\varepsilon} &\simeq \inf_c \left(\fint_Q \Big||f(y)|^\varepsilon - |c|^\varepsilon\Big| dy\right)^{1/\varepsilon} \\ &\lesssim \inf_c \left(\fint_Q \Big|f(y)-c\Big|^\varepsilon dy\right)^{1/\varepsilon} \end{aligned}$$

The key point of using the operator $M_\varepsilon^\sharp$ is that if $0<\varepsilon<1$ there exists a constant c depending on T and ε such that for all f,

$$M_\varepsilon^\#(Tf)(x) \le c\, Mf(x).$$

This result was shown in [1] and, as it was mentioned above, it allows to provide a simpler proof of the main result in the classical and celebrated paper [5], with the additional advantage that it can be extended to the multilinear case (see [23]). We remit to [3] for a recent extension of this property to many other situations.

This philosophy leads to establish the following result that provides a better dependence on the $A_\infty - A_1$ constant than the one obtained in Theorem 8.2.2.

Theorem 8.2.3 *Let T be a Calderón–Zygmund operator, w be a weight and $f\in L^p$ for some $p\in[1,\infty)$ a function such that $|f|\lesssim w$ a.e. Then for any $0<\varepsilon<1$ and any cube Q with sides parallel to the axes, one has*

$$\inf_{c}\left(\fint_{Q}\left|Tf(y)-c\right|^{\varepsilon}dy\right)^{1/\varepsilon}\leq C_{T,\varepsilon}\left\|\frac{f}{w}\right\|_{L^{\infty}}\operatorname*{ess\,inf}_{Q}Mw, \tag{8.2.8}$$

where $C_{T,\varepsilon}$ is a constant depending on the (regularity of the) kernel K and on ε. Hence

$$\left\|\frac{M_{\varepsilon}^{\sharp}(Tf)}{Mw}\right\|_{L^{\infty}}\leq C_{T,\varepsilon}\left\|\frac{f}{w}\right\|_{L^{\infty}}. \tag{8.2.9}$$

As a consequence if $w\in A_1$

$$\left\|\frac{M_{\varepsilon}^{\sharp}(Tf)}{w}\right\|_{L^{\infty}}\leq C_{T,\varepsilon}\,[w]_{A_1}\left\|\frac{f}{w}\right\|_{L^{\infty}}. \tag{8.2.10}$$

Also in (8.2.10) is sharp in the sense that $[w]_{A_1}$ cannot be replaced by $\psi([w]_{A_1})$ with $\psi(t)=o(t)$ as $t\to\infty$.

8.3 Some Definitions and Key Results

In this section, we gather some results and definitions that will be fundamental for the proofs of our main results.

For $1<p<\infty$, we say that a locally integrable function $w\geq 0$ belongs to the Muckenhoupt A_p class if

$$[w]_{A_p}:=\sup_{Q}\left(\frac{1}{|Q|}\int_{Q}w\right)\left(\frac{1}{|Q|}\int_{Q}w^{1-p'}\right)^{p-1}<\infty,$$

where p' is such that $\frac{1}{p}+\frac{1}{p'}=1$. We call $[w]_{A_p}$ the A_p constant. If $p=1$ we say that $w\in A_1$ if there exists a constant $\kappa>0$ such that

$$Mw(x)\leq\kappa w(x)\qquad \text{a.e. } x\in\mathbb{R}^n. \tag{8.3.1}$$

We define the A_1 constant or characteristic $[w]_{A_1}$ as the infimum of all κ such that (8.3.1) holds. It is also a well-known fact that the A_p classes are increasing, namely, that $p\leq q\Rightarrow A_p\subset A_q$. We can define in a natural way the A_∞ class as $A_\infty=\bigcup_{p\geq 1}A_p$. Associated to this A_∞ class, it is also possible to define an A_∞ constant as

$$[w]_{A_\infty}:=\sup_{Q}\frac{1}{w(Q)}\int_{Q}M(w\chi_Q)dx.$$

This constant was essentially introduced by Fujii [11] and rediscovered by Wilson [28].

Another basic tool for us is the following reverse Hölder inequality with optimal bound, as obtained in [18] (see also [19] for a different proof).

Lemma 8.3.1 *Let $w \in A_\infty$. There exists $\tau_n > 0$ such that for every $\delta \in \left[0, \frac{1}{\tau_n [w]_{A_\infty}}\right]$ and every cube Q*

$$\left(\frac{1}{|Q|}\int_Q w^{1+\delta}\right)^{\frac{1}{1+\delta}} \leq \frac{2}{|Q|}\int_Q w.$$

We will also use the following well-known lemma.

Lemma 8.3.2 *Let w be a weight and Q a cube with center c_Q, then*

$$\int_{Q^c} \frac{|x - c_Q|}{|y - c_Q|^{n+1}} w(y)\, dy \leq 2^n \operatorname*{ess\,inf}_{y\in Q} Mw(y). \tag{8.3.2}$$

In particular if $w \in A_1$, then

$$\int_{Q^c} \frac{|x - c_Q|}{|y - c_Q|^{n+1}} w(y)\, dy \leq 2^n [w]_{A_1} \operatorname*{ess\,inf}_{y\in Q} w(y). \tag{8.3.3}$$

Proof Assume that Q has edge length R. One has

$$\begin{aligned}
\int_{Q^c} \frac{|x - c_Q|}{|y - c_Q|^{n+1}} w(y)\, dy &\leq \sum_{k=0}^{\infty} \int_{2^{k+1}Q\setminus 2^k Q} \frac{R/2}{(2^k R)^{n+1}} w(y)\, dy \\
&\leq \sum_{k=0}^{\infty} \frac{2^{n-1}}{2^k} \frac{1}{(2^{k+1}R)^n} \int_{2^{k+1}Q} w(y)\, dy \\
&\leq 2^{n-1} \sum_{k=0}^{\infty} \frac{1}{2^k} \operatorname*{ess\,inf}_{y\in 2^k Q} Mw(y) \\
&\leq 2^{n-1} \sum_{k=0}^{\infty} \frac{1}{2^k} \operatorname*{ess\,inf}_{y\in Q} Mw(y) \\
&= 2^n \operatorname*{ess\,inf}_{y\in Q} Mw(y).
\end{aligned}$$

We end the proof of the lemma observing that (8.3.3) follows from (8.3.2) by the definition of A_1.

We recall that a family of cubes $\mathcal{S}$ is η-sparse ($\eta \in (0, 1)$) if for every cube $Q \in \mathcal{S}$ there exists a measurable subset $E_Q \subset Q$ such that $\eta|Q| \leq |E_Q|$ and the sets E_Q are pairwise disjoint.

The idea of sparse family was implicit in the literature from a long time ago. It is implicit, for instance, in the proof of the reverse Hölder inequality (see [12]).

However, it was not until the rise of a big interest in the area for quantitative weighted estimates that it was deeply understood and widely developed. In modern time, the first use of the sparsity in the context of singular integrals for "smooth" kernels can be found in [8] where the A_2 "conjecture" was proved for these operators and many others. Another major development can be found in the paper by Lerner [21] where he proved that Calderón–Zygmund operators can be controlled in norm by sparse operators, namely, operators defined by

$$A_{\mathcal{S}}f(x) = \sum_{Q\in\mathcal{S}} \frac{1}{|Q|}\int_Q f(y)dy\chi_Q(x)$$

Simplifying the proof of the A_2 theorem that had been established by Hytönen [17] was the motivation for Lerner. Sparse domination was pursued after that work leading to the proof of the pointwise domination of Calderón–Zygmund operators by sparse operators obtained independently by Lerner and Nazarov [22] and Conde-Alonso and Rey [4]. After those results, a number of authors, which would be hard to mention here, have developed several papers to the study of problems in the theory of weights from the point of view of sparse domination.

To end with the preliminaries, we are going to borrow from [22] the following result that will be crucial for our purposes.

Theorem 8.3.3 *Let $f : \mathbb{R}^n \to \mathbb{R}$ be any measurable almost everywhere finite function such that for every $\varepsilon > 0$,*

$$|\{x \in [-R, R]^n : |f(x)| > \varepsilon\}| = o(R^n) \quad as \quad R \to \infty. \tag{8.3.4}$$

Then for every $\lambda \in (0, 2^{-n-2}]$, there exists a $\frac{1}{6}$-sparse family $\mathcal{S}$ depending on f such that

$$|f(x)| \le \sum_{Q\in\mathcal{S}} w_\lambda(f, Q)\chi_Q(x)$$

where

$$w_\lambda(f, Q) = \inf\{w(f, E) : E \subset Q |E| \ge (1-\lambda)|Q|\}$$

and $w(f, E) = \sup_E f - \inf_E f$.

Remark 2 At this point we would like to note that functions in $L^p(\mathbb{R}^n)$, $p > 1$ satisfy (8.3.4). Indeed, if $f \in L^p$ let $R > 0$ and $\varepsilon > 0$,

$$\begin{aligned}\left|\{x \in [-R, R]^n : |f(x)| > \varepsilon\}\right| &\leq \frac{1}{\varepsilon}\int_{[-R,R]^n} |f(x)|dx \\ &\leq \frac{1}{\varepsilon}\left(\int_{[-R,R]^n} |f(x)|^p dx\right)^{1/p} (2R)^{n/p'} \\ &\leq \frac{1}{\varepsilon}\|f\|_{L^p}(2R)^{n/p'}.\end{aligned}$$

This gives immediately (8.3.4).

In our case, condition (8.2.1) for an operator T yields by standard estimates that it is bounded on $L^p(\mathbb{R}^n)$, $p > 1$. Hence, this formula holds for the case Tf when f is smooth.

8.4 Proofs

Proof of Theorem 8.2.1. We start building a Rubio de Francia Algorithm

$$Rg = \sum_{k=0}^{\infty} \frac{S_\sigma^k(g)}{\|S_\sigma\|_{L^p(\sigma)}^k}$$

where

$$S_\sigma(g) = \frac{M(f\sigma)}{\sigma}$$

We observe that

$$g \leq Rg \qquad \|Rg\|_{L^p(\sigma)} \leq 2\|g\|_{L^p(\sigma)}$$

and also

$$[\sigma Rg]_{A_1} \leq \|M\|_{L^p(w)}.$$

At this point we need to borrow a result from [15, Corollary p. 395]. There exists $g \in L^p(\sigma)$ with $\|g\|_{L^p(\sigma)} \leq 2$ such that

$$\left\|\frac{f}{\sigma g}\right\|_{L^\infty} \leq \|f\|_{L^p(w)}.$$

We see that the properties of the Rubio de Francia algorithm allow us to write

$$\left\|\frac{f}{\sigma Rg}\right\|_{L^\infty} \leq \left\|\frac{f}{\sigma g}\right\|_{L^\infty}.$$

Assume that f is smooth with compact support. Then Tf is well defined, our condition (8.2.1) yields by standard estimates that it is bounded on $L^p(\mathbb{R}^n)$, $p > 1$, i.e., without weights, and we can apply Remark 2 to Tf. Hence, by duality there exists $\|h\|_{L^{p'}(w)} = 1$ such that

$$\left(\int_{\mathbb{R}^n} |Tf|^p w dx\right)^{\frac{1}{p}} = \int_{\mathbb{R}^n} Tfhw.$$

We apply now the decomposition formula from Theorem 8.3.3 and using the well-known fact that

$$\omega_\lambda(f, Q) \leq c \inf_\alpha \left((f-\alpha)\chi_Q\right)^*(\lambda|Q|) \leq c_\lambda \inf_c \left(\frac{1}{|Q|}\int_Q |f-c|^\delta dx\right)^{1/\delta}$$

we can argue, taking into account all the estimates above and since the family of cubes is sparse, as follows:

$$\begin{aligned}
\int_{\mathbb{R}^n} Tfhw &\leq \sum_{Q\in\mathcal{S}} \omega_\lambda(Tf, Q)\int_Q hw \\
&\lesssim \sum_{Q\in\mathcal{S}} \inf_c \left(\frac{1}{|Q|}\int_Q |Tf(x)-c|^\delta\right)^{\frac{1}{\delta}} \int_Q hw \\
&\lesssim \varphi([\sigma Rg]_{A_1}) \sum_{Q\in\mathcal{S}} \inf_{z\in Q}\{\sigma(z)Rg(z)\} \left\|\frac{f}{\sigma Rg}\right\|_{L^\infty} \int_Q hw \\
&\lesssim \varphi([\sigma Rg]_{A_1})\|f\|_{L^p(w)} \sum_{Q\in\mathcal{S}} \inf_{z\in Q}\{\sigma(z)Rg(z)\} \int_Q hw \\
&\lesssim \varphi([\sigma Rg]_{A_1})\|f\|_{L^p(w)} \sum_{Q\in\mathcal{S}} \frac{1}{|Q|}\int_{E_Q} Rg\,\sigma\, dx \int_Q hw\, dx \\
&\leq \varphi([\sigma Rg]_{A_1})\|f\|_{L^p(w)} \int_{\mathbb{R}^n} M(hw)\, Rg\,\sigma\, dx \\
&\leq \varphi(\|M\|_{L^p(w)})\|f\|_{L^p(w)} \left(\int_{\mathbb{R}^n} M(hw)^{p'}\,\sigma\, dx\right)^{\frac{1}{p'}} \left(\int_{\mathbb{R}^n} (Rg)^p\,\sigma\, dx\right)^{\frac{1}{p}} \\
&\lesssim \varphi(\|M\|_{L^p(w)})\|f\|_{L^p(w)}\|M\|_{L^{p'}(\sigma)}\|h\|_{L^{p'}(w)}\|g\|_{L^p(\sigma)} \\
&\leq \varphi(\|M\|_{L^p(w)})\|M\|_{L^{p'}(\sigma)}\|f\|_{L^p(w)}
\end{aligned}$$

and we are done.

Proof of Theorem 8.2.2. To simplify the presentation, we shall assume that our Calderón–Zygmund operator satisfies a Hölder–Lipschitz condition with $\delta = 1$. The same argument with minor adjustments holds for any CZO satisfying any other Hölder–Lipschitz condition or the Dini condition.

By the hypothesis we have imposed on f and w, it is clear that Tf is well defined and that $(Tf)_Q < \infty$ for every cube Q. Let us call $g = f\chi_{2Q}$ and $h = f - g$. Again Tg, $(Tg)_Q$, Th, and $(Th)_Q$ make sense. It will be enough to control separately the oscillations of Tg and Th in Q. We note that it is enough to deal with (8.2.2) since just taking into account the reverse Hölder property in Lemma 8.3.1 in the case of (8.2.3) and also from the definition of A_1 in the case of (8.2.4) those estimates follow from (8.2.2).

We begin with h. Using Lemma 8.3.2, we obtain

$$\begin{aligned}\int_Q |Th - (Th)_Q| &\leq 2\int_Q |Th| = 2\int_Q \left|\int_{\mathbb{R}^n\setminus 2Q} [K(x,y) - K(c_Q,y)]f(y)\,dy\right| dx \\ &\leq C\left\|\frac{f}{w}\right\|_{L^\infty} \int_Q\int_{\mathbb{R}^n\setminus 2Q} \frac{|x-c_Q|}{|y-c_Q|^{n+1}} w(y)\,dy\,dx \\ &\leq C\left\|\frac{f}{w}\right\|_{L^\infty} |Q| \operatorname*{ess\,inf}_{y\in Q} Mw(y). \end{aligned} \tag{8.4.1}$$

To deal with g, we use Jensen's inequality and that the operator norm of T on L^r is r' up to constants only depending on the kernel K. We obtain

$$\begin{aligned}\fint_Q |Tg - (Tg)_Q| &= \fint_Q |Tg - (Tg)_Q| \leq 2\fint_Q |Tg| \lesssim \left(\fint_Q |Tg|^r\right)^{1/r} \\ &\lesssim r'\left(\fint_{2Q} |f|^r\right)^{1/r} \lesssim r'\left\|\frac{f}{w}\right\|_{L^\infty}\left(\fint_{2Q} w^r\right)^{1/r} \\ &\lesssim r'\left\|\frac{f}{w}\right\|_{L^\infty} \operatorname*{ess\,inf}_{2Q} M_r w \lesssim \left\|\frac{f}{w}\right\|_{L^\infty} \operatorname*{ess\,inf}_{Q} M_r w.\end{aligned}$$

This proves (8.2.2).

To show the sharpness of the theorem, it is enough to consider (8.2.4). We will consider in the real line the Hilbert transform and the following weight for $\alpha \in (0,1)$ (actually we will consider α close to zero):

$$w(y) = |y|^{\alpha-1}.$$

Observe first that $[w]_{A_\infty} \lesssim [w]_{A_1} \simeq \frac{1}{\alpha}$

Since we assume that (8.2.4) holds for any interval Q and any function f, we consider $f = w\chi_{(-1,1)}$ and let $Q = (-1,1)$. We observe that Hf is odd, integrable in Q, and hence

$$\int_Q H(f)(x) = 0.$$

This yields

$$\int_Q |H(f)(x) - (Hf)_Q|dx = \int_Q |H(f)(x)|dx = 2\int_0^1 |H(f)(x)|dx \geq -2\int_0^1 H(f)(x)dx$$

Now, taking into account results in [14, p. 315],

$$-\int_0^1 H(f)(x)dx = \int_{-1}^1 |x|^{\alpha-1} H(\chi_{(0,1)})(x)\, dx = \int_{-1}^1 |x|^{\alpha-1} \log \frac{|x-1|}{|x|}\, dx$$

$$= \int_{-1}^1 |x|^{\alpha-1} \log|x-1|\, dx + \int_{-1}^1 |x|^{\alpha-1} \log \frac{1}{|x|}\, dx = I + II$$

and we have

$$II = 2\int_0^1 x^{\alpha-1} \log \frac{1}{x}\, dx = \frac{2}{\alpha^2}$$

and

$$I > -\int_0^1 x^{\alpha-1} \log \frac{1}{1-x}\, dx$$

but

$$\int_0^1 x^{\alpha-1} \log \frac{1}{1-x}\, dx = \int_0^{\frac{1}{2}} x^{\alpha-1} \log \frac{1}{1-x}\, dx + \int_{\frac{1}{2}}^1 x^{\alpha-1} \log \frac{1}{1-x}\, dx$$

$$< \log 2 \int_0^{\frac{1}{2}} x^{\alpha-1}\, dx + 2^{\alpha-1} \int_{\frac{1}{2}}^1 \log \frac{1}{|x-1|}\, dx < \frac{\log 2}{\alpha} + c.$$

Combining estimates, we have for α small enough,

$$⨍_Q |H(f)(x) - (Hf)_{(Q)}|dx > \frac{c}{\alpha^2}.$$

This finishes the optimality of the estimate.

Proof of Theorem 8.2.3. We use the same notation as in the proof of Theorem 8.2.2. We fix a cube Q and split $f = g + h$. Then to prove (8.2.8) it is enough to choose $c = |(Th)_Q|^{\varepsilon}$. Using again the numeric inequality $||a|^{\varepsilon} - |b|^{\varepsilon}| \leq |a-b|^{\varepsilon}$

$$\begin{aligned}
\fint_Q \left| |Tf|^\varepsilon - |(Th)_Q|^\varepsilon \right| &\leq \fint_Q \left| Tf(x) - (Th)_Q \right|^\varepsilon dx \\
&= \fint_Q \left| Tg(x) + Th(x) - (Th)_Q \right|^\varepsilon dx \\
&\leq \fint_Q \left| Tg(x) \right|^\varepsilon dx + \fint_Q \left| Th(x) - (Th)_Q \right|^\varepsilon dx \\
&\leq \fint_Q \left| Tg(x) \right|^\varepsilon dx + \left(\fint_Q \left| Th(x) - (Th)_Q \right| dx \right)^\varepsilon = I + II^\varepsilon.
\end{aligned}$$

To estimate II, we use (8.4.1)

$$II \leq C \left\| \frac{f}{w} \right\|_{L^\infty} \operatorname*{ess\,inf}_{y \in Q} Mw(y).$$

For I, we will be using the well-known Kolmogorov's inequality (see [12], p. 485 for instance), if $q \in (0, 1)$ and (X, μ) is a probability space

$$\|g\|_{L^q(X,\mu)} \leq c_q \, \|g\|_{L^{1,\infty}(X,\mu)}.$$

Now, since T is of weak type $(1, 1)$, $B := \|T\|_{L^1 \to L^{1,\infty}} < \infty$ and we have

$$\begin{aligned}
\left(\fint_Q |Tg|^\varepsilon \right)^{1/\varepsilon} &\leq 2^{1/\varepsilon} \left(\frac{1}{1-\varepsilon} \right)^{1/\varepsilon} B_T \left\| \frac{f}{w} \right\|_{L^\infty} \fint_{2Q} w \\
&\leq 2^{1/\varepsilon} \left(\frac{1}{1-\varepsilon} \right)^{1/\varepsilon} B_T \left\| \frac{f}{w} \right\|_{L^\infty} \operatorname*{ess\,inf}_{2Q} Mw.
\end{aligned}$$

This concludes the proof of (8.2.8). For the proof of the optimality, we remit to the proof of Theorem 8.5.1.

8.5 Further Remarks

Some of the estimates we have obtained in Theorems 8.2.2 and 8.2.3 can be slightly generalized. We say that $(u, v) \in A_1$ if

$$[(u, v)]_{A_1} = \left\| \frac{Mu}{v} \right\|_{L^\infty} < \infty.$$

Relying upon that definition, we can establish the following result.

Theorem 8.5.1 *Let T be a Calderón–Zygmund operator, w be a weight and $f \in L^p$ for some $p \in [1, \infty)$ a function such that $|f| \lesssim u$. Then, if $\varepsilon \in (0, 1)$,*

$$\left\| \frac{M^{\sharp}_{\varepsilon}(Tf)}{v} \right\|_{L^{\infty}} \leq C_T \; [(u,v)]_{A_1} \left\| \frac{f}{u} \right\|_{L^{\infty}}. \tag{8.5.1}$$

If in addition $u \in A_{\infty}$, then

$$\left\| \frac{M^{\sharp}(Tf)}{v} \right\|_{L^{\infty}} \leq C_T \; [u]_{A_{\infty}} [(u,v)]_{A_1} \left\| \frac{f}{u} \right\|_{L^{\infty}}. \tag{8.5.2}$$

The dependence of the constant $[(u,v)]_{A_1}$ in (8.5.1) is sharp in the same sense as in Theorem 8.2.3.

Proof It is straightforward to deduce (8.5.1) from (8.2.9) and (8.5.2) from (8.2.6). In order to settle the optimality of the linear dependence of the constants on $[(u,v)]_{A_1}$, we will show the following for H the Hilbert transform. If (u,v) are such that for each $f \lesssim u$ a.e. such that $f \in L^p$ for some $p \in [1,\infty)$

$$\inf_c \left(\fint_I \left| |Hf|^{\varepsilon} - c \right| \right)^{1/\varepsilon} \leq C^{\star} \left\| \frac{f}{u} \right\|_{L^{\infty}} \operatorname*{ess\,inf}_I v, \tag{8.5.3}$$

then $(u,v) \in A_1$ and $C^{\star} \gtrsim [(u,v)]_{A_1}$. Let us fix an interval J. We define $f = u\chi_J$ and $J^i = J + i|J|$ for $i \in \mathbb{Z}$. Note that for $x \in J^i$ $i \geq 2$ we have that

$$|Hf(x)| = \int_J \frac{u(y)}{y-x} dy.$$

Then it is clear that

$$\begin{aligned} \frac{u(J)}{3|J|} \leq |Hf(x)| \leq \frac{u(J)}{|J|} \qquad & x \in J^2 \\ \frac{u(J)}{6|J|} \leq |Hf(x)| \leq \frac{u(J)}{4|J|} \qquad & x \in J^5 \end{aligned}$$

This yields that for every $x \in J^2$

$$\left| |Hf(x)|^{\varepsilon} - (|Hf|^{\varepsilon})_{J^5} \right| \simeq \left(\frac{u(J)}{|J|} \right)^{\varepsilon} \tag{8.5.4}$$

We take $I = \bigcup_{i=0}^{5} J^i$. We observe that in view of (8.5.4),

$$\left(\fint_I \left| |Hf|^{\varepsilon} - (|Hf|^{\varepsilon})_{J^5} \right| \right)^{1/\varepsilon} \gtrsim \left(\fint_{J^2} \left| |Hf|^{\varepsilon} - (|Hf|^{\varepsilon})_{J^5} \right| \right)^{1/\varepsilon} \gtrsim \frac{u(J)}{|J|}.$$

So writing (8.5.3) for our choice of I and f yields

$$\frac{u(J)}{|J|} \lesssim \left(\fint_I \left| |Hf|^\varepsilon - (|Hf|^\varepsilon)_{J^5} \right| \right)^{1/\varepsilon}$$
$$\simeq \inf_c \left(\fint_I \left| |Hf|^\varepsilon - c \right| \right)^{1/\varepsilon} \leq C^\star \operatorname{ess\,inf}_I v \leq C^\star \operatorname{ess\,inf}_J v$$

and we are done since the choice of J is arbitrary.

Remark 3 Combining Theorems 8.2.1 and 8.2.3, we obtain the following result for Calderón–Zygmund operators T:

$$\|T\|_{L^p(w)} \leq c_{n,p,T} \, [w]_{A_p}^{1+\frac{1}{p-1}} \qquad w \in A_p.$$

The preceding estimate is not sharp. We wonder whether it would be possible to recover the sharp result using an extrapolation argument in the spirit of Theorem 8.2.1. In view of the proof of that result, it seems clear that we could replace (8.2.1) by the following condition. For every cube Q and some $\lambda \in (0, 2^{-n-2}]$,

$$w_\lambda(Tf, Q) \leq c_{n,\lambda} \varphi([u]_{A_1}) \inf_{z\in Q} u(z) \left\| \frac{f}{u} \right\|_{L^\infty}.$$

We do not know whether this condition could allow to obtain more precise estimates.

Acknowledgements The second and the third authors are supported by the Spanish Ministry of Economy and Competitiveness MINECO through BCAM Severo Ochoa excellence accreditation SEV-2013-0323 and through the project MTM2017-82160-C2-1-P and also by Basque Government through the BERC 2014-2017 program and the grant IT-641-13.

References

1. J. Alvarez, C. Pérez, Estimates with A_∞ weights for various singular integral operators. Bollettino U.M.I. (7)8-A, 123–133 (1994)
2. S. Bloom, A commutator theorem and weighted BMO. Trans. Amer. Math. Soc. **292**(1), 103–122 (1985)
3. E. Cejas, K. Li, C. Pérez, I. Rivera-Ríos, Vector-valued operators, optimal weighted estimates and the C_p condition, to appear Science China Mathematics
4. J.M. Conde-Alonso, G. Rey, A pointwise estimate for positive dyadic shifts and some applications. Math. Ann. **365**(3–4), 1111–1135 (2016)
5. R. Coifman, C. Fefferman, Weighted norm inequalities for maximal functions and singular integrals. Studia Math. **51**, 241–250 (1974)
6. R.R. Coifman, R. Rochberg, G. Weiss, Factorization theorems for Hardy spaces in several variables. Ann. of Math. (2) 103, 3, 611–635 (1976)
7. D. Cruz-Uribe, José M. Martell, C. Pérez, Weights, Extrapolation and the Theory of Rubio de Francia, Oper. Theory: Adv. Appl. 215. Birkhäuer/Springer Basel AG, Basel (2011)
8. D. Cruz-Uribe, J.M. Martell, C. Pérez, Sharp weighted estimates for classical operators. Adv. Math. **229**, 408–441 (2012)

9. O. Dragičević, L. Grafakos, C. Pereyra, S. Petermichl, Extrapolation and sharp norm estimates for classical operators on weighted Lebesgue spaces. Publ. Mat. **49**(1), 73–91 (2005)
10. J. Duoandikoetxea, Extrapolation of weights revisited: new proofs and sharp bounds. J. Funct. Anal. **260**, 1886–1901 (2015)
11. N. Fujii, Weighted bounded mean oscillation and singular integrals. Math. Japon. **22**, 529–534 (1977/1978)
12. J. García-Cuerva, J.L. Rubio de Francia, *Weighted Norm Inequalities and Related Topics*, North Holland Math. Studies 116, North Holland, Amsterdam (1985)
13. J. García-Cuerva, E. Harboure, C. Segovia, J.L. Torrea, Weighted norm inequalities for commutators of strongly singular integrals. Indiana Univ. Math. J. **40**(4), 1397–1420 (1991)
14. L. Grafakos, *Classical Fourier Analysis*. Third edition. Graduate Texts in Mathematics, 249. Springer, New York (2008)
15. E. Harboure, R.A. Macías, C. Segovia, Extrapolation results for classes of weights. Amer. J. Math. **110**, 383–397 (1988)
16. I. Holmes, M.T. Lacey, B.D. Wick, Commutators in the two-weight setting. Math. Ann. **367**(1–2), 51–80 (2017)
17. T.P. Hytönen, The sharp weighted bound for general Calderón-Zygmund operators. Ann. of Math. (2) **175**(3), 1473–1506 (2012)
18. T.P. Hytönen, C. Pérez, Sharp weighted bounds involving A_∞. Anal. PDE **6**, 777–818 (2013)
19. T.P. Hytönen, C. Pérez, E. Rela, Sharp Reverse Hölder property for A_∞ weights on spaces of homogeneous type. J. Funct. Anal. **263**, 3883–3899 (2012)
20. B. Jawerth, A. Torchinsky, Local sharp maximal functions. J. Approx. Theory **43**(3), 231–270 (1985)
21. A.K. Lerner, A simple proof of the A_2 conjecture. Int. Math. Res. Not. IMRN **14**, 3159–3170 (2013)
22. A.K. Lerner, F. Nazarov, *Intuitive Dyadic Calculus: the Basics*, to appear in Expo. Math
23. A. Lerner, S. Ombrosi, C. Pérez, R. Torres, R. Trujillo-Gonzalez, New maximal functions and multiple weights for the multilinear Calderón-Zygmund theory. Adv. Math. **220**, 1222–1264 (2009)
24. A.K. Lerner, S. Ombrosi, I.P. Rivera-Ríos, On pointwise and weighted estimates for commutators of Calderón-Zygmund operators. Adv. Math. **319**, 153–181 (2017)
25. A.K. Lerner, S. Ombrosi, I.P. Rivera-Ríos, *Commutators of singular integrals revisited*. Preprint
26. B. Muckenhoupt, R. Wheeden, Weighted bounded mean oscillation and the Hilbert transform. Studia Math. **54**, 221–237 (1976)
27. J.L. Rubio de Francia, Factorization theory and A_p weights. Amer. J. Math. **106**, 533–547 (1984)
28. J.M. Wilson, Weighted inequalities for the dyadic square function without dyadic A_∞. Duke Math. J. **55**(1), 19–50 (1987)

Chapter 9
L^q Dimensions of Self-similar Measures and Applications: A Survey

Pablo Shmerkin

Abstract We present a self-contained proof of a formula for the L^q dimensions of self-similar measures on the real line under exponential separation (up to the proof of an inverse theorem for the L^q norm of convolutions). This is a special case of a more general result of the author from Shmerkin (Ann Math, 2019), and one of the goals of this survey is to present the ideas in a simpler, but important, setting. We also review some applications of the main result to the study of Bernoulli convolutions and intersections of self-similar Cantor sets.

Mathematical Subject Classification: Primary: 28A75 · 28A80

9.1 Introduction

9.1.1 Self-similar Measures

The purpose of this survey is to present a special, but important, case of the main result of [19] concerning the smoothness properties of self-similar measures on the real line. Given a finite family $f_i(x) = \lambda_i x + t_i$, $i \in I$ of contracting similarities (that is, $|\lambda_i| < 1$) and a corresponding probability vector $(p_i)_{i \in I}$, there is a unique Borel probability measure μ on $\mathbb{R}$ such that

$$\mu = \sum_{i \in I} p_i f_i \mu,$$

Partially supported by Projects CONICET-PIP 11220150100355 and PICT 2015-3675 (ANPCyT).

P. Shmerkin (✉)
Department of Mathematics and Statistics, Torcuato Di Tella University, and CONICET, Buenos Aires, Argentina
e-mail: pshmerkin@utdt.edu
URL: http://www.utdt.edu/profesores/pshmerkin

A. Aldroubi et al. (eds.), *New Trends in Applied Harmonic Analysis, Volume 2*, Applied and Numerical Harmonic Analysis,
https://doi.org/10.1007/978-3-030-32353-0_9

where here and throughout the paper, if ν is a Borel probability measure on a space X and $g : X \to Y$ is a Borel map, then $g\nu$ is the push-forward measure, that is, $g\nu(B) = \nu(g^{-1}B)$ for all Borel B. We call the tuple $(f_i, p_i)_{i\in I}$ a *weighted iterated function system*, or WIFS, and μ the corresponding *invariant self-similar measure*.

Studying the properties of self-similar measures, and in particular quantifying their smoothness, is a topic of great interest since the 1930s. Let us define the *similarity dimension* of a self-similar measure μ (or, rather, the generating WIFS) by

$$dim_S(\mu) = \frac{\sum_{i\in I} p_i \log(1/p_i)}{\sum_{i\in I} p_i \log(1/|\lambda_i|)}.$$

The similarity dimension is one of the simplest instances of a very widespread expression in the dimension theory of conformal dynamical systems: it has the form

$$\frac{\text{entropy}}{\text{Lyapunov exponent}}.$$

Indeed, $\sum_{i\in I} p_i \log(1/p_i)$ is the entropy of the probability vector $(p_i)_{i\in I}$: a quantity that measures how uniform this vector is. For example, it attains its maximal value $\log |I|$ exactly at the uniform probability vector $(1/|I|, \ldots, 1/|I|)$. Lyapunov exponents quantify the average expansion or contraction of a dynamical system, and this is how the denominator $\sum_{i\in I} p_i \log(1/|\lambda_i|)$ should be interpreted.

It is well known that if $dim_S(\mu) < 1$, then μ is purely singular with respect to Lebesgue measure. In fact, even more is true. The Hausdorff dimension of a Radon measure ν on $\mathbb{R}$ is defined as

$$dim_H(\nu) = \inf\{dim_H(A) : A \text{ is Borel}, \nu(\mathbb{R} \setminus A) = 0\}.$$

For self-similar measures, it always holds that $dim_S(\mu) \leq dim_H(\mu)$, and it is clear from the definition that measures with Hausdorff dimension < 1 must be purely singular. We also note that one always has $dim_H(\mu) \leq 1$, even though it is possible that $dim_S(\mu) > 1$.

Two major problems in fractal geometry are (a) understanding when one actually has $dim_H(\mu) = dim_S(\mu)$ and (b) analyzing the properties of μ when $dim_S(\mu) > 1$; in particular, determining whether μ is absolutely continuous and, if so, characterizing the smoothness of its density. While both problems are still wide open in this generality, major progress has been accomplished in the last few years. The goal of this article is to present one of the several directions in which progress was achieved, following [19]. While the results of [19] concern a wider class of measures satisfying a generalized notion of self-similarity, here we focus on the proper self-similar case, both because it is an important class in itself and because it allows us to present some of the proofs of [19] in a technically simpler setting. In particular, some ergodic-theoretic concepts and tools are not required in the self-similar case.

9.1.2 The Overlaps Conjecture and Hochman's Theorem on Exponential Separation

Let $A = \mathrm{supp}(\mu)$. The set A is *self-similar*: it satisfies that $A = \bigcup_{i\in I} f_i(A)$ (here we assume that all the p_i are strictly positive; we can always remove the maps f_i with $p_i = 0$ to achieve this). When the pieces $f_i(A)$ are separated enough one does have an equality $dim_H(\mu) = dim_S(\mu)$. Indeed, this holds if the sets $f_i(A)$ are pairwise disjoint or, more generally, under the famous open set condition which allows the images $f_i(A)$ to intersect but in a very limited way.

On the other hand, there are two known mechanisms that force an inequality $dim_H(\mu) < dim_S(\mu)$. The first is if $dim_S(\mu) > 1$. The second is slightly less trivial but still quite simple. Suppose first that $f_i = f_j$ for some $i \neq j$. Then if we drop f_j from the WIFS and replace p_i by $p_i + p_j$ the invariant measure does not change. However, a simple calculation reveals that the similarity dimension of the new WIFS is strictly smaller than that of the original one, and hence the Hausdorff dimension of μ is strictly smaller than the similarity dimension derived from the original WIFS. Although the calculation is slightly more involved, the same argument shows that if the maps f_i do not freely generate a free semigroup or, in other words, if there exist different finite sequences $i = (i_1 \ldots i_k), j = (j_1 \ldots j_\ell)$ such that

$$f_{i_1} \circ \cdots \circ f_{i_k} = f_{j_1} \circ \cdots \circ f_{j_\ell},$$

then one also has $dim_H(\mu) < dim_S(\mu)$. In this case, we say that the WIFS has an *exact overlap*. We note that if this happens then it also happens for sequences of the same length, as we could replace i and j by the juxtapositions ij and ji.

A central conjecture in fractal geometry asserts that these are the *only* mechanisms by which a dimension drop $dim_H(\mu) < dim_S(\mu)$ can occur (we note that in higher dimensions this is not true, but there are related conjectures, see [13] for a discussion). This can be shortly stated in the form: if $dim_H(\mu) < \min(dim_S(\mu), 1)$, then there is an exact overlap. The conjecture has a long history. A version for sets was stated in print in [16], where it is attributed to K. Simon. We refer to M. Hochman's paper [12] for further background and discussion.

While the overlaps conjecture remains open, in [12] M. Hochman accomplished a decisive step toward it. Given a finite sequence $i = (i_1 \ldots i_k)$, we write $f_i = f_{i_1} \circ \cdots \circ f_{i_k}$ for short. Roughly speaking, Hochman proved that if $dim_H(\mu) < \min(dim_S(\mu), 1)$ then for all large k there must exist distinct pairs i, j of words of length k such that the maps f_i and f_j are super-exponentially close (as opposed to being identical, as the overlaps conjecture predicts). More precisely, given two similarity maps $g_j(x) = \lambda_j x + t_j, j = 1, 2$ we define a distance

$$d(g_1, g_2) = \begin{cases} |t_1 - t_2| & \text{if } \lambda_1 = \lambda_2 \\ 1 & \text{if } \lambda_1 \neq \lambda_2 \end{cases}.$$

(It may seem strange to define the distance to be 1 if $\lambda_1 \approx \lambda_2$ and $t_1 \approx t_2$, but it turns out that only the case in which $\lambda_1 = \lambda_2$ ends up being relevant.) Given a WIFS as above, we define the k-separation numbers Γ_k as

$$\Gamma_k = \inf\{d(g_i, g_j) : i = (i_1, \ldots, i_k), j = (j_1, \ldots, j_k), i \neq j\}. \tag{9.1.1}$$

We say that the WIFS has *exponential separation* if there exists $\delta > 0$ such that

$$\Gamma_k > \delta^k \quad \text{for infinitely many } k \in \mathbb{N}.$$

Note that this notion depends only on the similarity maps f_i and not on the weights p_i. We also remark that this condition is substantially weaker than the open set condition. We can now state Hochman's theorem.

Theoremsh 1.1 *If (f_i, p_i) is a WIFS with exponential separation and μ is the associated invariant self-similar measure, then*

$$dim_H(\mu) = \min(dim_S(\mu), 1).$$

Besides conceptually getting us closer to the overlaps conjecture, this theorem has some striking implications: it can be checked in many new explicit cases, and it can be shown to hold outside of very small exceptional sets of parameters in parametrized families of self-similar measures satisfying minimal regularity and non-degeneracy assumptions. Hochman's theorem (and its proof) has also underpinned much of the more recent progress in the area—we will come back to all these points in Sect. 9.4.

9.1.3 L^q Dimensions

Hochman's theorem is about the Hausdorff dimension of self-similar measures. In fractal geometry, and in particular in multifractal analysis, there is a myriad of other ways of quantifying the size of a (potentially fractal) measure. Of particular relevance is a one-dimensional family of numbers known as *L^q dimensions*, which we now define.

We introduce some further notation for simplicity. Let $\mathcal{P}$ denote the family of boundedly supported Borel probability measures on $\mathbb{R}$. The class of $\mu \in \mathcal{P}$ supported on $[0, 1)$ is denoted by $\mathcal{P}_1$. Given $m \in \mathbb{N}$, we let $\mathcal{D}_m$ denote the family of half-open dyadic intervals of side-length 2^{-m}, that is,

$$\mathcal{D}_m = \{[j2^{-m}, (j+1)2^{-m}) : j \in \mathbb{Z}\}.$$

Given $\mu \in \mathcal{P}$ and $q > 1$, the quantity $S_m(\mu, q) = \sum_{J \in \mathcal{D}_m} \mu(J)^q$ measures, in an L^q-sense, how uniformly distributed μ is at scale 2^{-m}. Using Hölder's inequality, one can check that if $\mu \in \mathcal{P}_1$, then

$$2^{(1-q)m} \leq S_m(\mu, q) \leq 1, \tag{9.1.2}$$

with the extreme values attained, respectively, when μ is uniformly distributed among the 2^m intervals $J \in \mathcal{D}_m$ contained in $[0, 1)$, and when μ gives full mass to a single interval in $\mathcal{D}_m$. This suggests that the decay rate of $S_m(\mu, q)$ as $m \to \infty$ may indicate the smoothness of μ at arbitrarily small scales, and this is precisely how the L^q dimensions $D_\mu(q)$ are defined:

$$\tau_\mu(q) = \liminf_{m\to\infty} \frac{-\log S_m(\mu, q)}{m},$$
$$D_\mu(q) = \frac{\tau_\mu(q)}{q-1}.$$

(Here and throughout the paper, the logarithms are to base 2.) We will sometimes write $\tau(\mu, q)$, $D(\mu, q)$ instead of $\tau_\mu(q)$, $D_\mu(q)$. The function $q \mapsto \tau_\mu(q)$ is called the *L^q spectrum* of μ. In light of (9.1.2), one always has $0 \leq D_\mu(q) \leq 1$ for $\mu \in \mathcal{P}_1$ and, indeed, the same inequality holds for $\mu \in \mathcal{P}$. Moreover, $D_\mu(q) = 0$ for purely atomic measures μ and $D_\mu(q) = 1$ if μ is Lebesgue measure on an interval or, more generally, if μ is absolutely continuous with an L^q density. These basic properties suggest that $D_\mu(q)$ is a reasonable notion of dimension.

We state two simple and well-known properties of L^q dimensions.

Lemmash 1 *The functions $q \mapsto D_\mu(q)$, $q \mapsto \tau_\mu(q)$ are, respectively, nonincreasing and concave on* $(1, \infty)$.

Proof Fix $0 < \lambda < 1$, $\mu \in \mathcal{P}$, $m \in \mathbb{N}$, $q_1, q_2 \geq 1$. It follows from Hölder's inequality applied with exponents $1/\lambda$ and $1/(1-\lambda)$ that

$$S_m(\mu, \lambda q_1 + (1-\lambda)q_2) \leq S_m(\mu, q_1)^\lambda \, S_m(\mu, q_2)^{1-\lambda}.$$

The concavity of τ is immediate from this. For the monotonicity of $D_\mu(q)$, suppose $1 < p < q$ and apply the above with $q_1 = q$, $q_2 = 1$ and $\lambda = (p-1)/(q-1)$.

So far we have dealt with a general measure $\mu \in \mathcal{P}$. We now turn to self-similar measures associated to a WIFS (f_i, p_i). We have seen that the similarity dimension is a "candidate" for the Hausdorff dimension of a self-similar measure μ, is always an upper bound for $dim_H(\mu)$, and is conjectured to equal $dim_H(\mu)$ under the terms of the overlaps conjecture. There is a natural L^q analog of the similarity dimension: first, we define $T(\mu, q)$ as the only number satisfying

$$\sum_{i\in I} p_i^q |\lambda_i|^{-T(\mu,q)} = 1,$$

and then let $dim_S(\mu, q) = T(\mu, q)/(q-1)$. The function T is a "symbolic" analog of the L^q spectrum, while dim_S is a version of similarity dimension for L^q dimensions.

A simple exercise shows that $\lim_{q\to 1+} dim_S(\mu, q) = dim_S(\mu)$. We also have that $q \mapsto dim_S(\mu, q)$ is a real-analytic, nondecreasing function of q; it is constant if and only if $p_i = |\lambda_i|^s$ for some s independent of i (in which case $dim_S(\mu, q) = s$ for all $q > 1$), and otherwise it is strictly decreasing.

Just like for Hausdorff dimension, it always holds that $D_\mu(q) \leq \min(dim_S(\mu, q), 1)$, and the only known mechanisms for a strict inequality are $dim_S(\mu, q) > 1$ and the presence of exact overlaps. A variant of the overlaps conjecture asserts that if μ is a self-similar measure then $D_\mu(q) = \min(dim_S(\mu, q), 1)$ unless there is an exact overlap. This conjecture is stronger than the Hausdorff dimension variant, since $D_\mu(q) \leq dim_H(\mu)$ for all $q > 1$ and $D_\mu(q) \to dim_H(\mu)$ as $q \to 1^+$ in the case of self-similar measures: see [20, Theorem 5.1 and Remark 5.2].

In [19], the author established the following variant of Hochman's Theorem 1.1 for L^q dimensions.

Theoremsh 1.2 *If (f_i, p_i) is a WIFS with exponential separation and μ is the associated invariant self-similar measure, then*

$$D_\mu(q) = \min(dim_S(\mu, q), 1) \quad \textit{for all } q > 1.$$

Again, this theorem is formally stronger than Theorem 1.1, since the latter can be recovered by letting $q \to 1^+$. While at first it may seem that the difference between Hausdorff and L^q dimensions is merely technical, the L^q dimension version has several advantages in applications, especially since it applies to *every* $q > 1$. It is useful to think of the difference between L^q and Hausdorff dimensions as being similar to the difference between L^q and L^1 functions (on bounded intervals). Both kinds of dimensions give information about the local behavior of a measure, but the L^q dimensions do so in a more quantitative fashion. If $dim_H(\mu) > s$, then it holds that

$$\mu(B(x, r)) \leq r^s \tag{9.1.3}$$

for μ-almost all x and all sufficiently small r (depending on x). On the other hand, if $\lim_{q\to\infty} D_\mu(q) > s$, then (9.1.3) holds *uniformly, for all* x and all sufficiently small r: see Lemma 13. For some applications of Theorem 1.2 beyond those described in this article, see [3, 10, 17].

Theorem 1.2 was originally featured in [19, Theorem 6.6]. In this article, we will present the proof of the special case in which the WIFS is *homogeneous*, that is, all of the scaling factors λ_i are equal. The homogeneous case of Theorem 1.2 is a particular case of [19, Theorem 1.1]. As indicated earlier, this particular case avoids an ergodic-theoretic part of the argument, and so we hope it will be more accessible. The proof borrows many ideas from Hochman's proof of Theorem 1.1, but there are also substantial differences.

One central element of the proof of Theorem 1.2 is an inverse theorem for the L^q norm of convolutions, which does not rely on self-similarity and may have other applications. This theorem is discussed and stated (without proof) in Sect. 9.2. Section 9.3 contains the proof of the homogeneous version of Theorem 1.2, starting with a sketch and proceeding to the details. In Sect. 9.4, we introduce some applications to Frostman exponents, self-similar measures generated by algebraic parameters, absolute continuity, and intersections of self-similar Cantor sets. We also briefly discuss some old and new results by other authors on Bernoulli convolutions, how they relate to ours, and a possible line for future research.

Although some of the applications in Sect. 9.4 have not been stated in this form in [18], both the results and the presentation of this survey are strongly based on [18].

9.2 An Inverse Theorem for the L^q Norms of Convolutions

Let $\mu, \nu \in \mathcal{P}$. The convolution $\mu * \nu$ is defined as the push-forward of the product measure $\mu \times \nu$ under the addition map $(x, y) \mapsto x + y$. Explicitly,

$$\mu * \nu(A) = (\mu \times \nu)\{(x, y) : x + y \in A\} \quad \text{for all Borel } A \subset \mathbb{R}.$$

Intuitively, one expects the convolution $\mu * \nu$ to be at least as smooth as μ. A natural question is then: if $\mu * \nu$ is not "much smoother" than μ, can we deduce any structural information about the measures μ and ν? Of course, this depends on the notion of smoothness under consideration, and on the precise meaning of "much smoother."

Here we will measure smoothness by the moment sums $S_m(\mu, q)$ (with $q > 1$ fixed, and m also fixed but very large). Nevertheless, we begin by discussing the situation for entropy. Let $\mu \in \mathcal{P}_1$. Its normalized level m entropy is

$$H_m(\mu) = \frac{1}{m} \sum_{J \in \mathcal{D}_m} -\mu(J) \log(\mu(J)),$$

with the usual convention $0 \log 0 = 0$. In [12, Theorem 2.7], Hochman showed that if $\mu, \nu \in \mathcal{P}_1$ satisfy

$$H_m(\nu * \mu) \leq H_m(\mu) + \mathrm{e},$$

where $\mathrm{e} > 0$ is small, then ν and μ have a certain structure which, very roughly, is of this form: the set of dyadic scales $0 \leq s < m$ can be split into three sets $\mathcal{A} \cup \mathcal{B} \cup \mathcal{C}$. At scales in $\mathcal{A}$, the measure ν looks "roughly atomic," at scales in $\mathcal{B}$ the measure μ looks "roughly uniform", and the set $\mathcal{C}$ is small. This theorem was motivated in part by its applications to the dimension theory of self-similar measures, as discussed above.

We aim to state a result in the same spirit, but with L^q norms in place of entropy. Given $m \in \mathbb{N}$, we will say that μ is a 2^{-m}-measure if μ is a probability measure supported on $2^{-m}\mathbb{Z} \cap [0, 1)$. Given $\mu \in \mathcal{P}_1$, we denote by $\mu^{(m)}$ the associated 2^{-m}-measure, given by $\mu^{(m)}(j2^{-m}) = \mu([j2^{-m}, (j+1)2^{-m}))$. Given a purely atomic measure ρ, we define the L^q norms

$$\|\rho\|_q = \left(\sum \rho(y)^q\right)^{1/q},$$

for $q \in (1, \infty)$ and also set $\|\rho\|_\infty = \max_y \rho(y)$. With these definitions, we clearly have

$$S_m(\mu, q) = \|\mu^{(m)}\|_q^q.$$

By the convexity of $t \mapsto t^q$, we have that $\|\mu * \nu\|_q \le \|\mu\|_q \|\nu\|_1$, for any $q \ge 1$ and any two finitely supported probability measures μ, ν (this is a simple instance of Young's convolution inequality). We aim to understand under what circumstances $\|\mu * \nu\|_q \approx \|\mu\|_q \|\nu\|_1$, where the closeness is in a weak, exponential sense. More precisely, we are interested in what structural properties of two 2^{-m}-measures μ, ν ensure an exponential flattening of the L^q norm of the form

$$\|\mu * \nu\|_q \le 2^{-\varepsilon m} \|\mu\|_q. \tag{9.2.1}$$

(Recall that, by definition, 2^{-m}-measures are probability measures, so that $\|\nu\|_1 = 1$.) One particular instance of this problem has received considerable attention. Given a finite set A, we denote $\mathbf{1}_A = \sum_{x \in A} \delta_x$. Then $\|\mathbf{1}_A * \mathbf{1}_A\|_2^2$ is the additive energy of A, a quantity of great importance in combinatorics and its applications. In particular, estimates of the form

$$\|\mathbf{1}_A * \mathbf{1}_A\|_2^2 \le |A|^{-\varepsilon} \|\mathbf{1}_A\|_2^2 \|\mathbf{1}_A\|_1 = |A|^{3-\varepsilon}$$

arise repeatedly in dynamics, combinatorics, and analysis: see, e.g., [1, 7] for some recent examples.

To motivate the inverse theorem, we discuss cases in which $\|\mu * \nu\|_q \approx \|\mu\|_q$ for 2^{-m}-measures μ and ν, where we are deliberately vague about the exact meaning of $\approx$. If $\nu = \delta_{k2^{-m}}$, then $\mu * \nu$ is just a translation of μ and so we have an exact equality. If ν is supported on a small number of atoms (say subexponential in m), then we still have $\|\mu * \nu\|_q \approx \|\mu\|_q$. Reciprocally, if λ denotes the uniform 2^{-m}-measure giving mass 2^{-m} to each atom $j2^{-m}$, then we also have $\|\lambda * \nu\|_q \approx \|\lambda\|_q$. The same holds if λ is replaced by a suitably small perturbation.

Furthermore, if $\nu = 2^{-\varepsilon m}\delta_0 + (1 - 2^{\varepsilon m})\lambda$ and μ is an arbitrary 2^{-m}-measure, then we still have $\|\mu * \nu\|_q \ge 2^{-\varepsilon m}\|\mu\|_q$. This shows that a subset of measure $2^{-\varepsilon m}$ is able to prevent exponential smoothening, so that in order to guarantee (9.2.1) we need to impose conditions on the structure of the measures inside sets of exponentially small measure. This is one significant difference with the case of entropy, since sets of exponentially small measure have negligible contribution to the entropy.

A naive conjecture might be that if (9.2.1) fails for a pair of 2^{-m}-measures, then either μ is close to uniform, or ν gives "large" mass to an exponentially small set of atoms. However, there are other situations in which $\|\mu * \nu\|_q \approx \|\mu\|_q$. Let $D \gg 1$ be a large integer and fix $\ell \gg D$. Given a subset $\mathcal{S}$ of $\{0, \ldots, \ell - 1\}$, let $\widetilde{\mu}$ be the distribution of an independent sequence of random variables $(X_1, \ldots, X_\ell)$ such that X_s is uniformly distributed in $\{0, 1, \ldots, 2^D - 1\}$ if $s \in \mathcal{S}$ and $X_s = 0$ if $s \notin \mathcal{S}$. Finally, let μ be the push-forward of $\widetilde{\mu}$ under the 2^D-ary expansion map. In other words, $\widetilde{\mu}$ is the $2^{-D\ell}$-measure such that

$$\mu\left(\sum_{s=1}^{\ell} X_s 2^{-Ds}\right) = \widetilde{\mu}(X_1, \ldots, X_s).$$

It is convenient to think about the structure of μ in terms of trees. Given a base 2^D and a non-empty closed subset A of $[0, 1)$, we may associate to A the family of all intervals of the form $[j2^{-Ds}, (j+1)2^{-Ds})$ (i.e., the 2^D-ary intervals) that intersect A. This family has a natural tree structure, where the interval $[0, 1)$ is the root and descendence is given by inclusion. In general, the tree associated to A is infinite, but in the case of $2^{-D\ell}$-sets we can think of a finite tree with ℓ levels. For the measure μ just defined, its support A has the following structure: vertices of level $s \in \mathcal{S}$ have a maximal number of offspring 2^D ("full branching"), while vertices of level $s \notin \mathcal{S}$ have a single offspring ("no branching"), corresponding to the leftmost interval. Moreover, μ is the uniform measure on A—it gives all points in A the same mass $1/|A|$.

The convolution $\mu * \mu$ has essentially the same structure, except that vertices of level $s \notin \mathcal{S}$ such that $s-1 \in \mathcal{S}$ have two offspring—due to the carries of the previous level. Using this structure, it is not hard to check that $\|\mu * \mu\|_q \approx \|\mu\|_q$ for all q. In similar ways, one can construct 2^{-m}-measures μ, ν supported on sets of widely different sizes, such that $\|\mu * \nu\|_q \approx \|\mu\|_q$.

The inverse theorem asserts that if (9.2.1) fails to hold then one can find subsets $A \subset \operatorname{supp}(\mu)$ and $B \subset \operatorname{supp}(\nu)$, such that A captures a "large" proportion of the L^q norm of μ and B a "large" proportion of the mass of ν, and moreover $\mu|_A, \nu|_B$ are fairly regular (they are constant up to a factor of 2). The main conclusion, however, is that A and B have a structure resembling the example above, and also the conclusion of Hochman's inverse theorem for entropy: if D is a large enough integer, then for each s, either B has no branching between scales 2^{-sD} and $2^{-(s+1)D}$ (in other words, once the first s digits in the 2^D-ary expansion of $y \in B$ are fixed, the next digit is uniquely determined), or A has nearly full branching between scales 2^{-sD} and $2^{-(s+1)D}$ (whatever the first s digits of $x \in A$ in the 2^D-adic expansion, the next digit can take "most" values). To formalize this, we introduce the following definition.

Definition 2 Given $D \in \mathbb{N}, \ell \in \mathbb{N}$ and a sequence $R = (R_0, \ldots, R_{\ell-1}) \in [1, 2^D]^\ell$, we say that a set $A \subset [0, 1)$ is (D, ℓ, R)-*regular* if it is a $2^{-\ell D}$-set, and for all $s \in \{0, \ldots, \ell-1\}$ and for all $J \in \mathcal{D}_{sD}$ such that $A \cap J \neq \varnothing$, it holds that

$$|\{J' \in \mathcal{D}_{(s+1)D} : J' \subset J, J' \cap A \neq \varnothing\}| = R_s.$$

In terms of the associated 2^D-ary tree, A is (D, ℓ, R)-regular if every vertex of level s has the same number of offspring R_s.

Before stating the theorem, we summarize our notation for dyadic intervals (some of it has already been introduced):

- $\mathcal{D}_s$ is the family of dyadic intervals $[j2^{-s}, (j+1)2^{-s})$.
- Given a set $A \subset \mathbb{R}$, we write $\mathcal{D}_s(A)$ for the family of intervals in $\mathcal{D}_s$ that hit A.
- Given $x \in \mathbb{R}$, we write $\mathcal{D}_s(x)$ for the only interval in $\mathcal{D}_s$ that contains x.
- We write aJ for the interval of the same center as J and length a times the length of J.

We also write $[\ell] = \{0, 1, \ldots, \ell - 1\}$.

Theoremsh 2.1 *For each $q > 1$, $\delta > 0$, and $D_0 \in \mathbb{N}$, there are $D \geq D_0$ and $\varepsilon > 0$, so that the following holds for $\ell \geq \ell_0(q, \delta, D_0)$.*

Let $m = \ell D$ and let μ and ν be 2^{-m}-measures with

$$\|\mu * \nu\|_q \geq 2^{-\varepsilon m}\|\mu\|_q.$$

Then there exist 2^{-m}-sets $A \subset \operatorname{supp}\mu$ and $B \subset \operatorname{supp}\nu$, numbers $k_A, k_B \in 2^{-m}\mathbb{Z}$, and a set $\mathcal{S} \subset [\ell]$, so that

(A1) $\|\mu|_A\|_q \geq 2^{-\delta m}\|\mu\|_q$.
(A2) $\mu(x) \leq 2\mu(y)$ *for all* $x, y \in A$.
(A3) $A' = A + k_A$ *is contained in* $[0, 1)$ *and is* (D, ℓ, R') *uniform for some sequence* R'.
(A4) $x \in \frac{1}{2}\mathcal{D}_{sD}(x)$ *for each* $x \in A'$ *and* $s \in [\ell]$.

(B1) $\|\nu|_B\|_1 = \nu(B) \geq 2^{-\delta m}$.
(B2) $\nu(x) \leq 2\nu(y)$ *for all* $x, y \in B$.
(B3) $B' = B + k_B$ *is contained in* $[0, 1)$ *and is* (D, ℓ, R'') *uniform for some sequence* R''.
(B4) $y \in \frac{1}{2}\mathcal{D}_{sD}(y)$ *for each* $y \in B'$ *and* $s \in [\ell]$.

Moreover

(5) for each s, $R''_s = 1$ if $s \notin \mathcal{S}$, and $R'_s \geq 2^{(1-\delta)D}$ if $s \in \mathcal{S}$.
(6) The set $\mathcal{S}$ satisfies

$$\log\|\nu\|_q^{-q'} - m\delta \leq D|\mathcal{S}| \leq \log\|\mu\|_q^{-q'} + m\delta.$$

Here, and throughout the paper, $q' = q/(q-1)$ denotes the dual exponent. We make some remarks on the statement.

(i) In the original version of the theorem in [19], both the convolution and the translations take place on the circle $[0, 1)$ with addition modulo 1. See [17, Theorem 2.2 and Remark 2.3] for this formulation.
(ii) The main claim in the theorem is part (5). Obtaining sets A, B satisfying (A1)–(A4) and (B1)–(B4) is not hard, and (6) is a straightforward calculation using (5).

(iii) The theorem fails for $q = 1$ and $q = \infty$. In the first case, there is an equality $\|\mu * \nu\|_1 = \|\mu\|_1$ for any 2^{-m}-measures, and in the second case there is always an equality $\|\mathbf{1}_A * \mathbf{1}_{-A}\|_\infty = \|\mathbf{1}_A\|_\infty \|\mathbf{1}_A\|_1$. On the other hand, the proof can easily be reduced to the case $q = 2$, with the remaining cases following by interpolation with the endpoints $q = 1$ and $q = \infty$.

The proof of Theorem 2.1 (including the proofs of the results it relies on) is elementary and, at least in principle, it is effective, although the value of e that emerges from the proof is extremely poor and certainly suboptimal. However, for the purposes of proving Theorem 1.2, the existence of any e > 0 is enough.

9.3 Proof of the Main Theorem

9.3.1 Homogeneous Self-similar Measures

We restate the particular case of Theorem 1.2 that we will prove.

Theoremsh 3.1 *Let $(f_i(x) = \lambda x + t_i)_{i\in I}$ be a homogeneous IFS with exponential separation. Then for any probability vector $(p_i)_{i\in I}$, if μ is the invariant self-similar measure for the WIFS (f_i, p_i), then*

$$D_\mu(q) = \min(dim_S(\mu, q), 1) = \min\left(\frac{\log \sum_{i\in I} p_i^q}{(q-1)\log|\lambda}|, 1\right) \tag{9.3.1}$$

for all $q > 1$.

We recall that "homogeneous" here refers to the fact that all scaling factors are equal. We may and do assume that $\lambda > 0$; if $\lambda < 0$, note that μ can also be generated by the WIFS $(f_if_j, p_ip_j)_{i,j\in I}$, for which the scaling factor is $\lambda^2 > 0$. This iteration of the IFS does not change the validity of exponential separation.

It is known that for self-similar measures the limit in the definition of L_q dimension always exists, see [15].

The key advantage of homogeneity is that, in this case, the self-similar measure μ has an infinite convolution structure: if $\Delta = \sum_{i\in I} p_i \delta_{t_i}$, then

$$\mu = *_{n=0}^{\infty} S_{\lambda^n}\Delta, \tag{9.3.2}$$

where $S_a(x) = ax$ rescales by a. Formally, this infinite convolution is defined as the push-forward of the countable self-product $\mu^{\mathbb{N}}$ under the series expansion map $(x_1, x_2, \ldots) \mapsto \sum_{n=0}^{\infty} x_n$; this is well defined since the series always converges absolutely. To verify that this is indeed the self-similar measure, one only needs to check that it satisfies the self-similarity relation

$$\mu = \sum_{i \in I} p_i f_i \mu.$$

In more probabilistic terms, μ can also be defined as the distribution of the random series $\sum_{n=0}^{\infty} \lambda^n X_n$, where X_n are IID random variables with distribution Δ. The well-known fact that the distribution of a sum of independent random variables is the convolution of the distributions (which extends to countable sums) then gives another derivation of (9.3.2).

9.3.2 *Outline of the Proof*

The overall strategy of the proof of Theorem 3.1 follows the broad outline of [12]. However, while Hochman's method is based on entropy, we need to deal with L^q norms and, as we will see, this forces substantial changes in the implementation of the outline.

The right-hand side in (9.3.1) is easily seen to be an upper bound for the left-hand side, so the task is to show the reverse inequality. Write $\tau = \tau_\mu$ and $D = D_\mu$. We want to show that if $D(q) < 1$ (or, equivalently, $\tau(q) < q - 1$) then $D(q) = dim_S(\mu, q)$ (under the hypothesis of exponential separation).

Recall that the L^q spectrum $\tau(q)$ is concave, so in particular it is continuous and differentiable outside of at most a countable set. Hence, it is enough to prove the claim above for a fixed differentiability point q. The advantage of this assumption is that the "multifractal structure" of a measure μ is known to behave in a regular way for points q of differentiability of the spectrum. In particular, we will see that if $\alpha = \tau'(q)$ then, for large enough m, "almost all" of the contribution to the sum $\sum_{J \in \mathcal{D}_m} \mu(J)^q$ comes from $\approx 2^{\tau^*(\alpha)m}$ intervals I such that $\mu(J) \approx 2^{\alpha m}$; here τ^* is the Legendre transform of τ (see Sect. 9.3.5 for the definition). Moreover, using the self-similarity of μ, we establish also a multi-scale version of this fact, see Proposition 11.

The following is the key estimate in the proof; as we will see in Sect. 9.4, it has other applications. Recall that $\mu^{(m)}$ is given by

$$\mu^{(m)}(j2^{-m}) = \mu([j2^{-m}, (j+1)2^{-m})) \tag{9.3.3}$$

and that, by definition, $\|\mu^{(m)}\|_q^q = S_m(\mu, q) \approx 2^{-m\tau(q)}$.

Theoremsh 3.2 *Let μ be a self-similar measure associated to a homogeneous WIFS (not necessarily with exponential separation) and let $q > 1$. Suppose $\tau_\mu(q) < q - 1$. Then for every $\sigma > 0$ there is $\mathrm{e} = \mathrm{e}(\sigma, q) > 0$ such that the following holds for all large enough m: if ρ is an arbitrary 2^{-m}-measure such that $\|\rho\|_q^{q'} \le 2^{-\sigma m}$, then*

$$\|\rho * \mu^{(m)}\|_q^q \le 2^{-\mathrm{e}m} \|\mu^{(m)}\|_q^q. \tag{9.3.4}$$

This theorem is proved by combining the inverse theorem for the L^q norm of convolutions (Theorem 2.1), together with the study of the multifractal structure of μ. We sketch the idea very briefly: suppose (9.3.4) fails. The inverse theorem then asserts that there is a regular subset A of $\text{supp}(\mu^{(m)})$ that captures much of the L^q norm of $\mu^{(m)}$. By part (5) of the inverse theorem, and since ρ is assumed to have exponentially small L^q norm, A must have almost full branching on a positive density set of scales in a multi-scale decomposition. But A itself does not have full branching (this will follow from the assumption $\tau(q) < q - 1$, which rules out $\mu^{(m)}$ having too small L^q norm). So there must also be a positive density set of scales on which A has smaller than average branching. The regularity of the multifractal spectrum discussed above rules this out, since it forces A to have an almost constant branching on almost all scales. For a detailed proof, see Sect. 9.3.6.

The conclusion of the proof of Theorem 3.1 from (9.3.4) tracks fairly closely the ideas of [12]. One consequence of self-similarity, as expressed by (9.3.2), is that

$$\mu = \mu_n * S_{\lambda^n}\mu,$$

where

$$\mu_n = *_{j=0}^{n-1} S_{\lambda^j}\Delta. \tag{9.3.5}$$

Note that the atoms of μ_n are the points of the form $f_{i_1} \circ \cdots \circ f_{i_n}(0)$. The exponential separation assumption implies that there exist $R \in \mathbb{N}$ such that all these atoms are distinct and λ^{Rn}-separated. Hence, for this value of R we have

$$\frac{\log \|\mu_n^{(Rm)}\|_q^q}{(q-1)n\log(1/\lambda)} = \frac{\log \|\mu_n\|_q^q}{(q-1)n\log(1/\lambda)} = \frac{n\log \|\Delta\|_q^q}{(q-1)n\log(1/\lambda)},$$

where $m = m(n)$ is chosen so that $2^{-m} \leq \lambda^n$ and $2^{-m} \sim \lambda^n$. It is easy to see that the right-hand side is equal to the right-hand side of (9.3.1). Hence, it remains to show that

$$\lim_{n\to\infty} \frac{\log \|\mu_n^{(Rm)}\|_q^q}{n\log(1/\lambda)} = \tau(q). \tag{9.3.6}$$

In other words, we need to show that the L^q norm of μ_n at scale $2^{-m} \approx \lambda^n$ (which is easily seen to be comparable to the L^q norm of μ at scale 2^{-m}, and hence is $\approx S_m(\mu, q)^{1/q}$) nearly exhausts the L^q norm of μ_n at the much finer scale 2^{-Rm} which, in turn, equals the full L^q norm of μ_n, by the exponential separation assumption.

To show (9.3.6), we recall that $\mu = \mu_n * S_{\lambda^n}\mu$, and use this to decompose

$$\mu^{((R+1)m)} = \sum_{J\in\mathcal{D}_m} \mu(J)\widetilde{\rho}_J * S_{\lambda^n}\mu,$$

where $\widetilde{\rho}_J$ is the normalized restriction of μ_n to J. Since the supports of $\widetilde{\rho}_J * S_{\lambda^n}\mu$ have bounded overlap, it is not hard to deduce that

$$\|\mu^{((R+1)m)}\|_q^q \approx \sum_{J\in\mathcal{D}_m} \mu(J)^q \|\rho_J * \mu^{(Rm)}\|_q^q,$$

where $\rho_J = S_{\lambda^{-n}}\widetilde{\rho}_J$. This is the point where we apply Theorem 3.2, to conclude that if on the right-hand side above we only add over those J such that $\|\rho_J\|_q \geq 2^{-\sigma q}$, where $\sigma > 0$ is arbitrary, then, provided n is large enough depending on σ, we still capture almost all of the left-hand side. This follows since (9.3.4) can be shown to imply that the contribution of the remaining J is exponentially smaller than the left-hand side (incidentally, this is the only step where it is crucial to use that $q > 1$). A similar calculation, now with $\mu_n^{((R+1)m)}$ in place of $\mu^{((R+1)m)}$ in the left-hand side, then shows that (9.3.6) holds, finishing the proof.

9.3.3 *Notational Conventions*

Throughout this section, μ denotes a self-similar measure associated to a homogeneous WIFS $\{\lambda x + t_i\}_{i\in I}$ with weights $(p_i)_{i\in I}$. We do *not* assume exponential separation until the very end, when we finish the proof of Theorem 3.1. We continue to denote

$$\Delta = \sum_{i\in I} p_i\,\delta_{t_i}.$$

Other measures, without any assumptions on self-similarity, will be denoted by ρ and ν, possibly with subindices.

We use Landau's $O(\cdot)$ and related notation: if X, Y are two positive quantities, then $Y = O(X)$ means that $Y \leq CX$ for some constant $C > 0$, while $Y = \Omega(X)$ means that $X = O(Y)$, and $Y = \Theta(X)$ that $Y = O(X)$ and $X = O(Y)$. If the constant C is allowed to depend on some parameters, these are often denoted by subscripts. For example, $Y = O_q(X)$ means that $Y \leq C(q)X$, where $C(q)$ is a function depending on the parameter q.

9.3.4 *Preliminary Lemmas*

In this section, we collect some standard lemmas for later reference. They are all of the form: bounded overlapping does not affect L^q norms too much. We refer to [19, Sect. 4] for the very short proofs.

Lemmash 3 *Let $(Y, \nu, \mathcal{B})$ be a probability space. Suppose $\mathcal{P}$, $\mathcal{Q}$ are finite families of measurable subsets of Y such that each element of $\mathcal{P}$ can be covered by at most M elements of $\mathcal{Q}$ and each element of $\mathcal{Q}$ intersects at most M elements of $\mathcal{P}$. Then, for every $q \geq 1$,*

$$\sum_{P\in\mathcal{P}} \nu(P)^q \leq M^q \sum_{Q\in\mathcal{Q}} \nu(Q)^q$$

Lemmash 4 *Let* $\nu = \sum_{i=1}^{\ell} \nu_i$*, where* ν_i *are finitely supported measures on a space* Y*, such that each point is in the support of at most* M *of the* ν_i*. Then*

$$\|\nu\|_q^q \leq M^{q-1} \sum_{i=1}^{\ell} \|\nu_i\|_q^q.$$

Lemmash 5 *For any* $q \in (1, \infty)$*, for any* $\nu_1, \nu_2 \in \mathcal{P}_1$ *and any* $m \in \mathbb{N}$*,*

$$\|(\nu_1 * \nu_2)^{(m)}\|_q^q = \Theta_q(1)\|\nu_1^{(m)} * \nu_2^{(m)}\|_q^q.$$

Recall the definition of μ_n given in (9.3.5). Let $m = n\lceil \log(1/\lambda) \rceil$.

Lemmash 6 *For any* $q \in (1, \infty)$*,*

$$\|\mu^{(m)}\|_q^q = \Theta_q(1)\|\mu_n^{(m)}\|_q^q.$$

9.3.5 Multifractal Structure

We turn to the multifractal estimates that will be required in the proof of Theorem 3.2. The Legendre transform plays a key role in multifractal analysis. Given a concave function $\zeta : \mathbb{R} \to \mathbb{R}$, its Legendre transform $\zeta^* : \mathbb{R} \to [-\infty, \infty)$ is defined as

$$\zeta^*(\alpha) = \inf_{q \in \mathbb{R}} \alpha q - \zeta(q).$$

It is easy to check that if ζ is concave and is differentiable at q, then

$$\zeta^*(\alpha) = \alpha q - \zeta(q) \text{ for } \alpha = \zeta'(q).$$

As indicated earlier, we will establish some regularity of the multifractal structure for those values of q such that τ is differentiable at q.

The next lemma is well known; we include the very short proof for completeness.

Lemmash 7 *If* τ *is differentiable at* $q > 1$*,* $\tau(q) < q - 1$*, and* $\alpha = \tau'(q)$*, then* $\tau^*(\alpha) \leq \alpha < 1$.

Proof Since $\tau(1) = 0$ (this is immediate from the definition) and $\tau(q) < q - 1$, we have $(\tau(q) - \tau(1))/(q - 1) < 1$. On the other hand, as τ is concave and differentiable at q, we must have $\alpha \leq (\tau(q) - \tau(1))/(q - 1) < 1$. Furthermore, $\tau^*(\alpha) \leq \alpha \cdot 1 - \tau(1) = \alpha$, so the lemma follows.

The following lemmas illustrate the regularity of the L^q spectrum for values q of differentiability of τ (or dually, points of strict concavity of τ^*). The proofs are similar to [14, Theorem 5.1]. The heuristic to keep in mind is that, whenever $\alpha = \tau'(q)$ exists, almost all of the contribution to $\|\mu^{(m)}\|_q^q$ comes from $\approx 2^{\tau^*(\alpha)m}$ intervals, each of mass $\approx 2^{-\alpha m}$.

Lemmash 8 *Suppose that* $\alpha_0 = \tau'(q_0)$ *exists for some* $q_0 \in (1, \infty)$.

Given e > 0, *the following holds if* δ *is small enough in terms of* e, q_0 *and* m *is large enough in terms of* e, q_0, *and* δ.

Suppose $\mathcal{D}' \subset \mathcal{D}_m$ *is such that*

(1) $2^{-\alpha m} \le \mu(J) \le 2 \cdot 2^{-\alpha m}$ *for all* $J \in \mathcal{D}'$ *and some* $\alpha \ge 0$.
(2) $\sum_{J\in\mathcal{D}'} \mu(J)^{q_0} \ge 2^{-(\tau(q_0)+\delta)m}$.

Then $|\mathcal{D}'| \le 2^{m(\tau^*(\alpha_0)+\mathrm{e})}$.

Proof Set $\eta := \mathrm{e}/(3q_0)$, and pick $\delta \le \eta^2/9$, and also small enough that, if $q_1 = q_0 - \delta^{1/2}$, then

$$\tau(q_0) - \tau(q_1) \le \delta^{1/2}\alpha_0 + \delta^{1/2}\eta. \tag{9.3.7}$$

On one hand, using (1) and the definition of $\tau(q)$, we get

$$2^{-(\tau(q_1)-\delta)m} \ge \|\mu^{(m)}\|_{q_1}^{q_1} \ge |\mathcal{D}'|2^{-\alpha q_1 m},$$

if m is large enough (depending on q_0, τ). On the other hand, by the assumptions (1)–(2),

$$|\mathcal{D}'|2^{-\alpha q_0 m} \ge 2^{-q_0}2^{(-\tau(q_0)-\delta)m} \ge 2^{(-\tau(q_0)-2\delta)m}$$

if $m \gg_{\delta,q_0} 1$. Eliminating $|\mathcal{D}'|$ from the last two displayed equations yields

$$\alpha q_0 - \tau(q_0) - 2\delta \le \alpha(q_0 - \delta^{1/2}) - \tau(q_0 - \delta^{1/2}) + \delta,$$

so that, recalling (9.3.7),

$$\delta^{1/2}\alpha \le \tau(q_0) - \tau(q_0 - \delta^{1/2}) + 3\delta \le \delta^{1/2}\alpha_0 + \delta^{1/2}\eta + 3\delta.$$

Hence $\alpha - \alpha_0 < 2\eta$, since we assumed $\delta \le (\eta/3)^2$. Using this, we get that if $m \gg_\varepsilon 1$, then

$$2^{(-\tau(q_0)+\mathrm{e}/3)m} \ge \|\mu^{(m)}\|_{q_0}^{q_0} \ge 2^{-q_0\alpha m}|\mathcal{D}'| \ge 2^{-q_0\alpha_0 m}2^{-(q_0 2\eta)m}|\mathcal{D}'|.$$

The conclusion follows from the formula $\tau^*(\alpha_0) = q_0\alpha_0 - \tau(q_0)$ and our choice $\eta = \mathrm{e}/(3q_0)$.

Lemmash 9 *Let* $q_0 > 0$ *be such that* $\alpha_0 = \tau'(q_0)$ *exists. Given* $\sigma > 0$, *there is* e $=$ e$(\sigma, q_0) > 0$ *such that the following holds for large enough* m *(in terms of* σ, q_0*):*

$$\sum\{\mu(J)^{q_0} : J \in \mathcal{D}_m, \mu(J) \ge 2^{-m(\alpha_0-\sigma)}\} \le 2^{-m(\tau(q_0)+\mathrm{e})}. \tag{9.3.8}$$

Proof Let $\eta \in (0, 1)$ be small enough that

$$\tau(q_0 + \eta) \geq \tau(q_0) + \eta\alpha_0 - \delta, \tag{9.3.9}$$

where $\delta = \eta\sigma/(4 + 2q_0)$.

Let $\alpha_j = \alpha_0 - \delta j$, and write $N(\alpha_j, m)$ for the number of intervals J in $\mathcal{D}_m$ such that $2^{-m\alpha_j} \leq \mu(J) < 2^{-m\alpha_{j+1}}$. For any fixed value of q, if $m \gg_q 1$, then

$$N(\alpha_j, m)2^{-mq\alpha_j} \leq \|\mu^{(m)}\|_q^q \leq 2^{-m(\tau(q)-\delta)}.$$

Applying this to $q = q_0 + \eta$, and using (9.3.9), we estimate

$$\begin{aligned} N(\alpha_j, m)2^{-mq_0\alpha_j} &\leq 2^{m\eta\alpha_j}2^{-m(\tau(q_0+\eta)-\delta)} \\ &\leq 2^{2\delta m}2^{-j\delta\eta m}2^{-\tau(q_0)m}. \end{aligned}$$

Let $\mathcal{S}$ be the sum in the left-hand side of (9.3.8) that we want to estimate. Using that $\delta = \eta\sigma/(4 + 2q_0)$, we conclude that

$$\begin{aligned} \mathcal{S} &\leq \sum_{j:\delta(j+1)\geq\sigma} N(\alpha_j, m)2^{-mq_0\alpha_{j+1}} \\ &\leq \sum_{j:\delta(j+1)\geq\sigma} 2^{\delta q_0 m}2^{2\delta m}2^{-j\delta\eta m}2^{-\tau(q_0)m} \\ &\leq \sum_{j\geq 0} 2^{-j\delta\eta m}2^{(2+q_0)\delta m}2^{-\eta\sigma m}2^{-\tau(q_0)m} \\ &\leq O_{\delta\eta}(1)2^{(\eta\sigma/2-\eta\sigma)m}2^{-\tau(q_0)m}, \end{aligned}$$

as claimed.

Lemmash 10 *Let $q_0 > 1$ be such that $\alpha_0 = \tau'(q_0)$ exists. Given $\kappa > 0$, there is* $\mathrm{e} = \mathrm{e}(\kappa, q_0) > 0$ *such that the following holds for large enough m (in terms of q_0, e).*

If $\mathcal{D}' \subset \mathcal{D}_m$ has $\leq 2^{(\tau^*(\alpha_0)-\kappa)m}$ elements, then

$$\sum_{J\in\mathcal{D}'} \mu(J)^{q_0} \leq 2^{-(\tau(q_0)+\mathrm{e})m}.$$

Proof Let $\sigma = \kappa/(2q_0)$. In light of Lemma 9, we only need to worry about those J with $\mu(J) \leq 2^{-m(\alpha_0-\sigma)}$. But

$$\begin{aligned} \sum\{\mu(J)^{q_0} : J \in \mathcal{D}', \mu(J) \leq 2^{-m(\alpha_0-\sigma)}\} &\leq 2^{(\tau^*(\alpha_0)-\kappa)m}2^{-(q_0\alpha_0-q_0\sigma)m} \\ &= 2^{-(\kappa-q_0\sigma)m}2^{-\tau(q_0)m}. \end{aligned}$$

By our choice of σ, $\kappa - q_0\sigma = \kappa/2 > 0$, so this gives the claim.

The results in this section so far hold for general measures. The following proposition, on the other hand, relies crucially on self-similarity. The second part was first proved in [15]. Since the claim of Theorem 3.1 is not affected by rescaling and translating μ (from the point of view of the IFS, this amounts to doing these operations on the translation parameters t_i), from now on we assume that μ is supported on $[0, 1)$.

Proposition 11 *Let $q > 1$ be such that $\alpha = \tau'(q)$ exists.*

(i) Given $\kappa > 0$, there is $\eta = \eta(\kappa, q) > 0$ such that the following holds for all large enough m: for any $s \in \mathbb{N}$, $J \in \mathcal{D}_s$, if $\mathcal{D}'$ is a collection of intervals in $\mathcal{D}_{s+m}(J)$ with $|\mathcal{D}'| \le 2^{(\tau^*(\alpha)-\kappa)m}$, then

$$\sum_{J\in\mathcal{D}'} \mu(J)^q \le 2^{-(\tau(q)+\eta)m}\mu(2I)^q.$$

(ii) Given $\delta > 0$, the following holds for all large enough m: for any $I \in \mathcal{D}_s$, $s \in \mathbb{N}$,

$$\sum_{J\in\mathcal{D}_{s+m}(I)} \mu(J)^q \le 2^{-(\tau(q)-\delta)m}\mu(2I)^q.$$

Proof We prove (i) first. Let n be the smallest integer such that $\lambda^n < 2^{-s-2}$. Let y_j be the atoms of μ_n such that $[y_j, y_j + \lambda^n] \cap I \neq \varnothing$, let $\widetilde{p}_j$ be their respective masses, and write

$$\mu_{n,I} = \sum_j \widetilde{p}_j\delta_{y_j}.$$

Then the support of $\mu_{n,I}$ is contained in the λ^n-neighborhood of I. Moreover, since $\delta_z * S_{\lambda^n}\mu$ is supported on $[z, z + \lambda^n]$, as we assumed that μ is supported on $[0, 1]$, it follows from the self-similarity relation $\mu = \mu_n * S_{\lambda^n}\mu$ and the definition of $\mu_{n,I}$ that $\mu|_I = (\mu_{n,I} * S_{\lambda^n}\mu)|_I$. Write

$$\widetilde{p} = \|\mu_{n,I}\|_1 = \sum_j \widetilde{p}_j \le \mu(2I),$$

using that the support of μ_n is contained in the λ^n-neighborhood of the support of μ, and that $4\lambda^n \le 2^{-s}$.

We can then estimate

$$\begin{aligned}
\sum_{J\in\mathcal{D}'} \mu(J)^q &= \sum_{J\in\mathcal{D}'} \left(\sum_j \widetilde{p}_j\delta_{y_j} * S_{\lambda^n}\mu(J)\right)^q \\
&= \sum_{J\in\mathcal{D}'} \left(\sum_j \widetilde{p}_j\mu(\lambda^{-n}(J-y_j))\right)^q \\
&\le \sum_{J\in\mathcal{D}'} \widetilde{p}^{q-1} \sum_j \widetilde{p}_j\, \mu(\lambda^{-n}(J-y_j))^q \\
&= \widetilde{p}^{q-1} \sum_j \widetilde{p}_j \sum_{J\in\mathcal{D}'} \mu(\lambda^{-n}(J-y_j))^q,
\end{aligned}$$

where we used the convexity of t^q in the third line. Now for each fixed j, each interval $\lambda^{-n}(J - y_j)$ with $J \in \mathcal{D}'$ can be covered by $O_\lambda(1)$ intervals in $\mathcal{D}_m$, and reciprocally each interval in $\mathcal{D}_m$ hits at most two intervals among the $\lambda^{-n}(J - y_j)$. We deduce from Lemmas 3 and 10 that, still for a fixed j,

$$\sum_{J \in \mathcal{D}'} \mu(\lambda^{-n}(J - y_j))^q \leq O_{\lambda,q}(1) 2^{-(\tau(q)+\varepsilon)m},$$

provided m is taken large enough, where $\varepsilon = \varepsilon(\kappa, q) > 0$ is given by Lemma 10. Combining the last three displayed equations yields the first claim with $\eta = \varepsilon/2$.

The second claim follows in the same way, adding over $\mathcal{D}_{s+m}(I)$ instead of $\mathcal{D}'$.

9.3.6 *Proof of Theorem 3.2*

In this section, we prove Theorem 3.2. A similar result, with smoothness measured by entropy rather than L^q norms, was proved by Hochman in [12, Corollary 5.5], using his inverse theorem for the entropy of convolutions. In Hochman's approach, a crucial property of self-similar measures is that their entropy is roughly constant at most scales and locations, a property that Hochman termed *uniform entropy dimension*, see [12, Definition 5.1 and Proposition 5.2] for precise details. Unfortunately, there is no useful analog of the notion of uniform entropy dimension for L^q norms. One of the key differences is that nearly all of the L^q norm may be (and often is) captured by sets of extremely small measure; while sets of small measure also have small entropy. Instead, we will use the regularity of the multifractal spectrum established in the previous section in the following manner: if the flattening claimed in Theorem 3.2 does not hold, then the inverse theorem provides a regular set A which captures much of the L^q norm of μ. The upper bound on $\|\rho\|_q$, together with (5)–(6) in the inverse theorem imply that A has nearly full branching for a positive proportion of 2^D-scales, so it must have substantially less than average branching also on a positive proportion of scales. On the other hand, we will call upon the lemmas from the previous section to show that, in fact, A must have nearly constant branching on nearly all scales (this is the part that uses the differentiability of τ at q), obtaining the desired contradiction.

Proof of Theorem 3.2 Suppose ρ is a 2^{-m}-measure with $\|\rho\|_q^{q'} \leq 2^{-\sigma m}$. In the course of the proof, we will choose many numbers which ultimately depend on σ and q only. To ensure that there is no circularity in their definitions, we indicate their dependencies: $\alpha = \alpha(q)$, $\kappa = \kappa(\alpha, \sigma)$, $\gamma = \gamma(q, \alpha, \kappa)$, $\delta' = \delta'(\alpha, \sigma, \kappa)$, $\eta = \eta(q, \kappa)$, $\delta = \delta(q, \delta', \gamma, \eta)$, $\xi = \xi(q, \delta', \eta, \gamma)$, $D_0 = D_0(q, \sigma, \delta)$, $D = D(q, \delta, D_0)$, $\varepsilon = \varepsilon(q, \delta, D_0)$. Moreover, at different parts of the proof we will require δ', δ, ξ to be smaller than certain (positive) functions of the parameters they depend on; in particular, all of the requirements can be satisfied simultaneously.

Finally, m will be taken large enough in terms of all the previous parameters (hence ultimately in terms of q and σ).

Write $\alpha = \tau'(q)$, and define κ as

$$\kappa = (1 - \tau^*(\alpha))\sigma/4. \tag{9.3.10}$$

Then $\kappa > 0$, thanks to Lemma 7, and the assumption $\tau(q) < q - 1$.

We apply Proposition 11 to obtain a sufficiently large D_0 (in terms of δ, σ, q, with δ yet to be specified) such that:

(A) For any $D' \geq D_0 - 2$, any $I \in \mathcal{D}_{s'}$, $s' \in \mathbb{N}$, and any subset $\mathcal{D}' \subset \mathcal{D}_{s'+D'}(J)$ with $|\mathcal{D}'| \leq 2^{(\tau^*(\alpha)-\kappa)D'}$,

$$\sum_{J' \in \mathcal{D}'} \mu(J')^q \leq 2^{-(\tau(q)+\eta)D'} \mu(2J)^q,$$

where η depends on κ and q, hence on σ, q only.

(B) For any $D' \geq D_0 - 2$ and any $J \in \mathcal{D}_{s'}$, $s' \in \mathbb{N}$,

$$\sum_{J' \in \mathcal{D}_{s'+D'}(J)} \mu(J')^q \leq 2^{-(\tau(q)-\delta)D'} \mu(2J)^q.$$

(C) $1/D_0 < \delta$.

Let $\varepsilon > 0, D \in \mathbb{N}$ be the numbers given by Theorem 2.1 applied to δ, D_0, and q. For the sake of contradiction, suppose

$$\|\rho * \mu^{(m)}\|_q \geq 2^{-\varepsilon m} \|\mu^{(m)}\|_q.$$

We will derive a contradiction from this provided $m = \ell D$ is large enough (if m is not of the form ℓD, we apply the argument to $\lfloor m/D \rfloor D$ instead; we omit the details). We apply Theorem 2.1 to ρ and $\mu^{(m)}$ to obtain (assuming m is large enough) a set $A \subset \mathrm{supp}(\mu^{(m)})$ as in the theorem, with corresponding branching numbers R'_s. Since translating ρ and $\mu^{(m)}$ does not affect their norms or the norm of their convolution, we assume for simplicity that the numbers k_A, k_B are both 0.

The key to the proof is to show, using the structure of A provided by Theorem 2.1, that

$$|\{s \in [\ell] : R'_s \leq 2^{(\tau^*(\alpha)-\kappa)D}\}| \geq \gamma \ell, \tag{9.3.11}$$

where $\gamma > 0$ depends on q, α, and κ only (and κ is given by (9.3.10)). We first show how to complete the proof assuming this. Consider the sequence

$$L_s = -\log \sum_{J \in \mathcal{D}_{sD}(A)} \mu(J)^q.$$

By (B) applied with $s' = sD + 2$ and $D' = D - 2$,

$$L_{s+1} \geq (\tau(q) - \delta)(D - 2) - \log \sum_{J \in \mathcal{D}_{sD+2}(A)} \mu(2J)^q.$$

But if $J \in \mathcal{D}_{sD+2}(A)$, then $2J$ is contained in a single interval in $\mathcal{D}_{sD}(A)$ by property (A4) from Theorem 2.1, and conversely $J' \in \mathcal{D}_{sD}(A)$ hits at most two intervals $2J$, $J \in \mathcal{D}_{sD+2}(A)$. We deduce that

$$L_{s+1} \geq L_s + (\tau(q) - \delta)(D - 2) - 1$$

for all $s \in [\ell]$. Likewise, by (A),

$$L_{s+1} \geq L_s + (\tau(q) + \eta)(D - 2) - 1,$$

whenever $R'_s \leq 2^{(\tau^*(\alpha)-\kappa)D}$. Recall that η depends on q, κ. In light of (9.3.11), and using also (C), we have

$$\begin{aligned} L_\ell &\geq (\tau(q) + \eta)\gamma\ell(D - 2) + (\tau(q) - \delta)(1 - \gamma)\ell(D - 2) - \ell \\ &\geq (\tau(q) + \eta\gamma - \delta(1 - \gamma))m - 2\delta(\tau(q) + \eta)m - \delta m. \end{aligned}$$

Hence, by choosing δ small enough in terms of $\tau(q)$, γ and η we can ensure that, for m large enough,

$$L_\ell = -\log \|\mu^{(m)}|_A\|_q^q \geq (\tau(q) + \eta\gamma/2)m.$$

On the other hand, by (A1) in Theorem 2.1, if $\xi > 0$ is a small number to be fixed later, then (always assuming m is large enough)

$$\|\mu^{(m)}|_A\|_q^q \geq 2^{-q\delta m}\|\mu^{(m)}\|_q^q \geq 2^{-q\delta m}2^{-(\tau(q)+\xi)m}.$$

From the last two displayed equations,

$$\eta\gamma/2 \leq q\delta + \xi.$$

Recall that $\eta = \eta(\kappa, q)$, $\gamma = \gamma(q, \alpha, \kappa)$ is yet to be specified, while δ so far was taken small enough in terms of $\tau(q)$, γ, and η, and no conditions have been yet imposed on ξ. By ensuring $q\delta < \eta\gamma/8$ and $\xi \leq \eta\gamma/8$, we reach a contradiction, as desired.

It remains to establish (9.3.11). The idea is very simple: Theorem 2.1 (together with the assumption that $\|\rho\|_q^{q'} \leq 2^{-\sigma m}$) implies that A has "nearly full branching" on a positive proportion of scales. On the other hand, Lemma 8 says the size of A is at most roughly $2^{\tau^*(\alpha)m} \ll 2^m$ (by Lemma 7), so there must be a positive proportion of scales on which the average 2^D-adic branching is far smaller than $2^{\tau^*(\alpha)D}$, which is what (9.3.11) says.

We proceed to the details. Using (A1), (A2) in Theorem 2.1, we get that (for $m \gg_\delta 1$) there is $\widetilde{\alpha} > 0$ such that $\mu^{(m)}(a) \in [2^{-\widetilde{\alpha}m}, 2^{1-\widetilde{\alpha}m}]$ for all $a \in A$, and

$$\sum_{J\in\mathcal{D}_m(A)} \mu(J)^q \geq 2^{-q\delta m} \sum_{J\in\mathcal{D}_m} \mu(J)^q \geq 2^{-(\tau(q)+q\delta+\xi)m}.$$

We let $\delta \leq \delta'$ and ξ be small enough in terms of δ' and q that, invoking Lemma 8,

$$|A| \leq 2^{(\tau^*(\alpha)+\delta')m}. \tag{9.3.12}$$

Let $\mathcal{S}' = [\ell] \setminus \mathcal{S}$, where $\mathcal{S} = \{s : R'_s \geq 2^{(1-\delta)D}\}$. Using (A3) in Theorem 2.1, we see that

$$|A| = \prod_{s=0}^{\ell-1} R'_s \geq 2^{(1-\delta)D|\mathcal{S}|} \prod_{s\in\mathcal{S}'} R'_s. \tag{9.3.13}$$

Let $m_1 = D|\mathcal{S}|$, $m_2 = D|\mathcal{S}'| = m - m_1$. Combining (9.3.12) and (9.3.13), and using that $\delta \leq \delta'$, we deduce

$$\prod_{s\in\mathcal{S}'} R'_s \leq 2^{-(1-\delta)m_1} 2^{(\tau^*(\alpha)+\delta')m} \leq 2^{-(1-\tau^*(\alpha)-2\delta')m_1} 2^{(\tau^*(\alpha)+\delta')m_2}. \tag{9.3.14}$$

Note that $1 - \tau^*(\alpha) > 0$ by Lemma 7. At this point, we take δ' small enough that $1 - \tau^*(\alpha) - 2\delta' > 0$. Using (6) in Theorem 2.1, and the assumption $\|\rho\|_q^{q'} \leq 2^{-\sigma m}$, we further estimate

$$(\sigma - \delta)m \leq m_1 \leq ((\tau(q) + \xi)/(q - 1) + \delta)\, m. \tag{9.3.15}$$

We can plug in the left inequality (together with $m_2 \leq m$) into (9.3.14), to obtain the key estimate

$$\log \prod_{s\in\mathcal{S}'} R'_s \leq \big(\tau^*(\alpha) + \delta' - (1 - \tau^*(\alpha) - 2\delta')(\sigma - \delta)\big)\, m_2.$$

Recalling (9.3.10), this shows that by making δ' (hence also $\delta \leq \delta'$) small enough in terms of α, σ, κ, we have

$$\log \prod_{s\in\mathcal{S}'} R'_s \leq (\tau^*(\alpha) - 2\kappa)m_2.$$

Let $\mathcal{S}_1 = \{s \in \mathcal{S}' : \log R'_s \leq (\tau^*(\alpha) - \kappa)D\}$. Recall that our goal is to show (9.3.11), i.e., $|\mathcal{S}_1| \geq \gamma(q, \alpha, \kappa)\ell$. We have

$$D|\mathcal{S}' \setminus \mathcal{S}_1| \leq \frac{1}{\tau^*(\alpha) - \kappa} \sum_{s\in\mathcal{S}'\setminus\mathcal{S}_1} \log R'_s \leq \frac{\tau^*(\alpha) - 2\kappa}{\tau^*(\alpha) - \kappa} D|\mathcal{S}'|,$$

so that, using the rightmost inequality in (9.3.15), and recalling that $D|\mathcal{S}'| = m - m_1$,

$$D|\mathcal{S}_1| \geq \frac{\kappa(m-m_1)}{\tau^*(\alpha)-\kappa} \geq \left(\frac{\kappa(1-(\tau(q)+\xi)/(q-1)-\delta)}{\tau^*(\alpha)-\kappa}\right) m.$$

By ensuring that δ, ξ are small enough in terms of q, the right-hand side above can be bounded below by

$$\left(\frac{\kappa(1-\tau(q)/(q-1))/2}{\tau^*(\alpha)-\kappa}\right) m,$$

confirming that (9.3.11) holds with $\gamma = \gamma(q, \alpha, \kappa)$.

9.3.7 *Proof of Theorem 3.1*

Theorem 3.1 will be an easy consequence of the following proposition, which relies on Theorem 3.2. It is an analog of [12, Theorem 1.4], and we follow a similar outline. We emphasize that exponential separation is not required for the validity of the proposition.

Proposition 12 *Let $q \in (1, \infty)$ be such that τ is differentiable at q and $\tau(q) < q-1$. Fix $R \in \mathbb{N}$. Then*

$$\lim_{n\to\infty} \frac{\log \|\mu_n^{(Rm(n))}\|_q^q}{n \log \lambda} = \tau(q),$$

where $m(n)$ is the smallest integer with $2^{-m(n)} \leq \lambda^n$.

Proof Fix $n \in \mathbb{N}$. We write $m = m(n)$ for simplicity and allow all implicit constants to depend on q only. Using the self-similarity relation $\mu = \mu_n * S_{\lambda^n}\mu$ and Lemma 5, we get

$$\begin{aligned}
\|\mu^{((R+1)m)}\|_q^q &\leq O(1)\|\mu_n^{((R+1)m)} * (S_{\lambda^n}\mu)^{((R+1)m)}\|_q^q \\
&= O(1)\Big\| \sum_{J\in\mathcal{D}_m} \mu_n(J)(\mu_n)_J^{((R+1)m)} * (S_{\lambda^n}\mu)^{((R+1)m)} \Big\|_q^q.
\end{aligned}$$

Here $(\mu_n)_J = \mu_n|_J/\mu_n(J)$ is the normalized restriction of μ_n to J (note that we are only summing over J such that $\mu_n(J) > 0$). Since the measures $(\mu_n)_J^{((R+1)m)} * (S_{\lambda^n}\mu)^{((R+1)m)}$ are supported on $J + [0, \lambda^n]$, the support of each of them hits the supports of $O(1)$ others. We can then apply Lemma 4 to obtain

$$\|\mu^{((R+1)m)}\|_q^q \leq O(1) \sum_{J\in\mathcal{D}_m} \mu_n(J)^q \|(\mu_n)_J^{((R+1)m)} * (S_{\lambda^n}\mu)^{((R+1)m)}\|_q^q$$

Let $\rho_J = S_{\lambda^{-n}}(\mu_n)_J$ (we suppress the dependence on n from the notation, but keep it in mind). Note that $S_a(\eta) * S_a(\eta') = S_a(\eta * \eta')$ for any $a > 0$ and measures η, η'. It follows from Lemmas 3 and 5 that

$$\|(\mu_n)_J^{((R+1)m)} * (S_{\lambda^n}\mu)^{((R+1)m)}\|_q^q \le O(1)\|\rho_J^{(Rm)} * \mu^{(Rm)}\|_q^q,$$

so that, combining the last two displayed formulas,

$$\|\mu^{((R+1)m)}\|_q^q \le O(1) \sum_{J\in\mathcal{D}_m} \mu_n(J)^q \|\rho_J^{(Rm)} * \mu^{(Rm)}\|_q^q. \tag{9.3.16}$$

On the other hand, using Lemma 3 again,

$$\|\mu_n^{((R+1)m)}\|_q^q = \sum_{J\in\mathcal{D}_m} \mu_n(J)^q \|(\mu_n)_J^{((R+1)m)}\|_q^q \ge \Omega(1) \sum_{J\in\mathcal{D}_m} \mu_n(J)^q \|\rho_J^{(Rm)}\|_q^q. \tag{9.3.17}$$

Fix $\sigma > 0$, and let $\mathcal{D}' = \{J \in \mathcal{D}_m : \|\rho_J^{(Rm)}\|_q^q \le 2^{-\sigma m}\}$. According to Theorem 3.2, there is $\mathrm{e} = \mathrm{e}(\sigma, q) > 0$ such that, if n is taken large enough, then

$$J \in \mathcal{D}' \implies \|\rho_J^{(Rm)} * \mu^{(Rm)}\|_q^q \le 2^{-(\tau(q)+\mathrm{e})Rm}.$$

Applying this to (9.3.16), we get

$$\begin{aligned}\|\mu^{((R+1)m)}\|_q^q &\le O(1)2^{-(\tau(q)+\mathrm{e})Rm} \sum_{J\in\mathcal{D}'} \mu_n(J)^q + O(1) \sum_{J\notin\mathcal{D}'} \mu_n(J)^q \|\mu^{(Rm)}\|_q^q \\ &\le O(1)2^{-(\tau(q)+\mathrm{e})Rm}\|\mu^{(m)}\|_q^q + O(1)\|\mu^{(Rm)}\|_q^q \sum_{J\notin\mathcal{D}'} \mu_n(J)^q\end{aligned}$$

using Young's inequality in the first line, and Lemma 6 in the second. On the other hand,

$$2^{-(\tau(q)+\mathrm{e})Rm}\|\mu^{(m)}\|_q^q \le 2^{-\mathrm{e}m/2}\|\mu^{((R+1)m)}\|_q^q$$

if n is large enough (depending on R). Inspecting the last two displayed equations, we deduce that if $n \gg_\sigma 1$, then

$$\sum_{J\notin\mathcal{D}'} \mu_n(J)^q \ge \Omega(1)\frac{\|\mu^{((R+1)m)}\|_q^q}{\|\mu^{(Rm)}\|_q^q} \ge 2^{-m(\tau(q)+\sigma)}.$$

Recalling (9.3.17), we conclude that

$$\begin{aligned}\|\mu_n^{((R+1)m)}\|_q^q &\ge \Omega(1) \sum_{J\notin\mathcal{D}'} \mu_n(J)^q \|\rho_J^{(Rm)}\|_q^q \\ &\ge \Omega(1)2^{-\sigma m} \sum_{J\notin\mathcal{D}'} \mu_n(J)^q \ge \Omega(1)2^{-2\sigma m}2^{-m\tau(q)}.\end{aligned}$$

The inequality $\|\mu_n^{((R+1)m)}\|_q^q \le \|\mu_n^{(m)}\|_q^q$ holds trivially, so that by Lemma 6

$$\|\mu_n^{((R+1)m)}\|_q^q \le \|\mu_n^{(m)}\|_q^q \le 2^{\sigma m} 2^{-m\tau(q)},$$

provided $n \gg_\sigma 1$. Since $\sigma > 0$ was arbitrary and $2^{-m} = \Theta(\lambda^n)$, this concludes the proof.

We can now conclude the proof of Theorem 3.1.

Proof of Theorem 3.1 We continue to write $m = m(n) = \lceil n \log(1/\lambda) \rceil$. To begin, we note that, for any $q \in (1, \infty)$,

$$\|\mu_n^{(m)}\|_q^q \ge \|\mu_n\|_q^q \ge \|\Delta\|_q^{qn}. \tag{9.3.18}$$

(The latter inequality is an equality if and only if there are no overlaps among the atoms of μ_n.) Since $\|\nu^{(m)}\|_q^{q'} \ge 2^{-m}$ for any probability measure ν, it follows from (9.3.18) and Lemma 6 that

$$D(q) \le \min(dim_S(\mu, q), 1).$$

Hence, the proof will be completed if we can show that for each $q \in (1, \infty)$, either $\tau(q) \ge q - 1$ (so that in fact $\tau(q) = q - 1$) or

$$\tau(q) = \log \|\Delta\|_q^q. \tag{9.3.19}$$

Since $\tau(q)$ is concave, it is enough to prove this for all q such that τ is differentiable at q. Hence, we fix q such that $\tau(q) < q - 1$ and τ is differentiable at q, and we set out to prove (9.3.19).

By the exponential separation assumption, the atoms of μ_n are λ^{Rn}-separated for infinitely many n and some $R \in \mathbb{N}$. We know from Proposition 12 that

$$\lim_{n\to\infty} \frac{\log \|\mu_n^{(Rm(n))}\|_q^q}{n \log \lambda} = \tau(q). \tag{9.3.20}$$

On the other hand, if n is such that the atoms of μ_n are λ^{Rn}-separated, then (since $\lambda^{Rn} \ge 2^{-Rm(n)}$)

$$\|\mu_n^{(Rm(n))}\|_q^q = \|\mu_n\|_q^q = \|\Delta\|_q^{qn}. \tag{9.3.21}$$

Combining Eqs. (9.3.20) and (9.3.21), we conclude that (9.3.19) holds, finishing the proof.

9.3.8 *About the Proof of Theorem 1.2*

In the proof of Theorem 3.1, the convolution structure played a crucial role. While a general self-similar measure does not have such a clean convolution structure, we can proceed as follows. Let $(\lambda_i)_{i\in I}$ be the scaling factors of the IFS generating μ (there may be repetitions). Given m, let

$$\Omega_m = \{(j_1 \ldots j_k) : \lambda_{j_1} \cdots \lambda_{j_k} \le 2^{-m} < \lambda_{j_1} \cdots \lambda_{j_{k-1}}\},$$
$$\Lambda_m = \{\lambda_{j_1} \cdots \lambda_{j_k} : (j_1 \ldots j_k) \in \Omega_m\}.$$

One can then check, using self-similarity, that

$$\mu = \sum_{\lambda\in\Lambda_m} \mu_{\lambda,m} * S_\lambda \mu,$$

where $\mu_{\lambda,m}$ are certain purely atomic measures constructed from the translations of the maps $f_{j_1} \cdots f_{j_k}$ with $\lambda_{j_1} \cdots \lambda_{j_k} = \lambda$. Thanks to the fact that $|\Lambda_m|$ is polynomial in m (even though $|\Omega_m|$ is exponential in m), the proof given in the homogeneous case can be adapted with minor technical complications. We refer to [19, Sect. 6.4] for the details.

9.4 Applications

In this section, we present several applications of Theorem 3.1.

9.4.1 *Frostman Exponents*

If μ is a finite measure on a metric space X, we say that μ has *Frostman exponent* s if $\mu(B(x, r)) \le C\, r^s$ for some $C > 0$ and all $x \in X$, $r > 0$. There is a very simple relation between L^q dimensions for large q and Frostman exponents.

Lemmash 13 *Let $\mu \in \mathcal{P}$. If $D_\mu(q) > s$ for some $q \in (1, \infty)$, then there is $r_0 > 0$ such that*

$$\mu(B(x, r)) \le r^{(1-1/q)s} \text{ for all } x \in \mathbb{R}, r \in (0, r_0].$$

Proof If $D(\mu, q) > s$, then there is $s' > s$ such that for all large enough m and each $J' \in \mathcal{D}_m$,

$$\mu(J')^q \le \sum_{J\in\mathcal{D}_m} \mu(J)^q \le 2^{-m(q-1)s'}.$$

Since any ball can be covered by $O(1)$ dyadic intervals of size smaller than the radius, we get that if r is sufficiently small then

$$\mu(B(x, r)) \leq C\, r^{(1-1/q)s'},$$

where C is independent of x and r. This gives the claim.

Theorem 3.1 together with the previous lemma immediately yields the following corollary.

Corollarysh 14 *Let μ be the self-similar measure associated to a homogeneous IFS $(\lambda x + t_i)_{i\in I}$ with exponential separation and the uniform probability weights $(1/|I|, \ldots, 1/|I|)$. Then μ has Frostman exponent s for every $s < \min(\log |I|/\log (1/\lambda), 1)$.*

9.4.2 Algebraic Parameters

We now discuss the special case in which the IFS has algebraic parameters, that is, both the contraction ratio λ and the translations t_i are algebraic numbers. Hochman [12, Corollary 1.5] proved that the overlaps conjecture holds in this case and, in the same way, we extend this to the L^q-dimension version of the overlaps conjecture. The deduction is based on the following classical lemma; see [13, Lemma 6.30] for a proof.

Lemmash 15 *Given algebraic numbers (over $\mathbb{Q}$) $\alpha_1, \ldots, \alpha_k$ and a positive integer h, there exists $\delta > 0$ such that the following holds: if $P \in \mathbb{Z}[x_1, \ldots, x_k]$ is a polynomial of degree n, all of whose coefficients are at most h in modulus, then either $P(\alpha_1, \ldots, \alpha_k) = 0$ or*

$$|P(\alpha_1, \ldots, \alpha_k)| \geq \delta^n.$$

Corollarysh 16 *Let μ be the self-similar measure associated to a homogeneous WIFS with algebraic coefficients (i.e., the contraction ratio and the translations are algebraic). Then either there is an exact overlap or*

$$D_q(\mu) = \min(\dim_S(\mu, q), 1) \quad \text{for all } q > 1.$$

Proof Note that for any pair of sequences $i = (i_1, \ldots, i_n), j = (j_1, \ldots, j_n)$, the difference $f_i(0) - f_j(0)$ can be written as $P_{i,j}(\lambda, t_1, \ldots, t_{|I|})$, where $P_{i,j} \in \mathbb{Z}[x_1, \ldots, x_{|I|+1}]$ has degree at most $n + 1$ and coefficients ± 1. Since we assume that there are no exact overlaps, $P_{i,j}(\lambda, t_1, \ldots, t_{|I|}) \neq 0$ for $i \neq j$. Lemma 15 then guarantees that the IFS has exponential separation, so that Theorem 3.1 yields the corollary.

Even if there are exact overlaps, the proof of Theorem 3.1 yields an expression for the L^q dimensions of μ. Recall that μ_n is the purely atomic measure given by

$$\mu_n = *_{j=0}^{n-1} S_{\lambda^j} \Delta = \sum_{u \in I^n} p_{u_1} \cdots p_{u_n} \delta_{f_u(0)}.$$

Corollarysh 17 *Let μ be the self-similar measure associated to a homogeneous WIFS with algebraic coefficients (i.e., the contraction ratio and the translations are algebraic). Define*

$$T_\mu = \lim_{n\to\infty} -\frac{1}{n} \log \|\mu_n\|_q^q.$$

Then the limit in this definition exists, and

$$D_q(\mu) = \min\left(\frac{T_\mu}{(q-1)\log(1/\lambda)}, 1\right).$$

Proof By Lemma 15, and arguing as in the proof of Corollary 16, there is $R \in \mathbb{N}$ such that any two distinct atoms of μ_n are λ^{Rn}-separated. Suppose $D_q(\mu) < 1$. By Proposition 12,

$$\lim_{n\to\infty} \frac{\log \|\mu_n^{(Rm(n))}\|_q^q}{n \log \lambda} = \tau(q).$$

But $\|\mu_n^{(Rm(n))}\|_q^q = \|\mu_n\|_q^q$ since $2^{m(n)} \le \lambda^n$, so the claim follows.

9.4.3 Parametrized Families and Absolute Continuity

Exponential separation holds outside of a small set of exceptions in parametrized families satisfying mild regularity and non-degeneracy assumptions.

Lemmash 18 *Let $J \subset \mathbb{R}$ be a compact interval, and let $\lambda : J \to (-1, 0) \cup (0, 1)$ and $t_1, \ldots, t_\ell : J \to \mathbb{R}$ be real-analytic functions. For a pair of $\{1, \ldots, \ell\}$-valued sequences i, j, define*

$$g_{i,j}(u) = \sum_{k=0}^{\infty} \lambda(u)^k t_{i_k}(u) - \sum_{k=0}^{n-1} \lambda(u)^k t_{j_k}(u).$$

Assume that if $i \neq j$ then $g_{i,j}$ is not identically zero. Then the IFS $\{\lambda(u)x + t_i(u)\}_{i=1}^{\ell}$ has exponential separation for all u outside of a set $E \subset J$ of zero Hausdorff (and even packing) dimension.

See [12, Theorem 1.8] for the proof and some further discussion. Now Theorem 3.1 shows that for parametrized families of WIFS satisfying the assumptions of Lemma 18, there is a zero-dimensional exceptional set of parameters outside of which the L^q dimensions of the self-similar measures have the value predicted by the overlaps conjecture (note also that the exceptional set is independent of the probability weights).

When $dim_S(\mu, q) > 1$, the overlaps conjecture predicts that (in the absence of exact overlaps) $D_\mu(q) = 1$, but in fact it is plausible that under the same assumptions the measure μ is absolutely continuous with an L^q density. While this is of course still open and appears to be even harder than the dimension version of the overlaps conjecture, we have the following result for parametrized families.

Theoremsh 4.1 *Let $J \subset \mathbb{R}$ be an closed interval, and let $\lambda : J \to (-1, 0) \cup (0, 1)$ and $t_1, \ldots, t_\ell : J \to \mathbb{R}$ be real-analytic functions. For a pair of $\{1, \ldots, \ell\}$-valued sequences i, j, define*

$$g_{i,j}(u) = \sum_{k=0}^{\infty} \lambda(u)^k t_{i_k}(u) - \sum_{k=0}^{\infty} \lambda(u)^k t_{j_k}(u).$$

Assume that if $i \neq j$ then $g_{i,j}$ is not identically zero. Then, there is a set $E \subset J$ of zero Hausdorff dimension such that the following holds for all $u \in J \setminus E$: if μ is a self-similar measure associated to the IFS $(\lambda(u)x + t_i(u))_{i=1}^{\ell}$ and a probability vector $(p_i)_{i=1}^{\ell}$, and if $dim_S(\mu, q) > 1$ for some $q \in (1, \infty)$, then μ is absolutely continuous and its Radon–Nikodym density is in L^q.

This theorem provides the correct range for the possibility of having an L^q density (up to the endpoint), since measures μ with an L^q density satisfy $D(\mu, q) = 1$; this follows from the inequality $(\int_J f)^q \leq |J|^{q-1} \int_J f^q$ for all intervals J, where f is the L^q density of μ. The proof of the theorem follows the ideas from [18, 20]; the only new element is the stronger input provided by Theorem 3.1.

Recall that the Fourier transform of a measure $\rho \in \mathcal{P}$ is defined as

$$\widehat{\rho}(\xi) = \int \exp(2\pi i x \xi)\, d\rho(x).$$

The following result asserts that convolving a measure of full L^q dimension and another measure with power Fourier decay results in an absolutely continuous measure with an L^q density; see [20, Theorem 4.4] for the proof.

Theoremsh 4.2 *Let $\nu, \rho \in \mathcal{P}$ be such that $D_\nu(q) = 1$ for some $q > 1$ and ρ satisfies the Fourier decay estimate*

$$|\widehat{\rho}(\xi)| \leq C|\xi|^{-\delta}$$

for some $C, \delta > 0$. Then the convolution $\nu * \rho$ is absolutely continuous and its Radon–Nikodym density is in L^q.

The proof of this theorem shows that, additionally, $\nu * \rho$ has fractional derivatives in L^q.

Proof Proof of Theorem 4.1 Fix a weight $(p_1, \ldots, p_\ell)$ for some $\ell \geq 2$. For $u \in J$, let μ_u be the self-similar measure associated with the WIFS $(\lambda(u)x + t_i(u), p_i)_{i=1}^{\ell}$. We also denote

$$\Delta(u) = \sum_{i=1}^{\ell} p_i \, \delta_{t_i(u)}.$$

Fix $k \in \mathbb{N}$. Using the convolution structure of μ_u, we decompose

$$\mu_u = \left(*_{k\nmid j} S_{\lambda(u)^j} \Delta(u)\right) * \left(*_{k\mid j} S_{\lambda(u)^j} \Delta(u)\right) =: \nu_u^{(k)} * \rho_u^{(k)}. \tag{9.4.1}$$

We can think of $\rho_u^{(k)}$ and $\nu_u^{(k)}$ as the measures obtained from the construction of μ_u by "keeping only every k-th digit" and "skipping every k-th digit," respectively. Both $\rho_u^{(k)}$ and $\nu_u^{(k)}$ are, again, self-similar measures arising from homogeneous IFSs. Indeed, $\rho_u^{(k)}$ is the invariant measure for the IFS $(\lambda(u)^k x + t_i(u), p_i)_{i=1}^{\ell}$. The WIFS generating $\nu^{(k)}$ is more cumbersome to write down: it consists of ℓ^{k-1} maps, indexed by sequences $i \in \{1, \ldots, \ell\}^{k-1}$. The maps and weights are given by

$$g_{u,i}(x) = \lambda(u)^k(x) + \sum_{j=0}^{k-2} t_{i_{j+1}} \lambda(u)^j,$$

$$p_i = p_{i_1} \cdots p_{i_{k-1}}.$$

A short calculation shows that, for any $q > 1$,

$$dim_S(\nu_u^{(k)}, q) = (1 - 1/k) dim_S(\mu_u, q). \tag{9.4.2}$$

On the other hand, it is easy to check that (for each k) the family of IFSs generating $\nu_u^{(k)}$ also satisfies the assumptions of Lemma 18. Hence, there are sets E'_k of zero Hausdorff dimension such that the WIFS generating $\nu_u^{(k)}$ has exponential separation for all $u \in J \setminus E'_k$. Letting $E' = \cup_k E'_k$ and applying Theorem 3.1, we deduce that E' has zero Hausdorff dimension, and if $u \in J \setminus E$ then

$$D(\nu_u^{(k)}, q) = \min((1 - 1/k) dim_S(\mu_u), q) \quad \text{for all } k \in \mathbb{N}.$$

Turning to the measures $\rho_u^{(k)}$, we claim that there are exceptional sets E''_k of zero Hausdorff dimension such that if $u \in J \setminus E''_k$, then $\rho_u^{(k)}$ has power Fourier decay, that is, there are $C(u,k), \delta(u,k) > 0$ such that

$$|\widehat{\rho_u^{(k)}}(\xi)| \leq C(u,k)|\xi|^{-\delta(u,k)}.$$

This follows by variants of an argument that goes back to Erdnős [9]. If the function $\lambda(u)$ is nonconstant then, by splitting J into finitely many intervals and reparametrizing, we may assume that $\lambda(u) = u$. This case is closer to Erdős original argument; see, e.g., [18, Proposition 2.3] for a detailed exposition. Suppose now that $\lambda(u) \equiv \lambda$. In this case, we must have $\ell \geq 3$. Indeed, suppose $\ell = 2$. Replacing $t_1(u)$ by 0 and $t_2(u)$ by 1 has the effect of rescaling and translating the measures μ_u, which does not affect the claim. If $|\lambda| < 1/2$, then $dim_S(\mu, q) < 1$ for any q and there is nothing to do, while if $|\lambda| \geq 1/2$, there are two sequences $i, j \in \{0, 1\}^{\mathbb{N}}$ such that $\sum_{k=0}^{\infty}(i_k - j_k)\lambda^k = 0$, and this implies that the non-degeneracy assumption fails. Hence, we assume that $\ell \geq 3$ from now on. In this case, the function

$$h(u) = \frac{t_3(u) - t_1(u)}{t_2(u) - t_1(u)}$$

is nonconstant and real-analytic outside of a finite set of $u \in J$ (where the denominator vanishes). Otherwise, if either the denominator or $h(u)$ itself were constant, the non-degeneracy condition would fail. As before, this shows that we may assume $h(u) = u$. The claim now follows from [20, Proposition 3.1]. Let $E'' = \cup_k E''_k$.

Set $E = E' \cup E''$ and fix $u \in J \setminus E$. If $dim_S(\mu, q) > 1$, then (9.4.2) ensures that $D(\nu_u^{(k)}, q) = 1$ provided k is taken large enough. Since also $\rho_u^{(k)}$ has power Fourier decay by the definition of $E'' \subset E$, the decomposition (9.4.1) together with Theorem 4.2 shows that μ_u is absolutely continuous with an L^q density, finishing the proof.

9.4.4 Bernoulli Convolutions

Given $\lambda \in (0, 1)$, we define μ_λ as the distribution of the random sum $\sum_{n=0}^{\infty} X_n \lambda^n$, where the X_n are IID and take values 0 and 1 with equal probability 1/2. In other words, μ_λ is the self-similar measure associated to the WIFS $(\lambda x, 1/2), (\lambda x + 1, 1/2)$. The measures μ_λ are known as *Bernoulli convolutions*.

When $\lambda \in (0, 1/2)$, the topological support of μ_λ is a self-similar Cantor set of dimension $\log 2/\log(1/\lambda) < 1$; in particular, μ_λ is purely singular (and $D(\mu_\lambda, q) = \log 2/\log(1/\lambda)$ for all q). For $\lambda = 1/2$, the Bernoulli convolution μ_λ is a multiple of Lebesgue measure on the interval $[0, 1/(1-\lambda)]$. Understanding the smoothness properties of μ_λ for $\lambda \in (1/2, 1)$ has been a major open problem since the 1930s. Although the problem is still very much open, dramatic progress has been achieved in the last few years. In this section, we briefly state the consequences of the results of the previous sections for Bernoulli convolutions and discuss their connections with other old and new results about them.

In two foundational papers, Erdős [8, 9] showed that μ_λ is singular if $1/\lambda$ is a Pisot number (an algebraic integer >1 all of whose algebraic conjugates are <1 in modulus), and that μ_λ has a density in C^k for almost all λ sufficiently close to 1 (depending on k). In the 1960s, Garsia [11] exhibited an explicit infinite family of algebraic numbers λ for which ν_λ is absolutely continuous. These remained the only explicit known parameters of absolute continuity until very recently when Varju [22], introducing several new techniques, exhibited a new large family of algebraic numbers very close to 1 for which μ_λ is absolutely continuous, with a density in $L \log L$.

In a celebrated paper, Solomyak [21] proved that μ_λ is absolutely continuous with an L^2 density for almost all $\lambda \in (1/2, 1)$. Much more recently, in another landmark paper [12] that we have already encountered several times, Hochman proved that $dim_H(\mu_\lambda) = 1$ for all λ outside of a set of λ of zero Hausdorff (and even packing) dimension. Building on that, the author [18] proved that μ_λ is absolutely continuous for all λ outside of a set of λ of zero Hausdorff dimension. As an immediate application of Theorem 4.1, we have the following corollary.

Corollarysh 19 *There exists a set $E \subset (1/2, 1)$ of zero Hausdorff dimension such that ν_λ is absolutely continuous and its density is in L^q for all $q \in (1, \infty)$, for all $\lambda \in (1/2, 1) \setminus E$.*

We underline that the information that the density is in L^q for $q > 2$ is new even for a.e. parameter. Note that Corollary 14 shows that μ_λ has Frostman exponent $1 - \text{e}$ for every $\text{e} > 0$ for every λ for which there is exponential separation. Although this is weaker than L^q density for all $q > 1$, exponential separation can be checked for some explicit parameters; in particular, it holds for all rationals in $(1/2, 1)$.

An active area of research concerns investigating the properties of μ_λ for algebraic values of λ. We only summarize some of the recent results in this area. The entropy of a purely atomic measure ν is defined as $H(\nu) = \sum_x \nu(x) \log(1/\nu(x))$. The *Garsia entropy* associated to μ_λ is defined as

$$h_\lambda = \lim_{n\to\infty} \frac{1}{n} H(\mu_{\lambda,n}),$$

where $\mu_{\lambda,n}$ is the nth step discrete approximation to μ_λ, that is, the distribution of the finite random sum $\sum_{j=0}^{n-1} X_j \lambda^j$. It is well known that the limit exists. The number h_λ can also be interpreted as the entropy of the uniform random walk generated by the similarities λx and $\lambda x + 1$.

It follows from Hochman's work [12] (see [6, Sect. 3.4] for a detailed argument) that if λ is algebraic, then

$$dim_H(\mu_\lambda) = \min\left(\frac{h_\lambda}{\log(1/\lambda)}, 1\right). \tag{9.4.3}$$

Breuillard and Varju [6, Theorem 5] gave bounds for h_λ in terms of the Mahler measure M_λ of λ (see, e.g., [6, Eq. (1.1)] for the definition of Mahler measure):

$$c \min(1, \log M_\lambda) \le h_\lambda \le \min(1, \log M_\lambda), \tag{9.4.4}$$

where $c > 0$ is a universal constant that they numerically estimate to be at least 0.44. Using this theorem, they uncover a connection between Bernoulli convolutions and problems related to growth rates in linear groups. Very roughly, the idea is that the worst possible rate occurs for the group generated by the similarities λx, $\lambda x + 1$, which can be easily realized as a linear group. An easy consequence of (9.4.3) and (9.4.4) is that, assuming Lehmer's conjecture that the Mahler measure M_λ is either 1 or bounded away from 1, the Hausdorff dimension of μ_λ is 1 for all algebraic numbers which are close enough to 1. In [5], further progress was obtained; among many other results, the authors show that if $dim_H(\mu_\lambda) < 1$ for some transcendental number λ, then λ can be approximated by algebraic numbers with the same property. Hence, conditional on the Lehmer conjecture, $dim_H(\mu_\lambda) = 1$ for all λ close to 1. Very recently, combining results from most of the papers mentioned in this section with a clever new argument, Varju [23] achieved another major breakthrough by proving that $dim_H(\mu_\lambda) = 1$ for all transcendental $\lambda \in (1/2, 1)$.

The formula (9.4.3) makes it important to be able to compute Garsia entropy. An algorithm for this was developed in [2]. Among other applications, this algorithm makes it possible to check that $dim_H(\mu_\lambda) = 1$ for specific (new) algebraic values of λ.

All of these recent advances depend on the formula (9.4.3), and hence apply only to Hausdorff dimension and not to L^q dimensions. However, Corollary 17 shows that the L^q version of (9.4.3) remains valid: for all algebraic $\lambda \in (1/2, 1)$,

$$D(\mu_\lambda, q) = \min\left(\frac{T_{q,\lambda}}{(q-1)\log(1/\lambda)}, 1\right),$$

where

$$T_{q,\lambda} = \lim_{n\to\infty} -\frac{1}{n}\log\|\mu_{\lambda,n}\|_q^q$$

is an L^q analog of Garsia entropy. Hence, it would be interesting to know if there are L^q versions of some of the results described above.

9.4.5 *Intersections of Cantor Sets*

To finish the paper, we show how Theorem 3.1 can be used to obtain strong bounds on the dimensions of intersections of certain Cantor sets. Indeed, a conjecture of Furstenberg about the dimensions of intersections of $\times 2$, $\times 3$-invariant closed subsets of the circle was the main motivation for the results of [19]. While the resolution of Furstenberg's intersection conjecture requires a more general version of Theorem 3.1 and is therefore beyond the scope of this survey, we will still be able to derive other intersection bounds.

In the following simple lemma, we show how Frostman exponents (and therefore, by Lemma 13, also L^q dimensions) of projected measures give information about the size of fibers. We recall the definition of upper box-counting (or Minkowski) dimension in a totally bounded metric space (X, d). Given $A \subset X$, let $N_\varepsilon(A)$ denote the maximal cardinality of an ε-separated subset of A. The upper box-counting dimension of A is then defined as

$$\overline{\dim}_B(A) = \limsup_{\varepsilon\downarrow 0} \frac{\log(N_\varepsilon(A))}{\log(1/\varepsilon)}.$$

Lemmash 20 *Let X be a compact metric space, and suppose $\pi : X \to \mathbb{R}$ is a Lipschitz map. Let μ be a probability measure on X such that $\mu(B(x,r)) \geq r^s$ for all $x \in X$ and all sufficiently small r (independent of x). If $\pi\mu$ has Frostman exponent α, then there exists $C > 0$ such that for all balls B_ε of radius ε in $\mathbb{R}$, any ε-separated subset of $\pi^{-1}(B_\varepsilon)$ has size at most $C\varepsilon^{-(s-\alpha)}$.*

In particular, for any $y \in \mathbb{R}$,

$$\overline{\dim}_B(\pi^{-1}(y)) \leq s - \alpha$$

Proof Let $(x_j)_{j=1}^M$ be an ε-separated subset of $\pi^{-1}(B_\varepsilon)$ with ε small. Then

$$\mu\left(\bigcup_{j=1}^M B(x_j, \varepsilon/2)\right) \geq M(\varepsilon/2)^s,$$

while the set in question projects onto an interval of size at most $O(\varepsilon)$. Hence, $M = O(\varepsilon^{\alpha-s})$, giving the claim.

We give one concrete application of Theorem 3.1 in conjunction with this lemma, and refer to [19, Sect. 6.3] for further examples. Let $p \geq 2$ be an integer, and let $D \subset \{0, 1, \ldots, p-1\}$ be a proper subset. Let $A = A_{p,D}$ be the set of $[0, 1]$ consisting of all points whose p-ary expansion has only digits from D. This is the self-similar set associated to the IFS $((x + j)/p : j \in D)$. For example, the middle-thirds Cantor set is the case $p = 3$, $D = \{0, 2\}$. We call such a set a *p-Cantor set*.

Corollarysh 21 *Let $A \subset [0, 1)$ be a p-Cantor set, $p \geq 2$. Then for every irrational number $t \in \mathbb{R}$ and any $u \in \mathbb{R}$,*

$$\overline{\dim}_B(A \cap (tA + u)) \leq \max(2dim_H(A) - 1, 0).$$

Proof Let $A = A_{p,D}$, and let μ be the uniform self-similar measure on A. Since the IFS generating A satisfies the open set condition, it is well known, and not hard to see, that $\mu(B(x, r)) = \Theta(r^s)$ for all $x \in A$, with the implicit constant depending only on p, D. Hence, the product measure $\mu \times \mu$ satisfies

$$(\mu \times \mu)(B(z, r)) = \Theta(r^{2s}) \tag{9.4.5}$$

for all $z \in A \times A = \text{supp}(\mu \times \mu)$.

Let $\Pi_t(x, y) = x + ty$. Then $\Pi_t(\mu \times \mu)$ is the uniform self-similar measure generated by the IFS

$$\left(p^{-1}(x + i + tj) : i, j \in D\right).$$

We claim that this IFS has exponential separation for all irrational t. Assuming the claim, the corollary follows by combining Corollary 14 and Lemma 20 (keeping (9.4.5) in mind).

The argument to establish exponential separation in this setting is due to B. Solomyak and the author, and was originally featured in [12, Theorem 1.6]. Fix $t \in \mathbb{R} \setminus \mathbb{Q}$. The separation number Γ_k associated to $\Pi_t \mu$ has the form $x_k + ty_k$, where x_k, y_k have the form $\sum_{j=0}^{k-1} a_j p^{-j}$ with $a_j \in D - D$. Moreover, x_k and y_k cannot be simultaneously 0 since this would imply an exact overlap in the IFS generating A. If either x_k or y_k are zero for infinitely many k, then $\Gamma_k \geq \min(1, t)p^{-k}$ for infinitely many k, and hence we are done. So assume $x_k y_k \neq 0$ for all $k \geq k_0$, and therefore

$$\left| \frac{\Gamma_k}{y_k} - \frac{\Gamma_{k+1}}{y_{k+1}} \right| = \left| \frac{x_k}{y_k} - \frac{x_{k+1}}{y_{k+1}} \right| = \left| \frac{z_k}{y_k y_{k+1}} \right|,$$

where $z_k = x_k y_{k+1} - x_{k+1} y_k$. If $z_k = 0$ for all $k \geq k_1$, then for all $k \geq k_1$ we have

$$\Gamma_k = |y_k(x_{k_1}/y_{k_1} + t)| \geq p^{-k-1}|x_{k_1}/y_{k_1} + t|$$

so, again using the irrationality of t, there is exponential separation. It remains to analyze the case $z_k \neq 0$ for infinitely many k. For any such k, the quotient $z_k/(y_k y_{k+1})$ is a nonzero rational number of denominator at most $4p^{2k+1}$. Since $|y_k| \leq 2$ for all k, we conclude that there are infinitely many k such that either $\Gamma_k \geq p^{-2k-1}/16$ or $\Gamma_{k+1} \geq p^{-2k-1}/16$. Thus exponential separation also holds in this case, finishing the proof.

For rational t, the behavior is completely different: it follows from [4, Theorem 1.2] that if $A = A_{p,D}$ is any p-Cantor set of dimension $> 1/2$, and $p \nmid |D|^2$ (in particular this holds if p is prime), then for every rational t there are many values of u such that

$$dim_H(A \cap tA + u) > 2dim_H(A) - 1.$$

More precisely, for a given t this holds for a typical u chosen according to the natural self-similar measure on $A \times A$.

References

1. C. Aistleitner, G. Larcher, M. Lewko, Additive energy and the Hausdorff dimension of the exceptional set in metric pair correlation problems. Israel J. Math. **222**(1), 463–485 (2017). With an appendix by Jean Bourgain
2. S. Akiyama, D.-J. Feng, T. Kempton, T. Persson, On the Hausdorff dimension of Bernoulli convolutions. Int. Math. Res. Not. IMRN (2018). arXiv:1801.07118
3. B. Bárány, M. Rams, K. Simon, On the dimension of triangular self-affine sets. Ergod. Theory Dyn. Syst. **39**(7), 1751–1783 (2019)
4. B. Bárány, M. Rams, Dimension of slices of Sierpiński-like carpets. J. Fractal Geom. **1**(3), 273–294 (2014)
5. E. Breuillard, P.P. Varjú, On the dimension of Bernoulli convolutions. Ann. Probab. **47**(4), 2582–2617 (2016)
6. E. Breuillard, P.P. Varjú, Entropy of Bernoulli convolutions and uniform exponential growth for linear groups. J. Anal. Math. (2018). arXiv:1510.04043

7. S. Dyatlov, J. Zahl, Spectral gaps, additive energy, and a fractal uncertainty principle. Geom. Funct. Anal. **26**(4), 1011–1094 (2016)
8. P. Erdős, On a family of symmetric Bernoulli convolutions. Am. J. Math. **61**, 974–976 (1939)
9. P. Erdős, On the smoothness properties of a family of Bernoulli convolutions. Am. J. Math. **62**, 180–186 (1940)
10. J.M. Fraser, T. Jordan, The Assouad dimension of self-affine carpets with no grid structure. Proc. Am. Math. Soc. **145**(11), 4905–4918 (2017)
11. A.M. Garsia, Arithmetic properties of Bernoulli convolutions. Trans. Am. Math. Soc. **102**, 409–432 (1962)
12. M. Hochman, On self-similar sets with overlaps and inverse theorems for entropy. Ann. of Math. **180**(2), 773–822 (2014)
13. M. Hochman, On self-similar sets with overlaps and inverse theorems for entropy in $\mathbb{R}^d$. Mem. Am. Math. Soc. (2017)
14. K.-S. Lau, S.-M. Ngai, Multifractal measures and a weak separation condition. Adv. Math. **141**(1), 45–96 (1999)
15. Y. Peres, B. Solomyak, Existence of L^q dimensions and entropy dimension for self-conformal measures. Indiana Univ. Math. J. **49**(4), 1603–1621 (2000)
16. Y. Peres, B. Solomyak, Problems on self-similar sets and self-affine sets: an update, in *Fractal Geometry and Stochastics, II (Greifswald/Koserow, 1998). Progress in Probability*, vol. 46 (Birkhäuser, Basel, 2000), pp. 95–106
17. E. Rossi, P. Shmerkin, On measures that improve L^q dimension under convolution. Rev. Mat. Iberoam (2018). arXiv:1812.05660
18. P. Shmerkin, On the exceptional set for absolute continuity of Bernoulli convolutions. Geom. Funct. Anal. **24**(3), 946–958 (2014)
19. P. Shmerkin, On Furstenberg's intersection conjecture, self-similar measures, and the L^q norms of convolutions. Ann. Math. **2**(2), 319–391 (2019)
20. P. Shmerkin, B. Solomyak, Absolute continuity of self-similar measures, their projections and convolutions. Trans. Am. Math. Soc. **368**(7), 5125–5151 (2016)
21. B. Solomyak, On the random series $\sum \pm\lambda^n$ (an Erdös problem). Ann. of Math. (2) **142**(3), 611–625 (1995)
22. P.P. Varjú, Absolute continuity of Bernoulli convolutions for algebraic parameters. J. Am. Math. Soc. **32**(2), 351–397 (2018)
23. P.P. Varjú, On the dimension of Bernoulli convolutions for all transcendental parameters. Ann. Math. (2) **189**(3), 1001–1011 (2019)

Chapter 10
Sample Paths Properties of the Set-Indexed Fractional Brownian Motion

Erick Herbin and Yimin Xiao

Abstract For $0 < H \leq 1/2$, let $\mathbf{B}^H = \{\mathbf{B}^H(t);\ t \in \mathbb{R}_+^N\}$ be the Gaussian random field obtained from the set-indexed fractional Brownian motion restricted to the rectangles of $\mathbb{R}_+^N$. We prove that $\mathbf{B}^H$ is tangent to a multiparameter fBm which is isotropic in the l^1-norm and we determine the Hausdorff dimension of the inverse image of $\mathbf{B}^H$ and its hitting probabilities. By applying the Lamperti transform and a Fourier analytic method, we show that $\mathbf{B}^H$ has the property of strong local nondeterminism (SLND) for $N = 2$. By applying SLND, we obtain the exact uniform and local moduli of continuity and Chung's law of iterated logarithm for $\mathbf{B}^H = \{\mathbf{B}^H(t);\ t \in \mathbb{R}_+^2\}$. These results show that, away from the axes of $\mathbb{R}_+^2$, the local behavior of $\mathbf{B}^H$ is similar to the ordinary fractional Brownian motion of index H.

Mathematical Subject Classification: 60G60 · 60G15 · 60G17 · 60G18 · 60G22 · 28A80 · 28A78.

10.1 Introduction

The set-indexed fractional Brownian motion was introduced in [12] as an extension of fractional Brownian motion. Let $\mathcal{T}$ be a metric space equipped with a Radon measure m and let $\mathcal{A}$ be a collection of compact subsets of $\mathcal{T}$, which forms an indexing collection (see [12, 14] for the definition). Then, for any constant $0 < H \leq 1/2$, the *set-indexed fractional Brownian motion (SI-fBm)* $\{\mathbf{B}_U^H;\ U \in \mathcal{A}\}$ of index H is a centered Gaussian process such that

E. Herbin
Département de Mathématiques, CentraleSupélec, Bâtiment Bouygues, 3 rue Joliot Curie, 91190 Gif-sur-Yvette, France
e-mail: erick.herbin@centralesupelec.fr

Y. Xiao (✉)
Department of Statistics and Probability, Michigan State University, 619 Red Cedar Road, East Lansing, MI, USA
e-mail: xiao@stt.msu.edu; xiaoy@msu.edu

A. Aldroubi et al. (eds.), *New Trends in Applied Harmonic Analysis, Volume 2*, Applied and Numerical Harmonic Analysis,
https://doi.org/10.1007/978-3-030-32353-0_10

$$\mathbf{E}\big[\mathbf{B}_U^H\mathbf{B}_V^H\big] = \frac{1}{2}\big[m(U)^{2H} + m(V)^{2H} - m(U \bigtriangleup V)^{2H}\big], \quad \forall U, V \in \mathcal{A}, \quad (10.1.1)$$

where $U \bigtriangleup V$ denotes the symmetric difference between the sets U and V. For $H > 1/2$, set-indexed fractional Brownian motion on a general indexing collection does not exist because, as pointed out by Herbin and Merzbach [12, p. 345], even for the simple case of $\mathcal{T} = \mathbb{R}_+^2$ equipped with the Lebesgue measure m, and the indexing collection $\mathcal{A} = \{[0, t];\ t \in \mathbb{R}_+^2\}$, the function on the right-hand side of (10.1.1) is not nonnegative definite if $H > 1/2$.

Various properties such as projections on flows, the stationarity of increments, and self-similarity properties of SI-fBm $\{\mathbf{B}_U^H;\ U \in \mathcal{A}\}$ have been studied in [13, 14]. The Hölder continuity and exponents, in the general framework of set-indexed processes of Ivanoff and Merzbach [16], have been studied in [15].

By taking the indexing collection to be $\mathcal{A} = \{[0, t];\ t \in \mathbb{R}_+^N\}$, the set of the rectangles in $\mathbb{R}_+^N$, and m the Lebesgue measure on $\mathbb{R}^N$, we obtain a Gaussian random field $\mathbf{B}^H = \{\mathbf{B}^H(t);\ t \in \mathbb{R}_+^N\}$, where $\mathbf{B}^H(t) = \mathbf{B}^H_{[0,t]}$ for all $t \in \mathbb{R}_+^N$, and for simplicity still call it set-indexed fractional Brownian motion (SI-fBm) of index H. The sample function of $\mathbf{B}^H$ is almost surely continuous. When $H = 1/2$, $\mathbf{B}^H$ is the Brownian sheet whose sample path properties have been studied extensively. We refer to [7, 18–22, 30] for further information. When $N = 1$, $\mathbf{B}^H$ reduces to the one-parameter fractional Brownian motion with index H. However, by checking the covariance functions, one can see that for $0 < H < \frac{1}{2}$ and $N > 1$, $\mathbf{B}^H$ is different from the two important fractional Gaussian fields with index H in the literature. These two fractional Gaussian fields with index H are the multiparameter fractional Brownian motion (or fractional Brownian field) $X^H = \{X^H(t),\ t \in \mathbb{R}^N\}$ and the fractional Brownian sheet $W^H = \{W^H(t),\ t \in \mathbb{R}^N\}$, respectively. The former is a centered Gaussian field with covariance function

$$\mathbf{E}\big[X^H(s)X^H(t)\big] = \frac{1}{2}\big(\|s\|^{2H} + \|t\|^{2H} - \|s - t\|^{2H}\big),$$

where $\|\cdot\|$ is the Euclidean norm on $\mathbb{R}^N$; and the latter is a centered Gaussian field with covariance

$$\mathbf{E}\big[W^H(s)W^H(t)\big] = \frac{1}{2^N}\prod_{j=1}^{N}\big(|s_j|^{2H} + |t_j|^{2H} - |s_j - t_j|^{2H}\big).$$

Note that both X^H and W^H are well defined for every $H \in (0, 1)$. Many authors have studied sample path properties of X^H and W^H. See, for example, [2–5, 23, 27–29, 32, 37, 40–43]. It is known that some fine properties of X^H and W^H such as Chung's laws of the iterated logarithm, Hölder conditions for the local times, and exact Hausdorff measure functions for their level sets are significantly different.

In this paper, we will focus on the case $0 < H \le \frac{1}{2}$ and $N > 1$. In this case, $\mathbf{B}^H$ loses some important properties of the Brownian sheet and fractional Brownian motion (e.g., independence of increments over disjoint intervals, or stationarity of increments) and there have only been a few papers that study its sample path properties. For instance, Herbin et al. [11] determined the Hausdorff dimension of the graph of $\mathbf{B}^H$ and Richard [33] studied recently the Chung-type law of the iterated logarithm at the origin.

The main purpose of this paper is to further study sample path properties of SI-fBm $\mathbf{B}^H$ and to compare them with those of fractional Brownian field X^H and the fractional Brownian sheet W^H.

The rest of paper is organized as follows. In Sect. 10.2, we apply the approach of Falconer [8] to study the tangent structure of SI-fBm $\mathbf{B}^H$. Our Theorem 1 shows that the tangent field of $\mathbf{B}^H$ at every point $t_0 \in (0, \infty)^N$ is a centered Gaussian field $Y = \{Y(t);\ t \in \mathbb{R}_+^N\}$ with covariance function

$$\mathbf{E}\big[Y(s)Y(t)\big] = \frac{1}{2}\left[\|s\|_1^{2H} + \|t\|_1^{2H} - \|s - t\|_1^{2H}\right], \quad \forall s, t \in \mathbb{R}_+^N,$$

where $\|.\|_1$ denotes the l^1-norm of $\mathbb{R}^N$. It is clear that the Gaussian field Y shares the properties of stationary increments and H-self-similarity with the ordinary fractional Brownian motion X^H, but it is not isotropic in the Euclidean metric (or, in other words, Y is not rotationally invariant). In Sect. 10.3, we determine the Hausdorff dimensions of the inverse images and the hitting probabilities of a d-dimensional SI-fBm $\mathbf{B}^H$. These results are similar to those of the d-dimensional analog of X^H and W^H. In Sect. 10.4 we take advantage of the multi-self-similarity (in the sense of [9]) of $\mathbf{B}^H$ and consider spectral analysis of the stationary Gaussian random field Z obtained via the Lamperti transform. In Sect. 10.5, we study the asymptotic behavior of the spectral measure of Z at infinity for the case of $N = 2$. In particular, we obtain precise asymptotic properties of the spectral density of Z at infinity. This later result, combined with Theorem 2.1 in [45], allows us to prove that the SI-fBm $\mathbf{B}^H$ has the property of strong local nondeterminism (SLND); see Theorem 8 With applications of the property of SLND, we establish the exact uniform and local modulus of continuity, Chung's law of the iterated logarithm of $\mathbf{B}^H$. Finally, at the end, Sect. 10.5, we remark that the SLND property can also be useful for determining the Fourier dimensions of the image and level sets of the d-dimensional SI-fBm $\mathbf{B}^H$, and for studying regularity properties of the local times of $\mathbf{B}^H$.

We mention that, in [33, 34], Richard proposed a more general framework for studying fractional Brownian fields including $\mathbf{B}^H$. He considered a fractional Brownian field indexed by the Hilbert space $L^2(m)$ in [34] and proved that it has the property of SLND. In [33], he derived a spectral representation for a large class of $L^2(m)$-indexed Gaussian processes and studied the small ball probability and the Chung-type law of the iterated logarithm of $\mathbf{B}^H$. Our arguments in Sects. 10.4 and 10.5 extend those in [39] and are different from those in [33, 34].

We end this section with some notations. An element $t \in \mathbb{R}^N$ is written as $t = (t_1, \ldots, t_N)$, or as $\langle t_i \rangle$. For any $s, t \in \mathbb{R}^N$, $s \prec t$ means such that $s_j < t_j$ $(j = 1, \ldots, N)$. If $s \prec t$, then $[s, t] = \prod_{j=1}^N [s_j, t_j]$ is called a closed interval (or a rectangle).

10.2 Tangent Structure of SI-fBm on Rectangles of $\mathbb{R}_+^N$

In the pioneering work [4, 31], the notion of *local asymptotic self-similarity* was introduced to characterize the evolution of local regularity and scaling structure of the multifractional Brownian motion. In [8], the more general issue of tangent structure of random fields was extensively studied.

For the set-indexed processes including SI-fBm $\{\mathbf{B}_U^H;\ U \in \mathcal{A}\}$, [12, 14] introduced the concept of increments over all sets in the semi-algebra

$$\mathcal{C} = \Big\{U \backslash \bigcup_{i=1}^{n} U_i \,:\, n \geq 1 \text{ and } U, U_i \in \mathcal{A} \text{ for all } 1 \leq i \leq n\Big\}.$$

They showed in Proposition 5.2 in [14] that $\{\mathbf{B}_U^H;\ U \in \mathcal{A}\}$ has m-stationary $\mathcal{C}_0$-increments. Namely, for all integers $n \geq 1$, $V \in \mathcal{A}$, and for all increasing sequences $\{U_i, 1 \leq i \leq n\}$ and $\{A_i, 1 \leq i \leq n\}$ in $\mathcal{A}$ that satisfy $m(U_i \backslash V) = m(A_i)$ for all $1 \leq i \leq n$, one has

$$\big(\Delta \mathbf{B}_{U_1\backslash V}^H, \ldots, \Delta \mathbf{B}_{U_n\backslash V}^H\big) \stackrel{d}{=} \big(\mathbf{B}_{A_1}^H, \ldots, \mathbf{B}_{A_n}^H\big),$$

where $\Delta \mathbf{B}_{U\backslash V}^H = \mathbf{B}_U^H - \mathbf{B}_{U\cap V}^H$ and $\stackrel{d}{=}$ means equality in distribution.

With our special choice of $\mathcal{A} = \{[0, t];\ t \in \mathbb{R}_+^N\}$, the family of increments of $\{\mathbf{B}_U^H;\ U \in \mathcal{A}\}$ includes all the increments of $\mathbf{B}^H = \{\mathbf{B}^H(t), t \in \mathbb{R}_+^N\}$ over rectangles defined as follows: For any $s, t \in \mathbb{R}^N$ such that $s_i \leq t_i$ for all $1 \leq i \leq N$,

$$\Delta_{s,t}\mathbf{B}^H := \sum_{r\in\{0,1\}^N} (-1)^{N-\sum_i r_i}\mathbf{B}^H(\langle s_i + r_i(t_i - s_i)\rangle). \tag{10.2.1}$$

Hence, the aforementioned m-stationarity of $\mathcal{C}_0$-increments of $\{\mathbf{B}_U^H;\ U \in \mathcal{A}\}$ in [12, 14] implies the stationarity of the increments of $\mathbf{B}^H = \{\mathbf{B}^H(t), t \in \mathbb{R}_+^N\}$ over rectangles.

In the following, we apply the approach of Falconer [8] to study the tangent structure of $\mathbf{B}^H$. We consider the increments of the form $\mathbf{B}^H(t) - \mathbf{B}^H(s)$ rather than increments over rectangles. The following result shows that $\mathbf{B}^H$ has a local behavior close to a multiparameter extension of fBm using the l^1-norm of $\mathbb{R}^N$. The tangent field Y in Theorem 1 is not rotationally invariant, but it can be regarded as isotropic in the l^1-norm. This result reveals some subtle local properties of $\mathbf{B}^H$.

For simplicity of presentation of Theorem 1, we only consider tangent field indexed by $[0, 1]^N$. There is no difficulty in replacing $[0, 1]^N$ by an arbitrary compact subset of $\mathbb{R}^N$ as in Falconer [8]. For the definition of weak convergence of probability measures on the space of continuous functions on $[0, 1]^N$ and its criteria, we refer to Khoshnevisan [18, pp. 193–201].

Theorem 1 *Let $\mathbf{B}^H = \{\mathbf{B}^H(t), t \in \mathbb{R}_+^N\}$ be a real-valued set-indexed fractional Brownian motion of index $H \in (0, 1/2]$. For any $t_0 \in (0, \infty)^N$ and $\rho > 0$, we define the Gaussian random field $X^{(\rho)} = \{X^{(\rho)}(u);\ u \in [0, 1]^N\}$ by*

$$X^{(\rho)}(u) = \frac{\mathbf{B}^H(t_0 + \rho\langle (t_0)_j\, u_j\rangle) - \mathbf{B}^H(t_0)}{\left(\rho \prod_{i=1}^N (t_0)_i\right)^H}; \quad (u \in [0, 1]^N).$$

Then, as ρ goes to 0, the random fields $X^{(\rho)}$ converge weakly to the centered Gaussian random field $Y = \{Y(u);\ u \in [0, 1]^N\}$ with the covariance function

$$\forall u, v \in [0, 1]^N; \quad \mathbf{E}\big[Y(u)Y(v)\big] = \frac{1}{2}\left[\|u\|_1^{2H} + \|v\|_1^{2H} - \|u - v\|_1^{2H}\right],$$

where $\|.\|_1$ denotes the l^1-norm of $\mathbb{R}^N$.

Proof We begin by the proving of the convergence in finite-dimensional distributions. For this goal, we consider for any fixed $u, v \in \mathbb{R}_+^N$ the behavior of

$$\begin{aligned}
&\mathbf{E}\left[X^{(\rho)}(u)X^{(\rho)}(v)\right] \\
&= \left(\rho \prod_{i=1}^N (t_0)_i\right)^{-2H} \mathbf{E}\left[\left(\mathbf{B}^H(t_0 + \rho\langle (t_0)_j\, u_j\rangle) - \mathbf{B}^H(t_0))(\mathbf{B}^H(t_0 + \rho\langle (t_0)_j\, v_j\rangle) - \mathbf{B}^H(t_0)\right)\right],
\end{aligned}$$

as ρ goes to 0. Let

$$\nu(\rho, t_0) = \left(\rho \prod_{i=1}^N (t_0)_i\right)^{2H} \quad \text{and} \quad \phi_{t_0}^{(\rho)}(u, v) = \left(\rho \prod_{i=1}^N (t_0)_i\right)^{2H} \mathbf{E}\left[X^{(\rho)}(u)X^{(\rho)}(v)\right].$$

We have

$$\begin{aligned}
\phi_{t_0}^{(\rho)}(u, v) =& \mathbf{E}\left[\mathbf{B}^H(t_0 + \rho\langle (t_0)_j\, u_j\rangle)\mathbf{B}^H(t_0 + \rho\langle (t_0)_j\, v_j\rangle)\right] + \mathbf{E}\left[\mathbf{B}^H(t_0)\right]^2 \\
&- \mathbf{E}\left[\mathbf{B}^H(t_0 + \rho\langle (t_0)_j\, u_j\rangle)\mathbf{B}^H(t_0)\right] - \mathbf{E}\left[\mathbf{B}^H(t_0 + \rho\langle (t_0)_j\, v_j\rangle)\mathbf{B}^H(t_0)\right] \\
=& \frac{1}{2}\Big[m([0, t_0] \triangle [0, t_0 + \rho\langle (t_0)_j\, u_j\rangle])^{2H} + m([0, t_0] \triangle [0, t_0 + \rho\langle (t_0)_j\, v_j\rangle])^{2H} \\
&- m([0, t_0 + \rho\langle (t_0)_j\, u_j\rangle] \triangle [0, t_0 + \rho\langle (t_0)_j\, v_j\rangle])^{2H}\Big].
\end{aligned}$$

In Lemma 3.1 in [11], it is proved that for any $s, t \in \mathbb{R}_+^N$,

$$m([0, s] \setminus [0, t]) = \prod_{i \notin I} |s_i| \sum_{J \subsetneq I} \left(\prod_{i \in J} |t_i| \prod_{i \in I \setminus J} |t_i - s_i|\right), \tag{10.2.2}$$

where $I = \{1 \leq i \leq N : t_i < s_i\}$. In (10.2.2), the product over an empty index set, $\prod_{i \in \emptyset} a_i$, is taken to be 1, and the sum over an empty set is taken to be 0.

Assuming $\rho > 0$, the set I is equal to $\{1, \ldots, N\}$ and then

$$\begin{aligned} m([0, t_0] \triangle [0, t_0 + \rho \langle (t_0)_j\, u_j \rangle]) &= \sum_{J \subsetneq \{1,\ldots,N\}} \left(\prod_{i \in J} (t_0)_i \prod_{i \in \{1,\ldots,N\} \setminus J} \rho (t_0)_i u_i \right) \\ &= \prod_{i \in \{1,\ldots,N\}} (t_0)_i \sum_{J \subsetneq \{1,\ldots,N\}} \rho^{N-\#J} \prod_{i \in \{1,\ldots,N\} \setminus J} u_i, \end{aligned}$$

using the fact that I in the formula (10.2.2) is equal to $\{1, \ldots, N\}$ here.

Since $0 < 2H < 1$, as ρ goes to 0, we only keep the first-order terms in ρ in the previous sum

$$\begin{aligned} \frac{m([0, t_0] \triangle [0, t_0 + \rho \langle (t_0)_j\, u_j \rangle])^{2H}}{\nu(\rho, t_0)} &= \left(\sum_{\substack{J \subsetneq \{1,\ldots,N\} \\ \#J = N-1}} \prod_{i \in \{1,\ldots,N\} \setminus J} u_i \right)^{2H} + O(\rho^{2H}) \\ &= \left(\sum_{i \in \{1,\ldots,N\}} u_i \right)^{2H} + O(\rho^{2H}). \end{aligned} \quad (10.2.3)$$

We also have

$$\frac{m([0, t_0] \triangle [0, t_0 + \rho \langle (t_0)_j\, v_j \rangle])^{2H}}{\nu(\rho, t_0)} = \left(\sum_{i \in \{1,\ldots,N\}} v_i \right)^{2H} + O(\rho^{2H}). \quad (10.2.4)$$

To deal with the term in $m([0, t_0 + \rho \langle (t_0)_j\, u_j \rangle] \triangle [0, t_0 + \rho \langle (t_0)_j\, v_j \rangle])$, we consider $I = \{1 \leq i \leq N : v_i > u_i\}$ and use the expression (10.2.2),

$$\begin{aligned} & m([0, t_0 + \rho \langle (t_0)_j\, v_j \rangle] \setminus [0, t_0 + \rho \langle (t_0)_j\, u_j \rangle]) \\ &= \prod_{i \notin I} (t_0)_i (1 + \rho v_i) \sum_{J \subsetneq I} \left(\prod_{i \in J} (t_0)_i (1 + \rho u_i) \prod_{i \in I \setminus J} \rho (t_0)_i (v_i - u_i) \right) \\ &= \prod_{i \in \{1,\ldots,N\}} (t_0)_i \sum_{\substack{J \subsetneq I \\ \#(I \setminus J) = 1}} \rho \prod_{i \in I \setminus J} (v_i - u_i) + O(\rho^2), \end{aligned}$$

keeping only the first-order terms in ρ. This implies

$$\begin{aligned} \frac{m([0, t_0 + \rho \langle (t_0)_j\, v_j \rangle] \setminus [0, t_0 + \rho \langle (t_0)_j\, u_j \rangle])}{\rho \prod_{i \in \{1,\ldots,N\}} (t_0)_i} &= \sum_{\substack{J \subsetneq I \\ \#(I \setminus J) = 1}} \prod_{i \in I \setminus J} (v_i - u_i) + O(\rho) \\ &= \sum_{i \in I} (v_i - u_i) + O(\rho). \end{aligned}$$

We also have

$$\frac{m([0, t_0 + \rho\langle(t_0)_j\ u_j\rangle] \setminus [0, t_0 + \rho\langle(t_0)_j\ v_j\rangle])}{\rho \prod_{i\in\{1,\dots,N\}}(t_0)_i} = \sum_{i\in I^c}(u_i - v_i) + O(\rho).$$

Therefore, we get

$$\frac{m([0, t_0 + \rho\langle(t_0)_j\ u_j\rangle] \bigtriangleup [0, t_0 + \rho\langle(t_0)_j\ v_j\rangle])^{2H}}{\nu(\rho, t_0)} = \Bigg(\sum_{i\in\{1,\dots,N\}} |u_i - v_i|\Bigg)^{2H} + O(\rho^{2H}). \tag{10.2.5}$$

The convergence in finite-dimensional distributions to the process Y follows from (10.2.3), (10.2.4), and (10.2.5).

In order to prove the weak convergence, it remains to prove the tightness of the family $\{X^{(\rho)}(u), u \in [0,1]^N\}$ $(\rho > 0)$. Since the processes $X^{(\rho)}$ are Gaussian, by Theorem 3.3.1 in [18, p. 198] and Dudley's metric entropy bound for the uniform modulus of continuity (cf. Theorem 1.3.5 in [1]), it suffices to prove that for some finite constant $C > 0$ we have

$$\mathbf{E}\big[|X^{(\rho)}(u) - X^{(\rho)}(v)|^2\big] \le C\, \|u - v\|^{2H} \quad \forall u, v \in [0,1]^N. \tag{10.2.6}$$

We have just proved that $\mathbf{E}\big[|X^{(\rho)}(u) - X^{(\rho)}(v)|^2\big]$ converges to $\|u - v\|_1^{2H}$, as ρ goes to 0. Equation (10.2.6) follows from this and the equivalence of the l^1-norm and the Euclidean norm of $\mathbb{R}^N$. This completes the proof of Theorem 1.

10.3 Fractal Dimension Properties of SI-fBm on Rectangles of $\mathbb{R}_+^N$

It is known that the study of many sample path properties of a Gaussian random field relies on its incremental variances; see, for example, [39, 46].

By definition, the real-valued, set-indexed fractional Brownian motion $\mathbf{B}^H$ on rectangles of $\mathbb{R}_+^N$ satisfies

$$\forall s, t \in \mathbb{R}_+^N, \quad \mathbf{E}\big[|\mathbf{B}^H(t) - \mathbf{B}^H(s)|^2\big] = m([0,s] \bigtriangleup [0,t])^{2H},$$

where m is the Lebesgue measure on $\mathbb{R}^N$.

Let us start with the following lemma, which is essentially proved in [11, Lemma 3.1].

Lemma 2 *For any vectors $0 \prec \varepsilon \prec T \in \mathbb{R}_+^N$, there exist two positive constants $m_{\varepsilon,T}$ (depending on ε and T) and M_T (only depending on T) such that*

$$\forall s,t \in [\varepsilon, T]; \quad m_{\varepsilon,T}\ \|t-s\|^{2H} \le \mathbf{E}\big[|\mathbf{B}^H(t)-\mathbf{B}^H(s)|^2\big] \le M_T\ \|t-s\|^{2H}. \tag{10.3.1}$$

Proof By Lemma 3.1 of [11] there exist two positive constants $m'_{\varepsilon,T}$ and $M'_{\varepsilon,T}$ such that

$$\forall s,t \in [\varepsilon, T]; \quad m'_{\varepsilon,T}\ d_1(s,t) \le m([0,s] \,\triangle\, [0,t]) \le M'_{\varepsilon,T}\ d_\infty(s,t)$$

where d_1 and d_∞ are the usual distances of $\mathbb{R}^N$ defined by

$$d_1 : (s,t) \mapsto \|t-s\|_1 = \sum_{i=1}^N |t_i - s_i|,$$
$$d_\infty : (s,t) \mapsto \|t-s\|_\infty = \max_{1\le i\le N} |t_i - s_i|.$$

In proof of this result, it is clear that the constant $M'_{\varepsilon,T}$ can be chosen independently of ε. Since the distances d_1 and d_∞ are equivalent to the Euclidean distance of $\mathbb{R}^N$, (10.3.1) follows.

The following lemma shows that $\mathbf{B}^H$ satisfies the property of *two-point local nondeterminism* (cf. [46]).

Lemma 3 *Let* $\mathbf{B}^H = \{\mathbf{B}^H(t);\ t \in \mathbb{R}_+^N\}$ *be the set-indexed fractional Brownian motion* ($0 < H \le 1/2$) *on rectangles of* $\mathbb{R}_+^N$.

For any $\varepsilon, T \in \mathbb{R}_+^N$ *with* $0 \prec \varepsilon \prec T$ *there is a constant* $c > 0$ *such that*

$$\mathrm{Var}(\mathbf{B}^H(t) \mid \mathbf{B}^H(s)) \ge c\ \|t-s\|^{2H}, \quad \forall s,t \in [\varepsilon,\ T]. \tag{10.3.2}$$

Proof We recall the following formula for the conditional variance for a mean zero Gaussian vector (U, V):

$$\mathrm{Var}(U \mid V) = \frac{(\rho_{U,V}^2 - (\sigma_U - \sigma_V)^2)((\sigma_U + \sigma_V)^2 - \rho_{U,V}^2)}{4\sigma_V^2}, \tag{10.3.3}$$

where $\rho_{U,V}^2 = \mathbf{E}[(U-V)^2]$, $\sigma_U^2 = \mathbf{E}[U^2]$ and $\sigma_V^2 = \mathbf{E}[V^2]$.

We consider the case $U = \mathbf{B}^H(t)$ and $V = \mathbf{B}^H(s)$, for any $s,t \in [\varepsilon, T]$. We proceed in two steps:

- The denominator of (10.3.3) is

$$4\ \sigma_{\mathbf{B}^H(s)}^2 = 4\ \mathbf{E}\big[|\mathbf{B}^H(s)|^2\big] = 4\ m([0,s])^{2H},$$

  which is bounded by $m([0,\varepsilon])^{2H}$ and $m([0,T])^{2H}$.
- In order to study the first factor of the numerator of (10.3.3), we consider the function $\Phi : [\varepsilon, T] \to \mathbb{R}_+$ defined by $\Phi(t) = m([0,t])^H$. Then $\Phi(t) = \prod_{i=1}^N t_i^H$, and therefore Φ is differentiable on (ε, T) and admits the first-order Taylor expansion

$$\sigma_{\mathbf{B}^H(t)} - \sigma_{\mathbf{B}^H(s)} = \Phi(t) - \Phi(s) = \sum_{i=1}^{N} \partial_i \Phi(s).(t_i - s_i) + o(\|t - s\|).$$

For all $1 \leq i \leq N$, we evaluate the ith partial derivative of Φ

$$\partial_i \Phi(s) = H\ s_i^{H-1} \prod_{j \neq i} s_j^H = \frac{H}{s_i}\ \Phi(s).$$

Then,

$$\sigma_{\mathbf{B}^H(t)} - \sigma_{\mathbf{B}^H(s)} = H\ \sigma_{\mathbf{B}^H(s)} \sum_{i=1}^{N} \frac{t_i - s_i}{s_i} + o(\|t - s\|),$$

and

$$\left(\sigma_{\mathbf{B}^H(t)} - \sigma_{\mathbf{B}^H(s)}\right)^2 = \left[H\ \sigma_{\mathbf{B}^H(s)} \sum_{i=1}^{N} \frac{t_i - s_i}{s_i}\right]^2 + o(\|t - s\|^2). \qquad (10.3.4)$$

But $H\ \sigma_{\mathbf{B}^H(s)}/s_i$ is bounded by $H\ m([0,T])^H/\varepsilon$ for all $i \in \{1, \ldots, N\}$, so that

$$\left[H\ \sigma_{\mathbf{B}^H(s)} \sum_{i=1}^{N} \frac{t_i - s_i}{s_i}\right]^2 \leq \left[\frac{H\ m([0,T])^H}{\varepsilon}\ d_1(s,t)\right]^2. \qquad (10.3.5)$$

Since the distance d_1 is equivalent to the Euclidean distance of $\mathbb{R}^N$, we deduce from (10.3.4) and (10.3.5) that

$$\left(\sigma_{\mathbf{B}^H(t)} - \sigma_{\mathbf{B}^H(s)}\right)^2 = O(\|t - s\|^2).$$

Since $0 < 2H \leq 1$, we get

$$\mathbf{E}\left[|\mathbf{B}^H(t) - \mathbf{B}^H(s)|^2\right] - \left(\sigma_{\mathbf{B}^H(t)} - \sigma_{\mathbf{B}^H(s)}\right)^2 \asymp \|t - s\|^{2H}.$$

- The second factor of the numerator of (10.3.3)

$$\underbrace{\left(\sigma_{\mathbf{B}^H(t)} + \sigma_{\mathbf{B}^H(s)}\right)^2}_{\geq 4m([0,\varepsilon])^{2H}} - \underbrace{\mathbf{E}\left[|\mathbf{B}^H(t) - \mathbf{B}^H(s)|^2\right]}_{\leq M_T \|t-s\|^{2H}}$$

is clearly bounded from below by a positive constant.

The result follows from the two steps above.

By Lemmas 2 and 3, we can derive from [5, 46] the following results on Hausdorff dimension of the level sets and hitting probability of a d-dimensional MpfBm $\mathbf{B}^H$

defined by

$$\mathbf{B}^H(t) = (\mathbf{B}_1^H(t), \ldots, \mathbf{B}_d^H(t)), \qquad \forall t \in \mathbb{R}_+^N, \tag{10.3.6}$$

where $\mathbf{B}_1^H, \ldots, \mathbf{B}_d^H$ are independent copies of $\mathbf{B}^H$.

Theorem 4 *Let $F \subseteq \mathbb{R}^d$ be a Borel set such that* $\dim F \geq d - \frac{N}{H}$. *Then for every rectangle $I \subseteq \mathbb{R}_+^N$ the following statements hold:*

(i) Almost surely

$$\dim \left(\mathbf{B}^H\right)^{-1}(F) \cap I \leq N - H\left(d - \dim F\right), \tag{10.3.7}$$

where dim *denotes Hausdorff dimension (cf. e.g., [18]). In particular, if* $\dim F = d - \frac{N}{H}$, *then* $\dim \left(\mathbf{B}^H\right)^{-1}(F) \cap I = 0$ *a.s.*

(ii) If $\dim F > d - \frac{N}{H}$, *then for every $\varepsilon > 0$,*

$$\dim \left(\mathbf{B}^H\right)^{-1}(F) \cap I \geq N - H\left(d - \dim F\right) - \varepsilon \tag{10.3.8}$$

on an event of positive probability (which may depend on ε).

(iii) If $N > Hd$, then for every $x \in \mathbb{R}^d$, with positive probability,

$$\dim \left(\mathbf{B}^H\right)^{-1}(x) = N - Hd. \tag{10.3.9}$$

Now we define a metric ρ on $\mathbb{R}^N \times \mathbb{R}^d$ by

$$\rho\big((s, x), (t, y)\big) = \max\{\|s - t\|^H, \|x - y\|\}.$$

For any $r > 0$ and $(s, x) \in \mathbb{R}^N \times \mathbb{R}^d$, let

$$B_\rho((s, x), r) = \{(t, y) \in \mathbb{R}^N \times \mathbb{R}^d : \rho((s, x), (t, y)) < r\}$$

denote the open ball in the metric space $(\mathbb{R}^N \times \mathbb{R}^d, \rho)$ centered at (s, x) with radius r.

For any $q > 0$ and any set $A \subseteq \mathbb{R}^N \times \mathbb{R}^d$, q-dimensional Hausdorff measure of A under the metric ρ is defined by

$$\mathcal{H}_q^\rho(A) = \lim_{\varepsilon \to 0} \inf \left\{ \sum_i (2r_i)^q : A \subseteq \bigcup_{i=1}^\infty B_\rho(u_i, r_i),\ r_i < \varepsilon \right\}, \tag{10.3.10}$$

where $B_\rho(u, r)$ is the open ball in the metric space $(\mathbb{R}^N \times \mathbb{R}^d, \rho)$ centered at u with radius r.

The Hausdorff dimension $\dim^\rho A$ of $A \subset \mathbb{R}^N \times \mathbb{R}^d$ under the metric ρ is defined by

$$\dim^\rho A = \inf\{q > 0 : \mathcal{H}_q^\rho(A) = 0\}. \tag{10.3.11}$$

This type of Hausdorff measure and Hausdorff dimension has been applied by Hawkes [10], Taylor and Watson [38], Wu and Xiao [41], and Xiao [46] to characterize the polar sets for stochastic processes and the heat equation.

For any compact sets $E \subset \mathbb{R}^N$ and $F \subset \mathbb{R}^d$, we let $\mathcal{P}(E \times F)$ denote the collection of all probability measures that are supported in $E \times F$. For any such $E \times F$ and $\mu \in \mathcal{P}(E \times F)$, define the energy of μ in the metric ρ by

$$I_d^\rho(\mu) \hat{=} \int_{\mathbb{R}^N \times \mathbb{R}^d} \int_{\mathbb{R}^N \times \mathbb{R}^d} \frac{1}{\rho^d\big((s,x),(t,y)\big)} d\mu(s,x) d\mu(t,y). \tag{10.3.12}$$

The capacity of $E \times F$ on $\mathbb{R}^N \times \mathbb{R}^d$ is defined by

$$\mathcal{C}_d^\rho(E \times F) = \left[\inf_{\mu \in \mathcal{P}(E \times F)} I_d^\rho(\mu) \right]^{-1}. \tag{10.3.13}$$

The following theorem is an extension of Theorem 2.1 in [5], whose proof follows from Lemmas 3.1 and 3.2 and the proof of Theorem 2.1 in [6]. We omit the details.

Theorem 5 *If $E \subseteq \mathbb{R}_+^N$ and $F \subseteq \mathbb{R}^d$ be Borel sets, then there exists a constant $c > 1$ such that*

$$c^{-1}\, \mathcal{C}_d^\rho(E \times F) \le \mathbf{P}\Big\{ \mathbf{B}^H(E) \cap F \neq \varnothing \Big\} \le c\, \mathcal{H}_d^\rho(E \times F). \tag{10.3.14}$$

In the above, $\mathcal{H}_q^\rho(E \times F) = 1$ whenever $q \le 0$.

10.4 Stationary Random Field Obtained via the Lamperti Transform

It follows from Proposition 3.12 of [12] that the real-valued SI-fBm $\mathbf{B}^H = \{\mathbf{B}^H(t), t \in \mathbb{R}_+^N\}$ and its d-dimensional analog in (10.3.6) have the following multi-self-similarity in the sense of [9]: For any constants $c_1, \ldots, c_N > 0$,

$$\{\mathbf{B}^H(c_1 t_1, \ldots, c_N t_N), t \in \mathbb{R}_+^N\} \overset{d}{=} \{(c_1 \cdots c_N)^H \mathbf{B}^H(t), t \in \mathbb{R}_+^N\},$$

where $\overset{d}{=}$ means equality of all finite-dimensional distributions.

By Proposition 2.1.1 in [9], the centered Gaussian process $Z = \{Z(t);\ t \in \mathbb{R}^N\}$ defined by

$$Z(t) = e^{-H(t_1 + \cdots + t_N)}\, \mathbf{B}^H(e^{t_1}, \ldots, e^{t_N}), \quad \forall t = (t_1, \ldots, t_N) \in \mathbb{R}^N \tag{10.4.1}$$

is stationary. By the multi-self-similarity of $\mathbf{B}^H$, it can be verified that the covariance function $r(t) = \mathbf{E}\big[Z(0)Z(t)\big]$ is symmetric, namely, $r(t) = r(-t)$ for all $t \in \mathbb{R}^N$.

The aim of this section is to study the covariance structure of Z, which will be useful for establishing the property of strong local nondeterminism of $\mathbf{B}^H$. This argument was applied for a one-parameter bi-fractional Brownian motion by Tudor and Xiao [39]. As will be seen, the case of $N > 1$ is more complicated.

The stationary random field Z has the following integral representation:

$$\{Z(t), t \in \mathbb{R}^N\} \stackrel{d}{=} \left\{ \int_{\mathbb{R}^N} e^{i\langle t,\xi\rangle}\, \mathbb{W}(d\xi),\ \forall t \in \mathbb{R}^N \right\}, \tag{10.4.2}$$

where $\mathbb{W}$ is a complex-valued Gaussian random measure with control measure F, which is related to the covariance function $r(t)$ by

$$r(t) = \int_{\mathbb{R}^N} e^{i\langle t,\xi\rangle}\, F(d\xi).$$

The measure F is a finite measure on $\mathbb{R}^N$ and is called the spectral measure of Z. It is known that many local properties of Z are determined by the asymptotic behavior of F at infinity, while long-term properties (such as long-range dependence) of Z are determined by the behavior of F at the origin $\xi = 0$.

Because of (10.4.1) and (10.4.2), we see that $\mathbf{B}^H$ has the following stochastic integral representation:

$$\mathbf{B}^H(t) = \left(\prod_{j=1}^N t_j^H\right) \int_{\mathbb{R}^N} e^{i\langle \log t,\xi\rangle}\, \mathbb{W}(d\xi), \quad \forall t \in (0,\infty)^N, \tag{10.4.3}$$

where $\log t = (\log t_1, \ldots, \log t_N)$.

10.4.1 Expression of r(t)

For all $t = (t_1, \ldots, t_N)$ in $\mathbb{R}^N$, we compute

$$\begin{aligned} r(t) = \mathbf{E}\big[Z(0)Z(t)\big] &= e^{-H(t_1+\cdots+t_N)}\, \mathbf{E}\left[\mathbf{B}^H(1,\ldots,1)\, \mathbf{B}^H(e^{t_1},\ldots,e^{t_N})\right] \\ &= \frac{1}{2}\, e^{-H(t_1+\cdots+t_N)} \left[m([0,1])^{2H} + m([0,e^t])^{2H} - m([0,1] \,\triangle\, [0,e^t])^{2H}\right], \end{aligned} \tag{10.4.4}$$

where m is the Lebesgue measure of $\mathbb{R}^N$ and we have used the notation $e^t = (e^{t_1}, \ldots, e^{t_N})$.

We have $m([0,1])^{2H} = 1$ and $m([0,e^t])^{2H} = e^{2H(t_1+\cdots+t_N)} = e^{2H|t|}$ with the usual notation $|t| = \|t\|_1 = t_1 + \cdots + t_N$.

We start by considering the case $t \in \mathbb{R}^N$ such that $t_i \geq 0$ for all $1 \leq i \leq N$. Then $m([0,1] \,\triangle\, [0,e^t]) = m([0,e^t] \setminus [0,1])$. In [11, Lemma 3.1], it is proved that for any $s,t \in \mathbb{R}^N$,

$$m([0,s] \setminus [0,t]) = \prod_{i \notin I} |s_i| \sum_{J \subsetneq I} \left(\prod_{i \in J} |t_i| \prod_{i \in I \setminus J} |t_i - s_i| \right),$$

where $I = \{1 \leq i \leq N : t_i < s_i\}$. As $t_i \geq 0$ for all $1 \leq i \leq N$, this formula leads to

$$\begin{aligned} m([0,1] \bigtriangleup [0,e^t]) &= \sum_{J \subsetneq \{1,\ldots,N\}} \left(\prod_{i \in J} 1 \prod_{i \in \{1,\ldots,N\} \setminus J} |e^{t_i} - 1| \right) \\ &= \sum_{J \subsetneq \{1,\ldots,N\}} \left(\prod_{i \in \{1,\ldots,N\} \setminus J} |e^{t_i} - 1| \right). \end{aligned} \tag{10.4.5}$$

But we can remark that

$$\begin{aligned} \sum_{J \subsetneq \{1,\ldots,N\}} \left(\prod_{i \in \{1,\ldots,N\} \setminus J} |e^{t_i} - 1| \right) &= \prod_{i \in \{1,\ldots,N\}} \left(1 + (e^{t_i} - 1)\right) - 1 \\ &= \prod_{i \in \{1,\ldots,N\}} e^{t_i} - 1 = e^{t_1 + \cdots + t_N} - 1. \end{aligned} \tag{10.4.6}$$

From Eqs. (10.4.5) and (10.4.6), we deduce that

$$m([0,1] \bigtriangleup [0,e^t]) = e^{|t|} - 1,$$

and

$$\begin{aligned} r(t) &= \frac{1}{2} e^{-H|t|} \left[1 + e^{2H|t|} - (e^{|t|} - 1)^{2H}\right] \qquad (10.4.7) \\ &= \frac{e^{H|t|} + e^{-H|t|}}{2} - \frac{1}{2} e^{-H|t|} (e^{|t|} - 1)^{2H} \\ &= \frac{e^{H|t|} + e^{-H|t|}}{2} - \frac{1}{2} \left[e^{-|t|/2} (e^{|t|} - 1)\right]^{2H} \\ &= \frac{e^{H|t|} + e^{-H|t|}}{2} - \frac{1}{2} \left(e^{|t|/2} - e^{-|t|/2}\right)^{2H}. \end{aligned}$$

The other cases of $t \in \mathbb{R}^N$ can be considered similarly. By using hyperbolic trigonometry functions, we get

$$r(t) = \cosh(H|t|) - 2^{2H-1} (\sinh(|t|/2))^{2H}, \quad \forall t \in \mathbb{R}^N. \tag{10.4.8}$$

10.4.2 Behavior of $1 - r(t)$ as t Goes to 0

Expression (10.4.8) allows to obtain a Taylor expansion of $r(t)$ as t goes to 0. First, we write the expansion of the hyperbolic cosinus and sinus

$$\cosh H|t| = 1 + \frac{H^2|t|^2}{2} + o(|t|^3) \tag{10.4.9}$$

and

$$\begin{aligned} \sinh\frac{|t|}{2} &= \frac{|t|}{2} + \frac{|t|^3}{48} + o(|t|^4) \\ &= \frac{|t|}{2}\left(1 + \frac{|t|^2}{24} + o(|t|^3)\right). \end{aligned} \tag{10.4.10}$$

The expansion of $(1+x)^\alpha$ and (10.4.10) give

$$\begin{aligned} \left(\sinh\frac{|t|}{2}\right)^{2H} &= \left(\frac{|t|}{2}\right)^{2H}\left(1 + \frac{|t|^2}{24} + o(|t|^3)\right)^{2H} \\ &= \left(\frac{|t|}{2}\right)^{2H}\left(1 + 2H\,\frac{|t|^2}{24} + o(|t|^2)\right). \end{aligned} \tag{10.4.11}$$

From (10.4.8), (10.4.9), and (10.4.11), we get the expansion of $r(t)$ in the neighborhood of 0:

$$r(t) = 1 - \frac{1}{2}\,|t|^{2H} + \frac{H^2}{2}\,|t|^2 + o(|t|^2). \tag{10.4.12}$$

Since $0 < 2H < 1$, the expansion (10.4.12) leads to

$$1 - r(t) \sim \frac{1}{2}\,|t|^{2H} \qquad \text{as } t \to 0. \tag{10.4.13}$$

10.4.3 Integrability of $t \mapsto R(t)$ on $\mathbb{R}^N$

Starting from (10.4.4), we study the difference

$$m([0, e^t])^{2H} - m([0,1] \triangle [0, e^t])^{2H}.$$

Since $0 < 2H < 1$, the study of $x \mapsto x^{2H}$ shows that $x^{2H} - y^{2H} \leq (x-y)^{2H}$ for all $0 \leq y \leq x$. Then, we have

$$m([0, e^t])^{2H} - m([0, e^t] \setminus [0,1]^{2H} \leq m([0,1])^{2H},$$

and therefore

$$\begin{aligned} r(t) &= \frac{1}{2}\, e^{-H|t|}\left[m([0,1])^{2H} + m([0,e^t])^{2H} - m([0,1] \bigtriangleup [0,e^t])^{2H}\right] \\ &\leq m([0,1])^{2H}\, e^{-H|t|} \end{aligned}$$

for all $t \in \mathbb{R}^N$. This inequality proves that $r \in L^1(\mathbb{R}^N)$.

10.4.4 Fourier Transform of $t \mapsto R(t)$

Since $r \in L^1(\mathbb{R}^N)$, the spectral density of the stationary Gaussian field Z can be expressed as

$$\begin{aligned} f(\xi) &= (2\pi)^{-N} \int_{\mathbb{R}^N} e^{-i\langle t,\xi\rangle} r(t) dt \\ &= \pi^{-N} \int_{\mathbb{R}^N} \cos(\langle t,\xi\rangle) r(t)\, dt \end{aligned} \tag{10.4.14}$$

by the symmetry of the covariance function $r(\cdot)$. As shown in [45, 46], the asymptotic behavior of $f(\xi)$ as $\|\xi\| \to \infty$ carries a lot of information on the sample path properties of the Gaussian random fields Z and $\mathbf{B}^H$.

10.5 Strong Local Nondeterminism and Fine Properties of MpfBm

In this section, we study the exact uniform and local moduli of continuity of the set-indexed fractional Brownian motion $\mathbf{B}^H = \{\mathbf{B}^H(t), t \in \mathbb{R}_+^N\}$. Our main technical tool is the property of strong local nondeterminism which is established by applying a Fourier analytic method (see [45]). For technical reasons, we only consider the case $N = 2$.

10.5.1 Property of Strong Local Nondeterminism (SLND)

In order to prove SLND for $\mathbf{B}^H$, we first consider the stationary Gaussian random field Z defined in (10.4.1).

Recall (10.4.7), we denote by $\tilde{r}(\rho)$ the positive function on $\mathbb{R}_+$ defined by

$$\tilde{r}(\rho) = \frac{1}{2}\, e^{H\rho}\left[1 + e^{-2H\rho} - (1 - e^{-\rho})^{2H}\right], \quad \forall \rho \in \mathbb{R}_+.$$

We start with some elementary properties of $\tilde{r}(\rho)$.

$$\tilde{r}'(\rho) = \frac{1}{2}\, He^{H\rho}\big[1 - e^{-2H\rho} - (1-e^{-\rho})^{2H-1}(1+e^{-\rho})\big].$$

Note that $\tilde{r}'(\rho) < 0$ for all $\rho > 0$ and it is elementary to verify that

$$|\tilde{r}'(\rho)| \le c\, e^{-\beta\rho} \tag{10.5.1}$$

for ρ large, where $\beta = \min\{H, 1-H\}$ and

$$\tilde{r}'(\rho) \sim -H\rho^{2H-1} \quad \text{as } \rho \to 0.$$

Similarly,

$$\begin{aligned}\tilde{r}''(\rho) = \frac{1}{2}\, He^{H\rho}\Big[&H(1+e^{-2H\rho}) + (H+1)(1-e^{-\rho})^{2H-1}(1+e^{-\rho})\\ &+ (1-2H)(1-e^{-\rho})^{2H-2}(1+e^{-\rho}) + (1-e^{-\rho})^{2H-1}e^{-\rho}\Big].\end{aligned} \tag{10.5.2}$$

Clearly $\tilde{r}''(\rho) > 0$ for all $\rho \ge 0$.

If we write $\tilde{r}''(\rho) = H(1-2H)\rho^{2H-2}L(\rho)$, then we can verify that $L(\cdot)$ is a slowly varying function at 0 and satisfies

$$\lim_{\rho\to 0+} L(\rho) = 1 \quad \text{and} \quad \lim_{\rho\to 0+} \frac{\rho L'(\rho)}{L(\rho)} = 0. \tag{10.5.3}$$

The following lemma describes the asymptotic properties of the spectral density $f(\xi)$ of Z as $|\xi| \to \infty$.

Lemma 6 *Let $f(\xi_1, \xi_2)$ be the spectral density of Z. Then $f(\xi_1, \xi_2)$ is symmetric in ξ_1 and ξ_2:*

$$f(\xi_1, \xi_2) = f(-\xi_1, \xi_2) = f(\xi_1, -\xi_2) = f(-\xi_1, -\xi_2), \quad \forall\, (\xi_1, \xi_2) \in \mathbb{R}^2. \tag{10.5.4}$$

Moreover, the following statement hold:

(i) As $|\xi_1| \to \infty$ and $|\xi_2| \to \infty$, we have

$$f(\xi_1, \xi_2) \sim \frac{c_1}{|\xi_1^2 - \xi_2^2|}\left|\frac{1}{\xi_2^{2H}} - \frac{1}{\xi_1^{2H}}\right|, \tag{10.5.5}$$

where

$$c_1 = \frac{4H(1-2H)}{\pi^2} \int_0^\infty \sin(\eta)\, \eta^{2H-2}\, d\eta.$$

(ii) For any constant $M > 0$, as $|\xi_1| \to \infty$, we have

$$f(\xi) \sim \frac{4H(1-2H)}{\pi^2(\xi_1^2-\xi_2^2)} \frac{1}{\xi_2^{2H}} \int_0^\infty \sin(\eta)\, \eta^{2H-2}\, L(\eta/\xi_2)\, d\eta. \tag{10.5.6}$$

uniformly for all $\xi_2 \in [-M, M]$. *The same conclusion holds if we switch* ξ_1 *and* ξ_2.

Consequently, there is a constant $M > 0$ *such that*

$$f(\xi) \geq \frac{c}{|\xi|^{2+2H}}, \quad \forall \xi \in \mathbb{R}^2 \text{ with } |\xi| \geq M. \tag{10.5.7}$$

Proof The spectral density $f(\xi)$ is given in (10.4.14)

$$f(\xi) = \pi^{-2} \int_{\mathbb{R}^2} \cos(\langle t, \xi \rangle) r(t)\, dt, \quad \xi \in \mathbb{R}^2.$$

Since $r(t)$ only depends on the l^1-norm $|t|$. Our basic idea is to make a change of variables using the "polar coordinates" in the l^1-norm $|\cdot|$. This can be done explicitly when $N = 2$.

Note that the unit circle in the L^1 norm is $S_1^2 = \{\theta \in \mathbb{R}^2 : |\theta| = 1\}$ consists of four line segments. The one in the first quadrant is $\theta_1 + \theta_2 = 1$ $(0 \leq \theta_1 \leq 1)$. We make the following change of variables:

$$t_1 = \rho\theta_1 \quad \text{and} \quad t_2 = \rho(1-\theta_1)$$

to get

$$\int_{\mathbb{R}_+^2} \cos(\langle t, \xi \rangle) r(t)\, dt = \int_0^1 d\theta_1 \int_0^\infty \cos\left[\rho(\theta_1\xi_1 + (1-\theta_1)\xi_2)\right] \rho\tilde{r}(\rho)\, d\rho.$$

Using the same argument to other three quadrants and adding them together, we derive

$$\begin{aligned} f(\xi) &= \left(\frac{2}{\pi}\right)^2 \int_0^1 d\theta_1 \int_0^\infty \cos(\rho\theta_1\xi_1) \cos(\rho(1-\theta_1)\xi_2) \rho\tilde{r}(\rho)\, d\rho \\ &= \frac{2}{\pi^2} \int_0^\infty \rho\tilde{r}(\rho)\, d\rho \int_0^1 \left[\cos(\rho(\xi_1+\xi_2)\theta_1 - \rho\xi_2) + \cos(\rho(\xi_1-\xi_2)\theta_1 + \rho\xi_2)\right] d\theta_1. \end{aligned}$$

Integrating $[d\theta_1]$ first and then using integration by parts twice to integrate $[d\rho]$, we obtain

$$
\begin{aligned}
f(\xi) &= \frac{2}{\pi^2}\int_0^\infty \left[\frac{\sin(\rho\xi_1)+\sin(\rho\xi_2)}{\xi_1+\xi_2} + \frac{\sin(\rho\xi_1)-\sin(\rho\xi_2)}{\xi_1-\xi_2}\right]\tilde{r}(\rho)\,d\rho \\
&= \frac{4}{\pi^2(\xi_1^2-\xi_2^2)}\int_0^\infty \big[\cos(\rho\xi_1)-\cos(\rho\xi_2)\big]\,\tilde{r}'(\rho)\,d\rho \qquad (10.5.8)\\
&= \frac{-4}{\pi^2(\xi_1^2-\xi_2^2)}\int_0^\infty \left[\frac{\sin(\rho\xi_1)}{\xi_1} - \frac{\sin(\rho\xi_2)}{\xi_2}\right]\tilde{r}''(\rho)\,d\rho.
\end{aligned}
$$

It is now clear that (10.5.4) holds.

By using (10.5.1) and the facts that $\tilde{r}'(\rho)$ is negative and strictly increasing, we can see that the function

$$
x \mapsto -\int_0^\infty \cos(\rho x)\,\tilde{r}'(\rho)\,d\rho \qquad (10.5.9)
$$

is continuous and takes positive values (write the integral as an alternative series). This will be used to prove (ii) below.

In order to prove the rest of the lemma, it is sufficient to consider $\xi \in \mathbb{R}^2_+$ only. Furthermore, without loss of generality, we may and will assume from now on that $0 < \xi_2 < \xi_1 < \infty$.

Now we prove (i). To this end, we rewrite $f(\xi)$ as

$$
\begin{aligned}
f(\xi) &= \frac{-4}{\pi^2(\xi_1^2-\xi_2^2)}\left[\int_0^\infty \frac{\sin(\rho\xi_1)}{\xi_1}\,\tilde{r}''(\rho)\,d\rho - \int_0^\infty \frac{\sin(\rho\xi_2)}{\xi_2}\,\tilde{r}''(\rho)\,d\rho\right] \\
&= \frac{-4}{\pi^2(\xi_1^2-\xi_2^2)}\left[\int_0^\infty \frac{\sin(\eta)}{\xi_1^2}\,\tilde{r}''(\eta/\xi_1)\,d\eta - \int_0^\infty \frac{\sin(\eta)}{\xi_2^2}\,\tilde{r}''(\eta/\xi_2)\,d\eta\right]
\end{aligned}
\qquad (10.5.10)
$$

by a simple change of variables. Writing the above as

$$
\begin{aligned}
f(\xi) = \frac{4H(1-2H)}{\pi^2(\xi_1^2-\xi_2^2)}\Bigg[&\frac{1}{\xi_2^{2H}}\int_0^\infty \sin(\eta)\,\eta^{2H-2}\,L(\eta/\xi_2)\,d\eta \\
&- \frac{1}{\xi_1^{2H}}\int_0^\infty \sin(\eta)\,\eta^{2H-2}\,L(\eta/\xi_1)\,d\eta\Bigg]
\end{aligned}
\qquad (10.5.11)
$$

and applying the dominated convergence theorem, we obtain that as $\xi_1 \to \infty$ and $\xi_2 \to \infty$,

$$
f(\xi) \sim \frac{4H(1-2H)}{\pi^2(\xi_1^2-\xi_2^2)}\left(\frac{1}{\xi_2^{2H}} - \frac{1}{\xi_1^{2H}}\right)\int_0^\infty \sin(\eta)\,\eta^{2H-2}\,d\eta.
$$

In the above, we have used the fact that the last integral is absolutely convergent. This proves (i).

Let $M > 0$ be a fixed constant. By (10.5.9), we see that the function

$$\xi_2 \mapsto \xi_2^{-2H} \int_0^\infty \sin(\eta)\, \eta^{2H-2}\, L(\eta/\xi_2)\, d\eta$$

in (10.5.10) attains its minimum on $[0, M]$ and the minimum is positive. This implies that as $\xi_1 \to \infty$

$$f(\xi) \sim \frac{4H(1-2H)}{\pi^2(\xi_1^2 - \xi_2^2)} \frac{1}{\xi_2^{2H}} \int_0^\infty \sin(\eta)\, \eta^{2H-2}\, L(\eta/\xi_2)\, d\eta.$$

uniformly for all $\xi_2 \in [-M, M]$. This proves (ii).

Finally, thanks to (i) and (ii), we take a constant $M > 0$ large enough such that for all $\xi \in \mathbb{R}_+^2$ with $\xi_1 + \xi_2 \geq M$, we have

$$f(\xi) \geq \frac{c}{2|\xi_1^2 - \xi_2^2|} \left| \frac{1}{\xi_2^{2H}} - \frac{1}{\xi_1^{2H}} \right| \geq \frac{Hc}{|\xi|^{2+2H}}.$$

This finishes the proof of Lemma 6.

The following result states that the stationary Gaussian field $Z = \{Z(t), t \in \mathbb{R}^2\}$ in (10.4.1) has the property of strong local nondeterminism.

Lemma 7 *For any compact interval $I \subset \mathbb{R}^2$, there exists a constant $c_2 > 0$ such that for all integers $n \geq 1$ and all $u, t^1, \ldots, t^n \in I$, we have*

$$\mathrm{Var}\Big(Z(u) \mid Z(t^1), \ldots, Z(t^n)\Big) \geq c_2 \min_{1 \leq k \leq n} \left|u - t^k\right|^{2H}. \tag{10.5.12}$$

Proof This follows from (10.5.7) and the Fourier analytic argument in the proof of Theorem 2.1 in [45] (see also [46]). We omit the details.

The main result of this section is the following theorem. We remark that in [34, Lemma 4.5], Richard proved a similar result for $\mathbf{B}^H = \{\mathbf{B}^H(t), t \in \mathbb{R}_+^N\}$ by using a different method.

Theorem 8 *The real-valued, set-indexed fractional Brownian motion $\mathbf{B}^H = \{\mathbf{B}^H(t), t \in \mathbb{R}_+^2\}$ has the following property of strong local nondeterminism: For any compact interval $T \subset (0, \infty)^2$, there exists a constant $c_3 > 0$ such that for all integers $n \geq 1$ and all $u, t^1, \ldots, t^n \in T$,*

$$\mathrm{Var}\Big(\mathbf{B}^H(u) \mid \mathbf{B}^H(t^1), \ldots, \mathbf{B}^H(t^n)\Big) \geq c_3 \min_{1 \leq k \leq n} \left|u - t^k\right|^{2H}.$$

Proof Let $T = [a, b]$, where $a, b \in (0, \infty)^2$. For any $t \in T$, we write $t^H = (t_1 t_2)^H$ and $\log t = (\log t_1, \log t_2)$.

Note that, by the definition of Z in (10.4.1), for every $t \in (0, \infty)^2$ we have

$$\mathbf{B}^H(t) = t^H Z(\log t). \tag{10.5.13}$$

Hence, for any integer $n \geq 1$ and any $u, t^1, \ldots, t^n \in T$, we apply (10.5.13) and Lemma 7 to derive

$$\begin{aligned} &\mathrm{Var}\Big(\mathbf{B}^H(u)\big|\mathbf{B}^H(t^1), \ldots, \mathbf{B}^H(t^n)\Big) \\ &= \mathrm{Var}\Big(u^H Z(\log u)\big|(t^1)^H Z(\log t^1), \ldots, (t^n)^H Z(\log t^n)\Big) \\ &= u^{2H}\, \mathrm{Var}\Big(Z(\log u)\big|Z(\log t^1), \ldots, Z(\log t^n)\Big) \\ &\geq c\, u^{2H} \min_{1\leq k\leq n} \big|\log u - \log t^k\big|^{2H} \\ &\geq c_3 \min_{1\leq k\leq n} \big|u - t^k\big|^{2H}. \end{aligned} \tag{10.5.14}$$

This proves Theorem 8.

10.5.2 Fine Properties of SifBm

With Theorem 8 in hand, we may investigate various fine properties of set-indexed fractional Brownian motion $\mathbf{B}^H = \{\mathbf{B}^H(t), t \in \mathbb{R}^2_+\}$. For example, the exact Hausdorff measure functions for its range, graph, and level sets can be studied using the methods in [25, 37, 43, 44] and sharp Hölder conditions for its local times can be derived from [43]. We will not work out the details here; just point out that we expect these properties to be different from those for the Brownian sheet in [7] and fractional Brownian sheets in [2, 41].

In this section, we will only consider the uniform and local moduli of continuity of $\mathbf{B}^H$. We mention that many authors have investigated uniform and local moduli of continuity of Gaussian random fields; see [3, 4, 26, 28, 30, 40]. The Chung-type laws of the iterated logarithm have also been proved for the Brownian sheet, fractional Brownian motion, and other Gaussian random fields in [23, 24, 27, 29, 33, 36, 43].

The following theorem gives the exact uniform modulus of continuity for SifBm $\mathbf{B}^H$, which is similar to that for ordinary fractional Brownian motion.

Theorem 9 *There is a constant $\kappa_1 \in (0, \infty)$ such that*

$$\lim_{\varepsilon\to 0+} \sup_{s,t\in[0,1]^2, |s-t|\leq\varepsilon} \frac{|\mathbf{B}^H(t) - \mathbf{B}^H(s)|}{|s-t|^H \sqrt{\log 1/|s-t|}} = \kappa_1 \quad a.s. \tag{10.5.15}$$

Proof By Lemma 2 and Theorem 8, $\mathbf{B}^H$ satisfies conditions (A1) and (A2) in [28]. Thus Theorem 4.1 in [28] implies that (10.5.15) holds a.s. provided $[0, 1]^2$ is replaced by $[a, 1]^2$, where $a \in (0, 1)$ is any given constant. This result, together with the 0–1 law in Lemma 7.1.1 in [26] and the following easily proven upper bound

$$\limsup_{\varepsilon\to 0+} \sup_{s,t\in[0,1]^2, |s-t|\le\varepsilon} \frac{|\mathbf{B}^H(t)-\mathbf{B}^H(s)|}{|s-t|^H\sqrt{\log 1/|s-t|}} \le c_4 \quad \text{a.s.}$$

for some finite constant c_4, gives the desired result (10.5.15).

In the following, we consider Chung's law of the iterated logarithm for SifBm $\mathbf{B}^H$. Compared with the results in [23, 24, 39], the following theorem only describes the local oscillation at a fixed point $t_0 \in (0,\infty)^2$, which is assumed to be away from the axes. In this case, (10.5.16) is similar to Chung's LIL for ordinary fractional Brownian motion [23, 29, 43]. As shown by Richard [33, Theorem 3], however, the Chung's LIL at the origin is different. See also [27, 36] for Chung's LIL for fractional Brownian sheets.

Theorem 10 *For every fixed $t_0 \in (0,\infty)^2$, there exists a positive and finite constant $\kappa_2 = \kappa_2(t_0)$ such that*

$$\liminf_{\varepsilon\to 0} \frac{\max_{|s|\le\varepsilon} |\mathbf{B}^H(t_0+s)-\mathbf{B}^H(t_0)|}{\varepsilon^H(\log\log 1/\varepsilon)^{-H/2}} = \kappa_2, \quad a.s. \tag{10.5.16}$$

Proof Let $Z = \{Z(t), t\in\mathbb{R}^2\}$ be the stationary Gaussian random field defined in (10.4.1). Consider the Gaussian field $X = \{X(t), t\in\mathbb{R}^2\}$ defined by $X(t) = Z(t) - Z(0)$. Then X has stationary increments and has the spectral density of Z as its spectral density. Thanks to Lemma 6, Chung's laws of the iterated logarithm in [23, 24] can be applied to $X = \{X(t), t\in\mathbb{R}^2\}$. In particular, there is a positive and finite constant c_5 such that for every fixed $u\in\mathbb{R}^2$,

$$\liminf_{\varepsilon\to 0} \frac{\max_{|v|\le\varepsilon} |Z(u+v)-Z(u)|}{\varepsilon^H(\log\log 1/\varepsilon)^{-H/2}} = c_5, \quad \text{a.s.}$$

From this and (10.5.13), we derive (10.5.16).

Finally, in this section, we consider two kinds of local moduli of continuity for $\mathbf{B}^H$. Theorem 11 is concerned with the local modulus of continuity measured in the most general way. Theorem 12 provides the exact local modulus of continuity in the L^1-norm $|\cdot|$. It should be noticed that the logarithmic factors in these two theorems are quite different. Note that, since their proofs do not use the property of local nondeterminism, we will state them for general $N\ge 2$.

Theorem 11 *For every fixed $t_0\in(0,\infty)^N$, there exists a positive and finite constant κ_3, which depends on t_0, such that*

$$\limsup_{\|\varepsilon\|\to 0+} \sup_{\langle|s_j|\rangle\le\langle\varepsilon_j\rangle} \frac{|\mathbf{B}^H(t_0+s)-\mathbf{B}^H(t_0)|}{|s|^H\sqrt{\log\log(1+\prod_{j=1}^N |s_j|^{-H})}} = \kappa_3 \quad a.s., \tag{10.5.17}$$

where $\langle\varepsilon_j\rangle = (\varepsilon_1,\ldots,\varepsilon_N)$.

Proof As in the proof of Theorem 10, we let $Z = \{Z(t), t \in \mathbb{R}^N\}$ be the stationary Gaussian random field defined in (10.4.1) and consider the Gaussian field $X = \{X(t), t \in \mathbb{R}^N\}$ defined by $X(t) = Z(t) - Z(0)$. Then X has stationary increments and, by (10.4.13), satisfies the condition of Theorem 5.1 in [28]. It follows that there is a constant $c_6 \in (0, \infty)$ such that for any fixed $u \in \mathbb{R}^N$,

$$\limsup_{\|\varepsilon\|\to 0+} \sup_{\langle |v_j| \rangle \le \langle \varepsilon_j \rangle} \frac{|Z(u+v) - Z(u)|}{|v|^H \sqrt{\log\log(1 + \prod_{j=1}^N |s_j|^{-H})}} = c_6 \quad \text{a.s.},$$

From this and (10.5.13), we can derive (10.5.17).

By the same argument, we derive from Theorem 5.6 in [28] the following law of the iterated logarithm.

Theorem 12 *For every fixed $t_0 \in (0, \infty)^N$, there exists a positive and finite constant $\kappa_4 = \kappa_4(t_0)$ such that*

$$\lim_{\varepsilon\to 0+} \sup_{s:|s|\le\varepsilon} \frac{|\mathbf{B}^H(t_0 + s) - \mathbf{B}^H(t_0)|}{|s|^H \sqrt{\log\log(1 + |s|^{-1})}} = \kappa_4 \quad a.s. \tag{10.5.18}$$

Remark 13 We conclude this section with the following remark.

- Our results show that SI-fBm $\mathbf{B}^H$ shares many properties such as fractal dimension and hitting probability results with fractional Brownian field X^H and the fractional Brownian sheet W^H. The Chung's laws of the iterated logarithm of $\mathbf{B}^H$ away from the axes (Theorem 10) and at the origin (Theorem 3 in [33]) are significantly different from those for W^H proved in [27, 36]. It is interesting to notice that some properties of $\mathbf{B}^H$ such as Chung's LIL away from the axes are closer to those of fractional Brownian field X^H than to those of W^H. This is due to the fact that, even if $\mathbf{B}^H$ does not have stationary increments as X^H, it shares the same property of strong local nondeterminism (see Theorem 8).
- Since $\mathbf{B}^H(t) = 0$ whenever $t \in \partial\mathbb{R}^N_+$, we expect that the local moduli of continuity of $\mathbf{B}^H(t)$ at the origin or $t_0 \in \partial\mathbb{R}^N_+$ are different from (10.5.17) and (10.5.18). For fractional Brownian sheets, [40] considered this problem in his Theorems 4.1 and 4.2. It would be of interest to compare the asymptotic behavior of SifBm $\mathbf{B}^H$ on $\partial\mathbb{R}^N_+$ with the results in [40].
- The method for proving Theorem 8 can also be extended for determining the Fourier dimension of the image set $\mathbf{B}^H(E)$, where $E \subset \mathbb{R}^2_+$ is a compact set, and show that it is almost surely a Salem set when $\dim E \le Hd$. We refer to [17, 22, 35] for definitions of Fourier dimension and Salem set, and their importance in Fourier analysis and connections to Gaussian random fields.

Acknowledgements The authors thank the referee and the editors for their very thoughtful comments which have led to improvements of the paper. The research of Y. Xiao is partially supported by NSF grants DMS-1612885 and DMS-1607089.

References

1. R.J. Adler, J.E. Taylor, *Random Fields and Geometry* (Springer, New York, 2007)
2. A. Ayache, D. Wang, Y. Xiao, Joint continuity of the local times of fractional Brownian sheets. Ann. Inst. H. Poincaré Probab. Statist. **44**, 727–748 (2008)
3. A. Ayache, Y. Xiao, Asymptotic properties and Hausdorff dimensions of fractional Brownian sheets. J. Fourier Anal. Appl. **11**, 407–439 (2005)
4. A. Benassi, S. Jaffard, D. Roux, Elliptic Gaussian random processes. Rev. Mat. Iberoam. **13**, 19–90 (1997)
5. H. Biermé, C. Lacaux, Y. Xiao, Hitting probabilities and the Hausdorff dimension of the inverse images of anisotropic Gaussian random fields. Bull. Lond. Math. Soc. **41**, 253–273 (2009)
6. Z. Chen, Y. Xiao, On intersections of independent anisotropic Gaussian random fields. Sci. China Math. **55**, 2217–2232 (2012)
7. W. Ehm, Sample function properties of multi-parameter stable processes. Z. Wahrsch. verw Gebiete **56**, 195–228 (1981)
8. K. Falconer, Tangent fields and the local structure of random fields. J. Theoret. Probab. **15**, 731–750 (2002)
9. M.G. Genton, O. Perrin, M.S. Taqqu, Self-similarity and Lamperti transformation for random fields. Stoch. Models **23**, 397–411 (2007)
10. J. Hawkes, Measures of Hausdorff type and stable processes. Mathematika **25**, 202–212 (1978)
11. E. Herbin, B. Arras, G. Barruel, From almost sure local regularity to almost sure Hausdorff dimension for Gaussian fields. ESAIM Probab. Stat. **18**, 41–440 (2014)
12. E. Herbin, E. Merzbach, The set-indexed fractional Brownian motion. J. Theoret. Probab. **19**, 337–364 (2006)
13. E. Herbin, E. Merzbach, The Multiparameter fractional Brownian motion, in *Math Everywhere* ed. by G. Aletti, M. Burger, A. Micheletti, D. Morale (Springer, 2006)
14. E. Herbin, E. Merzbach, Stationarity and self-similarity characterization of the set-indexed fractional Brownian motion. J. Theoret. Probab. **22**, 1010–1029 (2009)
15. E. Herbin, A. Richard, Local Hölder regularity for set-indexed processes. Israel J. Math. **215**, 397–440 (2016)
16. G. Ivanoff, E. Merzbach, *Set-Indexed Martingales* (Chapman & Hall/CRC, 2000)
17. J.-P. Kahane, *Some Random Series of Functions*, 2nd edn. (Cambridge University Press, Cambridge, 1985)
18. D. Khoshnevisan, *Multiparameter Processes: An Introduction to Random Fields* (Springer, 2002)
19. D. Khoshnevisan, Brownian sheet images and Bessel-Riesz capacity. Trans. Am. Math. Soc. **351**, 2607–2622 (1999)
20. D. Khoshnevisan, Z. Shi, Brownian sheet and capacity. Ann. Probab. **27**, 1135–1159 (1999)
21. D. Khoshnevisan, Y. Xiao, Images of the Brownian sheet. Trans. Am. Math. Soc. **359**, 3125–3151 (2007)
22. D. Khoshnevisan, D. Wu, Y. Xiao, Sectorial local non-determinism and the geometry of the Brownian sheet. Electron. J. Probab. **11**, 817–843 (2006)
23. W.V. Li, Q.-M. Shao, Gaussian processes: inequalities, small ball probabilities and applications, in *Stochastic Processes: Theory and Methods, Handbook of Statistics* ed. by C.R. Rao, D. Shanbhag, vol. 19 (North-Holland, 2001), pp. 533–597
24. N. Luan, Y. Xiao, Chung's law of the iterated logarithm for anisotropic Gaussian random fields. Stat. Probab. Lett. **80**, 1886–1895 (2010)
25. N. Luan, Y. Xiao, Spectral conditions for strong local nondeterminism and exact Hausdorff measure of ranges of Gaussian random fields. J. Fourier Anal. Appl. **18**, 118–145 (2012)
26. M.B. Marcus, J. Rosen, *Markov Processes, Gaussian Processes, and Local Times* (Cambridge University Press, Cambridge, 2006)
27. D.M. Mason, Z. Shi, Small deviations for some multi-parameter Gaussian processes. J. Theoret. Probab. **14**, 213–239 (2001)

28. M.M. Meerschaert, W. Wang, Y. Xiao, Fernique-type inequalities and moduli of continuity of anisotropic Gaussian random fields. Trans. Am. Math. Soc. **365**, 1081–1107 (2013)
29. D. Monrad, H. Rootzén, Small values of Gaussian processes and functional laws of the iterated logarithm. Probab. Theory Rel. Fields **101**, 173–192 (1995)
30. S. Orey, W.E. Pruitt, Sample functions of the N-parameter Wiener process. Ann. Probab. **1**, 138–163 (1973)
31. R.F. Peltier, J. Lévy-Véhel, Multifractional brownian motion: definition and preliminary results, in *Rapport de recherche INRIA* (1995), (RR-2645), p. 39
32. L.D. Pitt, Local times for Gaussian vector fields. Indiana Univ. Math. J. **27**, 309–330 (1978)
33. A. Richard, Some singular sample path properties of a multiparameter fractional Brownian motion. J. Theoret. Probab. **30**, 1285–1309 (2017)
34. A. Richard, A fractional Brownian field indexed by L^2 and a varying Hurst parameter. Stoch. Process. Appl. **125**, 1394–1425 (2015)
35. N.-R. Shieh, Y. Xiao, Images of Gaussian random fields: Salem sets and interior points. Studia Math. **176**, 37–60 (2006)
36. M. Talagrand, The small ball problem for the Brownian sheet. Ann. Probab. **22**, 1331–1354 (1994)
37. M. Talagrand, Hausdorff measure of trajectories of multiparameter fractional Brownian motion. Ann. Probab. **23**, 767–775 (1995)
38. S.J. Taylor, N.A. Watson, A Hausdorff measure classification of polar sets for the heat equation. Math. Proc. Camb. Philos. Soc. **97**, 325–344 (1985)
39. C.A. Tudor, Y. Xiao, Sample paths properties of bifractional Brownian motion. Bernoulli **13**, 1023–1052 (2007)
40. W. Wang, Almost-sure path properties of fractional Brownian sheet. Ann. Inst. Henri Poincaré Probab. Statist. **43**, 619–631 (2007)
41. D. Wu, Y. Xiao, Geometric properties of fractional Brownian sheets. J. Fourier Anal. Appl. **13**, 1–37 (2007)
42. D. Wu, Y. Xiao, On local times of anisotropic Gaussian random fields. Commun. Stoch. Anal. **5**, 15–39 (2011)
43. Y. Xiao, Hölder conditions for the local times and the Hausdorff measure of the level sets of Gaussian random fields. Probab. Theory Rel. Fields **109**, 129–157 (1997)
44. Y. Xiao, Hausdorff measure of the graph of fractional Brownian motion. Math. Proc. Camb. Philos. Soc. **122**, 565–576 (1997)
45. Y. Xiao, Strong local nondeterminism and the sample path properties of Gaussian random fields, in *Asymptotic Theory in Probability and Statistics with Applications*, ed. by T.L. Lai, Q. Shao, L. Qian (Higher Education Press, Beijing, 2007), pp. 136–176
46. Y. Xiao, Sample path properties of anisotropic Gaussian random fields, in *A Minicourse on Stochastic Partial Differential Equations* ed. by D. Khoshnevisan, F. Rassoul-Agha. Lecture Notes in Mathematics 1962, (Springer, 2009), pp. 145–212

Applied and Numerical Harmonic Analysis

(96 volumes)

1. A.I. Saichev, W.A. Woyczyński, *Distributions in the Physical and Engineering Sciences* (ISBN: 978-0-8176-3924-2)
2. C.E. D'Attellis, E.M. Fernandez-Berdaguer, *Wavelet Theory and Harmonic Analysis in Applied Sciences* (ISBN: 978-0-8176-3953-2)
3. H.G. Feichtinger, T. Strohmer, *Gabor Analysis and Algorithms* (ISBN: 978-0-8176-3959-4)
4. R. Tolimieri, M. An, *Time-Frequency Representations* (ISBN: 978-0-8176-3918-1)
5. T.M. Peters, J.C. Williams, *The Fourier Transform in Biomedical Engineering* (ISBN: 978-0-8176-3941-9)
6. G.T. Herman, *Geometry of Digital Spaces* (ISBN: 978-0-8176-3897-9)
7. A. Teolis, *Computational Signal Processing with Wavelets* (ISBN: 978-0-8176-3909-9)
8. J. Ramanathan, *Methods of Applied Fourier Analysis* (ISBN: 978-0-8176-3963-1)
9. J.M. Cooper, *Introduction to Partial Differential Equations with MATLAB* (ISBN: 978-0-8176-3967-9)
10. A. Procházka, N.G. Kingsbury, P.J. Payner, J. Uhlir, *Signal Analysis and Prediction* (ISBN: 978-0-8176-4042-2)
11. W. Bray, C. Stanojevic, *Analysis of Divergence* (ISBN: 978-1-4612-7467-4)
12. G.T. Herman, A. Kuba: *Discrete Tomography* (ISBN: 978-0-8176-4101-6)
13. K. Gröchenig, *Foundations of Time-Frequency Analysis* (ISBN: 978-0-8176-4022-4)
14. L. Debnath, *Wavelet Transforms and Time-Frequency Signal Analysis* (ISBN: 978-0-8176-4104-7)
15. J.J. Benedetto, P.J.S.G. Ferreira, *Modern Sampling Theory* (ISBN: 978-0-8176-4023-1)

A. Aldroubi et al. (eds.), *New Trends in Applied Harmonic Analysis, Volume 2*, Applied and Numerical Harmonic Analysis,
https://doi.org/10.1007/978-3-030-32353-0

16. D.F. Walnut, *An Introduction to Wavelet Analysis* (ISBN: 978-0-8176-3962-4)
17. A. Abbate, C. DeCusatis, P.K. Das, *Wavelets and Subbands* (ISBN: 978-0-8176-4136-8)
18. O. Bratteli, P. Jorgensen, B. Treadway, *Wavelets Through a Looking Glass* (ISBN: 978-0-8176-4280-80
19. H.G. Feichtinger, T. Strohmer, *Advances in Gabor Analysis* (ISBN: 978-0-8176-4239-6)
20. O. Christensen, *An Introduction to Frames and Riesz Bases* (ISBN: 978-0-8176-4295-2)
21. L. Debnath, *Wavelets and Signal Processing* (ISBN: 978-0-8176-4235-8)
22. G. Bi, Y. Zeng, *Transforms and Fast Algorithms for Signal Analysis and Representations* (ISBN: 978-0-8176-4279-2)
23. J.H. Davis, *Methods of Applied Mathematics with a MATLAB Overview* (ISBN: 978-0-8176-4331-7)
24. J.J. Benedetto, A.I. Zayed, *Sampling, Wavelets, and Tomography* (ISBN: 978-0-8176-4304-1)
25. E. Prestini, *The Evolution of Applied Harmonic Analysis* (ISBN: 978-0-8176-4125-2)
26. L. Brandolini, L. Colzani, A. Iosevich, G. Travaglini, *Fourier Analysis and Convexity* (ISBN: 978-0-8176-3263-2)
27. W. Freeden, V. Michel, *Multiscale Potential Theory* (ISBN: 978-0-8176-4105-4)
28. O. Christensen, K.L. Christensen, *Approximation Theory* (ISBN: 978-0-8176-3600-5)
29. O. Calin, D.-C. Chang, *Geometric Mechanics on Riemannian Manifolds* (ISBN: 978-0-8176-4354-6)
30. J.A. Hogan, *Time–Frequency and Time–Scale Methods* (ISBN: 978-0-8176-4276-1)
31. C. Heil, *Harmonic Analysis and Applications* (ISBN: 978-0-8176-3778-1)
32. K. Borre, D.M. Akos, N. Bertelsen, P. Rinder, S.H. Jensen, *A Software-Defined GPS and Galileo Receiver* (ISBN: 978-0-8176-4390-4)
33. T. Qian, M.I. Vai, Y. Xu, *Wavelet Analysis and Applications* (ISBN: 978-3-7643-7777-9)
34. G.T. Herman, A. Kuba, *Advances in Discrete Tomography and Its Applications* (ISBN: 978-0-8176-3614-2)
35. M.C. Fu, R.A. Jarrow, J.-Y. Yen, R.J. Elliott, *Advances in Mathematical Finance* (ISBN: 978-0-8176-4544-1)
36. O. Christensen, *Frames and Bases* (ISBN: 978-0-8176-4677-6)
37. P.E.T. Jorgensen, J.D. Merrill, J.A. Packer, *Representations, Wavelets, and Frames* (ISBN: 978-0-8176-4682-0)
38. M. An, A. K. Brodzik, R. Tolimieri, *Ideal Sequence Design in Time-Frequency Space* (ISBN: 978-0-8176-4737-7)
39. S.G. Krantz, *Explorations in Harmonic Analysis* (ISBN: 978-0-8176-4668-4)
40. B. Luong, *Fourier Analysis on Finite Abelian Groups* (ISBN: 978-0-8176-4915-9)

41. G.S. Chirikjian, *Stochastic Models, Information Theory, and Lie Groups*, vol. 1 (ISBN: 978-0-8176-4802-2)
42. C. Cabrelli, J.L. Torrea, *Recent Developments in Real and Harmonic Analysis* (ISBN: 978-0-8176-4531-1)
43. M.V. Wickerhauser, *Mathematics for Multimedia* (ISBN: 978-0-8176-4879-4)
44. B. Forster, P. Massopust, O. Christensen, K. Gröchenig, D. Labate, P. Vandergheynst, G. Weiss, Y. Wiaux, *Four Short Courses on Harmonic Analysis* (ISBN: 978-0-8176-4890-9)
45. O. Christensen, *Functions, Spaces, and Expansions* (ISBN: 978-0-8176-4979-1)
46. J. Barral, S. Seuret: *Recent Developments in Fractals and Related Fields* (ISBN: 978-0-8176-4887-9)
47. O. Calin, D.-C. Chang, K. Furutani, C. Iwasaki, *Heat Kernels for Elliptic and Sub-elliptic Operators* (ISBN: 978-0-8176-4994-4)
48. C. Heil: *A Basis Theory Primer* (ISBN: 978-0-8176-4686-8)
49. J.R. Klauder, *A Modern Approach to Functional Integration* (ISBN: 978-0-8176-4790-2)
50. J. Cohen, A.I. Zayed, *Wavelets and Multiscale Analysis* (ISBN: 978-0-8176-8094-7)
51. D. Joyner, J.-L. Kim, *Selected Unsolved Problems in Coding Theory* (ISBN: 978-0-8176-8255-2)
52. G.S. Chirikjian, *Stochastic Models, Information Theory, and Lie Groups*, vol. 2 (ISBN: 978-0-8176-4943-2)
53. J.A. Hogan, J.D. Lakey, *Duration and Bandwidth Limiting* (ISBN: 978-0-8176-8306-1)
54. G. Kutyniok, D. Labate, *Shearlets* (ISBN: 978-0-8176-8315-3)
55. P.G. Casazza, P. Kutyniok, *Finite Frames* (ISBN: 978-0-8176-8372-6)
56. V. Michel, *Lectures on Constructive Approximation* (ISBN: 978-0-8176-8402-0)
57. D. Mitrea, I. Mitrea, M. Mitrea, S. Monniaux, *Groupoid Metrization Theory* (ISBN: 978-0-8176-8396-2)
58. T.D. Andrews, R. Balan, J.J. Benedetto, W. Czaja, K.A. Okoudjou, *Excursions in Harmonic Analysis*, vol. 1 (ISBN: 978-0-8176-8375-7)
59. T.D. Andrews, R. Balan, J.J. Benedetto, W. Czaja, K.A. Okoudjou, *Excursions in Harmonic Analysis*, vol. 2 (ISBN: 978-0-8176-8378-8)
60. D.V. Cruz-Uribe, A. Fiorenza, *Variable Lebesgue Spaces* (ISBN: 978-3-0348-0547-6)
61. W. Freeden, M. Gutting, *Special Functions of Mathematical (Geo-)Physics* (ISBN: 978-3-0348-0562-9)
62. A.I. Saichev, W.A. Woyczyński: *Distributions in the Physical and Engineering Sciences, Volume 2: Linear and Nonlinear Dynamics of Continuous Media* (ISBN: 978-0-8176-3942-6)
63. S. Foucart, H. Rauhut, *A Mathematical Introduction to Compressive Sensing* (ISBN: 978-0-8176-4947-0)
64. G.T. Herman, J. Frank, *Computational Methods for Three-Dimensional Microscopy Reconstruction* (ISBN: 978-1-4614-9520-8)

65. A. Paprotny, M. Thess, *Realtime Data Mining: Self-Learning Techniques for Recommendation Engines* (ISBN: 978-3-319-01320-6)
66. A.I. Zayed, G. Schmeisser, *New Perspectives on Approximation and Sampling Theory: Festschrift in Honor of Paul Butzer's* 85th *Birthday* (ISBN: 978-3-319-08800-6)
67. R. Balan, M. Begue, J. Benedetto, W. Czaja, K.A. Okoudjou, *Excursions in Harmonic Analysis, Volume 3* (ISBN: 978-3-319-13229-7)
68. H. Boche, R. Calderbank, G. Kutyniok, J. Vybiral, *Compressed Sensing and its Applications* (ISBN: 978-3-319-16041-2)
69. S. Dahlke, F. De Mari, P. Grohs, D. Labate, *Harmonic and Applied Analysis: From Groups to Signals* (ISBN: 978-3-319-18862-1)
70. A. Aldroubi, *New Trends in Applied Harmonic Analysis* (ISBN: 978-3-319-27871-1)
71. M. Ruzhansky, *Methods of Fourier Analysis and Approximation Theory* (ISBN: 978-3-319-27465-2)
72. G. Pfander, *Sampling Theory, a Renaissance* (ISBN: 978-3-319-19748-7)
73. R. Balan, M. Begue, J. Benedetto, W. Czaja, K.A. Okoudjou: *Excursions in Harmonic Analysis*, vol. 4 (ISBN: 978-3-319-20187-0)
74. O. Christensen, *An Introduction to Frames and Riesz Bases*, 2nd edn. (ISBN: 978-3-319-25611-5)
75. E. Prestini, *The Evolution of Applied Harmonic Analysis: Models of the Real World, 2nd edn.* (ISBN: 978-1-4899-7987-2)
76. J.H. Davis, *Methods of Applied Mathematics with a Software Overview*, 2nd edn. (ISBN: 978-3-319-43369-1)
77. M. Gilman, E.M. Smith, S.M. Tsynkov, *Transionospheric Synthetic Aperture Imaging* (ISBN: 978-3-319-52125-1)
78. S. Chanillo, B. Franchi, G. Lu, C. Perez, E.T. Sawyer, *Harmonic Analysis, Partial Differential Equations and Applications* (ISBN: 978-3-319-52741-3)
79. R. Balan, J. Benedetto, W. Czaja, M. Dellatorre, K.A. Okoudjou, *Excursions in Harmonic Analysis*, vol. 5 (ISBN: 978-3-319-54710-7)
80. I. Pesenson, Q.T. Le Gia, A. Mayeli, H. Mhaskar, D.X. Zhou, *Frames and Other Bases in Abstract and Function Spaces: Novel Methods in Harmonic Analysis*, vol. 1 (ISBN: 978-3-319-55549-2)
81. I. Pesenson, Q.T. Le Gia, A. Mayeli, H. Mhaskar, D.X. Zhou, *Recent Applications of Harmonic Analysis to Function Spaces, Differential Equations, and Data Science: Novel Methods in Harmonic Analysis*, vol. 2 (ISBN: 978-3-319-55555-3)
82. F. Weisz, *Convergence and Summability of Fourier Transforms and Hardy Spaces* (ISBN: 978-3-319-56813-3)
83. C. Heil, *Metrics, Norms, Inner Products, and Operator Theory* (ISBN: 978-3-319-65321-1)
84. S. Waldron, *An Introduction to Finite Tight Frames: Theory and Applications.* (ISBN: 978-0-8176-4814-5)
85. D. Joyner, C.G. Melles, *Adventures in Graph Theory: A Bridge to Advanced Mathematics.* (ISBN: 978-3-319-68381-2)

86. B. Han, *Framelets and Wavelets: Algorithms, Analysis, and Applications* (ISBN: 978-3-319-68529-8)
87. H. Boche, G. Caire, R. Calderbank, M. März, G. Kutyniok, R. Mathar, *Compressed Sensing and Its Applications* (ISBN: 978-3-319-69801-4)
88. A.I. Saichev, W.A. Woyczyński, *Distributions in the Physical and Engineering Sciences, Volume 3: Random and Fractal Signals and Fields* (ISBN: 978-3-319-92584-4)
89. G. Plonka, D. Potts, G. Steidl, M. Tasche, *Numerical Fourier Analysis* (ISBN: 978-3-030-04305-6)
90. K. Bredies, D. Lorenz, *Mathematical Image Processing* (ISBN: 978-3-030-01457-5)
91. H.G. Feichtinger, P. Boggiatto, E. Cordero, M. de Gosson, F. Nicola, A. Oliaro, A. Tabacco, *Landscapes of Time-Frequency Analysis* (ISBN: 978-3-030-05209-6)
92. E. Liflyand, *Functions of Bounded Variation and Their Fourier Transforms* (ISBN: 978-3-030-04428-2)
93. R. Campos, *The XFT Quadrature in Discrete Fourier Analysis* (ISBN: 978-3-030-13422-8)
94. M. Abell, E. Iacob, A. Stokolos, S. Taylor, S. Tikhonov, J. Zhu, *Topics in Classical and Modern Analysis: In Memory of Yingkang Hu* (ISBN: 978-3-030-12276-8)
95. H. Boche, G. Caire, R. Calderbank, G. Kutyniok, R. Mathar, P. Petersen, *Compressed Sensing and its Applications: Third International MATHEON Conference 2017* (ISBN: 978-3-319-73073-8)
96. A. Aldroubi, C. Cabrelli, S. Jaffard, U. Molter, *New Trends in Applied Harmonic Analysis, Volume II: Harmonic Analysis, Geometric Measure Theory, and Applications* (ISBN: 978-3-030-32352-3)

For an up-to-date list of ANHA titles, please visit http://www.springer.com/series/4968

GPSR Compliance
The European Union's (EU) General Product Safety Regulation (GPSR) is a set of rules that requires consumer products to be safe and our obligations to ensure this.

If you have any concerns about our products, you can contact us on

ProductSafety@springernature.com

In case Publisher is established outside the EU, the EU authorized representative is:

Springer Nature Customer Service Center GmbH
Europaplatz 3
69115 Heidelberg, Germany

www.ingramcontent.com/pod-product-compliance
Ingram Content Group UK Ltd.
Pitfield, Milton Keynes, MK11 3LW, UK
UKHW021833270726
14058UKWH00001B/125

* 9 7 8 3 0 3 0 3 2 3 5 5 4 *